GERMAN-ENGLISH DICTIONARY

for

ELECTRONICS ENGINEERS

and

PHYSICISTS

WITH A PATENT-PRACTICE VOCABULARY

BY BERNARD R. REGEN
Researcher and Translator

AND RICHARD R. REGEN
Physicist and Field Engineer
Magnetic Analysis Corporation

•

FIRST EDITION

1946

J. W. EDWARDS ∽ ANN ARBOR, MICHIGAN

Lithoprinted by Edwards Brothers, Inc.
Ann Arbor, Michigan, U.S.A.

To
Nelson M. Cooke
and
John Markus,

Authors of
Electronics Dictionary

PREFACE

"L'avenir de électricité, c'est dans le vide." (The future of electricity lies in the vacuum.) This remark, made before the Académie a generation ago by a famous French physicist, has proved to be prophetic. For the world has witnessed amazing developments in that branch of electricity which is based upon the employment of a vacuum: ELECTRONICS. Yet in spite of these great developments (or possibly because of them), and in spite of an increasingly urgent need for foreign-language dictionaries, students of electronics and allied problems have been forced to proceed without a lexicographical tool which brings together the terminologies of the two languages most vital to their work, German and English.

The authors believe that this volume will be welcomed by all electronics engineers and all physicists, notwithstanding certain limitations in size and scope which have been imposed by the war. It contains nearly 21,000 entries, representing most of the terminology current in 1939 and an appreciable number of terms introduced during the war years.

Owing to a lack of earlier German-English dictionaries in the general field of physics, the task of compilation has proved unusually difficult. The only outstanding reference books hitherto available for consultation have been Sattelberg's Wörterbuch der elektrischen Nachrichtentechnik (1926) covering the signal art in general, and Austin M. Patterson's famous German-English Dictionary for Chemists (1935). The bulk of the material in the present volume has been drawn from magazines, scientific journals, books, and patent literature. The definitions gleaned from other books have in most instances been amplified and re-worded.

A somewhat unorthodox policy has been followed by frequently adding explanatory annotations to the English-language expressions. This has been done in order to supply related terminology for the benefit of the reader who may be unfamiliar with a given subject. Ample cross-references have been inserted throughout. And for those who desire fuller information, source and reference books--including a number of outstanding works of British origin--have been listed in a bibliography. The authors recommend purchase of some of the items included in this list, particularly the Electronics Dictionary by Nelson M. Cooke and John Markus.

Indeed, the authors are deeply indebted to these authors and their book for much valuable information.

We also wish to express our gratitude to all those who have aided us in one way or another, especially to E. R. Weinmann, Arthur Worischeck, Walter Brunberg and Dr. G. Ornsen who assisted in the proofreading. We shall be equally grateful to those who bring to our attention such errors, imperfections, and omissions as are inevitable in the first printing of a work which is a pioneer in its field. A revised and amplified edition is being planned for a future date, and correspondence with the authors is solicited from those wishing to co-operate in the compilation of a revised text.

The Twelve Acres,
Stone Ridge, N.Y.
and
55 W. 42nd Street,
New York, 18, N.Y.

Bernard R. Regen
Richard R. Regen

August, 1946

BIBLIOGRAPHY

Andretto, P. C. Principles of Aeronautical Radio Engineering. 1942.

Chamber's Technical Dictionary. 1944.

Cooke, Nelson M. & Markus, John. Electronics Dictionary 1945.

Drake's Cyclopedia of Radio and Electronics. 1943.

Gherardi, A. A. Modern Radio Servicing. 1936.

Jacobs, Donald H. Fundamentals of Optical Engineering. 1936.

Keen, R. Wireless Direction-finding. 1938.

Manly, Harold P. Radio and Electronics Dictionary. 1931.

Olson, Harry F. Elements of Acoustical Engineering. 1940.

Patterson, Austin M A German-English Dictionary for Chemists 1935.

Roget's Dictionary of Electrical Terms. 1943.

Zworykin, V. K., and Morton, G. A. Television. 1940.

Ardenne, Manfred von Handbuch der Funktechnik. 3 vols. 1935-1938.

Kohlrausch, F. Praktische Physik. 1944.

Sattelberg, Otto. Wörterbuch der elektrischen Nachrichtentechnik 1926.

Schröter, F. Fernsehen. 1937.

Magazines covering radio, sound-film, television and related physical sciences and technological arts, foreign-literature reviews regularly appearing in some of them. Abstract journals such as Science Abstracts (A & B), Physikalische Berichte, etc. Patents.

ABBREVIATIONS

a-c, a.c.	alternating current
ac.	acoustics
a-m	amplitude modulation
AF; a-f	audio frequency
avc	automatic volume control
c.-r.; CR t.	cathode-ray tubes
cer.	ceramics
chem.	chemistry
cr.	crystallography
d-c; d.c.	direct current
d. f.	direction-finding
el. engg.	electrical engineering
el. opt.	electron optics
f-m; FM	frequency modulation
ldspk.	loudspeaker
magn.	magnetism
micr.	microscopy
m.p.	moving pictures
mus.	music
obs.	obsolete
opt.	optics
phon.	phonograph
phot.	photography
phys.	physics
q.a.v.c.	silent automatic volume control
RF, r-f	radio frequency
spec.	specifically
spectr.	spectroscopy
telev.	television.

A

a.a.O. in the place cited, loco citato, quoted reference.
Abänderung change, alteration, modification, amendment.
Abänderungspatent re-issue patent.
abarten vary, degenerate.
Abbau decomposition, disintegration, degradation, attrition. **Frequenz-** frequency division, f.submultiplication. **-mittel** disintegrant. **-verhinderndes Mittel** anti-disintegrant, stabilizer.
abbiegen bend off or away, bend angularly; turn off.
abbilden image, focus.
abbildende Ringelektrode focusing ring (el.opt.).
Abbildung imaging, imagery, image formation; figure(Fig.), illustration, drawing, sketch, cut.
Abbildung, geteilte split focus (Walton method of scanning). **-, komatische** comatic image. **-, konforme** conformable representation. **-, punktförmige** point image. **-, punktuelle** point focal vision or imagery.
Abbildungs-fehler defect of image, aberration(opt.). **-gegenstand** imaging object. **-optik** imaging "optic", focusing o. **-vermögen** resolving power (opt., micr.).
abblättern scale (off), shell, chip or peel off, descale, defoliate, exfoliate.
Abblendekapotte dimming cap.
abblenden (cf.abdecken) mask off, stop down (with masking diaphragm), occult, diaphragm, gate, blank, dowse (a beam), wipe, dissolve, fade or black out.
Abblender, Strahl- means to mask, stop, dowse, occult or eclipse rays or a beam, blanking or blanketing means.
Abblendung, Iris- irising out.
abbreiten stretch out, flatten.
abbremsen decelerate, retard, brake (electrons, etc.).
Abdeck-aufzeichnung masking recording, matting. **-blende** shutter, mask, shutter m. **-doppelzackenspur** duplex variable -area track **-einfachzackenspur** unilateral variable-area track.
abdecken (cf.abblenden) diaphragm, mask, eclipse (beam or pencil, by shutter, etc.), occult (a lens); uncover; cover; provide a resist coat (in metal etching).
Abdeck-flügel masking or shutter blade or vane. **-scheibe** shutter disk with spiral slot, stop disk with spiral slot cyclically co-operating with quadruple-spiral scanning disk apertures (telev.). **-seite** masked side (m.p.). **-streifen** squeeze unit, barrier (m.p. film).
Abdeckung, Einfachzackenschrift mit unilateral variable-area track made with single-vane shutter.
Abdeckverfahren shutter mask method.
abdichten seal, pack, provide glands
Abdichtung tight seal, packing, hermetic plumbing.
Abdrift drift (off course, of air plane, etc.). **-visier** drift meter.
abdrosseln throttle, choke.
Abdruck copy, impression, print, cast, counterpart, replica.
Abdruck, Kontakt- contact print (phot.). **Probe-** proof, p.print, p.sheet.

abdrucken print, reproduce or copy (by printing or impression), stamp.
abdunkeln darken, cut off or dim (light), occult, eclipse, obscure, blanking or gating of spot (in telev.).
Aberkennung abrogation, abjudication, deprivation.
Aberration, Flächen- zonal aberration.
aberregen de-energize, de-activate.
Aberregung, Relais- de-energization of a relay.
abfachen classify, partition, form compartments.
Abfall drop, fall, droop, decline, descent; waste, scrap, refuse, discard.
abfallen fall or drop down or off, slope down, droop, release (a relay).
Abfallprodukt by-product, waste p.
abfangen capture, catch, intercept; pull (an airplane) out of a dive and level off.
abfasen chamfer, face.
abfasern lose fibers, fuzz.
abfassen frame, draft, prepare (text), write, draw up (a specification, etc.).
abfiltern (cf.Filter, ausfiltern) filter or strain off (by sieve, etc.).
Abfindungssumme indemnity, money paid in settlement (of claims, etc.).
abflachen flatten, smooth down, broaden, make oblate.
Abflachung bevel(ing), rounding off (a curve), leveling, flattening, smoothing, oblateness(at poles).
Abflachungsdrossel smoothing coil, s.choke.
Abfliegen outward course, flight away from airport.
abfliessen leak off, flow or drain off or away.
abfluchten align, line out.
Abflug, Ziel- flight away from object.
Abflugszeit outward flying time (away from airport, in ZZ landing).
Abfluss leak path (for grid current).
-hahn discharge cock, drain c.
-ventil discharge valve, drain v.
abfressen corrode, eat away or up.
Abfühlnadel pecker, selecting needle.
Abfuhr outlet (e.g., in water-cooled tube).
Abführung, Wärme- heat dissipation, h.loss, h."abduction", effect of carrying or conducting away heat.
abfuttern case, line.
Abgabe import duty, tax, fee, royalty; emission (of radiations, electrons, etc.).
Abgabefläche, Elektronen- electron emitting area.
Abgangswinkel angle of departure.
abgeflachte Spannung flat-topped potential(wave).
abgeglichen balanced, equalized, aligned, flush.
abgehender Strom outgoing current.
abgehende Wirbelbahnen trailing vortices(airfoil).
abgekantet chamfered, beveled, canted.
abgeklungene Emanation emanation with diminished or decaying radioactivity.
abgelagert aged, seasoned.
abgeleitete Grösse derived or subsidiary quantity or magnitude.
abgeschirmte Stelle dead spot, radio shadow.
abgeschlossen, in sich selbst self-contained.
abgeschmolzene Röhre sealed off tube.
abgeschrägt beveled, chamfered, canted.
abgesetzt offset.
abgestimmt, nicht- untuned, non-resonant, aperiodic, not in syntony or resonated.
abgestumpft truncated, blunt-ended, with frustum.
abgewickelte Linie evolute.
abgezweigte Anmeldung divisional application (portion divided off

from original a.).

abgleichen equalize, compensate, balance, trim, align, level.

abgleichend, selbst- self-balancing.

Abgleich-fehler unbalance, balancing or alignment error, misalignment. **-kondensator** trimming condenser or capacitor, aligning c. **-platte** tuning spade, tuning plate, t. wand.

Abgleichung, Fehl- misalignment, faulty alignment.

Abgleichvorrichtung, Phasen- phase balancer, p. shifter, p. changer, p. compensator network.

Abgreifen der Entfernung measuring map range with dividers, etc.

abgrenzen mark off, bound, demarcate, define (the limits of), delimit, circumscribe.

Abgrenzung und Einteilung, Wellen- frequency and wave-band designation: very low frequency (vlf) 10-30 kc, low f. (lf) 30-300 kc; medium f. (mf) 300-3000 kc, high f. (hf) 3-30 mc, very high f. (vhf) 30-300 mc; ultra-high f. (uhf) 300-3000 mc; super high f. (shf) 3000-30000 mc.

Abgriff tap.

abhängen depend on; lift off (receiver).

abhängig, druck- oder temperatur- be a function of pressure or temperature, pressure-or temperature-dependent or responsive.

Abhängigkeit, Frequenz- frequency dependence, having f. effect, being affected by, or being a function of, frequency.

Abheber plow (to peel film from teeth).

Abhebestange damper rod (piano).

Abhilfe remedy, redress, relief, cure, obviating a difficulty.

Abhörbox monitoring box.

Abhorchen (von Flugzeugen) auditory locating, spotting, detecting, or fixing position (of aircraft, etc.) by sound locator devices.

Abhorchgerät monitor, check or pick-up microphonic device, standby circuit means.

abhören intercept, listen in; monitor.

Abhören, Maschine zum - von Tonfilmstreifen machine for editing and rehearsing sound film.

Abhören der Negative direct sound reproduction from negatives.

Abhör-gerät (Flugzeug, etc.) airplane, etc., sound locator or detector; radio locator, Radar (using uhf waves and echo principle). **-lautsprecher** monitoring loudspeaker. **-raum** mixing room, monitoring booth, "box". **-tisch (und Schneidetisch)** sound film editing machine.

Abhör-, Klebe- und Schneidetisch movieola.

Abisolierer wire skinner.

abkanten level, take off edge, chamfer, cant.

abkappen lop off (peaks or crests), clip, limit amplitude.

Abkapper amplitude or output limiter.

abkehren turn away or aside, divert; sweep or brush off.

abklappen swing or hinge down, out or away, let down.

abklingender Kurvenzug decay train.

Abkling-kurve decay curve. **-zeit** decay period, die-out p.; quench p. (in super-regeneration). **-zeitmesser(Fluoreszenz)** fluorometer.

abkneifen pinch off, nip off.

abknicken bend or deviate (a ray, at sharp angle).

Abkommen sight graticule or reticle (opt.), point of aim; agreement, convention, contract, arrangement; deviation.

Ablagerung deposit, sediment.

Ablagerungen, inselartige "island" deposits.

Ablass, Wasser- draining or drainage of water.

Ablauf expiry, expiration or lapse (of a period of time), rotation or flight (of electrons); film feed (from magazine spool). **-bahn** runway, landing ramp (of airplanes).

ablaufen run down (of a spring), return (of dial switch); run off, drain; expire (rights, contracts, etc.).

ablaufen lassen release, let go or return (dial switch).

ablaufend(e) Kante trailing edge. **-(er) Kurs** course or direction away from airport (in ZZ landing), outward course.

Ablaufzeit return time (of a dial switch).

ablegen(Zeugnis) bear testimony or witness.

ablehnen decline, reject, refuse, challenge, dismiss.

ableiten derive, deduce, turn aside, draw off, let escape, lead or carry off, dissipate or abduct (heat, etc.), shunt, drain to earth.

Ableiter arrester, means to let a charge leak or drain away.

Ableiter, Kathodenfall- valve arrester.

Ableitung leakance (line constant), shunt conductance, derivation (with respect to), differential coefficient (math.); downlead, lead-in (of an antenna).

Ableitung, dielektrische dielectric leakance, conductance. **dritte, der Charakteristik** third differential of characteristic. **Antenne mit -** shunt-excited antenna.

Ableitung, Oberflaechen- surface leakage. **Wärme-** heat dissipation, h. evacuation, loss or abduction, the carrying away or conducting away of thermal energy. **Wasser-** draining or drainage of water.

Ableitungs-dämpfung leakance loss. **-glied** shunt element. **-strom** leakage or stray current. **-widerstand** leak resistance, resistance leak (of a tube).

ablenken deflect, deviate, inflect, diffract; turn off or aside.

Ablenk-gerät deflection or deflector means, time-base. **-kammer** deflection chamber, d. space (in electron microscope). **-platte** deflection plate, deflector p. **-spannungen, symmetrische** equal and opposite or symmetric sweep voltages or deflection potentials. **-spule** deflection coil, deflector yoke.

Ablenkung, horizontale horizontal or line scan, sweep or deflection.

Ablenkung, Bild- frame or picture scan or sweep. **Bord-** aircraft error, distortion of bearings (due to ship's field, etc.). **Horizontal-** horizontal or line scan, sweep or deflection. **Strahl-** ray deviation, r. deflection; r. refraction. **Zeilen-** line scan, line sweep. **Zeit-** time base.

Ablenkungs-empfindlichkeit deflection sensitivity (reciprocal of d. factor). **-generator** deflection generator, sawtooth g., time-base g., sweep oscillator. **-spannung** sweep voltage. **-spule** deflecting yoke, sweeping y., s. coil. **-wechsel, Vertikal-** vertical deflection cycle, v. or frame d. or scan. **-weite** sweep amplitude. **-winkel** angle of deviation, a. of refraction.

Ablese-fernrohr reading telescope. **-lupe** reading lens. **-marke** index line, i. mark.

ablesen (Karte) read (off from) map or chart.

Ablesescheibe (Peilen) dial of direction-finder.

ableugnen deny, abnegate.

abliegen be spaced apart or away, differ from.

Ablösung, Phasen- phase switching (in polyphase operation of ignitron).

Ablösungs-arbeit, -energie des Elektrons total internal work function (amount of energy required by electron to break away from incandesced or other metal surface), excitation energy.

Abmachung agreement, arrangement, convention, stipulation, contract.

Abmass tolerance.

Abnahme-elektrode output electrode, collector e. **-magnet** pickup magnet. **-stelle, Ton-** sound gate.

abnehmen tap, transfer (potential, etc.), take off, pick up.

Abnehmer-sprechkopf sound pickup head, reproducing h. **-stelle, Ton-** sound gate, s. pickup.

Abnutzung abrasion, wear and tear, attrition.

abplatten flatten, oblate.

abprallen rebound, be reflected.

Abprallwinkel angle of ricochet.

abrechnen make an accounting, settle an account.

Abrede, in - stellen deny.

Abreiss-effekt (Schwingungen) discontinuity effect (oscillations). **-kontakt** arcing contact (of a switch). **-methode** adhesion method (used in surface-tension testing). **-spannung** stopping potential.

abrichten fit, adjust, true, trim, dress.

Abriss abridgment, abstract, synopsis.

abrollen (über) ride (on), roll on or off, roll in contact, unroll.

abrunden round out, r. off.

Abrundungsradius contour radius.

Abrutschen, seitliches side slip (of an airplane).

Absatz wear and tear (of film), offset, reduction,recess, shoulder; subsection (in text).

absatzweise intermittently, discontinuously, progressively, gradually, by steps, stages or degrees, with intervals, stepwise.

absaugen (treat with a) squeegee.

Absaug-feld positive field, suction f., f. designed to draw away, drive or concentrate electrons. **-ring** ring-shaped or annular collector.

abschaltbar disconnectible, severable, disengageable; adapted to be disconnected, switched off or cut off.

Abschaltleistung (eines Schalters) rupturing, circuit-opening or circuit-breaking capacity (of a switch).

abschattieren shade off.

Abschattierungsstörsignal spurious signal, shading s. (in iconoscope operation, known also as "black spot", "tilt and bend","shading", etc.).

abschichten separate into layers, stratify.

abschiessen (des Getters) set off, cause vaporization (of a getter).

Abschirm-becher shield(ing) can. **-koeffizient** absorption coefficient (screening function).

Abschleifmesser grinding blade or knife.

Abschlusswand (Kondensator) end plate.

abschmirgeln abrade (with emory), grind off.

Abschneiden (von Amplitudenspitzen) clipping or lopping off of amplitude peaks or crests (in a limiter). **-(von Silben und Worten, in Telephonie)** clipping, obliteration or mutilation of speech.

abschrägen bevel, skew, scarf, slope, slant, taper, cant.

Abschrift, beglaubigte certified or legalized copy.

abschuppen scale or flake off.

Abschützung shielding, screening.

abschwächen reduce (contrast, phot.).

Abschwächer attenuator.

Abschwächungslösung reducer solution.

Abschwörung denial under oath, abjuration.

absedimentieren sediment out.

absehen (von) dispense with, disregard, let alone.

Absehen sight graticule.

Absenkung in den Höhen de-accentuation or attenuation of high frequencies, using treble de-emphasizer means.

absetzen lay off or plot (on a map or chart), shape course for; deposit, put down, interrupt.

absitzen lassen allow to settle, deposit or clarify.

Absorptionskante, K- K edge. **Umkehr-** reverse absorption edge or limit.

Absorptionskonstante absorption coefficient. **-küvette** absorption cell. **-modulation** absorption or Heising modulation method. **-rohr** absorption column or tube. **-sprung** absorption discontinuity. **-streifen** absorption bands, a. spectrum.

abspalten cleave (cryst.).

abspannen relieve of tension or strain, relax, release, stay, strain back, dead-end or terminate (a wire).

Abspann-isolator terminal or strain insulator. **-seil** guy or span rope or cable. **-stange** stay terminal, strain pole or mast.

absperren (cf sperren) shut off, cut off, confine, stop, stopper, seal off.

Absperrklappe, Drossel- throttle clack valve.

Abspieldose electric phonograph soundbox.

abspielen, direktes direct playback.

Abspielnadel phonograph needle.

Abspulen (cf. Ablauf) feeding film from magazine or feed spool.

Abstand nehmen (von etwas) refrain or desist from, dispense with.

Abstand, Augen- interocular distance. **Betrachtungs-** viewing distance. **Frequenz-** interference guard band, tolerance frequency. **Geräusch-** signal: noise ratio. **Hauptebenen-** distance between principal planes or points. **Haut-** (focal) skin distance. **Knoten-** inter-nodal distance. **Netzebenen-** grating constant (spectrometer), interlattice plane distance. **Pfeifpunkt-** singing, whistling or stability margin. **Wellen-** interference guard band, tolerance frequency. **Zeilen-** line separation, inter-line distance, pitch (telev.).

Abstand-halter spacer. **-isolator** stand-off insulator.

Abstands-messer, Augen- interpupillary distance gage. **-taste** spacer key, blank k.

absteifen strut, prop, stiffen, reinforce, support, shore, truss, stay.

Absteller automatic disconnector or cut-off device (for soldering irons, etc.)

Abstieg, Ballon- descent of a balloon.

Abstimm-änderung tuning drift (due to temperature effects). **-anzeigeröhre** tuning indicator tube, magic eye. **-bereichüberstreichung** scanning a tuning range. **-einrichtung** tuning device or means, tuning control, selector.

abstimmen tune, syntonize (circuits), shade off (phot.), harmonize.

abstimmen, auf Resonanz resonate, tune to resonance.

Abstimm-glimmröhre tuning indicator glow-tube, flashograph. **-kette** tuning arrangement comprising a plurality of oscillation circuits. **-kreis** tuned or tuning circuit, tank c. **-melder** tuning indicator (e.g., of c-r. tube or magic eye type). **-platte** tuning spade or plate, t. wand. **-schärfe** sharpness of tuning or of resonance condition, selectivity.

Abstimmung, feine sharp or fine tuning. **-, fühlbare** lazy tuning (condenser sensibly detained on passing carrier wave). **-, leise** silent tuning, quiet (avc) tuning. **-, rohe** broad tuning, coarse tuning, flat tuning (with little selectivity). -, **scharfe** sharp, fine or precise tuning. -, **schlechte** mistuning. -, **ungenaue** off-resonance tuning. -, **unscharfe** broad or flat tuning.

Abstimmung, Falsch- mistuning. **Primär-** pre-selector means.

Abstimmungs-aggregat tuning set, tuning unit (L and C). **-kreuz** tuning indicator (magic eye, etc.). **-stromkreis, Still-** silent (q.a.v.c) tuning circuit.

Abstoppen von Zeitdauern stopwatch measurement of time intervals.

abstossend, wasser- water-repellant, non-hygroscopic.

Abstossung, Coulomb- Coulombian repulsion. **Eigen-, der Elektronenwolken** repelling action of electron charges, negative space charge formation.

Abstossungs-kraft repelling force, repulsion. **-potential** repulsion potential.

Abstrahl reflected ray.

abstrahlen blast (with shot or sand), radiate, reflect.

Abstrahler radiator (ac., antenna).

Abstrahlfläche radiating, radiation or radiant surface (loudsp.), reflecting surface or area.

Abstrahlung, Schall- sound radiation or projection.

Abstrakte tracker, prolonge (organ).

Abstrebekraft centrifugal force.

abstreichen scrape or strike off, skin, strickle, lute (clay in brick mold).

abstreifen strip, wipe (off).

Abstrich downstroke or trailing edge (of an impulse).

abstufen grade (phot.), graduate, step.

Abstufung, empfindungsgerechte natural or subjective grading or spacing (in chromaticity or chromatic scale, etc.).

Abstufung, Licht- gradation of light. **Schwärzungs-** density graduation. **Ton-** grading of tones, graduation of intensity.

abstumpfen neutralize, saturate (acids), truncate, blunt; dull, deaden.

Abstumpfungsfläche truncating face (cryst.).

absuchen search, sweep, scan, explore.

Abtast-band scanning belt. **-blende (Farnsworth, etc.)** scanning aperture or hole (in Farnsworth dissector, Nipkow disk, etc.). **-dose** electric or pickup soundbox.

abtasten scan, explore, sweep, pick up, probe, search.

Abtaster, Elektronen- electronic scanner, beam sweep, spot or lever electron brush or pencil. **Fernseh-** televisor, video scanner. **Film-** film scanner (in intermediate film system). **Kathodenstrahl-** electron beam scanner. **Lichtstrahl-** spotlight scanner, scanning brush or pencil, sweep, flying or exploring spot scanner. **Linsenkranz-** lens drum scanner. **Oberflächen-** surface analyzer (to record irregularities). **Personen- (mit wanderndem Licht)** spotlight scanner for persons.

Abtast-frequenz scan frequency (= lines x frame f.). **-geschwindigkeit** pickup velocity, pickup rate (phon.). **-scheibe (cf. Spirallochscheibe)** Nipkow scanning or exploring disk, spiral disk. **-spalte, Ton-** sound scanning slit. **-stelle, Bild-** picture or film gate. **-strahl, Elektronen-** electron spot, e. beam, e. brush (iconoscope).

Abtastung scanning, scansion, exploration, sweeping out (of picture area).

Abtastung nacheinanderfolgender Bildpunkte sequential or progressive scanning of elementary areas. -, **quadratische** square-law scanning. -, **sprungweise**

interlaced scanning. **-mit unsichtbaren ultraroten Strahlen** noctovisor scan (using infra-red rays).
Abtastung, Fernseh-, von Kinofilm telecine scan of moving-picture film. **Film-** film scanning. **Geschwindigkeits-** variable-speed scanning. **Schwingspiegel-** vibrating mirror scanning (in Mihaly Telehor). **Strahl-** electron beam scan, spotlight sweep or scan. **Teil-** partial or fractional scan, coarse scan. **Ton-** pickup of sound. **Vielfach-** multiple scan.
Abtastzeile scanning line, s. strip.
Abteilung, Berufungs- appeal division or department. **Beschwerde-** appeal dept., board of appeals. **Nichtigkeits-** annulment dept. (of G.P. Office).
abtonen, abtönen shade off, tint, tone down.
abtragen plot or lay off.
Abtragung disintegration, attrition, degradation.
Abtreibschmelze refining melt.
Abtrenner, Impuls- synchronizing separator, amplitude s., clipper.
Abtrennung severance, separation, division (of a patent application).
abtreten assign, cede, convey, transfer.
Abtretender assignor.
Abtretungserklärung declaration of assignment.
Abtrift drift (of ships or airplanes). **-messer** drift meter.
Abtropf-schale jar, dish, drip, drainer. **-ständer** draining stand.
Abwägeschaltung comparator circuit organization.
abwalzen (cf. abrollen))ride on, roll.
abwandeln vary, change, alter, modify.
Abwärtstransformator step-down transformer.
abwassern (Flugzeug) rise or take off from water (airplane).
abwegig reden argue beside the point, tergiversate, dodge the issue.
Abwehr, Lärm-, Schall- noise abatement, n. suppression, n. attenuation, combating of n.
abweichen differ, vary, deviate, diverge, depart from; soak off.
Abweichung departure, difference, disparity, divergence, deviation, variation, declination, anomaly, swing, drift. **-, seitliche sphärische** transverse spherical aberration.
Abweichung, Farben- chromatic aberration. **Flächen-** zonal aberration. **Frequenz-** frequency departure, swing, deviation or drift, lilt (slow fluctuation). **Kurs-** yaw, deviation from course. **Linsen-** lens aberration. **Nullpunkt-** zero error. **Rest-** residual aberration. **Seiten-** lateral deviation.
Abweichungskorrektur achromatization (opt.).
abweisen (eine Klage) dismiss (a suit), non-suit.
Abweisungsmaterial anticipatory material or references, anticipations (in prior art).
abwenden avert, avoid.
abwerfen release (a keeper).
Abwerg waste tow or oakum.
Abwesender absentee.
Abwickelachse feed spindle.
abwickelbar developable.
abwickeln wind off, unwind, reel off, rectify (math.).
Abwickelspule feed spool, f. reel.
Abwickler, Film- magazine drum, feed reel.
Abwind down current.
abzapfen draw off, tap.
abziehbares Papier transfer paper.
abziehen print, copy; trim or true (a wheel); draw off or tap.
Abzug print, copy.
Abzug, Korrektur- proof sheet.
Abzüge machen take prints, print,

copy, multigraph.
Abzugsrohr vent stack (of arc-lamp); waste or drain pipe or tube, vent p., outlet.
abzweigen (von Strahlen) split off or deviate (rays).
Abzweigpunkt tap, distributing point, branch-off point.
Abzweigung tap, branch-off, tie line, ramification; division (of a patent application).
Abzweigwiderstand leak resistance.
Achromasie achromatism, achromatization, degree of achromatic correction.
Achse, ausser extra-axial, abaxial. **-, grosse** major axis. **-, kleine** minor axis.
Achse, Abwickel- feed spindle. **Dreh-** axis of rotation, center of r., pivot, fulcrum, hinge. **Drill-** rotator (quantum theory). **Flugzeuglängs-** fore-aft line, major axis of airplane. **Gegen-** anti-axis. **Halb-** semi-axis. **Haupt-** prime or major axis, vertical axis (of rhombic crystal). **Kristall-** crystal axis, crystallographic a. **Lande-** correct bearing or line for approach or landing. **Längs-** longitudinal axis, fore-aft line (of air and water craft). **Linsen-** optical axis (of a lens). **Makro-** macro-axis (cryst.). **Neben-** lateral axis (of rhombic crystal), secondary axis. **Säulen-** prismatic axis. **Spulen-** bob or core of a spool (film).
Achsenankerrelais axial-armature relay.
achsenentfernt abaxial, off axis.
Achsenkreuz coordinate system.
Achsenrichtung, Kristall- crystallographic axis orientation.
achsensymmetrisch axially symmetric, in axial symmetry.
Achsenwinkel axial angle.
achsig, ein- uniaxial. **ein-und-ein-** orthorhombic (cryst.). **zwei-** biaxial.
achsparallel axis-parallel, paraxial (e.g., rays beside, or parallel to, axis).
Achs-schenkel axle journal. **-stand** wheel base.
achteckig octagonal, eight-cornered.
Achtelkreis octant.
achtelkreisförmige Komponenten octantal error components (df).
achteraus, Backbord- port aft. **recht-** right astern. **Steuerbord-** starboard aft.
Achterbündel eight-wire core, quadruple twin.
achterliche Kurslinie astern course, direct line away from transmitter, "come-from" indication.
Achter-schraube octet ring. **-spule** figure "8" coil.
Acht-flach octahedron. **-polröhre** octode, eight-electrode type of tube or valve.
adaptiertes Auge, dunkel- dark adapted eye.
Adaption, Dunkel- dark adaptation, scotopia.
Adaptionszustand, Augen- (state of) retinal adaptation.
Adatom adsorption atom.
Adcocksystem (U-System) (spec.) vertical aerial or Adcock system (U-system), also spaced aerial df system.
additive Modulation upward modulation, negative m.
Ader, Zähl- pilot wire, marked w., meter w.
AEF (Ausschuss für Einheiten und Formeln) Committee on Units and Formulae, Standards or Standardization Committee.
"Aeo" Aufzeichnungslampe Aeo light, alkaline-earth oxide lamp, Aeo lamp.
Aeols-glocke, Glas- glass Aeolian bell. **-harfe** Aeolian harp.
affinieren refine.
After-kristall pseudomorph. **-kugel** spheroid.

Aggregat, Lampen- bank or pillar of lamps, battens (used on stages). **Molekül-** molecular cluster.

Aggregatzustand state of aggregation.

Agraffe stud (piano).

Ah (Amperestunde) ampere hour.

Ähnlichkeits-gesetz, -theorie similarity principle, similitude theory.

Akkomodations-entspannung relaxing of adaptation. **-fähigkeit** power of accommodation.

akkomodationsloses Auge unaccommodated or fixed eye.

Akkordion-balg accordion bellows. **-griff** keyboard.

Akkordzitter auto harp.

Akkumulator, kochender agitated or gassing accumulator (just disconnected from charger).

Akten, Anmeldungs- file wrapper, application papers.

Aktendalarm reel-end signal.

Akten-zeichen serial number, file n., docket n., reference n. **-zeichnung** drawing incorporated in docket or case records.

aktinisch undurchlässig adiactinic.

Aktiniumemanation actinon.

aktive Masse active paste.

Aktivator, Schwermetall- phosphorogen (e.g., Mn, promoting phosphorescence).

Aktivierung, Neu- reactivation, rejuvenation (of a thoriated cathode, etc.).

Akustik s.a. Schall, Ton, Klang, Geräusch.

Akustik, Bau- architectural acoustics, enclosure, space or room a.

akustisch(e) Blendung aural dazzle. **-(e) Empfindung** acoustic perception. **-(e) Gegenwirkung** acoustic reactance. **-(er) Kurzschluss** acoustic feedback or short-circuit. **-(e) Ortung** sound location, acoustic orientation. **-(e) Rückkopplung** acoustic feedback. **-(er) Widerstand** acoustic resistance, a. screen.

Alarm, Aktend- reel-end signal. **Luft-** air-raid alarm. **-ende, Luft-** all-clear signal.

Albedo albedo (reflection factor of diffusely reflecting surfaces, say, a planet).

d'Alembert'sche Wellengleichung D'Alembertian wave equation.

Alhidade alidade, sight rule.

Alkalimetall, Erd- alkaline earth metal.

Alkalizelle alkali metal photo-emissive cell.

Allein-berechtigung sole or exclusive right. **-inhaber** sole owner or holder.

allitieren coat with aluminum, aluminize.

Allstromempfangsgerät universal receiver set.

Alphabet, Funfer- oder Fünfströme- five-unit code.

Alt alto, counter tenor.

Alt, tiefer contralto.

Altersrank priority.

alterssichtig presbyopic.

Altposaune alto trombone.

Aluminium-knetlegierung malleable aluminum alloy. **-töne** aluminum reeds (accordion).

Ametall non-metal.

Ampere-stunde(Ah) ampere hour. **-windungszahl** number of ampere-turns(ats).

Amplitude, Bewegungs-, Geschwindigkeits- velocity amplitude. **Membranschwing-** diaphragm excursion amplitude. **Stör-** noise level.

Amplituden-abschneider peak limiter, clipper, lopper. **-anzeiger** glow-tube amplitude, tuning or resonance indicator, flashograph t.i. **-begrenzer** peak limiter, clipper, lopper.

amplitudengetreu of equal amplitude, with a. fidelity.

Amplituden-glimmröhre glow-tube amplitude, tuning or resonance indicator, flashograph t.i. **-schrift**

variable-area (v.-width) recording or sound track. -selektion, -sieb amplitude filter, a. separator or selector, a. discrimination selector. -spitzen, Abschneiden der clipping or lopping off of amplitude peaks or crests, by limiter means.

Amt, Eich- gage or calibration office, standards bureau. **Not-** temporary or emergency exchange or office.

amtlicher Bescheid Patent Office action, decision, official notice or notification.

Amtsblatt official gazette or journal.

Amtverkabelung central wiring, office wiring.

Analysator, Suchton- heterodyne analyzer.

anarbeiten attach, join or work onto.

anätzen etch or cauterize superficially or slightly.

anbacken bake on, fix by firing, bake superficially.

anberaumen, Termin zur Verhandlung einer Sache set or fix date for hearing a case.

Anblaston Aeolian tone.

anbohren tap, bore, drill.

Anbruch fracture.

Änderungsgeschwindigkeit remaining velocity (ballistics).

Androhung warning, caution, giving notification (of a threatened action).

Andruck-fenster pressure plate, aperture p. (m.p.). **-rolle, Gummi-** rubber-tired pressure or pad roll (m.p.). **-schiene** pressure pad, p. guide.

aneinandergesprengte Prismen broken contact (uncemented) prisms (closely fitted together).

anerkennen acknowledge, admit, recognize.

Anerkennungsurteil consent judgment.

Anfachraum generator or working space (beam tube).

Anfachung continuous excitation of waves (or point where oscillating begins); fanning, kindling. **Schwingungs-** wave excitation, w. generation.

Anfachzeit building-up period (time required for oscillating to start).

Anfälligkeit, Stör- susceptibility to trouble, interference or noise.

Anfangs-drift initial drift (of photo-cell). **-geschwindigkeit** initial velocity, muzzle v. (ballistics). **-konzentration** (cf. **Vorsammellinse und Vorkonzentration**) pre-focusing. **-strom** starting, initial, incipient or original current. **-wert** initial or early value. **-zustand** initial state or condition.

anfärben stain (superficially).

anfechten, Gültigkeit eines Patentes challenge validity of a patent, argue non-validity.

Anfechtungsprozess contested action.

anfeuchten wet, moisten or dampen (superficially).

Anflug approach (flight), coming f.

Anflug, Ziel- flight towards object or target.

Anflug-grundlinie approach of center of sector. **-kegel** sector of approach. **-manoever** approach procedure. **-richtung** direction of approach, homing direction (of an airplane). **-richtungsfunkbake** runway localizing beacon (marker). **-schneise** approach sector, corridor of approach. **-sektor** sector of approach. **-zeit** time for flight to airdrome.

Anflussgeschwindigkeit velocity of approach, afflux v. (of fluids).

anfressen corrode (superficially or slightly), eat away or eat at, erode, pit, attack, stain.

Anfressung, Rost- corrosion, honeycombing, pitting, tubercular or

channel-shaped corrosion.
anführen quote, cite, mention.
Angeber, Ton-, chromatischer chromatic pitch pipe.
Angeklagter defendant, respondent.
angelagerte Elektronen trapped electrons.
Angelegenheit case, matter, affair, in re.
angepasste Leitung matched line.
-, nicht- mismatched line.
angeregte Gruppe exciton (in dielectric breakdown).
angeschärften Enden, Verbindung mit scarfed joint.
angezapfte Wicklung tapped winding.
-, in der Mitte center-tapped winding, with midpoint tap.
angiessen cast on, pour on; color (by coat of clay).
Angleich, Farb- color matching, c. comparison.
angleichen equate, correct, compensate.
Angleichung adjustment, adaptation, (approximate) matching (of brightness, color, etc.).
angrenzend border on, adjacent, adjoining, contiguous.
Angrenzung contiguity, adjacency, adjoining.
Angriffspunkt point of engagement, attachment, action, application or attack (say, of a force).
Anguss lug (cast on); feedhead.
anhaftend adherent, inherent, be attached to.
Anhallen building-up of acoustic energy.
Anhaltspunkt clue, criterion, index; stopping point, fulcrum.
anharmonische Terme anharmonic terms or constants.
Anhäufung aggregation, accumulation, collection, damming up, concentration.
Anhäufung, Korn- grain aggregation.
Anhäufungszone zone of accumulation (in ultrasonic waves)
anheben (von Tönen, Bändern, etc.) emphasize, accentuate, underscore or favor (certain sounds or frequency bands, by tone-control or tonalizer means).
Anhebung, Bass- bass boosting, b. compensation, tone-control means to accentuate low-pitch or b. range, b. emphasis.
anheizen heating up, initial h. (of a tube); kindle.
Anheizzeit cathode heating time (till stable temperature is reached), thermal time constant (of a cathode), warming-up period.
anisotropische Verzeichnung anisotropic distortion, shear d., d. of orientation.
Anker armature, keeper (of permanent magnet); anchor, stay, guy; escapement.
Anker, Doppel-T- shuttle armature, H-type a. **Dreh-** pivoted armature. **Eichkatzen-** squirrel-cage armature. **Schwing-** oscillating armature.
Anker-gesperre anchor escapement. **-relais, Schneide-** knife-edge relay. **-ruhestellung, Relais mit mittlerer neutraler** neutral relay, relay with differentially moving armature. **-umschlag** armature travel, a. transit, armature excursion.
Anklang concord, accord, sympathy, tuned state.
Anklang finden find acceptance, become popular.
Anklingen gradual rise or waxing (of vibrations, etc.), initial sounding, intonation.
Anklingzeit starting or onset time (of a tone).
ankohlen char or carbonize (superficially, slightly or partially).
ankommender Strom incoming current.
anlagern combine, become added to, adsorbed or attached.
Anlagerungs-reaktion addition reaction. **-verbindung** addition compound.

Anlass-anode starting anode (in arc rectifier). **-beständigkeit** tempering resistance.
anlassen start, turn on, let in, temper, anneal (glass, metal, etc).
Anlasstemperatur tempering temperature, annealing t., drawing t.
Anlauf pickup, start, acceleration. **-beständigkeit** resistance to tarnish, t.-proofness. **-charakteristik** transfer characteristic.
anlaufen become coated with oxide or moisture, tarnish, become dull or dim; start, buck against (a potential).
anlaufendes Flugzeug approaching airplane.
Anlauf-farbe temper color. **-stromgebiet** range or region of incipient current flow (say, at grid). **-winkel** angle of approach.
Anlege-kassette clamp-on dark slide. **-punkt** aiming point.
anlenken pivot, fulcrum, hinge.
Anlenkungspunkt pivotal point, fulcrum, hinge.
annetzen moisten, dampen (superficially).
Anmassung, Patent- usurpation of patent rights.
anmelden (ein Patent) apply for letters patent, file an application.
Anmelder applicant.
Anmeldung, abgezweigte divisional application (part divided off from the original case or docket).
Anmeldungsakten file wrapper, application papers.
Anmerkung annotation, memorandum, remark, comment, (foot, marginal) note.
Anmerkung, Rand- marginal note or annotation, marginalia. **Streit-** annotations of litigation.
Annahme acceptance, acceptation, adoption; assumption, supposition, presumption, conjecture, surmise.
Anode, Anlass- starting anode (to strike initial arc in Hg rectifier). **Auffang-** collecting or gathering anode (of multiplier); catcher (of klystron). **Beschleunigungs-** gun anode. **Diaphragma-** aperture anode, anode stop. **Fang-** collecting anode, gathering a., (Farnsworth telev., multiplier). **Lochscheiben-** apertured disk anode, ring a. **Nachbeschleunigungs-** second gun anode. **Netz-** plate fed from power pack. **Prall-** target anode, impactor a, reflecting electrode or dynode (of electron multiplier). **Sammel-** collector anode, gathering a. **Voll-** unsplit anode (in magnetron). **Vor-** first anode (c-r. tube), screen grid connected with cathode g. and acting as "fore-anode" in power pentode (Pierce circuit).
Anoden-absinkkurve anode or plate current curve. **-ansatzpunkt** spot on anode (of an arc). **-aussortierung** anode sorting. **-blende** (anode) diaphragm or partition (in cathode-ray tube), first anode, defining aperture (of telev. tube). **-fleck** anode spot. **-gleichrichtung** plate-current rectification or detection, anode **bend r.** **-loch (cf. Anodenblende)** gun (in cathode-ray tube). **-rückwirkung** anode or plate feedback, plate reaction. **-schale** anode segment (of magnetron). **-schutznetz** screen grid, plate shield. **-spannungsaussteuerung** plate voltage excursion. **-strahlen** anode rays, canal r., positive r. **-stromaussteuerung** plate current excursion. **-unruhe** fluctuation of plate or anode feed current. **-verlustleistung** anode dissipation.
Anomalie, Nullpunkt- origin distortion.
an Ort und Stelle in situ, in the site, on the spot.
anpassen match, adapt, accommodate, adjust, fit, suit.
Anpassung, Augen- adaptation (to change in distance or focus), accommodation (to change in light intensity). **Dunkel-** dark adapta-

tion, scotopia. Falsch- Fehl- mismatching, defective focus. Hell- light adaptation, photopia. Über- overmatching. Unter- under-matching.

Anpassungs-dämpfung matching attenuation, non-reflection a. **-fähigkeit** adaptability, accommodation powers (of eye). **-netz** matching network. **-schaltung** circuit means designed to form or shape synchronizing signals. **-transformator** matching transformer.

anpeilen take bearings from or on (a radio station).

Anregung excitation, stimulation, stimulus, response, sounding (as of a Kundt glass tube).

Anregung, Kern- nuclear excitation.

Anregungs-energie excitation or stimulation energy (e.g., potential). **-faktor** excitation factor. **-spannung** exciting or stimulating potential, pre-ionization p., p. to raise an electron to a higher level out of its atomic bond. **-welle** exciton (strictly: an electron in a crystal moving in the field of a positive hole).

anreichern enrich, strengthen, concentrate.

Anreicherungsfaktor concentration factor.

anreihen arrange in, or attach to, a series or row, align.

anreissen pluck (a string or chord).

Anriss cracking, critical break.

ansässig domiciled, residing at.

Ansatz side arm or appendage (of a bulb or tube), adapter (phot.), loss, extension, tail, ear, lug, nipple, shoulder, deposit, sediment.

Ansatz einer Gleichung laying down a formula or equation.

Ansatz, Lichtton- sound-film head. **Nadelton-** sound-on-disk attachment or head.

Ansatzfläche des Objektivgewindes screw shoulder, screw collar or seat of objective amount.

Ansatz-punkt contact point (airplane landing); spot (of arc, etc., on cathode or anode), point of attachment or insertion. **-stück** extension piece, adapter piece. **-tubus** extension tube.

anschichten pile in layers, stratify.

anschiessen shoot into crystals, crystallize.

Anschlag limitation or buffer, stop, dog.

Anschlag, Auslösungs- trip dog, trigger means. **Begrenzungs-** back stop, limiting means. **Finger-** finger stop (dial switch).

Anschlagbund stop collar.

anschlagen touch (piano), strike (a string, in percussion instruments), sound or tap (a tuning fork).

Anschlag-linie striking line (piano). **-stelle** striking point.

Anschliff ground and polished surface (under microscopic examination).

Anschluss, Kopfhör- telephone jacks. **Sammel-** p.b.x. line. **Vollnetz-** (**Radioempfanger**) all-electric set (with mains supply).

Anschluss-berufung cross appeal. **-teil** terminal part, connector. **-teilnehmer, Sammel-** subscriber with several lines.

anschmelzen melt, fuse or join onto, join by fusion, begin to melt.

Anschmiegung, enge intimate or snug adhesion, adherence, engagement or fit, close contact, hug.

anschmoren scorch.

anschneiden intersect, determine a fix, obtain a bearing.

Anschnitt gating (foundry work).

Anschuss, Kristall- crop of crystals, crystallographic growth, crystallization.

Anschwingzeit build-up period (time required for oscillations to start).

Anschraubring lens flange.

ansetzen crystallize, effloresce, make up (a solution), prepare,

mix; start or strike (an arc).
an sich per se, basically or fundamentally speaking, in or by itself.
Ansicht, Gesamt- total view, general v., assembly v. **Hinter-** rear elevation, posterior view. **Seiten-** side or lateral elevation. **Vorder-** front elevation, anterior view.
Ansichtskopie release print.
anspannen strain, stretch, bend, tension, tense.
anspielen auf make allusion to, hint at.
Ansprache response; inception, onset.
ansprechen in freier Luft response in free space or air (of a microphone).
Ansprech-konstante sensitivity constant (microphone), m. responsiveness. **-wahrscheinlichkeit (eines Zählrohres)** efficiency of counter tube (likelihood to respond or register).
Anspruch (eines Patentes) claim (of letters patent).
Anspruch, Unter- sub-claim.
Anspruch geltend machen advance, prefer, raise or plead a claim or point.
Anstalt, Forschungs- research institution.
anstechen pierce, tap, open.
Anstellwinkel angle of attack; a. of pitch, blade a.
ansteuern take bearings from, keep on course, head for.
Ansteuerungsfeuer line of approach beacon; directional beacon.
Anstoss nehmen take exception to.
anstossen butt against, collide with, knock against, impulse (a circuit).
Anstrebekraft centripetal force.
Anstreichen stroking (a string), (initial) bowing (of a violin).
Anstrengung, Augen- eye strain.
Anströmrichtung direction of air flow.
Anströmungssgeschwindigkeit afflux velocity, approach v.
Anteil constituent, portion, share, component.
anteilig proportionate, in proportion, pro rata.
Anteilziffer. percentage share.
Antenne Siehe auch "Luftdraht"
Antenne, abgeschirmte shielded (downlead) antenna, anti-static antenna, screened a. **- mit Ableitung** shunt-excited antenna. **-, ausgekurbelte** reeled out, lowered or paid out antenna or aerial (through fairlead of airplane). **-, eingebaute** built-in antenna. **-, eingefahrene** retracted or drawn-in antenna. **- einziehen** reel in, haul in or retract antenna wire. **-, künstliche** artificial, mute or phantom antenna. **-, vom Sender entkuppelte** balanced antenna. **- mit seitlichem Minimum** directive antenna. **-, ungerichtete** non-directional antenna.
Antenne, Baken- beam antenna, beacon a., radio range a. **Band-** tape antenna. **Beggerow-Zeppelin** Zeppelin antenna. **Behelfs-** temporary or makeshift antenna. **Boden-** ground or earth antenna, underground or buried antenna. **Dach-** gable roof antenna. **Doppelkegel-, Doppelkonus-** cage antenna, double-cone a., sausage a. **Doppel-V-** double-V antenna, doublet a. **Einfach-** plain antenna. **Erd-** ground or earth antenna, underground or buried a. **Ersatz-** artificial antenna, mute a., phantom a. **Fächer-** fan antenna, harp a. **Fischbauch-** fishbelly antenna (mast antenna in which maximum cross-section and minimum characteristic impedance are in middle). **Flächen-** sheet antenna. **Flossen-** skid-fin antenna (on airplane wing). **Frei-** outdoor antenna. **Gegengewicht-** screened aerial. **Gemeinschafts-** block antenna, communal or party a., centralized or shared a. **Gitter-** parasol-type

antenna. **Harfen-** harp antenna, fan a. **Hoch-** open outdoor antenna (installed at, or reaching, a point high above ground). **Innen-** indoor antenna. **Käfig-** cage antenna, sausage a., hoop antenna. **Kreuzrahmen-** crossed-coil antenna. **Kurbel-** reel antenna, trailing-wire a. **Lichtnetz-** light-socket or -circuit antenna. **Linear-** plain or straight one-wire antenna. **Mikrophon-** microphone boom or outrigger. **Nahschwund-** low-angle or short-range fading antenna. **Netz-** light- or power-circuit antenna, lamp-socket a.; a. adapter, a. eliminator. **Peil-** directional antenna. **Rahmen-** frame antenna, loop a., coil a. **Rauten-** rhombic antenna. **Reusen-** cage antenna, sausage a., hoop a. **Richt-** directional or directive antenna, unilateral a. **Richtungsgeber-** director. **Rundstrahl-** polydirectional antenna, omni-directional a., omni aerial. **Schlepp-** trailing-wire antenna. **Schwundverminderungs-** anti-fading antenna. **Spiral-** (flat) spiral aerial, extensible spiral or coiled antenna, s . loop. **Stab-** rod antenna, whip antenna, upright antenna stick. **Tannenbaum-** christmas tree antenna. **Unterwagen-** under-car antenna. **Weihnachtsbaum-** christmas tree antenna. **Wellen-** wave antenna, Beverage a.

Antennen-ableitung antenna lead (in), downlead. **-effekt** antenna effect (error in df). **-ei, -endgewicht** fish (of trailing antenna). **-gebilde** antenna array. **-nahfeld** vicinity or short-range field of antenna. **-ringleitung** multiple-receiver connection to antenna. **-schwingung** swaying or swinging of antenna. **-system, Peil-** directional antenna system. **-verkürzungskondensator** antenna (shortening) condenser, aerial series cond., padding condenser. **-verlängerungsspule** serial-capacity or loading coil. **-zuleitung** antenna lead, downlead, feeder line.

Antikathode target, anti-cathode (of X-ray tube).

Antragsteller applicant, petitioner.

antreten, Beweis offer evidence. **"Zum Tanz"-** phase focusing, formation of groups of electrons in beam tube, bunching.

Antrieb, schlupffreier (cf. zwangsläufig) non-slip drive, geared d., positive d. **Druckluft-** pneumatic drive.

Antriebs-element motor element (of loudspeaker). **-zahntrommelwelle** drive sprocket shaft.

antworten answer, respond, plead, reply.

Antwort-geber revertive signal means, check-back or monitoring position indicator (of measuring instruments). **-schrift** answer.

A.N. Verfahren A.N. method (of instrument landing).

anvisieren sight, aim at.

Anwaltsgebühren counsel fee.

anwassern "land", alight or descend on, or make contact with, the water (airplane).

Anweisung, Fehl- incorrect indication or reading, misreading.

Anwuchs growth, increase, increment, accrescence, accretion.

anzapfen tap.

Anzapf-transformer (center) tapped or split transformer. **-turbine** bleeder turbine.

Anzapfung, Mittel- center tapping, mid-tap.

Anzeige reading,indication, notice, notification, advertisement.

Anzeige, gedämpfte deadbeat, non-ballistic or aperiodic reading (measuring instruments). **-, ungedämpfte** ballistic reading or indication (measuring instruments).

Anzeige, Richtungs-, eindeutige unidirectional direction-finding (with sense finding). **Stand-** deadbeat indication or reading (measuring instruments). **Zuck-** ballistic reading, flash or kick r. (of instruments).

Anzeigemarke, schwarze black dot pointer or mark (on dials) to observe or read bearings.

Anzeiger s.a. MELDER

Anzeiger, Amplituden- flashograph (and similar neon types of indicators of resonance and volume). **Aussteuerungs-** volume indicator (in sound recording). **Fern-** telemetric device (e.g., metameter of GE Co. using impulse carriers). **Fernblitz-** Keraunophone. **Glimmlicht-** flashograph, Tune-a-lite, neon tube, etc., tuning indicators. **Höchstwert-** crest indicator, peak i. **Mindest-, Minimum-** minimum (voltage) indicator. **Resonanz-** resonance indicator, tuning i. **Richtungs-** direction-finder, df indicator. **Schatten-** shadow-type tuning indicator, shadowgraph. **Seitenabrutsch-** side slip indicator. **Spannungsspitzen-** crest or peak voltage indicator, crest voltmeter. **Stand-** bearing indicator, direction-finder dial. **Verhältnis-** exponent (math.). **Wanderwellen-** surge indicator, klydonograph. **Wellen-** cymoscope. **Windrichtungs-** wind-direction indicator (say, of oscillatory-vane type).

Anzeige-röhre, Abstimm- tuning indicator tube (e.g., magic eye, etc.). **-vorrichtung** indicator device, visual reading instrument

anziehen der Rückkopplung tighten feedback, make regenerative coupling closer.

anziehend, wasser- attracting water or moisture, hygroscopic.

Anziehung, Flächen- surface attraction, adhesion. **Haarrohr-** capillary attraction. **Massen-** gravitation.

Anziehungskraft attraction, attractive power, adhesive force, gravitation.

Anziehungskraft, Erd- gravitational force.

Anzielen des Zielpunktes sighting of target.

Anzugsdrehmoment starting torque.

anzünden ignite, kindle, light, start, strike (an arc).

anzweifeln call in question, doubt.

Aeols-glocke, Glas- glass Aeolian bell. **-harfe** Aeolian harp.

Apertur, Zwitter- split focus (Walton method of scanning).

Aperturblende aperture stop or diaphragm.

Apparate-gamma apparatus gamma. **-zuleitung** instrument lead, apparatus l.

Apparat-geräusch system noise (sound recorder). **-patent** apparatus patent.

Apparaturgamma apparatus gamma.

Äquipotentialfläche isopotential or equi-p. surface or plane.

Arbeit, Ablöse- total internal work function (energy acquired by electron to break away from incandesced surface). **Reibungs-** frictional work, magnetic hysteresis.

Arbeits-eichkreis working reference system. **-fähigkeit** capacity for doing work, energy. **-gang** working, operating or procedural step or stage, run, course of manufacture, working operation or schedule. **-kennlinie** operating curve, working c. or characteristic, performance characteristic. **-kontakt** marking contact, make or operating c. (closure occurs upon energization of relay). **-kopie** studio copy, s. print. **-leistung** work, task, performance, efficiency, output. **-linie und -punkt** load line and operating point. **-punkt, Ruhe-** Q point, quiescent p. (of a valve). **-raum (Triftröhre)** gener-

ating space, working s. (of drift or beam tube). **-stellung** operating position, working p. **-strom** watt current, working c. **-strombetrieb** open-circuit operation.
Architektur, Film- settings and composition.
arglistig fraudulent.
argumentierbar arguable.
arm (z.B. verlust-armes Material) low- (e.g., low-loss material).
Arm, Kontakt- wiper arm. **Lampen-** lamp bracket. **Längs-** line or series arm (network). **Quer-** cross arm, shunt arm (network). **Tast-** wiper (arm). **Verhältnis-** ratio arm (Wheatstone bridge).
Armatur, Mast- pole hardware or fittings, accessories.
armiert, stahlband- steel tape armored (cable).
Art (Besselfunktion) kind (of a Bessel function; e.g., J_n denotes Bessel function of first kind and order n).
Atelier studio, stage, teletorium, telestudio. **Synchronisier-** scoring stage. **Ton-** sound stage, studio.
Atelier-direktor studio manager. **-kamera** studio camera. **-szenen** stage scenery, s. sets.
Atlas, Farb- color chart, c. scale (e.g., Munsell).
atmen breath, respire, throb, inhale, exhale.
Atmen des Films in and out of focus effect. **-des Kohlemikrophons** breathing of microphone (slow changes in resistance and response of carbon m.). **-des Störgeräusches** fluctuations, or waxing and waning of noise.
atmende Membran flexible or non-reflecting diaphragm.
atmendes Mikrophon breathing microphone.
Atmosphärendynamo (Stewart) atmospheric dynamo.
Atmosphärilien atmospheric agencies, the elements.
atmosphärische Störungen atmospherics, statics, strays, "X's".
Atom, Rückstoss- recoil atom.
atombindende Kraft atomic combining power, valence.
Atom-formfaktor atom form factor, atomic scattering or structure factor. **-gewicht** atomic weight. **-gramm** gram atom. **-hülle** atom shell. **-modell (Rutherford-Bohr)** R-B. atom model or conception. **-strahlmethode** molecular or atomic ray method. **-verrückung** atom displacement. **-zahl** atomic number, nuclear charge number. **-zerfall** disintegration of atom, atomic d. **-zertrümmerung** atom splitting, nuclear transmutation, a. fission.
Atü (Atmosphärenüberdruck) atmospheric excess pressure.
ätzen corrode, etch, cauterize, treat with mordants.
ätzen, tief intaglio etching (whole metal coated with resist, pattern cut out with scriber).
Ätz-figur etch pattern, etching figure. **-grund** resist, protective coating, "ground". **-schliff** ground section prepared for etching of metals.
Audienz hearing, reception.
Audiogramm audiogram (gives hearing loss vs. frequency).
Audion, rückgekoppeltes regenerative grid-current detector, ultraudion. **-, selbstschwingendes** autodyne.
Audion, Bremsfeld- electron oscillation detector or rectifier (negative anode connected with leak resistance and condenser). **Gegentakt-** push-pull grid detector. **Kraft-** power (grid-current) detector. **Schwing-** oscillating detector, autodyne.
Aufbau construction, structure, assembly, assemblage, mount, mounting, superstructure, synthesis or building up (line by line, of a telev. picture).
Aufbau, Ionisierungsspiel- Townsend structure or build-up. **Mikroskop-**

microscope mounting.
Aufbauten superstructural parts; set (filmwork).
Aufbauzeit, Funkenentladungs- formation time of spark discharge.
aufbereiten prepare, work up, dress (ore), separate.
aufblähen swell (up), bulge out, inflate, belly.
aufblättern exfoliate.
aufblenden dissolve or fade in (film projection).
Aufblendung, Fächer- fan fade-in. **Iris-** irising-in. **Kulissen-** side-curtain fade-in. **Vorhang-** curtain fade-in. **Winkel-** angle fade-in.
aufblitzen flash (up).
aufdämpfen deposit by evaporation, smoke or vaporize upon, distil (a coat or film) onto.
aufdrücken impress (a potential upon).
aufdrucken imprint, stamp upon.
Aufenthaltsort residence, domicile.
auffallendes Licht incident light.
Auffanganode (cf. Auffänger) collecting or gathering anode (of Farnsworth device), target (in a multiplier); catcher (of Klystron).
auffangen pick up (a signal by antenna), collect, intercept, catch or capture (electrons).
Auffänger collector electrode (in Farnsworth dissector), target (of cyclotron), catcher (in beam tube or Klystron).
Auffänger-dicke target thickness. **-käfig** collector electrode (in Farnsworth dissector). **-platte** target plate, impactor p. (of multiplier), collector (electrode). **-plattenverlustleistung** collector dissipation (of multiplier).
auffasern separate into fibers, unravel.
Auffassung, indeterministische indeterministic conception or interpretation.
aufflackern flare or flash up.
aufflammen flame up, blaze, deflagrate.
Auffrischen re-processing, renewing, reviving, restoring, regeneration.
Auffrischer replenisher (film development).
Aufführung, Privat- preview, private show.
aufgeben abandon (an application or a patent), forfeit, discontinue, relinquish, renounce, waive, surrender.
aufgelöste Linienbilder resolved line patterns.
aufgeschrumpft shrunk-fitted.
aufgeteilte Abbildung split focus (scanning method).
aufgewundener Exponentialtrichter twisted or coiled exponential loudspeaker.
Aufgussverfahren infusion process.
Aufhängebügel loop hanger.
Aufhängung, Mikroskop- microscope mounting. **Schneiden-** knife-edge suspension. **Spitzen-** pivot suspension, point s.
aufheben lift, raise; preserve, keep; abolish, discontinue, terminate, revoke, abate, cancel, quash, reverse (a judgment), compensate; reduce (a fraction) to its lowest terms.
Aufhebung abrogation, abolition, suspension.
aufhellen spotlight.
Aufheller Klieg light.
Aufhellung schwarzer Stellen blooming, excessive brightness, loss of contrast (in picture).
Aufhellungsorgan intensity or brilliance modulation or control electrode (of c-r. tube).
aufkeilen key on.
aufklappbar capable of being swung open or raised (on hinges).
aufklappen open up on hinges, lift up, drop (a leaf).
Aufkohlung carburization.
Auflager abutment, seat, bearing surface, support.
Auflaufen des Films take-up (by spool) of film strip. **- der Gleichlauffehler** compounding or integration of differences of synchronism.

Auflaufrolle take-up reel.
Aufleuchtdauer illuminated or light period.
aufleuchten light up, flash up, shine, glow.
Auflicht azimuthal or top illumination, light from vertical sources.
auflockern loosen (up), relax, slacken, make less rigid or less compact.
auflösen resolve (into), solve, analyze, clear out, release, open (connections).
auflösen (eine Gleichung nach n) solve an equation with respect of n.
auflösende Kraft resolving power (opt.).
Auflösung, Bild- picture resolution, p. definition; scan.
Auflösungsvermögen resolving power (micr.), solvent power. **Minimum-** minimum separabile (of eye).
auflöten solder on, unsolder.
Aufmachung scenery, settings, composition.
Aufnahme reception, detection; plotting (of graphs), pickup (telev. and film work), recording (sound), shooting or taking (of pictures), uptake or absorption (say, of a gas), sorption.
Aufnahme, weitwinklige magnascopic picture.
Aufnahme, Augenblicks- instantaneous photograph, snapshot. **Aussen-** outdoor shot, exterior shot or take, outdoor scene; phosphor screen pattern photographed with outside camera (in oscillograph work). **Ball-** bulb exposure. **Bild-, Kofferapparat für** box equipment, trunk unit. **Blindstrom-** drawing or taking wattless, idle or reactive current. **Doppel-** duplex shot. **Fahr-** running, follow or traveling shot. **Farbenfilm-** color photograph. **Farbraster-** color screen photograph. **Fern-** telephotograph, telephotographic work; long shot (m.p.). **Flieger-, Flugzeug-** aerial or airplane picture or photograph. **Freilicht-** outdoor or exterior shooting or shot. **Funkbeschikkungs-** calibration of direction-finder. **Gas-** occlusion of gas, absorption of gas (in metals, etc.). **Gitterstrom-** drawing or taking of current by grid. **Gleichtakt-** in-phase recording (on film). **Gross-** (big) close-up (m.p.). **Hoch-** vertical or upright picture. **Innen-** indoor or studio shot; oscillogram directly recorded on photographic emulsion inside vacuum space (in oscillograph work). **Kombinations-** combination exposure, composite shot (made by mask method). **Kurbel-** hand-driven shot. **Lichtton-** photographic sound-film recording. **Luftbild-** air or aerial photography. **Nach-** re-take. **Panorama-** panning shot, panoramic picture or view, pam or pan shot (studio slang). **Probe-** test picture, trial shot. **Quer-** horizontal picture. **Raff-** time-lapse photography (low-speed take, high-speed projection). **Roentgenbeugungs-** X-ray diffraction exposure. **Roentgen (strahl)-** Roentgenography, X-ray picture, skiagraph. **Rundblick-** panoramic photograph, p. view, panning shot, "pam" or "pan" (studio slang). **Schall-** sound pickup, s. recording (in form of a track). **Schallband-** magnetic steel tape recording. **Sprach- Sprech-** speech, voice, vocal or dialog recording. **Stahlband-** magnetic steel tape recording. **Stahldraht-** magnetic steel wire recording. **Strom-** current input, the taking of current (say, by a grid), current absorption. **Taucher-** submarine shooting. **Ton-, Kofferapparatur für** box equipment, trunk unit. **Verfolge-** running shooting. **Wieder-** re-take (m.p.); re-opening (of law case). **Zeitlupen-, Kamera für** camera for

slow-motion or retarded-action picture projection work. **Zeitraff-** time-lapse photography (low-speed shooting, high-s. projection).

Aufnahmefähigkeit pickup performance (of antenna); absorbability, absorptiveness, absorptivity.

Aufnahme-fähigkeit, magnetische magnetic susceptibility. **-gelände** location (film work). **-kamera, Zeitlupen-** camera for slow-motion or retarded-action picture projection work. **-kamera, Zeitraff-** time-lapse camera (low-speed shooting, high-speed projection). **-lampen** studio lights. **-leiter** chief camera man, first c. **-magnet** recording or pickup magnet (in steel band or wire sound recording). **-platte** platter (16" electric transcription record). **-raum** studio (of broadcast station), teletorium, telestudio. **-stelle** location (film shooting); receiving or translating place (telegraphy). **-techniker** camera technician. **-verfahren** pickup method. **-verstärker** recording amplifier. **-wagen, Fernseh-** television or electron camera truck, video bus, pickup camera truck, televising car. **-winkel** shooting angle.

aufnehmen trace or plot (graphs); take (current); collect or pickup (waves, vibrations); record (sound); take or shoot (pictures); copy (messages); receive, accommodate, lodge.

Aufnehmer, Schall-, Ton- sound collector (microphone), pickup; recorder.

Aufprall (cf. Prall) impact, bombardment, impingement, striking (of electrons, etc.); bound, rebound.

Aufpunkt "space" point.

aufquellen swell (up), well up.

aufrauhen granulate, roughen or grind superficially, knurl, rag, produce a nap.

aufrecht upright, erect.

aufrechterhalten maintain, keep alive (an application or patent), uphold.

aufreiben ream, broach.

Aufrichtung, Bild- erection of an image, rectification of i., inversion (when left and right, top and bottom are simultaneously interchanged).

Aufrichtungs-blende erector stop. **-prisma** erecting prism.

Aufriss elevation, vertical section, sketch.

aufrührbar stirrable, agitable.

Aufsatz any part put or set on, attachment, head-piece, neck, top, dome.

Aufsatz, Geschütz- optical instruments for artillery. **Peil-** bearing plate (in direction-finder). **Photometer-** photometer head.

Aufschaukeln resonant rise, build-up or increase of amplification (of oscillations, in proper phase).

Aufschaukelzeit time-constant of resonant amplification.

aufschichten pile up, stack up, stratify, layer.

Aufschiebung postponement, adjournment, suspension, arrest (of judgment).

Aufschlemmung solution, suspension.

Aufschlemmverfahren floatation method.

Aufschlussraum part of prospected volume subject to current or displacement lines.

Aufschrumpfung shrunk-fit.

Aufschub adjournment, time-extension, grace of time, stay, postponement, prolongation, deferment, respite, suspension, delay, moratorium.

Aufsetzpunkt (Flugzeug) landing point, ground-contact point.

Aufsicht plan or top view.

Aufsichts-betrachtung direct or frontal viewing (of television pictures). **-sucher** view finder.

Aufspaltung separation, split-up, cleavage, division, dissociation.

Aufspaltung, Dubletten- doublet separation. **Spinbahn-** spin orbit splitting.
Aufspaltungs-bilder splitting patterns (opt.). **-faktor** splitting factor.
aufspannen stretch or spread out, unfold.
aufstapeln pile up, stack up, store.
aufstäuben (einen Überzug) spray, dust or atomize (a coat, film, etc.) upon.
Aufsteck-glas slip-on lens segment. **-glas, Okular-** eyepiece correcting lens. **-kappe** slip-on cap. **-spule** plug-in coil.
Aufsteller, Bühnen- set dresser.
Aufstellpunkt location.
Aufstellung, Gitter-, Rowlandsche (Rowland's) mounting for concave grating.
Aufstieg, Ballon- balloon ascent.
aufstreichen spread on, brush on, stain (paper).
Aufstrich leading edge or upstroke (of an impulse).
Aufsuchen, Ziel- detection of target or object (by reflection of radar or sonar pulses).
Aufteilungs-filter crossover network (sound production). **-verhältnis** potentiometer ratio.
auftragen apply, charge (furnace), plot (curves), protract (geometry).
Auftrag-geber mandator, principal, client, customer. **-rollen** application rollers, inking r.
Auftreff-geschwindigkeit striking velocity. **-winkel** angle of impact.
auftreiben, das Loch enlarge, expand, broach, ream (a hole).
Auftreiber reamer.
auftrennen sever, rip up or open.
Auftrieb upward thrust, up-thrust, lift, buoyancy, ascendency.
Auftrieb, Luft- air buoyancy.
Auftriebmethode float method, buoyant-force m. (hydrometer).
auf "Vordermann stehen" be or lie in the shadow of.
Aufwand, beträchtlicher, von Mitteln elaborate means or circuit organization, appreciable outlay.
Aufwand, Leistungs- power input, p. expenditure, energy dissipation. **Mehr-** additional or extra expenditure, outlay or means (used for circuits, apparatus, etc.), complement (of tubes).
Aufwärtstransformator step-up transformer.
Aufweiterung, Gitter- increase or expansion of lattice spacing.
Aufweitung expansion, bulge-out.
aufwickeln wind up, wrap up or upon, take-up (film); unwind, unwrap.
Aufwickelspule take-up reel.
Aufwind anabatic wind, upslope w.
aufzählen enumerate, list, itemize.
Aufzehrung, Gas- gettering, gas absorption, outgasing.
aufzeichnen (Stromlinien, etc.) map or plot (tubes or lines of force, etc., often in a tank).
Aufzeichner, Schall- sound recorder. **Schallwellen-** phonodeik (recording on film). **Ton-, magnetischer** magnetic sound recorder (magnetophone, Blattnerphone, Poulsen telegraphone). **Schwingungs-** vibrograph (to record mechanical vibrations).
Aufzeichnung, Abdeck- masking recording, matting. **Bild-** picture recreation, p. delineation, p. tracing, p. synthesis. **Vielkanal-** multi-channel recording.
Aufzeichnungs-kammer recording chamber (in electron microscope). **-kontrolle** monitoring (of sound recording or sound tracks). **-lampe, "Aeo"-** Aeo light, alkaline-earth oxide lamp, Aeo lamp (variable-density sound picture recording work). **-objektiv** recording objective. **-schlitz, -spalt,** recording slit or aperture. **-träger** record or track support (sound on film, disk, etc.); record sheet. **-umfang** re-

cording range.

aufziehen mount (a photograph), wind (a dial switch).

Aufziehhebel lever ratchet.

Augapfel eyeball, bulbus.

Auge, akkomodationsloses unaccommodated eye, fixed eye. **-, alterssichtiges** presbyopic eye. **-, berichtigtes** corrected eye. **-, bewaffnetes** aided eye. **-, dunkel adaptiertes** dark adapted eye. **-, entspanntes** relaxed eye. **-, fehlsichtiges** defectively sighted (ametropic) eye. **-, fernakkomodiertes** relaxed eye. **-, fernsichtiges** far-sighted eye, hyperopic eye. **-, kurzsichtiges** near- or short-sighted eye, myopic eye. **-, nachtblindes** night-blind eye, nyctalopic eye. **-, nahsichtiges** near-sighted eye, myopic eye. **-, rechtsichtiges** normal-sighted eye, emmetropic eye. **-, unbewaffnetes** unaided eye, naked eye. **-, weitsichtiges** far- or long-sighted eye, hyperopic eye.

Augen-abstand interocular distance. **-abstandsmesser** inter-pupillary distance gage. **-adaptationszustand** state of retinal adaptation. **-anpassung** eye adaptation (to luminosity or wave-length of light), eye accommodation (to changes of distance or focus). **-anstrengung** eye strain. **-blicksaufnahme** instantaneous photograph, snapshot. **-brechungs-messer** skiascope, retinoscope. **-eigenlicht** self-light, intrinsic light of eye or retina, eye's own l. (associated with photopsia). **-empfindlichkeit** intensity discrimination, contrast sensitivity (embodied in Weber-Fechner law). **-farbenempfindlichkeit** color or spectral sensitivity of eye, spectral response of eye. **-höhle** socket of eye, orbit, orbital cavity. **-kreis** exit pupil. **-lehre** ophthalmology. **-lichtempfindlichkeit** luminosity response of eye. **-linse** eye lens. **-linsenfaserung** fibrillation of lens. **-messer** ophthalmometer, optometer. **-muschel** eye cup. **-punkt** exit pupil, eyepit. **-spiegel** ophthalmoscope, skiascope. **-stäbchen** cones (of nerve endings). **-täuschung** optical illusion. **-zäpfchen** rods (of nerve endings). **-zerstreuungsbilder** blur circles of eye.

ausätzen cauterize, destroy by caustic.

Ausbau demounting, taking apart or out, dismantling.

ausbauen disassemble, dismantle, demount.

ausbauchen bulge, swell, hollow out, belly.

ausbeulen take out dents (in a metal part), swell out, round out.

Ausbeute yield, crop, output, gain, returns, efficiency.

Ausbeute, photoelektrische photoelectric emissivity or yield. **Quanten-** quantum yield or efficiency.

ausbiegen bend out, turn out.

ausbilden fashion, form; develop, educate, train.

Ausbildung, Fach- technical education, schooling or training.

ausbleien provide with a lead lining or coating.

ausblenden mask with apertured diaphragm or stop, limit the field (of beam, pencil, etc.), diaphragm out, cut down aperture of a stop, use a field stop, iris out, circle out.

Ausblend-mittel limiting aperture, screening, masking or stopping means (for rays and radiations). **-steuerung** obturator or cut-off modulation.

Ausblickstutzen objective lens socket (opt.).

ausbohren bore or drill out.

ausbreiten spread (out), extend, expand, diffuse, propagate (waves).

Ausbreitprobe hammering test (specimen).

Ausbreitung, Licht- distribution of

exposure (on film). **Schall-, vertikale** vertical spread or diffusion of sound (should be angle subtended by audience at loudspeaker).
Ausbreitungs-geschwindigkeit velocity of propagation (of waves, etc.). **-strom** dispersion current. **-vorgang** propagation phenomenon or action. **-widerstand** diffusion resistance.
ausbringen yield, produce, bring out, take away.
Ausbruch, Gas- eruption, (violent) liberation or escape of gas. **Sonnen-** solar eruption.
ausdehnbar expansible, extensible, ductile.
Ausdehnung dimension, extent, extension, expansion, latitude, scope, range.
Ausdehnung, Langen- linear expansion.
Ausdehnungsmesser dilatometer, extensometer.
Ausdeutung interpretation, evaluation, explanation.
Ausdruck expression, term (of an equation), denotation (of a factor).
Ausdruck, Fach- technical term, t. expression.
ausdünnen thin (out).
auseinander-gehende Werte divergent values, discrepant v. **-ziehen** spread apart, draw a. or asunder.
Ausent-wicklung full development.
ausexponiert fully exposed.
Ausfall, elektrischer electric shock. **Intensitäts-** fading, fade-out (of radio signals). **Strahlen-** emergence of rays.
ausfallen precipitate, deposit.
ausfallender Lichtstrahl emergent light ray.
Ausfällapparat precipitator, precipitron (electronic).
Ausfallwinkel angle of emergence.
Ausfertigung office copy, official c.
ausfiltern filter out, exclude.
ausfindig machen ascertain, trace, discover.
ausfliessen flow out, issue, discharge, emanate.
Ausflockung (de)flocculation, coagulation.
Ausfluss outflow, efflux, discharge, emanation; mouth, outlet, drain.
Ausfluss-einschnitt contractio venae. **-rohre** outflow tube, discharge pipe. **-ton** jet tone, slit tone.
ausfransen fray.
ausfrasen notch or recess (in milling machine).
ausfrieren freeze out (e.g., in liquid-air trap).
Ausfrier-gerät freeze-out device. **-tasche** liquid-air trap (vacuum pump).
Ausführbarkeit practicability, feasibility.
Ausführung lead-in (wire).
Ausführungs-beispiel exemplified embodiment, form of construction (cited by way of example in patent specifications). **-zwang** compulsory working (of a patent).
ausfuttern (ausfüttern) line, case, bush, pad, upholster.
Ausgängen, Schalter mit 5 five-point switch.
Ausgangs-funktion trial function. **-gleichung** starting or initial equation. **-leitwert** output admittance, output conductance. **-material** starting, raw, initial or parent material. **-pupille** exit pupil. **-schlitz** exit slit (spectrograph). **-stellung** home position (of dial switch).
ausgefahrene Meterzahl paid-out or reeled-out yardage (of airplane antenna).
ausgefressene Kontakte pitted or worn contacts.
ausgeglichener Verstärker balanced amplifier.
ausgehen go out, become extinguished, originate, start, emanate, issue from.

ausgepeilte Richtung bearing.
ausgeprägt distinguished, peculiar, characteristic, singular, particular, marked, decided, significant, excellent; salient (re: poles).
ausgespannte Saite stretched or tensioned chord or string.
ausgesteuerter Tonstreifen, schwach low-modulation sound-track.
ausgetuchtes Loch bushed hole.
ausgezackte Linie, ausgezahnte Linie jagged line, serrated l., notched l., dented l.. dentated l,
ausgezeichnet distinguished, marked, excellent, singular, significant, particular.
ausgezeichnet(en) Lichtweg, Satz vom law of extreme path. **-(e) Punkte** cardinal points (opt.).
ausgezogene Linie solid line, full line, unbroken line (in graphs, drawings, etc.).
Ausgiebigkeit productiveness, fertility, abundance, yield(iness).
ausglätten (Kurven) smooth, flatten (graphs).
Ausgleich, Farb- color balancing (in color film). **Ladungs-** charge equalization. **Pegel-** level equalization, equalizing l. **Schrumpfungs-** shrinkage compensation. **Schwund-** fading compensation, volume control.
ausgleichen balance, compensate, equalize, align, bias out (by counterforce), counterbalance.
Ausgleichen (von Schallwellen) neutralization of sound waves (prevented by baffles).
Ausgleichentzerrung complementary recording.
Ausgleicher, Dämpfungs- equalizing network, compensating n.
Ausgleich-fläche (der Erdkruste) isostatic surface. **-getriebe** differential gear. **-schwungscheibe** rotary stabilizer (m.p. projector), impedance wheel. **-spannung** transient voltage, compensating v. **-strom** compensating current; balancing current (of a bridge). **-transformator** hybrid transformer, balanced t., differential t., 3-winding t.
Ausgleichung, Ladungs- charge equalization. **Schallwellen-** neutralization of sound waves (prevented by baffles).
Ausgleich-vorgang transient. **-widerstand** compensating resistance, balancing resistance, ballasting r., building-out r.
Ausguss lip, spout, outlet, drain, sink; delivery, effusion. **-schnauze** pouring lip, nozzle or mouth.
aushärten harden, set, indurate (thoroughly).
ausheizen heat, anneal, subject to (thorough) thermal treatment.
auskehlen channel, flute, groove.
auskeimen germinate; cease germinating.
auskellen ladle (out).
ausklagen (einer Geldstrafe) sue for a penalty.
ausklappbar capable of being swung out or dropped (on hinges).
auskleiden line, clothe.
ausklinken release, trip, trigger, disengage, throw out (of gear).
Ausknickung buckling, bending (at a sharp angle).
Auskopierpapier printing-out paper.
Auskoppelfeld output field, absorbing f., uncoupling f., delivery f. (in Hollmann's inverted cyclotron or electron turbine).
auskoppeln tune or balance out, neutralize, decouple (to suppress or lessen feedback).
auskurbeln (Antenne) reel out, lower, pay out the aerial (on airplane, through fairlead).
ausladend projecting, outrigging.
Ausladung projection, overhang, reach, length of action, working radius.
auslagern adsorb (at points of attachment).
Auslassventil escape valve, delivery v.

Auslauf von der Rolle point where film strip leaves drum.
Ausläufer tail, extension, streamers, off-shoot, satellite line.
Auslauf-kurve coasting curve, deceleration c. **-spitze** discharge tip.
ausleeren empty out, evacuate, drain
auslegen (für jedermann zur Einsicht) lay open (say, a patent application for public inspection), design, plan, lay out, interpret, construe.
Auslegung (eines Gesetzes) interpretation or construction (of an act or a law). **-, falsche** misinterpretation, erroneous interpretation or construction.
Auslenkhärte deflection hardness (force required to deflect phonograph needle point 100 μ).
Auslesefähigkeit selectivity.
ausleuchten illuminate, flash (an image).
Ausleuchtung, gleichmässige even illumination.
Ausleuchtung und Tilgung extinction or quenching (of fluorescence), evanescence.
Ausleuchtung, Spalt- slit illumination.
auslöschen erase, extinguish, wipe out, quench, blot, obliterate, cancel, put out, (cause to) evanesce.
Auslöschmagnet obliteration magnet, obliterating pole piece (to wipe out sound record,on wire or tape, for re-use).
Auslöschung, Fluoreszenz- fluorescence quenching, evanescence (say, by poison).
Auslöschzone dead spot, region of silence.
Auslöse-arbeit (cf. Ablösearbeit) total internal work function, excitation energy. **-daumen** releasing cam, resetting cam. **-hebel** trip lever, detent l.
auslösen emit, give off (electrons, etc.); trip, trigger, disengage, render operative, release.
Auslöser, Ball- ball release.
Auslöserknopf jack or escapement button, set-off button (piano).
Auslösespule trip coil.
Auslösung, Selbst- automatic release means, a. tripping action.
Auslösungs-anschlag trip dog. **-feder** escapement spring, release or trip spring. **-knopf** jack, escapement or set-off button or knob.
Ausmass amplitude, dimension, size, extent, quantity.
Ausmessung, Flächen- planimetering.
ausmultiplizieren eliminate by multiplication (math.).
Ausmündung orifice, outlet.
ausnehmen von except, exempt from.
Ausnehmung recess.
Ausnützungskoeffizient utilization coefficient.
auspressen press or squeeze out, express.
auspumpen pump out, exhaust, evacuate.
ausquetschen squeeze, crush or ring out, squeegee.
Ausrede plea, evasion, excuse, pretext.
Ausregelzeit decline period of control potential.
ausrichten orient, orientate, straighten, adjust, set (right).
Ausrichtung, falsche misalignment (of image or track).
ausrücken disengage, ungear, unmesh, throw out (of gear).
ausrunden round off.
Ausrundung fillet.
Aussage affidavit, allegation, declaration, assertion, affirmation, deposition, testimony, evidence, averment.
Aussagender declarant, deponent, affiant, witness.
Aussalzungskoeffizient salting-out coefficient.
ausscheiden (Teile einer Anmeldung) divide (an application for letters patent).
Ausscheidung precipitation, separation, elimination, deposit, efflorescence.

Ausscheidungs-effekt filtering, separating or excluding action or effect. **-härtung** precipitation hardening. **-punkt** point of separating out.

Ausschlag throw, excursion, deflection, deviation, kick, travel (re a coil, diaphragm, instrument needle, beam, etc.); exudation, efflorescence, scum.

Ausschlag, Schall- displacement (of a particle) (ac.).

ausschleudern centrifuge, spin out.

ausschliessen exclude, preclude, bar.

Ausschliesslichkeit uniqueness.

Ausschliessungs-gründe reasons for exclusion. **-prinzip** Pauli's equivalence principle (obs.), exclusion p.

ausschmelzen melt or fuse out, render, liquate, purify by melting.

ausschmieden forge, hammer out.

Ausschnitt cutout, notch, hole, opening, cutaway portion, louver, aperture, recording window. **Bild-** frame (of a camera); image area (on scanning disk). **Blenden-** aperture, slit (of diaphragm or stop).

Ausschuss für Einheiten und Formeln (AEF) Committee on Units and Formulae, Standards or Standardization C.

Ausschwingstrom decay current, decaying c. (of transients).

Ausschwingung decay, dying out (of an oscillation).

Ausschwingungs-verzerrung decay (transient) non-linear distortion, facsimile transient d. (in form of tailing, or as overthrow or underthrow distortion). **-verzug** hangover, tailing, excessive prolongation of decay of wave tail, in facsimile.

aussedimentieren sediment out

Aussehen, verschwommenes foggy, bleary or blurred appearance (of an image).

Aussenaufnahme outdoor or exterior picture or shot, o. scene, (c-r.) screen pattern photographed with outside camera (oscillography).

aussenden transmit, send (out), emit, give off, issue, dispatch, disseminate (of standard frequencies).

Aussen-gewinde male screw-thread, outside screw thread. **-leiter** "outer" (conductor or wire). **-linie** contour (line). **-mass** overall dimension, outside d.

aussenmittig eccentric, off center.

Aussen-schenkel outer leg, o. limb. **-taster** outside calipers. **-welt** external world, ambient.

ausser Achse extra-axial, abaxial.

ausserachsiale Strahlen extra-axial rays.

ausserachtlassen disregard, neglect, leave out of consideration.

Ausserfokusbild out-of-focus picture.

ausser-gerichtlich extra-judicial, out of court. **-irdische Störungen** extra-terrestrial or inter-stellar space noise or disturbances. **-ordentliche Komponente** extraordinary component (of rays). **-Phase** dephased, out of p. **-Tritt fallen** fall out of step or synchronism.

Äusserung, Kraft- manifestation or effect of force.

aussetzende Belastung intermittent load.

Aussetzer failure in ignition, slips (ignitron).

Aussetzung adjournment, discontinuance, deferment, stay (of proceedings), arrest (of judgment).

Aussichtsdichte reflection or specular density.

aussickern trickle out, ooze out, percolate.

aussieben filter out, exclude (by filter action); select, sift.

Aussiebung, Impuls- synchronizing separator (for sync and video signal separation), amplitude separator, clipper (in telev.).

Aussortierung, Anoden- anode sorting. **Phasen-** phase focusing, bunching (in klystron and other beam tubes).

aussparen recess, remove material from; set aside (say, marginal portions for sound track on film).
ausspringend salient, protruding, projecting, jutting.
ausspritzen squirt (out), wash by squirting, inject.
Aussprungwinkel angle of reflection.
Ausstattung scenery, settings, set, composition.
ausstellen (einer Urkunde, Vollmacht, etc.) execute (a document, power of attorney, etc.).
Aussteuerung, prozentuale percentage modulation.
Aussteuerung, Gitter- grid swing, g. sweep, g. excitation. **Lautsprecher-** operating condition in which input suffices to fully utilize loudspeaker power and performance, excursion of diaphragm. **Lichtstrahl-, maximale** clash point of light valve (in sound recording). **Röhren-** operating condition when signal (a.c.) voltage impressed on grid suffices in amplitude to swing plate current from zero to saturation, grid swing, g. excitation, amplitude of signal (a.c.) voltage. **Schirm-** full utilization of screen (up to its very border, in cathode-ray tube operation), sweeping out of screen.
Aussteuerungs-anzeiger volume indicator (sound recording). **-bereich** drive range (of a grid in an amplifier tube), grid swing, grid base, range below point where overload begins, that is, in straight portion of characteristic (maximum limit is 100 per cent modulation). **-grad** modulation percentage, depth of m., percentage m.; carrier amplitude (in telemetric or teletransmission work). **-intervall** control range, drive range. **-kontrolle mit Neonröhre** neon tube volume indicator. **-kontrollgerät** modulation meter, load indicator.
Ausstrahlung radiation, irradiancy, emission of rays, radiations, oscillations, vibrations and waves (in general, including heat and sound).
ausstreichen cancel, strike out, cross out, delete.
ausströmen emanate, emit, flow out, stream out, effuse.
Ausströmerscheinungen effusion phenomena.
austasten cut off or gate (pencil or beam, on flyback), blackout (by blanking signal); key off (a carrier).
Austastimpuls blanking pulse.
austauschen exchange, interchange.
Austausch-integral exchange integral. **-kräfte** exchange forces, e. energies. **-operator, Ladungs-** charge exchange operator. **-reaktion** substitute reaction, exchange r. **-werkstoff** substitute material.
austragen deliver, distribute, carry out, discharge.
austreiben expel, drive out, outgas (gas or vapor), cleanup, getter.
Austreibung, Gas- gas cleanup, outgasing, degasing, gettering.
austretendes Licht emergent light.
Austritts-arbeit (der Elektronen) work function (of electrons). **-blende** exit slit. **-geschwindigkeit** exit velocity, muzzle v., (re: guns); emergence v. **-pupille** exit pupil.
Austuchung bushing.
Ausübung (eines Gedankens, etc.) carrying into practice or effect, execute (an idea, invention, etc.).
Auswahl-prinzip, -regel selection principle, selection rule (in electron transition).
Auswandern shift of beam or beacon course.
auswechselbar exchangeable, interchangeable, replaceable, capable of substitution.
Auswechslung, Optik- optical system or assembly (comprising interchangeable intermediate lens barrel).

Auswertung evaluation, valuation, assay, analysis, appraisal, computation, interpretation; plotting (map).

Auswertung, punktweise point by point evaluation.

auszacken indent, notch, jag, serrate, dentate.

Auszeichnen, scharfes, einer Fläche durch Objektiv sharp focusing by lens of image in angle of field.

ausziehbar telescoping, extensible.

Ausziehtubus drawtube (re: microscope).

Auszug abridgment, extract, excerpt, synopsis, abstract.

Auszug, Balg- bellows extension. **Blau-** blue record. **Farb-** color record. **Kamera-** camera extension. **Posaunen-** telescoping or extension means (for tuning Lecher-wire or ultra-h-f systems, by short-circuiting bridge sliding along wires).

Auszugfilter, Farben- selective filter, s. screen.

Auto-elektronenemission auto-electronic emission, field or cold emission of electrons. **-frettage** cold drawing, auto-frettage. **-matenlegierung** machining alloy stock. **-matik** automatic (sharp) tuning means; any automatically operable mechanism. **-matik, Fading-, Schwund-** automatic volume control means (avc).

automatisch(e) Gittervorspannung automatic or self-biasing of grid. **-(e) Kurssteuerung** automatic or mechanical piloting. **-(er) Verschluss** automatic shutter.

A-Verstärker class A amplifier.

A W Zahl (Amperewindungszahl) number of ampere-turns (ats).

Ayrton'scher Nebenschluss Ayrton shunt, universal shunt (box).

Azetil-film acetate film stock. **-zellulose** cellulose acetate.

Azid azide.

Azimuth, Funk-, missweisender magnetic bearing. **Funk-, rechtweisender** true bearing.

Azimuthwinkel azimuth angle.

Azoverbindung azo compound.

Backbord achteraus port aft. **-voraus** port bow. **-kurve** port (left-hand) curve (in airplane manoeuvring and landing procedure).

Backe jaw, bit, die.

Backe, Einspann- clamping or gripping jaw, chuck jaw, grip (of a tester). **Schweiss-** welding die.

backen, zusammen- cake together, clinker.

Bahn path, orbit, shell (of an electron), trajectory, way, track, ecliptic (apparent path of sun), web (of paper, cloth, etc.).

Bahn, Ablauf- runway, landing ramp. **Blitz-,verschlungene** tortuous path of lightning flash. **Elektronen-** electron path, e. orbit, e. trajectory, flight of electrons. **Erd-** earth's orbit, terrestrial o. **Fahr-** track, runway (airfield), travel beam. **Film-** film track, f. channel. **Flächenstrom-** current sheet. **Gleit-** chute (film), slide (way), shoot **Herz-** cardioid-shaped path (of electrons). **Kreis-** circular path, orbit. **Lande-** runway (for landing of airplanes). **Leit-** transit path. **Papier-** paper web. **Roll-** cycloidal path (of electrons); roller type conveyor, runway (airport). **Rosetten-** (der Elektronen) rosette-shaped path (of electrons). **Rückstoss-** recoil track. **Schwung-** sound take-off drum (m.p.). **Spin-** spin orbit. **Start-** runway. **Tauch-(der äussersten Na-Elektronen)** dip orbit (of peripheral Na electrons). **Wirbel-,abgehende** trailing vortices (airfoil)

Bahn-aufspaltung, Spin- spin orbit splitting. **-durchmesser** diameter of orbit. **-elektronen** orbital electrons.

bahnen beat, smooth, clear (a way or path)

Bahn-ende, Mesonen- mesotron track end. **-impuls** orbital moment. **-kreis, Roll-** cycloidal path of motion (of electrons). **-schling** orbital loop. **-spur** track. **-verlauf** ray path, r.tracing, trajectory.

Bajonettverschluss bayonet joint, b union (as used in bayonet-type lamp holders).

Bake, Gleitweg- glide path beacon.

Bakenantenne beam antenna, beacon a.,radio range aerial.

Balancier beam.

Balg(en) bellows

Balgenauszug bellows extension.

Balgfalten bellows folds.

Ballast-röhre absorber tube or valve (operative during spacing periods). **-widerstand** ballast resistance, b.resistor.

Ballaufnahme bulb exposure. **-auslöser** ball release.

ballen form into balls, conglomerate, cake.

ballig crowned, cambered.

Ballon carboy, balloon (flask).

Ballon, Säure- acid carboy.

Ballon-abfüller pump-equipped balloon flask. **-abstieg** balloon descent. **-aufstieg** balloon ascent. **-variometer** balloon statoscope.

Ball-sender re-broadcast station, repeating station, s. belonging to a network.

Ballungsfähigkeit coalescing, balling or caking property, conglomerableness, spheroidizing property (cryst.).

Balsam, Kanada- Canada balsam.

Bananenstecker split (banana-shaped) plug, b.plug.

Band ribbon, tape, band, strip, tie, strap, bond, belt, hoop.

Band, besetztes full band. **am laufenden-fabrizieren** conveyor-belt manufacture, large-scale production.

Band, Führungs- control strip (of a printer). **Lautschrift-** sound recording band, tape or strip (magnetic sound recording). **Loch-** scanning belt (telev.). **Übertragungs-** signal band.

Bandabschattierung, nach rot oder violett degradation of bands to red or to violet.

bandagieren fit with tires, tire; bandage, wrap or tie with ribbon or tape.

Band-antenne tape antenna. **-aufnahme, Schall-, -aufzeichnung, Schall-** magnetic steel tape recording of sound. **-breiteregler** band-width control.

Bändchen-galvanometer twisted strip or band galvanometer. **-mikrophon** ribbon microphone, velocity m.

Band-einengung squeezing or compression of band. **-eisen** hoop iron, strip i., band i.

Banden, Ober- overtone absorption bands. **Schwanz-** tail bands (in afterglow)

Banden-besetzungsgrad degree of filling of bands. **-kopf** band-head, band edge (spectral analysis). **-zug** band progression. **-zweig** branch, part of series of lines forming a band.

Bandfilter band selector (circuit), preselector.

Band-filter von grosser Lochweite broad band filter. **-förderer** band or belt conveyor **-führung, Film-** aperture guide. **-haken** bridle wire, tie hook. **-kante** band edge, b.head. **-mikrophon** ribbon microphone, band m. **-mitte** mid band. **-öse** bridge guide. **-sperre** band-exclusion filter. **-umkehrung** speech band inversion (secret telephony).

Bank, optische optical bench.

Bank, Prüf- (lens) test bench (opt.)

Bär ram (of a press).

Barkhausensprung Barkhausen effect or jump (abrupt changes in magnetization).

Barometer, registrierendes barograph.

Barometer, Dosen-. (Feder-) aneroid barometer. **Gefäss-** cistern barometer. **Gefässheber-** combination cistern and syphon barometer. **Heber-** syphon barometer.

Barometer-korrektur barometric check (df). **-säule** barometric column.

barometrische Reglerdose control aneroid.

Baryumazid barium azide.

Basenbildner base former, basifier, basifying agent.

Basisentfernungsmesser base range-finder.

Bass-anhebung bass boosting, b.compensation, tone control means to accentuate, emphasize or underscore b. or low-pitch notes, b. emphasis. **-balken** bass bar (string instrument). **-kasten** bass box (musical instruments).

Bassin tank, cistern, reservoir, basin, bowl.

Bastler amateur, fan, "ham".

Batterie-glas battery jar. **-klinke** battery jack.

Bauakustik architectural acoustics, theater and auditorium acoustics.

Bauch (einer Schwingung) loop or antinode, internode (of an oscillation).

bauchig bellied, bulgy,

Baufehler, Kristall- crystal defects.

Baukunst, Film- setting and composition art.
baumähnlich tree-like, arborescent, dendriform.
bauschen swell out, bag, bulge, puff up; refine (tin).
Bauteil construction part, structural part or element.
Bauten, Auf- superstructural parts.
beabsichtigen intend, plan, contemplate.
Beamter, Funk- radio operator, station official.
Beanspruchender claimant.
Beanspruchung stress, strain, load; claim.
Beanspruchung, Prioritäts- priority claim. **Schlag-** shock load or stress, blow l., impact l.
beanstanden object, reject, criticize adversely.
beantragen apply (for), petition, make a motion.
Beantwortung, Klage- reply, answer, defence (plea).
bearbeitbar workable, machinable, processable, treatable.
Bearbeitbarkeit, spanabhebende free cutting machinability.
bearbeiten work (on), tool, finish, dress, machine.
bearbeiten, kalt cold-work. **-,warm** hot-work.
Beaufschlagung electric stress (placed on an insulator link or unit), acoustic action or sound pressure (brought on microphone), impacting (by electrons).
Beauftragter attorney, agent, delegate, commissioner, mandatory.
Bebenmesser seismometer, seismograph (device to measure and record tremors).
Becher, Abschirm- shielding can, can, shield. **Zink-** zinc case or cylinder (of dry cell).
Becher-block encased fixed tubular paper or mica wound or wrapped condenser, solid-dielectric fixed capacitor. **-glas** beaker. **-glaskolben** Erlenmeyer flask.
Becken cymbals. **Tamborin-** tambourine jingles.
Bedämpfung (introduction of) resistance, damping, attenuation.
Bedämpfung, Eingangs- input resistance, damping.
Bedarf need, demand, requirement.
Bedarf, Leistungs- power absorption, p.dissipation, p.requirements.
Bedarfpegel, Leistungs- power reference level.
bedeckt mit Cäsium caesiated.
Bedeckung surface coverage. **-des Netzes** coverage factor.
Bedeckung, Wandstoff- (cf **Bekleidung**) wall draping (ac.).
Bedeckungsfaktor, Pol- pole arc, p. pitch percentage.
Bedeutung import, importance, meaning, sense, significance.
bedienbar, leicht easily manipulable or operable.
Bedienbarkeit manipulability, manoeuvrability.
Bedienungsvorschrift service or working instruction, directions for use.
bedingen stipulate, contract for, condition.
Bedingungsgleichung equation of condition.
Bedrahtungsplan wiring diagram.
Beeidigung confirmation by oath.
Beeinflussung influence, control action, modulation, interference. **-,gegenseitige** mutual action, interaction. **Gitter-** grid control, grid modulation. **Sprach-** voice control, modulation by voice.
Beeinflussungsröhre modulating tube, modulator t.
beeinträchtigen impair, injure, prejudice, affect adversely, encroach upon, infringe, detract from.
Befehl, Einhalts- injunction, interdict. **Vollziehungs-** writ of execution, warrant.

Befehls-rohr (cf Folgerohr) pilot or master thyratron (fires first, in a trigger circuit). **-stelle** control unit (airport). **-übertragung zwischen Flugzeugen** inter-aircraft voice or command communication.

befestigen secure, fasten, attach, fix; fortify, strengthen.

Befestigungsring ring fastener, securing ring.

befeuchten moisten, dampen, wet, water.

Befeuerung, Flughafen- airport beacon service. **Mast-** mast beacon. **Nachtstrecken-** route beacon for night flying service.

beflecken spot, stain, soil, blur.

befördern forward, further, promote, convey, aid, transport.

begehen inspect or patrol (lines, on foot).

begiessen coat (film), wet, moisten, water, irrigate.

beglaubigte Abschrift certified, legalized, authentic or attested copy.

Begleit- concomitant, accompanying, attendant.

Begleiten zum Fernsehbild sound or audio action accompanying television or video program.

Begleiter associate, satellite, congener (if similar).

Begleiter, Eisen- elements accompanying iron, iron associates, congeners of iron (if similar).

Begleiterscheinung attendant, accompanying, or secondary phenomenon or action, satellite effect.

Begleit-körper accompanying substance or body, congener (if similar); impurity, foreign substance or admixture. **-linien** satellite lines. **-saiten** accompaniment strings. **-stoff** accompanying body or substance, congener (if similar); impurity, foreign substance.

Begrenzer, Amplituden- peak limiter, p.clipper, p.lopper. **Verstärkungs-** output suppressor or limiter, gain spoiler.

Begrenzungsanschlag back stop, limiting means.

Begriff concept, conception, idea, notion, precept.

Begriff, Ober- preamble, introductory portion of German patent claim setting forth prior art.

begründen establish, found, constitute, substantiate, create.

Begründung (eines Einspruches) argumentation (in support of an opposition).

Begründung, Berufungs- arguments or grounds for appeal.

Begründungsfrist period for giving grounds or to argue a case.

Begussmasse slip, engobe (cer.).

behaftete Oberfläche, gas- gas-contaminated surface (gas adhering to or adsorbed by surface).

Behaglichkeitswert comfort value, c.factor.

behandelbar manipulable, tractable, workable, processable, treatable.

Behandlung treatment, processing, manipulation.

Beharrungs-moment moment of inertia, rotational i. **-punkt** center of inertia, c. of mass, centroid. **-vermögen** inertia. **-zustand** steady-state condition, permanence, persistence, freedom from transients, resistance (of a machine).

behaupten contend, allege, state, hold, affirm, assent, maintain.

Behelfsantenne makeshift, temporary or emergency antenna or aerial.

beherrschen master, control, override, rule, overcome, dominate.

behinderte Rotation hindered or inhibited rotation (of molecules).

Behinderung hindrance, hindering, inhibition, restraint, impediment, obstacle.

Behinderung, raumliche steric hindrance. **Fliess-** inhibition of plastic deformation.

Behmlot fathometer, echo depth sounder; Behm sound ranging altimeter, sonic a.

beidäugig binocular.

beidrehen auf Anfluggrundlinie turn to or steer for center of approach sector.

Beikreis epicycle.

Beilage annex, schedule, supplement, enclosure.

Beilauf, Jute- cable filler, jute f. or compound.

beimengen, beimischen admix, add, intermix, intermingle.

Beintaste bone key.

Beisitzer aid, aide, assessor.

Beispiel, Ausführungs- exemplified embodiment (re:patent specifications).

beispielsweise by way of example, exemplified.

beistimmen accede, assent or agree to, acquiesce in.

Beiwert constant, coefficient, parameter, factor, co-ordinate.

Beiwert, Schluck- absorptivity.

Beiwerte (Peilung) correction values (to compensate bearing errors).

beizen corrode, mordant, cauterize, etch, pickle.

Beiz-brüchigkeit acid brittleness. **-mittel** mordant, corrosive, caustic. **-tonung** mordant toning.

bejahenden Sinnes in the affirmative (sense).

Bekämpfung, Geräusch-, Lärm- noise abatement, n.suppression, combating of n.

Bekanntmachung über Erteilung eines Patentes announcement of grant or allowance of a patent.

Bekanntmachungsgebühr lay-out fee, announcement f.

Beklagter defendant, respondent. **Berufungs-** appellee. **Wider-** cross- or counter-claim defendant.

bekleiden clothe, cover, coat, line, face, box.

Bekleidung, Wand-, schallschluckende baffle blanket, sound-absorbent wall draping or lining, tormentor, gobo (consisting of portable wall covered with absorbent material, used in m.p. studios).

Bekräftigung, Wahrheits- affirmation (of testimony).

beladen, eine Oberfläche mit Wasserstoff, etc. activate, sensitize or charge a surface with h, form an H skin.

Beladung, Vor- pre-charge, preliminary c.(of sorptive in sorbent).

Belag, Prioritäts- priority proof. **Spiegel-** mirror coating, reflecting film, silvering. **Strom-** current coverage or distribution (amperes per centimeter of periphery).

Belagverlust surface leakage loss (dielectric).

belangen, gerichtlich bring legal action against.

Belastbarkeit, Hoch- high load or current-carrying capacity.

belasten load, charge, weight, burden.

belasten, punktförmig lump-load.

belastet, stark heavily or highly loaded.

Belastung, aussetzende intermittent load. **Querspulen-** leak load.

Belastungsspitze peak load, crest l., maximum l.

beledert leather covered, leather-lined.

Belegung coat (of condenser), plating, deposit, film, covering, density (of electrons in beam).

Belegung, magnetische magnetic induction, m.charge, m.field, seat of m.flux. **Kondensator-, bewegliche** rotor plate.

Beleuchter light electrician, scaffold or top light controller.

Beleuchtung illumination, illuminance, irradiance (in lux or meter-candle units).

Beleuchtung, gerade direct illumination. **-, senkrechte** scaffold lighting, overhead l., top l. **-, spaltfreie** apertureless illumination.

Beleuchtung, Dunkelfeld- dark ground illumination. **Effekt-** effect light, spot or fancy l. **Gegen-** background illumination. **Hellfeld-** bright ground illumination. **Hinter-** Rembrandt illumination, half-back i. **Rampen-** footlights. **Zweizweck-** dual-purpose illumination.

Beleuchtungs-brücke traveling light crane. **-linse** condenser lens, illuminating l., bull's eye l. **-messer** illumination meter, i.photometer, illuminometer. **-mittel** illuminant. **-stärke** illumination (ratio luminous flux to area of element of surface, in lux units, etc.). **-stärkemesser** illuminometer, illumination photometer. **-trommel** exposure drum, recording d.(on which film is exposed to modulated light·beam).

Belichtung, Logarithmus der log. exposure. **Mehrfach-** superimposing, multiple exposure. **Nach-** post-exposure. **Ruhe-** average lighting, steady l., unmodulated or no-sound l. (in sound picture recording). **Vor-** priming illumination (of a cell), pre-exposure, preliminary exposure, or i.

Belichtungs-bereich, Unter- toe gradient of characteristic H. & D. (Hurter & Driffield) curve, toe range of under-exposure. **-fenster** aperture (in a.plate), gate, photocell window. **-kanal** film track, f.channel (camera). **-lampe** exciter lamp. **-mass** exposure level (of Decoutes). **-messer** exposure meter (e.g. with barrier-layer cell, used in sound recording, photography, etc.), cinophot (m.p. film pocket photometer); brightness meter, turbidity m. **-rolle** printing drum (in p.machine), sound recording drum; scanning point, translation p.(in sound film reproduction). **-schieber** exposure lid, e.shutter. **-spielraum** latitude or range of exposure, e.range. **-stelle** sound gate. **-tabelle** exposure time table. **-teil, Unter-** toe region of under-exposure (film). **-trommel** exposure drum, recording d.. (sound film). **-zahntrommel** printing sprocket, main s.

Belieben discretion, (at) random, selection, at will, ad libitum.

beliebig haphazard, (at) random, arbitrary, any...at all, any... whatever, optional.

Bemessung dimensioning, proportioning, choosing; size, dimensions or proportions.

Benachteiligter aggrieved party.

Benachteiligung der hohen Frequenzen de-accentuation, de-emphasizing, slighting or partial suppression of high frequencies, attenuation of treble or high-pitched frequencies.

benetzbar wettable, capable of being wetted or moistened.

Benetzungswärme heat of wetting.

Benutzung eines Patentes working or reduction to practice of a patent.

Beobachtungs-fehler error of observation. **-fenster** bezel, observation window, peep-hole. **-rohr** observation tube (in electron microscope).

Beratung advice, consultation, counsel.

berechenbar calculable, computable, appraisable.

Berechnung, Bildkraft-, halbklassische semi-classical image-force calculation.

Beregnungsversuch rain test, wet t. (of insulator).

Bereich, Einfluss- sphere of influence, radius of action. **Entzerrungs-** frequency range of equalization. **Klangintensitäts-** dynamic range. **Lautstärke-, der menschlichen Stimme** human voice intensity range. **Mitnahme-** range of forced oscillation, entrainment range, coherence r., pull-in-step range. **Schwärzungs-** density range (Y) (say, from .3 to 3). **Sperr-** suppression range. attenuation r., exclusion band.

Bereich-filter, Frequenz- band-pass filter. **-melder, Wellen-** wave-band indicator. **-schalter, Wellen-** wave-band switch.

Bereinigung des Minimums (cf **Enttrübungsrahmen**) zero clearing (df).

Berg, Energie- energy hill, e. barrier. **Potential-** potential barrier. **Wellen-** peak, crest or hump of a wave.

Berg-fett mountain tallow, ozocerite. **-fleisch** mountain flesh (asbestos). **-fluss** colored quartz. **-glimmer** margarite. **-harz** mineral pitch. **-kristall** rock crystal. **-öl** petroleum. **-pech** mineral pitch, asphalt. **-wachs** ozocerite, mineral wax.

Bericht, zusammenfassender survey report, review, summarizing article.

berichtigen correct, amend, adjust, rectify, redress.

berichtigendes Glas correcting lens.

berichtigtes Auge corrected eye.

Berichtigungs-gerät compensation means (cam, etc., in df). **-walze** adjustment drum (range finder). **-wert** correction value, c.factor.

berieseln cause to flow or trickle over, water, irrigate, douche, spray, scrub, wash, sprinkle.

Berieselungsapparat sprinkler.

Bernstein amber.

Berstdruck breaking limit, bursting l., explosion or cracking l.

Berufshandkamera professional hand camera.

Berufung appeal, brief for appeal. **-einlegen gegen eine Entscheidung bei höherer Instanz** take an appeal from a decision to a higher court.

Berufungs-abteilung appeal division, a.department. **-begründung** arguments and grounds for appeal, confirmation of appeal. **-beklagter** appellee, defendant in appeals court, respondent in an appealed cause. **-instanz, zuständig sein als** to have appellate jurisdiction. **-klage** appeal. **-kläger** appellor, appealer, appellant.

beruhen auf based on, due to, derived from, conditioned by, predicated upon.

beruhigen quiet, steady, stabilize, smooth, calm, kill (melt).

Beruhigung stabilizing (weight).

Beruhigungs-flügel intermediate blade, anti-flicker b. **-frequenz** fusion frequency, critical no-flicker frequency (m.p.). **-kapazität** smoothing capacitor **-spule., Eisenkern-** smoothing choke, filter c. **-vorrichtung** arc-silencer (suppresses hum and frying noise). **-widerstand** steadying resistance, ballasting r.

berührend, doppelt- bitangent.

Berührende tangent.

Berührungs-ebene tangential plane. **-gefahr** risk or hazard of electric shock. **-linie** tangent. **-punkt** contact point, touching p. p.of tangency or osculation. **-schutz** protection against electric shock hazard.

berührungssicherer Stecker shock-proof plug.

Berührungs-spannungsschutz protection against electric shock hazard. **-stelle** place of contact, osculation point (between two curves).

besäumen hem, edge, border, shear, trim, square (sheets).

Beschallung radiate or cause sound o r ultra-sound waves to act or impinge upon, acoustic irradiation.

Beschaltung wiring.

Bescheid, abschlägiger adverse action or decision, refusal, disallowing action of a Patent Office Examiner.

bescheinigen certify, attest.

beschichten coat (a film with emulsion), form a layer or a film, apply emulsion.

beschichteter Film, doppelt- double-coated film, sandwich f.

beschicken load (phot.), charge, feed.

Beschicker, Funk-, automatischer automatic radio quandrantal-error compensator or corrector means, cam compensator.

Beschickung correction or compensation (in df work), radio compass calibration.

Beschickung auf gleiche Temperatur reduction to equal temperature.

Beschickungsaufnahme, Funk- calibration of radio direction-finder.

beschiessen bombard (with electrons, etc.).

Beschlag, Oxyd- coating or film of oxide.

Beschläge fittings (of metal), fixtures.

Beschlagnahme seizure, attachment, arrest, distraint, distress.

beschleunigen accelerate, expedite, hasten, speed up.

Beschleuniger, Resonanz- magnetic resonance accelerator, cyclotron, induction electron a., betatron, rheotron. **Vielfach-** cyclotron (type of) tube with two dees and spiral beam accelerating electrons or ions in stages.

Beschleunigung, Erd-, Fall- gravitational acceleration, acceleration due to gravity. **Nach-** post-acceleration (by second lens) (in c.-r. tube).

Beschleunigungs-anode gun anode. **-einrichtung** high-potential transformer and vacuum tube (for nuclear disintegration). **-elektrode** accelerator electrode. **-elektrode, Nach-** second or additional gun anode or accelerator electrode, post-accelerator, after-accelerator (c.-r. t.). **-gitter** accelerator grid (c.-r. tube); screen-grid (of tetrode) **-linse aus zwei Lochelektroden** double aparture accelerator lens (c.-r. tube). **-messer** accelerometer. **-spannung** gun potential, accelerator p., beam voltage. **-stufe** target stage, accelerator s. (of multipler). **-system** gun system (c.-r. tube).

Beschluss action (of Patent Office), decision, decree, order (of a court of law).

Beschneideglas trimming glass, print trimmer.

Beschneidung clipping (loss of initial or final speech sounds).

Beschreibung specification (of a patent), description, outline, sketch, characterization. **-, weitgehende** broad specification. **-von Bildern auf Schirm** picture tracing, re-creation or delineation on screen.

Beschriften making a record or sound track upon, impressing a film or blank with acoustic actions.

Beschriftung recording (on film).

Beschussprobe shooting or bombardment test.

Beschwerde-abteilung appeals department, Board of Appeal. **-einrede** rejoinder (in an appeal). **-gericht** court of appeals.

-schrift appellatory plaint, appeal papers. **-verfahren** appeal action or procedure.

beschweren load, weight, burden; complain.

Beschwerung impedance.

beschworen take an oath, sign an affidavit, declare under oath.

Beseitigung, Geräusch- (an Klebestellen) blooping elimination.

besetzen populate.

besetztes Band full band.

Besetzung cast (m.p.)

Besetzungs-grad der Banden degree of filling of bands. **-verhältnisse** size of audience in a playhouse. **-zahl (der Elektronen in einem Niveau)** extent to which level is populated by electrons. **-zahl, Höchst-** maximum number of electrons in shell.

Besitzurkunde title deed.

Besonnung insolation.

Bespannung baffle cloth.

besponnen covered, braided.

Besprechung mit Eisendrossel modulation by voice action with magnetic modulator.

Besprechungs-anlage sound pickup outfit. **-mikrophon** sound pickup microphone, s. collector. **-punkt** point where sound action is impressed. **-raum** sound studio.

besprochener Modulator voice-impressed or voice-actuated modulator.

Bespulung provide with winding, coil or loading means.

Besselfunktion erster Art Bessel function of the first kind.

beständig stable, constant, durable, permanent, steady, fast (colors), resistant, proof; continuous, invariable.

beständig, feuer- resistant to fire or heat, fireproof, refractory. **flammen-** flame-proof, flame-resistant. **frost-** freeze-proof, resistant to frost. **hitze-** stable or resistant to heat, heat-proof, thermostable, refractory. **koch-** resistant, stable or fast to boiling (dyes, etc.). **lichtbogen-** arc-proof, arc-resistant. **säure-** stable to acid action, fast to a. **wasser-** stable in or towards action of water.

Beständigkeit, Anlauf- tarnish-proofness, resistance to t. **Raum-** volume stability. **Temperatur-** temperature stability, unaffected by temperature (changes).

Bestandteil constituent, ingredient, part, component. **Fremd-** foreign or extraneous matter or substance.

Bestätigungsurteil confirmatory decision.

Besteck dead reckoning, ship's position; set of instruments or utensils. **Funk-** radio fix.

Besteckungskosten trimming expense (re :arc-lamps).

Bestimmung determination, definition, clause, provision, stipulation, ascertainment.

Bestimmung, Mengen- quantitative determination, q. analysis.

Bestimmungs-gleichung defining equation. **-grösse** determinant. **-stück** factor, parameter, operating datum.

Bestrahlung, Schall-, hohe high radiation (distribution) efficiency. **Sonnen-** insolation, irradiation by solar rays.

bestreichen wipe (over), sweep out, scan, explore (say, the picture area or screen with beam, pencil or spot); cover (a range), contact, stroke (a magnet).

bestreitbar disputable, contestable.

bestreuen strew or sprinkle (say, powder) over.

Bestwert optimum value, optimal v., most favorable v.

Betätigungsspule working coil, tripping c., differential c.

Beteiligter party concerned, interested or involved.

Betonung accentuation, emphasis, action or stimulus (brought upon).

Betrachtung, unmittelbare direct viewing. **Aufsichts-, Direkt-** direct or frontal viewing (of television pictures).

Betrachtungs-abstand viewing distance. **-apparat** viewing apparatus, kinetoscope. **-richtung** sight-line, line or direction of regard. **-tisch** film viewing machine. **-zeichen** reference letter, r. numeral, symbol.

betreiben run, work, operate, exploit.

betreiben mit Arbeitsstrom operate on open circuit. **-mit Ruhestrom** operate on closed circuit.

Betrieb, Schwer- heavy-duty operation or service.

Betrieblautsprecheranlage works or plant public address system (loudspeakers distributed throughout a factory, etc.).

betrieblich operational, operative, functional, working, concerning actual use in service.

Betriebsbuch station log.

betriebsfähig operable, workable, in (good) working order.

Betroffener aggrieved party.

betupfen dab, dip, tip, spot, stipple.

Beugungs-aufnahme, Roentgen- X-ray diffraction exposure or pattern. **-bild, -figur** diffraction pattern, picture or image. **-fransen** diffraction fringes. **-gitter** diffraction grating. **-scheibchen** diffraction disk. **-winkel** diffraction angle.

beurkunden authenticate, legalize, record.

Beutel-filter bag filter. **-sieb** bolting sieve, bolter.

Bevollmächtigter attorney, agent, proxy, assignee, mandatory, mandatary.

bevorzugte Kristallachsenrichtung preferred crystallographic axis orientation. **-Orientierung** non-random orientation, privileged o.

bevorzugte, nicht- Orientierung randomly oriented, with non-preferential or non-privileged orientation.

Bevorzugung (von NF oder HF) accentuation or emphasizing (of AF or of RF and de-accentuation or de-emphasizing of the other).

Bevorzugung, energetische energy preference.

bewaffnetes Auge aided eye.

bewähren prove useful or good, give satisfactory results, prove (itself).

bewegliches, entfesseltes Mikrophon following microphone.

Beweglichkeit mobility, fluidity (being the reciprocal of dynamic viscosity).

Bewegung motion, movement, stir, travel, action (in film projection).

Bewegung, fortschreitende progressive motion. **-, hin- und hergehende** reciprocating movement, to-and-fro m., rocking m., oscillating m., shuttling motion. **-, rein sinusförmige** plain harmonic or sine movement. **Brown'sche** Brownian movement. **Kreis-, Kreisbahn-** circular motion, gyration, movement in an orbit, orbital m. **Leitbahn-** transit path motion. **Rollbahnkreis-** cycloidal path motion. **Rotations-** rotational motion, rotary m.

Bewegungs-amplitude velocity amplitude. **-eindruck** sensation of motion, illusion of m. **-elektrizitätslehre** electrokinetics. **-empfänger** velocity microphone, pickup or detector of motion. **-energie** kinetic energy, motional e. **-gleichung** equation of motion. **-grösse** kinetic or motional quantity or magnitude, impulse, momentum. -impedanz motional impedance. -kraft motive force, m. power. -lehre mechanics (re: machines), kinetics (re: motion of bodies and forces acting thereon), kinematics (re: motion in the abstract), dynamics (re: force action on bodies). -mikrophon velocity microphone, pressure-gradient m. -rohr mobility tube. -stufe stage of motion, phase of progress. -übergang motional transient. -vorgang motional action, cinematographic action.

bewehrt, stahlband- steel-tape armored.

Bewehrung re-inforcement, armoring, fittings (hardware, of an insulator), sheathing.

Beweis, der Gültigkeit eines Patentes Abbruch tut evidence impeaching, or prejudicial to validity of, a patent.

Beweis, Neben- secondary proof, collateral evidence, circumstantial e.

Beweis-beschluss direction or order for evidence. **-kraft** conclusiveness, evidential value.

Beweislast onus of proof, burden of p.

Beweis-material (documentary) evidence, evidential material, testimony, proofs. **als- vorlegen** tender in evidence. **-recht** law of evidence. **-stück** exhibit, document, proof. **-wert** evidential value.

bewerten weight.

Bewertungsfilter weighting network.

Bewetterung air conditioning.

Bewilligung grant, allowance, permission.

Bezeichnung marking, denotation, notation, sign, symbol, designation labeling (for identification), color coding (of conductors).

bezichtigen accuse of, charge with.

Bezieher, Piano- piano stringer.

Beziehung, Wechsel- interrelation, correlation, reciprocal or mutual relationship.

beziehungsweise and/or;...or...or both, and; or; as the case may be, respectively, or rather.

Bezirksgerichtshof district court.

bezogene Farben related colors.

Bezugsdämpfung volume loss, volume equivalent. **Rückhör-** side-tone reference equivalent.

Bezugs-ebene datum level, d. plane, reference l., fiducial l. **-punkt** reference point, fiducial p., datum p. **-stromkreis** reference circuit. **-system** reference system, standard or norm (used for comparison). **-wert** relative value. **-zeichen** reference symbol (numeral or letter).

bezw. s. **beziehungsweise**

bezweifeln call in question, doubt.

Bibliothekar librarian.

Bidipentode duodiode tetrode with suppressor grid; duodiode pentode (with bifilar filament).

Biegekristall "bender" crystal.

biegen bend, flex, curve, warp; diffract; refract.

Biegemoment flexural moment, bending m., flexural torque.

Biege-probe, Schlag- shock, blow or impact bending test piece or specimen. **-versuch, Dauer-** bending fatigue test. **-wechselfestigkeit** alternating bending strength.

biegsam flexible, pliant, pliable, bendable, ductile, supple. **-machen** ductilize (as a tungsten wire).

Biegungsschwingung flexural vibration.

Bienen-harz bee glue, propolis. **-korblampe** beehive neon lamp.

Bifilardraht bifilar wire, twisted pair. **-wicklung** bifilar winding, Ayrton-Perry winding.

Bild, aufgerichtetes, aufrechtes

erected image, erect i., rectified i. -,**detailreiches** contrasty picture (rich in details and contrasts). -,**flaches** very soft picture, flat picture (lacking contrast). -,**flaues** non-contrasty picture or image (presenting a limy condition or fuzziness, contrast often impaired by white tint permeating whole picture). -,**hartes** contrasty, crisp or harsh picture or image. -,**hochzeiliges** high-definition (television) image or picture. -,**lichtschwaches** low-luminosity picture, p. with low brightness. -,**reelles, umgekehrtes** real reversed image. -,**ruhendes** still, still picture, unanimated p., ordinary photograph. -,**scharf eingestelltes** sharp picture, sharply focused image. -,**tönungsreiches** picture with proper shading values and contrast. -,**umgekehrtes** reversed or inverted image. -,**unbewegtes** still picture, unanimated p. -,**unscharfes** fuzzy picture, ghosty p. -,**unverzeichnetes**, -,**unverzerrtes** orthoscopic image, undistorted picture. -,**weiches (zwischen flau und normal)** weak or soft picture (between flat and normal), uncontrasty p.

Bild, Stehen des-(es) steadiness of image. -,**mangelhaftes Stehen** jumping or unsteadiness of picture or image.

Bild, Aufspaltungs- splitting pattern (opt.). **Ausserfokus-** out-of-focus picture. **Beugungs-** diffraction pattern. **Durchsichts-** (**cf Durchprojektion**) transparent picture, glass transparency, diapositive, lantern slide; translux p. **Durchstrahler-** transmission image. **Eigenstrahler-** image obtained by self-emissive method (electron microscope). **Einzel-** unit frame, individual picture (in pictorial sequence). **Empfangs-** recorded copy (facisimile). **Ersatz-** equivalent circuit diagram. **Feld-** field pattern, f. configuration or map (made in electrolytic tank). **Fenster-** window transparency. **Fern-** television picture, televised p., telephoto. **Flugbahn-** trajectory diagram. **Formel-** structural formula. **Funk-** photoradio (transmission), radio picture, wired p., photo-radiogram, photo-telegram. **Gegen-** counterpart, antitype. **Geister-** ghost image. **Halbton-** half-tone or mezzo-tinto picture (in black and white, with grey shading values). **Interferenz-** interference figure. **Klang-** sound pattern, acoustic p. **Klatschen-** beat picture. **Komik-** funny picture, "funnies," cartoon. **Koma-** comatic image. **Kontroll-** monitoring picture (telev.). **Koordinaten-** pattern of co-ordination. **Ladungs-** charge pattern or image (re : iconoscope). **Lauf-** animated, moving or motion picture (in a sequence or series of pictures or pictorial actions). **Licht-** photo-image, photograph. **Linien-, unaufgelöstes** unresolved line pattern. **Luft-** aerial image (formed in space). **Mehrfach-** multiple image, double i., ghost i. **Moment-** instantaneous picture, snapshot. **Nach-** after-image; copy, imitation, replica. **Neben-** ghost image, multiple or double i. (telev.). **Netzhaut-** retinal image. **Öffnungs-** aperture image. **Panorama-** panoramic view or picture, panning shot, "pan" or "pam" (studio slang). **Raff-** time-lapse picture, fast-motion p. **Raum-** space diagram; stereoscopic picture, plastic p. **Rechen-** nomogram. **Reihen-** sequence of pictorial actions or pictures, motion p. **Schatten-** silhouette.

Schau- diagram, graph. Schwarzweiss- black and white (facsimile) picture or phototelegraphic transmission. Sende- subject or outgoing copy (in facsimile). Sinn- symbol. Spalt-, optisches slit image, optical slit. Spiegel- mirror or speculum image, flare ghosts or spots (camera), double or reflected (ghost) i. Stand-, Steh- still (film) picture, non-animated p., still. Stimmungs- key picture, sentiment p. -,dunkles low-key picture. -,überhelles high-key picture. Sucher- seeker or monitoring picture. Teil- (Farbenauszug) compound picture (color record). Ton- tonal pattern or spectrum. Trick- trick shot, t. picture. Ursprungs- original subject copy. Weitwinkel- wide-angle picture, "widescope" p. Zeitlupen- slow-motion picture. Zerr- distorted picture (made with anamorphotic lens). Zerstreuungs- image formed by divergent lens. Zerstreuungs-, des Auges blur circles of eye.

Bild-ablenkung frame or picture scan. -abtaster televisor, television scanning device. -abtasterröhre, doppelseitige two-sided mosaic pickup tube with separate image anode, iconoscope with two-sided mosaic screen. -abtaster-Röhre mit mechanischer Blende Farnsworth dissector of electron camera, Dieckmann magnetic scanner. -abtaststelle picture or film gate (m.p.). -abtastung frame scan, picture s. -anode barrier grid mosaic. -aufbau picture synthesis, build-up of picture (line by line). -auffänger, elektronischer electronic scanning device. -auflösung resolution, picture definition, scan. -aufnahme, Kofferapparat für box equipment, trunk unit. -aufnahme, Luft- (cf Luftbild) aerial photography, airplane picture. -aufnahmevorbereitung lining up for shooting pictures.

bildaufrichtendes Mikroskop image-erecting microscope.

Bildaufrichtung erection of image, rectification of i., inversion of i. (reversing in both axes simultaneously, right and left, top and bottom interchanged at the same time). -aufzeichnung picture re-creation, tracing or delineation. -ausleuchtung "flashing" of image. -ausschnitt frame (of camera), image area (on scanning disk). -beschreibung (auf Schirm) picture tracing, re-creation or delineation (on screen). -bühne aperture or film trap. -bühneneinstellung framing, racking (by framing device). -charakter key of a picture or image. -drehung image rotation. -einstellampe framing lamp. -einstellung centering control, height c., horizontal c., framing (in telev., often with rack and pinion for focusing). -empfänger picture receiver, p. reconstructor, video receiver, telev. receiver, picture viewing tube, p. reproducing device. -empfangsstelle picture or facsimile (telegraphic) receiving station or office, recording point. -entwerfung formation of an image, imagery, imaging. -entzerrung, Licht- rectification of aerial photos.

Bilder-kasten-, Laternen- lantern slide box. -masken, Laternen- lantern masks.

Bild-erzeugung formation of an image, imagery, picture (re-) creation, p. tracing. -fänger video pick-up camera, television c. -fangröhre, Sender- pickup tube, iconoscope, dissector tube (of Farnsworth) -fehler picture distortion, image defects. -feinheit detail, (degree of) definition of p.

Bildfeld frame, picture area, image field. **-ebnung** flattening of image field, elimination of curvature (opt.). **-krümmung,-wölbung** curvature of image field. **-zerleger** picture scanner, p. dissector (Farnsworth), p. exploring means (telev.). **-zerlegung** image (field) definition.

Bild-fenster film gate, aperture or trap, picture g. (m.p.). **-fenster mit 2 Ausschnitten** film gate with two frames. **-fläche** image area, picture field, image plane (opt.). **-flachheit** flat, very soft or non-contrasty condition or quality of picture. **-flauheit** flat, soft or non-contrasty quality of a picture (contrast impaired by white tint permeating p.), limy condition, fuzziness. **-flieger** aerial photographer. **-folge** sequence or series of optic or pictorial actions (m.p.). **-format** picture format, frame size or format. **-frequenz** picture frequency, frame (repetition) f., repetition rate (telev.), video f., visual f. **-frequenzverstärker** amplifier for frame sawtooth time-base. **-funk** photo-radio, wired picture, facsimile transmission. **-geber** pickup camera. **-gerät, Spalt-** photographic soundhead (m.p.). **-geräusch** frame noise (m.p.). **-greifer** in-and-out-claw (of threading mechanism). **-grösse** size of picture, format of p. **-grösse und Brennweite, Linse für veränderliche** zooming lens for variable magnification and variable focus, "vario-objective". **-güte** (figure of) merit or quality of picture. **-helligkeit, Mittel-** average or background illumination, mean picture brightness. **-helligkeitssignale** video signals, shading-value signals (telev.). **-hintergrund** picture background. **-höhe** picture or image height (opt.), frame height (of a camera). **-inhalt** picture content, p. subject-matter, "information". **-kadre, Normal-** standard picture frame. **-kantenschärfe** resolution on border. **-kippeinsatz** incipient frame flyback or beam return (telev.). **-kipper** frame time-base. **-kippschwingung** frame or picture time-base oscillation or impulse. **-kraft (zwischen Metall und Elektron)** image force, image potential (between metal and electron). **-kraftberechnung, halbklassische** semi-classical image-force calculation. **-linie** focal line, image line.

bildlos amorphous, non-crystalline.

Bild-meister director of photography, picture chief or director. **-nachleuchten** afterglow, phosphorescence of picture. **-nachstellung** framing, phasing, racking (m.p., telev.), centering control (telev.). **-nachstellung, falsche** misframing, out-of-frame condition.

Bildner, Basen- base former, basifier, basifying agent. **Korrosions-** corrosive, corrodent, corroding or rusting agent. **Schleifen-** loop setter.

Bildnis portrait, picture, photograph.

Bildplastik plastic effect, relief e., stereoscopic e.; distortion of telev. image in form of multiple contours of diminishing intensity, plastic effect.

Bildpunkt image point, picture p., picture unit or elementary area, point image (telev.). **-, dunkelster** shadow. **-, hellster** highlight. **-, nachleuchtender** phosphorescent picture point. **-signal** video signal, electrical impulse resulting from elementary area. **-verlagerung** spot shift (causing plastic effect, in telev.). **-verteiler** picture scanner (at receiving end).

Bild-raster (cf Raster) scanning field, picture raster (telev.). -raum image space. -reihe sequence of pictorial actions, frames or pictures (m.p.), strip mosaic (aerial phot.). -röhre, Elektronen- electron image tube.

bildsames Gleiten plastic deformation.

Bildsamkeit plasticity, capability of being kneaded, fashioned, formed or molded (by pressure and/or heat application), formability, moldability, fictility.

Bild-schale image shell. -schallplatte sound and picture on disk. -schaltung feed or movement of frame (m.p.), frame ratcheting, frame timebase (telev.). -schärfe picture definition. -scherung shearing of an image. -schicht, Film- film emulsion or coat. -schirm projection screen, picture s. -schreibröhre televisor tube, viewing tube, picture reproducing tube. -schreibröhre, Projektions- projector-type television receiver. -schritt frame gage. -schwingungen picture or video impulses or signals. -seite image side (of a lens). -seitenverhältnis picture aspect ratio. -sender video transmitter; facsimile t. -signal video or picture signal. -spannung, Untergehen der swamping of video or picture signal. -speicherröhre image storing tube, charge or signal storage tube, normal iconoscope. -speicherröhre, Bildwandler- super-iconoscope. -sprung picture repetition frequency, picture cycle, break or shift of vision. -spurzeit Watkins factor, development f. -stelle picture scanner.

Bildstrich-einstellung phasing or framing of picture. -einstellung, falsche misframing, out-of-frame condition. -geräusch frame line noise.

Bildstrom video current, picture c. -stürzung toppled condition of pictures (turned an angle in relation to one another).

Bildsynchronisierungs-impuls frame synchronizing impulse, f. sync. pulse, low sync i. -impulsverstärker frame synchronizing impulse amplifier, f. sync pulse a. -lücke synchronizing gap (between end of field or frame traversal and beginning of next known also as underlap or interstitial period during which sync signal is introduced).

Bild-tanzen unsteadiness or jumping of picture. -teile verschiedener Gliederung picture portions of dissimilar nature, composition or make-up. -telegraphie picture telegraphy, wired or radio phototelegraphy, radio or wire transmission of pictures, facsimile, wire-photo. -transportrolle picture feed roller. -Übertragung (elektromechanische und elektrochem.) picture, photo or facsimile transmission by electro-mechanical or electrochemical means, writing or copying telegraphy, by telautograph, telectrograph, Bartlane device, etc. -umkehrung image reversion, image inversion (when in two axes, i.e., interchange of left and right, and top and bottom of image); solarization (reversal of gradation sequence due to overexposure to light).

Bildungs-energie energy of formation. -wärme heat of formation, enthalpy.

Bild-unschärfe lack of picture definition. -verbreiterung image spread. -verfahren, Steh- lantern slide projection method. -verschiebung slow drift or hunting of image. -, seitliche lateral shift of image. -verstärkung image intensification. -verstellungsknopf framing knob. -verstellungswelle framing shaft.

-verwackelung frame unsteadiness. -verzerrung image or picture distortion, jiggers (in facsimile). -vorführung, Raum- stereoscopic picture projection. -vorschubbewegung frame sweep scansion. -walze picture cylinder (in Bakewell app.). -wand projection screen. -wand, Ton- transoral screen (m.p). -wandler image converter, picture transformer, transducer (changes optic into electron image). -wandler-Bildspeicherröhre super-iconoscope. -wandleuchtdichte brightness or luminous density of screen.

Bildwechsel picture cycle. -, stetiger continuous, steady or non-intermittent feed or motion of film. -frequenz, -zahl frame or picture frequency, repetition frequency (telev.), number of frames per second. -zeit feeding time, picture cycle, moving period (m.p.).

bildweise Filmschaltung intermittent film feed, discontinuous film movement.

Bild-weite distance between screen and lens, image intercept. -wellenspannung video signal wave potential. -werfer picture projector. -wiederaufbau reconstruction (e.g., of a facsimile picture). -winkel angle of image, a. of view. -wirkung, plastische (cf Bildplastik) illusion of depth (of picture), stereoscopic effect. -wölbung curvature of image, c. of field. -wurf picture projection. -wurfelektrode target electrode. -zahl frame or picture frequency. -zähler frame counter, f. indicator. -zeile picture line, p. strip (telev.). -zeilendurchlauf line traversal. -zerdehnung radial image distortion. -zerdrehung rotational, tangential or orientational distortion, twist of image. -zerleger picture scanner or explorer, dissector (of Farnsworth). -ziehen photographic travel ghost. -zittern unsteadiness of picture, jumping of p. -zusammensetzvorrichtung picture reproduction means, p. recreator or delineator, p. scanner.

Billigkeitsverfahren equity suit.

Billigung approval, approbation, consent, assent, acquiescence.

Bimetallstreifen bimetallic strip (consisting of two strips of dissimilar metals with different expansion coefficients), bimetallic trip or tripping gear (of a circuit breaker).

Binantenschachtel box with a two-segment electrometer system.

bindend, wasser- hydrophylic, water-absorbent.

Bindeton ball clay.

Bindung binding, bond, union, association, affinity.

Bindung, Dreifach- triple bonding. Gas- gettering. Neben- secondary union or bond, second-order linkage.

Bindungs-energie (Elektronen-Konzentration) binding or bond energy. -kraft (chemische) chemical linkage force. -stärke bonding strength.

Binode diode-triode (rectifier and AF or RF amplifier in one bulb, with joint cathode).

Binokularspule binocular (form of closed-field) coil.

Bitter'sche Streifen Bitter (magnetic powder) patterns, Bitter bands.

Bkw reactive power, r. voltampere, var.

Blähungsgrad expanding or swelling property, degree of inflation or imbibition.

Blank, Zahlen- figure blank, figure space,

Blankfilm film base. -, gelatinebeschichteter gelatine-coated film base. -, unbeschichteter plain, uncoated film base.

Blank-seite cellulose face or side

(of film). -verdrahtung bare wiring.

Blasdüse blasting nozzle, gun, discharge n.

Blase, Dampf- steam-heated still, bubble of vapor or steam. **Glas-** (generally) bubble in glass, seed (when small), air bell (when of irregular shape). **Guss-** flaw or defect in a casting. **Schweiss-** arc crater.

blasen blast, blow, smelt (blast furnace), inject (steam).

Blasen, Stahlkugel- steel shot blast.

Blasenbildung blister formation, blistering (metal, phot., etc.); cavitation, occlusion of gases.

Bläser, Flamm- blowing magnet.

Blasinstrument wind instrument. **Blech-** brass instrument.

Blatt leaf, foil, sheet, page, folio, blade (of knife switch), lamination (of a magnetic core).

Blättchen lamina, lamella, small leaf, flake.

blättchenartig lamelliform, laminiform.

Blattelektrometer leaf electroscope.

Blätter-bürste laminated brush. **-kondensator** foil condenser, tubular capacitor.

Blatt-feder leaf, plate or flat spring, reed, blade. **-grün** chlorophyll. **-kurve** folium.

Blauauszug blue record.

blauempfindlich blue sensitive.

Blaufilter blue filter, viewing filter (phot.).

Blech plate, sheet, foil, lamination of metal.

Blechblasinstrument brass wind instrument.

blecherner Klang tinny sound.

Blech-gestell chassis (of receiver set). **-instrument** brass instrument (mus.).

Bleicherde Fuller's earth.

Bleiglätte litharge.

Blende light-stop, diaphragm, aperture (stop), slit; shield (with scanning aperture, in Farnsworth dissector); shutter (e.g., iris diaphragm, focal-plane blind, disk with cut-out sector for exposing film or plate, in m.p. work).

Blende, Abdeck- mask, shutter, shutter mask. **Abtast-(Farnsworth)** scanning aperture, hole (of dissector tube of Farnsworth). **Anoden-** anode diaphragm or partition (re:cathode-ray tube). **Apertur-** aperture stop, a. diaphragm. **Aufhell-** reflecting screen. **Aufrichtungs-** erector stop. **Austritts-** exit slit. **Blickfeld-** field stop. **Einhange-** inset diaphragm. **Einsatz-** interchangeable diaphragm. **Einschnürungs-** crossover aperture. **Eintritts-** input diaphragm or capacity disk (of klystron). **Fächer-** fan fading shutter. **Feld-** field stop. **Fenster-** apertural stop, a. diaphragm. **Gesichtsfeld-** field stop. **Halb-** half stop, scale. **Klang-** tone-shading means, tone control, tone switch, tonalizer. **Kombinations-** Goerz effect shutter. **Kreis-** iris stop or diaphragm. **Licht-** light stop, diaphragm, light shield (in sensitometer). **Loch-** stopping aperture, apertured partition, diaphragm, shield (in cathode-ray tube). **Noiseless-** bias or noise-reduction stop, shutter vane or gate. **Okular-** eyepiece diaphragm, eyepiece stop. **Raster-** scan hole, scan aperture (Farnsworth dissector), raster screen aperture. **Reinton-** bias or noise-reduction stop, shutter vane or gate. **Schlitz-** slit stop. **Schluss-** final diaphragm. **Schnürspur-** mask for making squeeze track. **Sehfeld-** field diaphragm, field stop. **Sonnen-** sunshade, black screen. **Spalt-** aperture stop, slit. **Sperr-** (blocking) light stop. **Steuer-** modulating electrode, modulation shield or grid (cathode-ray tube). Ton- tone-control means, tonalizer; fader. Trommel- drum with scan apertures or scan holes, scanning d. Vorhang- curtain fading shutter. Wechsel- stop disk with spiral slot cyclically cooperating with quadruple scanning disk holes, auxiliary rotary shutter disk with spiral slot. Zacken- triangular aperture (in film recording), vane with serrated or triangular edge. Zerleger- dissector aperture. Zonen- stop to diaphragm out a certain zone or area. Zusammensetz- picture receiver or p. reproducer scanning hole.

Blenden-ausschnitt aperture, slit. -ebene diaphragm plane. -linse, Loch- aperture disk lens. -loch anode aperture (cathode-ray tube). -nachstellung phasing of shutter. -öffnung aperture of diaphragm or stop. -rohr screening tube. -scheibe horizontal edge (film recording); auxiliary rotary shutter disk with spiral slot (for quadruple scanning). -schirm gobo (for sound absorption, in studio). -träger diaphragm support. -trommel scanning drum.

Blend-glas moderating glass. -scheibe stop. -schirme gobos (for sound absorbing, m.p.).

Blendung glare, blinding, dazzle (of eye). -, akustische aural dazzling.

Blendverfahren diaphragm method (for pencil modulation).

Blickfeld field of vision, range of v., field of view (opt.). -, beidäugiges binocular field of view.

Blick-feldblende field stop. -linie line of vision, l. of sight. -punkt point of view. -richtung sight-line, direction of sight, d. of regard. -winkel angle of view.

Blind-buchse dummy socket or jack. -komponente wattless, idle, quadrature, reactive, or reactance component. -landeachse blind or instrument landing line or path. -landedienst beacon service (to assist in blind or instrument landing). -leistung reactive power, r. voltampere, var. -leitwert susceptance. -schwanz arrangement to make resistance alternately zero and infinite, in dipole feeder operation, stub. -stecker dummy plug. -stromaufnahme drawing or taking of wattless, idle, or reactive current. -verbrauchszähler reactive voltamp.-hour meter, sine m., wattless component m., var-hour m. -voltampere var (unit of reactive power). -widerstand, induktiver inductive reactance, positive r., inductance. -widerstand, kapazitiver capacitive reactance, negative r., condensance.

Blinker, Zeit- chronographic camera, time recording c.

Blink-lampe flash lamp. -schaltung blinking arrangement, ratchet circuit scheme (in relaxation or sawtooth wave generator with R-C time-base and neon lamp).

Blitz (Film) statics (film).

Blitz, rücklaufender return lightning stroke.

Blitz, Kapsel- flashlight capsule. Kugel- flashbag, globular lightning, ball l. Licht- pulse or flash of light, scintillation.

Blitz-anzeiger, Fern- keraunophone. -bahn, verschlungene tortuous path of lightning flash. -lampe flashlight, flashlamp. -licht flashlight, flash bulb, magnesium light.

Block, Gitter- grid condenser; staircase-like structure of echelon grating. Kopplungs- fixed coupling capacitor.

Block-empfang party reception, block r., communal r. -hahn

stopcock.
Blockierung, Gitter- negative biasing of grid, g. cut-off.
Block-kondensator fixed capacitor, blocking c., stopping c., insulating c. **-kondensator, Wickel-** tubular capacitor, paper c., Mansbridge condenser (of fixed value). **-zeichnung** schematic or diagrammatic illustration, block diagram.
Blut-farbemesser plethysmograph (monitors color changes, etc.), electroarteriograph. **-körperchenzahler** haemacytometer, blood counter.
Bock, Lager- pedestal of a bearing.
Boden, gewölbter arched back (mus.).
Boden, Resonanz- sounding board. **Schrift-** recording base.
Boden-antenne ground antenna, earth a.; underground or buried antenna. **-flugleiter** ground control operator. **-funkstelle** ground direction-finding station, aeronautical g. radio s. **-funkstellen-Rufzeichen** ground station call sign or signal. **-geschwindigkeit** ground speed (of an airplane). **-peilapparat** ground station direction-finder. **-peilstelle** ground direction-finder station. **-welle** ground wave, direct w. (which hugs earth's curvature).
Bogen bow, arc, curve, curvature, bend, arch, clip.
Bogen, Flammen- flame arc, electric arc. **Grad-** graduated arc, bow, protractor. **Hochstrom-** heavy-current arc, large-current arc. **Kreis-** arc of circle, circular arch. **Pol-** pole arc. **Schwingungs-** amplitude of oscillations. **Voll-** semi-circle rail (re: microscope).
Bogenabstand pitch or distance along an arc (of holes in Nipknow disk).
bogenbildend, nicht- non-arcing (metal).
Bogen-frosch bow-frog (string instrument). **-gitter, Kreis-** arcuate grid. **-lampe mit breitwinklig gestellten Kohlen** scissors-type arc-lamp. **-lampe, Freifall-** gravity-feed arc-lamp. **-lampe, Wolfram-** tungsten arc-lamp, pointolite. **-lampenabzugsrohr** vent stack of arc-lamp. **-linie** arc line, arcuate l., curved l. **-mass** radian measure. **-rückschlag** arc-back. **-spektrum** arc spectrum.
Bohreinsatz, -eisen bit.
Bohrer, Zwick- twist drill.
Bolzen, Knebel- toggle bolt. **Ösen-** eyebolt. **Schlüssel-** clutch key.
Bolzen-mutter nut. **-scheibe** washer.
Bomben-sauerstoff tank oxygen. **-flugbahn** bomb trajectory. **-richtgerät, -visier** bombsight.
Bord-ablenkung distortion of bearing on site (by re-radiation, etc., or by ship's field), aircraft error. **-effekt** board effect, ship field error (in df work), quadrantal error (due to structural parts, metal, etc.).
bördeln bead over, edge, flange (an edge).
Bördelrand flanged or beaded edge.
Bordsprechgerät, -telephon intercommunication telephone, interphone.
Bottich vat, vessel, tub, tun, tank.
Bottich, Entwickler- developing tank.
Box, Abhör- monitoring box.
Bramme, Roh- slab, ingot.
Brand-messer pyrometer. **-stein** brick.
Bratpfannengeräusch frying noise.
Bratsche viola, tenor violin.
brauchbar usable, useful, serviceable, suitable.
Brause rose, douche.
brechbar frangible, fragile, breakable, brittle, refrangible, refractable.
brechend, doppelt- birefringent, birefractive. **einfach-** unirefringent.
brechende Brennfläche diacaustic (opt.).

Brechkraft refractive power (opt.).

Brechung, Strahl- refraction or splitting up of rays.

Brechungs-brennfläche diacaustic. **-einheit** diopter, dioptry (unit of focal length). **-exponent** refractive index, refractivity. **-gesetz, Snellius'sches** Snell's law of refraction. **-indexbestimmung** refractometry. **-messer** refractometer. **-messer, Augen-** diascope, rétinoscope. **-vermögen, Licht-** optical refractive power, refrangibility (of light). **-winkel** angle of refraction, refracting a.

Brech-wert, Hauptpunkt- principal point refraction. **-winkel** angle of refraction, refracting a.

Breitbandempfänger broad-band receiver, wide-band r.

Breite width, breadth, latitude. **-, nördliche** northern latitude.

Breite, Doppler- Doppler broadening.

Breiten-effekt latitude effect (cosmic rays). **-höhe** meridian altitude. **-minute** one minute of latitude. **-wert** azimuth value (of vision).

breitwinkliger Stellung der Kohlen, Bogenlampe mit scissors-type arc-lamp.

Bremsdynamometer Prony brake, brake dynamometer.

Bremse, Freilauf- free-wheeling brake, coaster b.

Bremselektrode reflecting electrode, retarding-field e. (in positive-grid tube).

bremsen brake, retard, check (motion), decelerate.

Bremsfeld reflecting field, retarding f. (in electron-oscillation tube). **-audion** positive-grid detector, electron-oscillation detector or rectifier (negative anode associated with leak resistance and condenser). **-elektrode** reflecting electrode, retarding-field e. (in positive-grid tube). **-generator** negative transconductance generator or oscillator, positive-grid or retarding-field o. **-röhre** retardation valve, Barkhausen-Kurz retarding-field tube. **-schaltung** retarding-field or positive-grid circuit scheme (operating with grid maintained at a higher potential than plate), Barkhausen-Kurz circuit organization, oscillating-electron or electron-oscillation scheme with reflecting electrode.

Brems-gitter suppressor or cathode grid (in a pentode). **-kreis** negative-anode or negative-plate circuit. **-moment (Zähler)** brake or retarding torque (of a meter, etc.). **-spannung** negative-anode potential (in electron-oscillation tube). **-strahlen** rays caused by particle retardation.

Bremsung der Elektronen reflection (in Barkhausen-Kurz tube); deflection (in magnetron); deceleration or retardation of electrons.

Brems-vermögen stopping power (of photographic emulsion); braking or retarding power. **-wirkung der negativen Raumladung** repelling action of electron charge, negative space charge. **-zylinder, Relais mit** dashpot relay.

Brennebene, hintere back focal plane. **-, vordere** front focal plane.

brennen, dicht vitrify (cer.).

Brenner, Fischschwanz- fishtail burner, batswing b. **Schlitz-** batswing burner, slit b. **Schneid-** cutting torch. **Zirkon-** zirconium burner, z. filament.

Brennfläche caustic surface (opt.). **-, brechende** diacaustic. **-, reflektierende, -, rückstrahlende** catacaustic.

Brenn-fleck (focused) spot. **-glas** burning glass, convex lens. **-linie** focal line, image l.

-punkt focus, focal point. -punkt, vorderer first focal point. -schwindung shrinkage in firing (cer.). -spannung constant glow potential (is slightly lower than firing or striking p.), normal running p. (of glow tube, etc.). -spiegel reflector sheet. -strahlhärtung flame hardening. -weite focal distance, f. length, convergence d. -weite, hintere back focal length. -weite und Bildgrösse, Linse für veränderliche zooming lens for variable focus and variable magnification, "vario-objective", zoom lens.

Brennweitenmesser focometer.

brennweitige Linse, kurz- short-focus lens. lang- long-focus lens.

Brett, Misch- mixing panel, monitoring p. Objektiv- lens panel, objective board. Schall- baffle (board), sounding b. Verstärker- amplifier rack, a. panel.

Brillanz (akustische Wiedergabe) "bounce" (in sound reproduction).

Brille, Zweistärken- bifocal spectacles.

Brillengläser, durchbogene meniscus spectacle glasses. -, punktuell abbildende point-focal glasses.

Brisanz shattering power.

Bristolkarton Bristol board.

bröckelig brittle, friable, crumbly, fragile.

brodeln boiling noise.

Brom-öldruckpapier bromoil printing paper. -silberpositiv positive in silver bromide. -zahl bromine number.

Brown'sche Bewegung Brownian movement.

Brown Zungentelephon Brown reed-type telephone receiver.

Bruch break, breakage, fracture, rupture, failure, fraction, fragment; scrap, waste.

Bruch, Längen- longitudinal fracture. Vertrauens- breach of confidence.

Bruch-dehnung total extension or elongation (to fracture), breaking tension, stretch (of paper). -festigkeit ultimate strength. -fläche surface of fracture.

brüchig brittle, short, friable, fragile, cracky.

Brüchigkeit, Beiz- acid brittleness.

Bruch-methode, Ketten- method of infinite continued fractions (math.). -platten bursting disks. -punkt brittle point (of glass). -spannung ultimate stress, breaking s. -stücke fission products, fragments, fracture or rupture pieces. -teil fraction, fractional part, aliquot part. -versuch, Rot- hot breaking test.

Brücke bridge, platform; rider (piano).

Brücke, Beleuchtungs- traveling light crane. Kapazitäts- capacitance bridge, Wien b. Querleitungs- transconductance bridge. Thermokreuz- thermo-couple or thermo-junction, thermal cross. Verhältnisarme der ratio arms of bridge. Widerstands- resistance bridge, Wheatstone b.

Brücken-ausgleichstrom bridge balancing current. -gleichgewicht bridge balance. -H-Schaltung, Entzerrungskette mit H-type attenuator network (balanced). -Scheitel bridge apex. -T-Schaltung, Entzerrungskette mit T-type attenuation network (unbalanced). -Übertrager differential transformer. -verhältnisarme ratio arms of bridge.

Brumm-kasten bass keys. -spannung hum potential, ripple potential. -spur, -streifen buzz track; hum (broad bands on screen of c.-r. tube). -tastenfeder spring for bass keys (accordion). -ton, Netz- motor boating (of a.c.), a.c. pickup.

Brust-fernsprecher head and chest set. -mikrophon breastplate microphone.

Buch, Dreh- shooting script.
Bücherei library.
Buchgold leaf gold.
Buchse sleeve, bush, socket.
Buchse, Blind- dummy socket, dummy jack.
Büchse, Fernrohr- telescope tube, barrel. **Kompass-** compass bowl, c. kettle. **Lager-** bearing bush.
Buchse, Rad- wheel bushing. **Schalt-** jack.
Büchse, Widerstands- resistance box.
Bucht, Verstärker- amplifier rack, repeater rack or bay. **Wähler-** bay of selectors, selector rack, switch frame.
Büffelleder buffalo skin.
Bug bow.
Bügel clamp, clip, strap, staple, bow; yoke (of a magnet); loop (of a magnetron); harness (of headphones); saddle bracket (microscope); bout (string instrument).
Bügel, Aufhänge- loop hanger. **Kopfhörer-** harness, headband or strap (of headphone or headset). **Stirn-** head rest (of magnifier).
Bügelhaken, Flaschenzug- block tackle.
Bugwelle (eines Geschosses) nose wave (of a projectile).
Bühne, Bild- apexture or film trap
Bühnen-aufsteller set dresser. **-einstellung, Bild-** framing.
Bund, Anschlag- stop collar.
Bündel beam, pencil, bunch, cone (of rays); wave packets.
Bündel, Achter- eight-wire core, quadruple twins. **Strahlen-** bundle, bunch, beam or pencil of rays or radiations.
Bündelung bunching (of electrons in Klystron), focusing, beaming, concentration, insuring beam or directive antenna effect.
bündig succinct, brief, concise, summarized, short, valid; flush, snug.
Bunker (für Verstärker) bay or rack (of amplifiers).
Bunsenphotometer Bunsen photometer, grease-spot p.
bunte Farben hue colors, chromatic c.
Bürde burden, load.
Bürste, Blätter- laminated brush. **Kontakt-** contact-brush, wiper.
Bürsten-entladung brush discharge (between glow discharge and sparking). **-maschine, Zwischen-** metadyne, metadynamo.
Büschel tuft, bunch, cluster, brush (discharge).
Bussole, Sinus- sine galvanometer.
Bütte tub, vat, tank, chest.
Büttenrandpapier deckle edged paper.
Butzenscheibenlinse bull's eye lens.
B Verstärker class B amplifier.

C

C Verstärker class C amplifier.

CW Schaltung capacity-resistance circuit scheme.

Callierfaktor (cf **Kontraststeigerung**) Callier factor (of print contrast).

cartesische Fläche Cartesian surface.

caesiumbedeckt caesiated.

Centradian centraradian (one-hundredth radian).

Charakter, Bild- key of a picture.

Charakteristik s.a. **Diagramm** & **Kurve**.

Charakteristik, fallende drooping characteristic, falling c. **-, quadratische** square-law characteristic. **Anlauf-** transfer characteristic. **Herz-** cardioid characteristic or space pattern. **Maschinen-** speed-load characteristic. **Nachglüh-** persistence characteristic. **Richtungs-** directive or directional characteristic or diagram, space pattern.

Chargengang heat run.

chargieren charge, feed, load.

Chiffrierung encipherment, coding.

Chor assembly of strings (mus.).

chörig, drei- trichord.

chromatischer Fehler chromatic aberration. **-Schaum** chromatic soft focus or bleeding.

Chromatisierung, Ortho- orthochromatic processing.

conphas in-phase (condition), in phase coincidence.

cotg α cot α, cotangent α.

Coulombabstossung Coulombian repulsion.

C Verstärker class C amplifier.

Cycloidhöhe (**Magnetron**) cycloidal height.

Cyclotron (cf **Resonanzbeschleuniger**) cyclotron, magnetic resonance accelerator using two dees and a spiral beam.

D Wert quadrantal error due to airplane fuselage, etc.
Dach-kapazität top (loading) capacity. **-kohle** upper coal.
d'Alembert'sche Wellengleichung d'Alembertian wave equation.
Damm margin or narrow strip of material between adjacent tracks or grooves of phonograph disks, barrier, land.
Dammarharz Dammar resin or gum.
dämmen stop up, dam up, curb.
Dämmerungs-effekt night effect. **-schalter** twilight switch.
Dämmung, Schall- sound reduction or attenuation, s. deadening or absorption, silencing (of noise).
Dämmzahl, Schall- sound transmission characteristic.
Dampf, direkter live steam. **-, indirekter** latent steam.
Dampf-auflösung volatilization. **-blase** bubble of vapor or steam; steam-heated still.
dampfdicht steam tight, vapor tight.
Dampf-dichtemesser dasymeter. **-dichtung** steam packing. **-druckmesser** manometer, steam gage, tensimeter. **-druckzünder** vapor-pressure igniter.
dampfecht fast to steam.
dämpfen attenuate, damp (out), dissipate, absorb (sound in air, etc.); muffle, deaden, silence, quench, die away, decay, mute (mus. instnument).
dämpfend, schall- sound absorbing, s. attenuating, s. damping or deadening.
Dämpfer mute (in string instrument), damper (of piano). **-, rotierender** rotary stabilizer (film feed). **Licht-** silk, light softener (m.p.). **Schall-** silencer, muffler, sound absorber, gobo (sound absorbing sheet) (m.p.).
Dämpferwicklung amortisseur winding, damping w.
Dampf-hülle steam jacket; vaporous envelope or sheath, vapor shroud. **-hunger** steam avidity. **-kolben** steam piston.
Dampflampe, Quecksilber-, mit Heizkathode phanotron, tungar tube (with argon filling). **-, Quecksilber, mit Steuerelektrode** mercury-vapor tube with control grid (of thyratron, ignitron, etc., type).
Dampf-messer manometer, steam gage, tensimeter. **-schale** evaporating dish or basin.
Dämpfung attenuation, damping, loss (of energy), dissipation, drop; impedance. **-, kilometrische** attenuation per kilometer of line. **-, mechanische** mechanical resistance, m. impedance.
Dämpfung, Ableitungs- leakance loss. **Bezugs-** volume loss, v. equivalent. **Fehler-** balance attenuation, return loss. **Flüssigkeits-** liquid or fluid damping or cushioning, dashpot action. **Kenn-** iterative attenuation constant. **Klirr-** correction of non-linear harmonic distortion. **Leitungs-** line loss, l. dissipation, l. attenuation. **Nebensprech-** crosstalk transmission equivalent, c. attenuation. **Nutz-** damping resulting in useful radiation or signal power, effective transmission equivalent. **Öl-** oil damping means (re: sound recording vibrators); oil dashpot. **Pseudo-** loss of selectivity by

parallel internal resistance. **Rückfluss-** structural return loss. **Rückhörbezugs-** side-tone reference equivalent. **Schall-** sound absorption, s. attenuation, s. deadening. **Schwingungs-** damping of oscillations or vibrations; shock absorption. **Sperr-** stop band attenuation. **Strahlungs-** radiation damping, r. resistance. **Übersprech-** crosstalk transmission equivalent. **Verlust-** damping resulting in outright loss or dissipation of energy (by ohmic resistance). **Vierpol-** image attenuation constant. **Wirk-** (cf **Wirkwiderstand**) transducer loss.

Dämpfungs-ausgleicher equalizing network, compensating network. **-dekrement, scheinbares logarithmisches** equivalent logarithmic decrement. **-entzerrer** equalizing network, compensating n. **-Frequenzkurve** attenuation-frequency curve. **-gang** frequency response of attenuation, attenuation-frequency characteristic. **-glied** attenuation network or mesh. **-klotz** damping plate. **-konstante, Wellen-** wave attenuation constant (real part of transfer c). **-mass** unit of attenuation (in decibels or in nepers); attenuation standard (denotes also real component of image transfer constant). **-messer** transmission (efficiency) measuring device, decremeter. **-mittel, Schall-** sound absorbing, attenuating, insulating, damping or deadening means or material. **-periode** quench period. **-pol** damping pole (point of infinite attenuation or suppression); point of attenuation. **-reduktion** regeneration, gain, de-attenuation. **-spule** damping coil, amortisseur c. **-verminderung** regeneration, gain, de-attenuation. **-verzerrung** frequency distortion; tone d. **-vorrichtung** damper, dashpot. **-wert** decrement, value of damping or of attenuation. **-wicklung** damping or amortisseur winding. **-widerstand** non-reactive, effective, dissipative or ohmic resistance; attenuator. **-widerstand, Kenn-** iterative impedance, characteristic i. (for uniform line). **-winkel** phase angle difference, loss a. **-wirkung** quench action.

Dampfwassertopf steam trap.

Darbietung entertainment (radio, acoustic or visual), program (material), offering.

Darmsaite gut string.

darren kiln, dry, kiln-dry; liquate (copper).

darstellen represent, prepare, produce, make, manufacture, exhibit, display, describe.

Darsteller actor, character player.

Darstellung, vektorielle vectorial representation. **Spektral-** representation by spectra.

Darstellungsweise method of preparation; manner of representation, style.

darunterliegend subjacent, underlying.

Dauer, Leucht- illuminated period, light p., time of luminescence or fluorescence.

Dauer-biegeversuch bending fatigue test. **-bruch** fatigue fracture.

Dauerfestigkeit fatigue strength. **Kerb-** notch fatigue strength. **Korrosions-** corrosion fatigue strength.

dauerhaft durable, lasting, permanent, stable, fast.

Dauer-haftigkeit durability, keeping quality, stability, permanence, endurance. **-haltbarkeit** fatigue endurance. **-haltbarkeit, Verdrehungs-** endurance under torsion stress. **-prozess** d-process (in phosphorescence decay). **-registrierung** long-period recording. **-riss** fatigue crack. **-schlagprobe** continuous or fatigue shock or impact test (specimen). -schlagwerk continuous or fatigue impact test machine. -standfestigkeit (long-time) creep strength, creep resistance, static fatigue stress, creep limit. -standfestigkeitsprüfung (long-time) creep test. -standversuch creep test, fatigue test. -strich long dash, permanent note (N and A interlocking, in df). -strom persistent current. -ton permanent note, tone, sound or signal. -zustand steady state.

Daumen mechanical operating part resemblung a thumb, as a cam.

Daumen, Hebe- cam, tappet, finger, disk with lobe. Steuer- sequence switch cam.

Daumen-mutter thumb nut. -rad sprocket (wheel). -scheibe cam (wheel), disk with lobes.

Däumling cam knob.

dazwischentreten intervene.

Debye-Scherrer-Ring D.-S. ring or circle, powder pattern, Hull r.

Decke belly (of string instrument).

deckeln provide with lid or cover.

Deckenprojektor ceiling projector (to determine cloud height).

Deck-fähigkeitsmesser (Farbe) cryptometer (measures concealing power of paint). -glas einer Linse anterior surface of a lens. -gläschen cover glass (of microscope). -kraft density (phot.), opacity, hiding, concealing or covering power (of paints). -kraftmesser cryptometer. -mittel covering material, resist, protective coat (in etching work), "ground". -peilung alignment bearing. -plättchen cover slides.

Deckung registry, cover, covering, coincidence, congruence (geom.); intensity or coverage (of a negative, opt.), stereoscopic contact (in range finder).

deformierte Fläche deformed surface, figured s.

dehnbar machen ductilize (wire).

Dehnbarkeitsmesser ductilimeter, dilatometer, extensometer.

Dehner expander (in volume-range control).

Dehnung, Bruch- total extension or elongation, breaking tension, stretch (paper). Rück- damping capacity (re metals). Zeit- high-speed camera shooting; time scale increase.

Dehnungs-diagramm, Spannungs- stress-strain diagram, stress-deformation d. -fuge expansion joint. -grenze yield point. -messer dilatometer, extensometer, ductilimeter. -modulus modulus of extension, (often) Young's modulus of elasticity. -schlitz slip joint (Martin furnace). -welle dilational wave. -zahl coefficient of expansion or extension.

dekadische Logarithmen common logarithms.

dekonzentriert de-focused, debunched, dispersed, out of focus.

Dekoration set (m.p.).

Dekrement, Dämpfungs-, scheinbares logarithmisches equivalent logarithmic decrement.

Dekrementmesser decremeter.

Delta-Verfahren "toe" negative method (phot.).

Demodulation demodulation, detection, rectification.

demontieren demount, disassemble, dismantle, take apart, knock down.

Densitometer, Kugel-, integrierendes integrating spherical densitometer.

Densogramm, Mikro- microdensitometer record.

derb solid, compact, dense, firm, rough, strong.

Derbheit ruggedness, robustness, solidity, strength.

Desakkomodation, zeitliche magnetic ageing (irreversible decrease of permeability with time).

detailreiches Bild picture rich in detail and contrasts, contrasty p.
Detektor, linearer linear detector or rectifier. **quadratischer** square-law detector or rectifier. **quantitativ arbeitender** integrating detector **Fest-** contact detector, fixed crystal d. (with non-adjustable cat whisker). **Hebel-** contact detector or crystal d. with cat whisker adjustable by lever action **Pinsel-** contact or crystal detector with cat whisker. **Wellen-** cymoscope, wave or oscillation detector, coherer.
Determinante, Gleichungs-, für Leitungen determinant for line equations.
detonierender Verstärker heterodyne amplifier.
Deutlichkeit, Konsonanten- consonant articulation. **Sprach-** articulation, intelligibility of voice, clarity of speech audition. **Vokal-** vowel articulation.
Deuton deuteron.
Deutsche Industrie-Normen (DIN) German industrial standards.
Deutung explanation, interpretation, evaluation.
Dewargefäss Dewar vessel.
Dexel adze (cutting tool).
Dezentrierung decentration, decentering, eccentricity.
Dezimalstelle decimal place.
Dezimeterwelle micro wave (for w. below 20 cm), decimeter wave.
Diagonalbeziehungen diagonal relationships (in periodic system).
Diagramm see also **Charakteristik, Kurve.**
Diagramm, Doppelkreis- figure "8" diagram or space pattern. **Kreis-** circle diagram, circular loci. **Orts-** locus or circle diagram. **Pulver-** powder pattern. **Rückstrahl-** back reflection photogram. **Strahlungs-** radiation diagram or characteristic, space or r. pattern. **Zeiger-** vector diagram. **Zustands-** phase diagram.
Dialogpegel dialog level.
Diaphonie (cf **Übersprechen**) cross-talk; babble (when from a great number of disturbing channels).
Diaphragma diaphragm, stop, baffle, mask, membrane, shutter. **-anode** apertured anode, a. stop.
Diapositiv, Strich- line diapositive, line transparency.
Diapositiv-projektion lantern slide projection. **-sender** film transmitter.
Diaprojektion lantern slide projection.
dichroitischer Schleier dichroic or dichromatic fog, silver or red fog.
dicht dense, close, tight, firm, compact, thick, impervious, sealed, hermetic.
dicht, zu overdense.
dicht brennen vitrify (ceramics).
dicht, dampf- steam or vapor tight. **licht-** opaque, (sealed) light-tight or light-proof, not transmitting l. **luft-** hermetic(ally sealed), airtight, excluding air. **vakuum-** vacuum-tight, sealed to insure vacuum.
Dichte, Einsatz- number of stations (satisfactorily operable) in a given band and their closeness. **Leucht-** luminous brightness, intrinsic brilliance, intensity or brightness of illumination or light in terms of stilb (sb) or apostilb (asb) units. **Lichtstrom-** lumirous flux density (in lumens). **Massen-** mass density.
Dichte-gitter density grid, (charge) d. control or regulating g. **-messer, Flüssigkeits-** liquid density meter, densimeter, hydrometer, areometer. **-messer, Gas-** dasymeter. **-messung, photogr.** densitometry, photodensitometry, photographic density measurement. **-modulation** density modulation, charge-density m.
dichten make tight, lute, pack, caulk; condense, compact, densify.

dichtend, selbst- self-sealing, tightening or packing.
Dichte-regulíergitter density grid, (charge) density regulator or control g. **-schrift** variable-density sound track. **-skala** density scale. **-steuerung, Ladungs-** charge-density modulation.
dichtgepackt close-packed (cryst.).
Dichtung, Labyrinth-, Turbinen- turbine labyrinth seal, t. gland. **Vakuum-** vacuum plumbing, sealing or packing.
Dichtungs-ring gasket ring. **-stoff** luting, calking, packing, plumbing material (vacuum work).
Dicken-messer thickness gage, calipers, gage c. **-schwingungen** thickness vibrations (cryst.). **-taster, Faden-** calipers with jaws to measure thread diameter capacitively.
Dickte thickness.
dielektrische Verschiebung dielectric displacement.
Dielektrizitätskonstante specific inductive capacity, dieletric constant, permittivity.
Dienstvorschriften service or working rules or regulations.
Dietrich skeleton key, picklock.
Differenzfrequenz difference frequency (re·beat, heterodyning).
Differenzialrelais differential or discriminating relay.
differenzierbar differentiable.
Differenzton difference tone, differential t.
diffundieren diffuse.
Diffusionswärme diffusion heat (gas).
Diktiermaschine dictograph.
D I N (Deutsche Industrie Normalien) German industrial standards.
Dinaston Dinas clay.
Ding object. **-punkt** object point.
dingseitig on the object side.
Diodengleichrichter, Zweifach- duodiode rectifier.
Diopter alidade, diopter.
Dioptrik dioptrics (now covering refraction branch of optics).
Dipole dipoles, doublets.
Dipolzeile dipole array.
direktanzeigender Peiler direct-reading direction-finder.
direkt(es) Abspielen direct playback. **-(er) Dampf** live steam. **-(er) Gegenbeweis** rebutting testimony or evidence. **-(e) Kopplung** direct coupling (by resistance, inductance or capacity).
Direktionskraft directing force, directive f., versorial f.
Direktor, Atelier- studio manager.
Direktverstärker straight-ahead amplifier.
Diskant treble. **-posaune** soprano trombone.
Diskontinuität, Elektronenfluss- shot effect.
Dispersion, Drehungs- rotatory dispersion.
dissoziierbar dissociable.
Distanz, Kern- inter-nuclear distance.
Distanz-halter spacer, separator, spacing means. **-isolator** spacing insulator, stand-off i. **-relais** distance relay. **-stück** spacer, separator piece.
DK (dielektrische Konstante) dielectric constant.
Dochtkohle cored carbon.
Döckchen stud.
docken wind (yarn, etc.).
Dockenfeder poppet spring.
Dolle nick.
Dompfaffenorgel bird organ.
Donnereffekt peculiar effect in sound recording when white and ultra-violet light is used; (non-linear) photographic distortion in variable-area recording known as "donner" or thunder effect.
Doppelaufnahme duplex shot.
doppelbeschichteter Film double-coated film stock, sandwich f.
Doppel-blickzielfernrohr double

direct sighting telescope **-brechung** double refraction, birefringence. **-diode** duodiode, double diode.

doppeleindeutig one to one.

Doppel-fernglas field glass, binoculars. **-fernrohr** binocular telescope. **-gängeraufnahme** dual-role work. **-gestaltung** dimorphism. **-gitterröhre** bigrid valve, twin-grid tube, four-electrode tube (tetrode, pliodynatron, etc.). **-gleichrichter** full-wave rectifier.

doppelhörig binaural.

Doppel-impulsverfahren double-impulse method. **-kegelantenne** cage antenna, double-cone antenna. **-klang** echoed sound, hearing sound twice (by electric and by air transmission). **-knoten** binode. **-konusantenne** cage antenna, double-cone atenna. **-kreisdiagram̄** figure-8 diagram, tangent circle pattern. **-linse** double lens, doublet. **-lötungen (Dewar-Gefäss)** double seals (Dewar vessel).

doppeln duplicate, copy.

Doppeln, Film- film duplication.

Doppel-pedalharfe double-action harp. **-platten** twin plates or crystals, composite crystal, biquartz. **-plattenkristall** bimorph crystal, divided-plate crystal, biquartz. **-quarzkeil** double quartz wedge. **-ring-Oberflächenspannungsmesser** twin-ring type surface tensiometer. **-röhre** twin-tube, two-system t., dual amplification t., reflex tube. **-röhrenschaltung** tandem tube (push-pull) circuit organization.

doppelsageförmige Spannung double sawtooth potential.

Doppel-schichtfilm sandwich film, double-coated f. **-schichtlinse** double-layer lens. **-schnürspur, Gleichtaktsprossen-** variable-density double-squeeze sound track.

doppelseitige Bildabtaströhre two-sided mosaic pickup tube (with separate image anode). **-Mosaikröhre** image multiplier iconoscope.

doppelsinnig ambiguous, equivocal, uncertain, double meaning, "double entendre".

Doppelsperrklinke (cf **Klinke**) double dog, d. pawls, d. detent.

doppelspuriger Tonstreifen double-edged variable-width sound track or record.

Doppelsteuerrohre power pentode in which cathode grid is connected with control g. rather than with cathode.

Doppel-T-Anker shuttle armature, h-type a.

doppelt-berührend bitangent. **-brechend** birefringent, birefractive. **-hochrund** convexo-convex. **-hohl** concavo-concave.

Doppel-umkehrprisma inversion prism (reverses or exchanges in both axes of image). **-V-Antenne** double-V-antenna, doublet. **-weggleichrichter** full-wave rectifier. **-wegschaltung** double or full-wave rectifier circuit organization. **-wendelsystem** biplane filament system. **-zackenschrift mit Abdeckung** bilateral track recording with double-vane shutter, double-edged variable-width track, duplex v.-area t. **-zackenschrift, Viel-** multiple double-edged variable-width track. **-zackenspur, Abdeck-** duplex variable-area track. **-zünder** time and percussion fuse, combination f. **-zungenpfeife** double-tongued flute. **-zweipolröhre** duodiode. **-zweipol-Dreipolröhre** duodiode-triode tube. **-zweipol-Vierpolröhre** duodiode-tetrode tube.

Dopplerbreite Doppler broadening (causing D. width).

Dorn mandrel, spine, pin, tongue.

Dorn, Lehr- inner cylinder or plug gage.

Dose, Abspiel- electric phonograph soundbox. **Abtast-** electric pick-up or soundbox. **Wiedergabe-** sound pickup head.

Dosen-barometer aneroid barometer. **-fernhörer** watchcase telephone receiver.

dosiert in measured quantities, quantitatively regulated, dosed (pharmacology, etc.).

Dosierung, Dosis dosage (re:rays, etc.).

Dosismeter, Roentgen- X-ray dosimeter.

Doubeln duplicating.

Dptr. diopter, dioptry.

Drachen, Karton- box kite.

Draht, durchgehender through-wire. **-, schwingender** filament swing (magnetron, etc.). **-, spannungsführender** live wire, charged w., "hot" w. **-, stromführender** current-carrying wire, live w.

Draht-, Durchgangs- through-wire. **Fasson-** profile wire, shaped or fashioned w. **Gitterwickel-** grid coil wire. **Haar-** Wollaston wire, capillary w. **Klingel-** bell wire, annunciator w. **Litzen-** stranded wire, Litz wire.

Draht-gaze wire gauze, w. mesh, w. cloth. **-gebilde, Luft-** aerial or radiating array. **-gewebe** wire gauze, w. mesh, w. cloth. **-kreuzung** wire crossing, transposition of (line) wires, cross-over. **-netz, Erd-** ground mat. **-schirm** wire screen. **-schleife, Funkbeschickungs-** QE (quadrantal error) correcting, clearing or compensating loop. **-walzwerk** wire rod mill. **-wellentelegraphie** wired radio, w. wave telegraphy.

Drall twist, spiral thread, torsional force, moment of momentum (phys.), spin (of electrons), rifling (gun).

drallfrei non-kinking, non-twisting; unrifled.

Drallwinkel angle of twist (rifling).

Dreh-achse axis of rotation, center of r., pivot, fulcrum, hinge. **-anker** pivoted armature (re:loudspeaker, etc.). **-ausgleich** torque compensation.

drehbar gelagerter Hebel pivoted lever; fulcrumed l. (when with two arms).

Dreh-buch shooting script, screen play. **-dose** crank box.

drehen, plan- face planing. **über-** super-speed operation (m.p.). **unter-** low-speed operation.

Dreh-federwage torsion balance. **-fehler, Nord-** northerly turning error. **-feld** rotating field, rotary f. **-feldschwingungen** rotating field oscillations. **-funkfeuer** rotating radio beacon or radio range, omnidirectional b. **-impuls, Spin-** spin angular momentum, spin moment, moment of momentum of electron. **-knopf** rotary (tuning) knob. **-lichtsignal** revolving beacon. **-ling** crank.

Drehmoment torque, moment of rotation, twisting or torsional moment, angular momentum (re: atomics). **Anzugs-** starting torque. **-ausgleich** torque compensation.

Dreh-platte turntable (re:phonograph). **-polarisation** rotatory polarization, optical rotation. **-punkt** fulcrum, pivot, center of rotation, c. of motion. **-resonanz** torsional oscillation resonance. **-scharnier** swivel hinge. **-scheibe** dial (re:automatic tuning, etc.); turntable (part of m.p. studio equipment). **-schwingversuch** oscillation or alternating torsion test. **-späne** turnings, borings, drillings. **-spiegelung** rotatory reflection, combined rotation and reflection (cryst.). **-spule, mit einem Lager versehene** unipivotal moving coil. -stabfeder torsion bar spring. -tisch rotating stage, rotator (opt.).

Drehung rotation, revolution, turn, angular motion, gyration, twist, torsion. -,magneto-optische magneto-optical rotation, magnetic r., Faraday effect. -,optische optical rotation, rotatory polarization. -,optischen, Umkehrung oder Wechsel der mutarotation.

Drehung, Bild- image rotation. Links- levogyration, left-handed polarization, counter-clockwise rotation. Phasen- phase rotation, p. shift. Rechts- dextrorotation, right-handed polarization, clockwise rotation.

Drehungsdispersion rotatory dispersion.

drehungsfrei irrotational, rotation-free.

Drehungs-kristall "twister" crystal. -schwingung torsional vibration. -vermogen rotatory power. -vermögens, Umkehrung oder Wechsel des mutarotation. -winkel angle of rotation; a. of deflection.

Dreh-wage torsion balance. -wähler, Heb- vertical and rotary selector. -zapfen pivot, trunnion, journal (of an axle).

Dreibein tripod.

dreichöriges Piano trichord piano, three-stringed p.

Dreidimensionalität tridimensionality, three-dimensionality, stereoscopic property (opt.).

dreieckig triangular, three-cornered.

Dreieckrechner triangulator (for trigonometric calculations).

Dreiecks-keil triangular wedge. -lehre trigonometry. -rechnung, Wind- air-drift triangulation.

Dreiecksternschalter delta-star switch.

Dreier-alphabet three-unit code.

-stösse triple collisons, three-fold c.

Dreifachbindung triple bond(ing).

dreifach(e) Oberharmonische triple harmonic. -(er) Streifenfilter tricolor banded filter.

dreifachrechtwinklig trirectangular.

Dreifach-röhre three-unit or three-purpose tube (three independent valves in one bulb). -stecker three-pin plug, triplug.

Dreifarben-leuchtdichten tricolor RD (reflection density) values. -trennung three-color separation. -verfahren tricolor method.

dreifarbig trichromatic, three-colored.

Dreifingerregel right-hand rule, thumb rule, Fleming's r.

dreiflächig three-faced, trihedral.

Dreifussring tripod ring.

dreigliedrig trinomial (math.). -(e) Farbgleichung three-color equation, three-stimulus e.

Drei-kreisempfänger three-circuit receiver. -punktschaltung Hartley circuit scheme (terminals including center tap of coil connected with three electrodes of tube), potentiometer circuit scheme. -schenkeldrossel three-legged reactor or choke coil.

drei-schenkelig three-legged, three-limbed. -seitig three-sided, trilateral, triangular. -stellig three-place.

Dreistoffsystem ternary system.

dreiteilig three-part, tripartite. -(er) Stecker three-point plug, three-way p.

Drei-teilung trisection, triequipartition. -wegehahn three-way tap or cock. -wegeschalter three-way switch. -wegverbindung three-way connection. -wicklungs-transformator hybrid transformer.

dreizählig trigonal (cryst.); three-fold, triple, ternary.

Drillachse rotator (quantum theory).

drillen drill, turn, twist.

Drillung torsion couple, twisting c.

Drillungs-kristall "twister" crystal. **-modulus** torsion modulus, shear m., rigidity m. **-schwingungen** torsional vibrations.

dritter Ordnung, Kurve cubic curve.

Drittwicklung tertiary winding (of a transformer).

Drossel choke coil, choke, inductor, reactance coil, impedance c. kicking c., tickler c. (regeneration).

Drossel, eisenfreie air-core coil, air-c. choke. **Abflachungs-** smoothing coil, s. choke. **Dreischenkel-** three-legged reactor coil or choke coil. **Eisen-** iron-cored choke coil. **Entzerrungs-** anti-resonant coil. **Kipp-** cut-off choke (regulating battery charging). **Klang-** tone-control choke. **Netz-** hum-bucking coil, hum eliminator c. **Steuer-** magnetic modulator, modulator choke, Heising modulator. **Tast-, Telephonie-** magnetic modulator coil. **Tonbeseitigungs-** hum eliminator choke, h. bucking coil.

Drossel-absperrklappe throttle clack valve. **-kette** low-pass filter. **-satz, Niederfrequenz-** low-pass filter. **-spule, Funk-** air-gap reactor. **-wirkung** attenuation (filter, etc.), damping, choking or throttling action.

Druck pressure, head; compression; impression, print(ing).

Druck, Gegen- counter-pressure, back-pressure, reaction. **Hoch-** high pressure; relief print. **Licht-** photo-mechanical printing, photographic p.; radiometer vane effect. **Lichtfarben-** photochemical color printing. **Rück-** reaction pressure. **Schall-** sound or acoustic pressure. **Schweiss-** welding upset. **Strahlungs-** radiation pressure. **Tief-** low pressure; intaglio printing process. **Um-** reprint. **Vor-** first impression, proof.

druckabhängig pressure-responsive, pressure dependent, being a function of p.

Druck-amplitude pressure amplitude. **-eigenspannung** residual compressive stress. **-empfang** typescript reception, printing r. **-empfänger** (cf **Bewegungsempfänger**) pressure (-responsive) microphone. **-entlastung** pressure relief.

Drucker, Fern- teleprinter, teletype, stock ticker.

Druck-erhöher intensifier, pressure step-up means. **-feder** compression spring. **-fenster** pressure plate or gate (m.p.). **-filter** pressure filter. **-gasschalter** gas blast switch, (cross) airblast s., auto-pneumatic circuit breaker. **-gebiet** compressed or pressure region or area. **-und Gefrierpunktsmesser** manocryometer. **-gleichgewicht, osmotisches** isosmotic equilibrium. **-kammerlautsprecher** pressure-chamber loudspeaker, pneumatic l. (without diaphragm). **-knopfempfänger** push-button or press-button receiver set, receiver with automatic tuning means. **-kolben** pressure flask, p. piston, piston ram plunger; digestion flask. **-körper** indentor (rubber). **-kufe** pressure pad (m.p.). **-lack, Gummi-** gum printing varnish. **-luftantrieb** pneumatic drive. **-luftschalter** (cross) airblast switch, auto-pneumatic circuit breaker.

Druckmesser pressure gage, piezometer, liquid manometer, fluid m. **-, osmotischer** osmometer. **Dampf-** steam gage, manometer, tensimeter. **-& Gefrierpunktsmesser** manocryometer.

Druck-minderer pressure reducing or relief valve. **-platte** platen (of a printing press). **-probe, Kugel-**

ball pressure test, Brinell ball impression or indentation t. -rahmen film trap (projection work). -raster printer's screen. -regler pressure regulator, barostat, piezostat. -rolle presser or pad roll (film feed). -schiene pressure shoes (film proj.). -steg pressure or harmonic bar (piano). -verbreiterung pressure broadening (re:gas spectrum). -verschiebung pressure shift (re· spectral lines). -wage piston manometer. -windkessel compressed airtank, blast pressure tank. -Zug-Verstärker push-pull amplifier. -zünder pressure igniter.

D-Spannung Dee voltage (cyclotron).

Duanten duants (cyclotron), dees.

dubbeln, dubben duplicate.

Dublette doublet.

Dubletten-Aufspaltung doublet separation (spectrum).

duff dead, dull.

Dulling dulling (suppression of AF by dubbing or scoring, m.p.).

dumpf all bottom, tubby, boomy; dull

dunkel, Modulierung auf modulation to dark condition.

Dunkel-adaptation, -anpassung dark adaptation, scotopia. -entladung dark discharge, dark current. -feldbeleuchtung dark ground illumination. -kammer dark room, camera obscura. -methode observation in darkness. -pause dark period, cut-off or obscuring period (m.p.). -raum, Aston'scher Aston dark space, primary d.s. -raum, Faraday'scher Faraday dark space. -,Kathoden-Crookes dark space, cathode d.s. -stösse background counts (in Geiger-Müller tube). -strahlung obscure radiation, heat r. (from a light source), invisible (actinic)r. beyond the violet. -strom dark current (of photoelectric cell flowing in absence of illumination). -wertsteuerung adjustment to value of darkness. -widerstand dark resistance. -zeit dark interval or period.

dunkles Stimmungsbild low-key picture.

dünn thin, tenuous, slender, dilute, feeble, rare, weak.

dünnwandig thin-walled.

Dunst, Fernen- distant fog.

Dunst-abzug hood (for fumes). -abzugsrohr vent pipe. -kreis atmosphere. -messer atmometer. -rohr ventilating pipe.

Dup-emulsion dupe emulsion. -negativ duplicating negative. -positivfilm duplicating positive stock, master p.

Dur (Tonart) major (musical pitch or mode).

durchbelichten fully expose or irradiate.

Durchbiegung deflection, flexure, bending, coflexure, sag, dip (of line wire).

Durchblickvorrichtung dioptric device.

durchbohren perforate, bore, drill, punch, pierce, penetrate.

durchbohrt perforated, punctured, punched, bored, apertured.

durchbrochen apertured, perforated.

Durchbruch burst (discharge); fracture, break. -stoss burst pulse.

durchdringen pervade, permeate, penetrate.

durchdringend (inter) penetrating (e.g., two crystal lattices), penetrant.

Durchdringungs-frequenz penetration frequency, critical f. -vermögen penetrating power, penetrativeness.

Durcheinander confusion, mixup, bedlam.

durchfallen fall or drop through (a potential).

durchfallendes Licht transmitted light, transcident l.

Durch-flussmesser flow meter. -flutung circulation (magnetic

potential); flow (of RF, etc.).
Durchführung wall entrance, bushing, duct, lead-in. **Innenraum-** indoor bushing.
Durchführungs-isolator, Wand- wall lead-in or bushing insulator. **-stromwandler** bushing-type current transformer.
Durchgangs-draht through-wire. **-frequenz** penetration frequency. **-richtung** forward or low-resistance direction (rectifier). **-widerstand** volume resistance (exclusive of surface r.), insulation r.
durchgeben filter, strain.
durchgehend(er) Draht through-wire. **-(es) Licht** transmitted light, transcident l.
Durchgriff grid controllance, g. penetration factor, grid transparency, transgrid action, gain reciprocal 1:μ
Durchgrifflinse (cf **Lochblendenlinse**) aperture lens (through which field or anode may act or draw electrons).
Durchhang sag, slack. **-verfahren** "toe" method of recording (underexposure of film, Tobis method using toe region of Hurter & Driffield curve).
durchkopieren print through.
Durchlass clearance, lumen. **-breite** band-width, pass band.
durchlassen filter, strain, transmit (light, etc.), let through or pass.
Durchlass-frequenz pass wave. **-grad (von Wasser)** transmission factor (of water), radiation through water.
durchlässig, halb- semi-permeable, semi-transparent.
Durchlässigkeit transmittancy, transmissivity, transmitting property or power (for rays, electrons, etc.), transparency, perviousness, penetrability, permeability (magnetic).
Durchlässigkeit für ultrarote Strahlen diathermancy, property of being diathermanous.
Durchlässigkeit, Filter- transmittance (ratio of filtered to unfiltered current). **Kopie-** transmission of picture. **Licht-** light transmittance, transmittancy (spec., of unit thickness). **Luft-** air penetrability, a. perviousness. **Ultrarot-** infrared transmittance or transmittancy, diathermancy.
Durchlass-kurve selectivity curve (of a direction-finding receiver). **-richtung** forward direction, low-resistance d. (of a rectifier). **-widerstand** forward resistance (of rectifier).
Durchlauf running through or passage (of a film). **Bildzeilen-** line traversal.
durchlaufen pass through, traverse, cover, sweep (a line).
Durchlaufkopiermaschine continuous film printer.
durchleuchtend transparent, diaphanous.
Durchleuchtung trans-illumination, transmission of light; radioscopy. **-mit Roentgenstr.** (diagnostic) radioscopy, radiography.
Durchleuchtungsapparat fluoroscope, cryptoscope.
durchlöchert punctured, perforated, pierced, punched, apertured, having holes or pits. **-(er) Bügel** beehive shelf (gas drying).
Durchmesserwicklung full-pitch winding.
Durchmesserwicklung des Senders operating transmitter at high modulation percentage.
durchmustern survey, review, catalog or count (stars).
durchnässen wet (thoroughly), soak, steep.
Durch-perlungselektrode bubbling-type electrode. **-projektion** back projection, diaprojection.

-projektionswand translucent screen.

durchrühren stir or agitate (thoroughly, to the bottom).

Durchsatz throughput (furnace work).

durchscheinend translucent, partially transparent, diaphanous, shining through.

Durchschlag breakdown, rupture, puncture (of insulation); punch, drift; filter, strainer; carbon copy. Wärme- breakdown due to thermal instability.

Durchschlagen (von Nachbarsender) interference by neighboring station in volume, neighboring stations "come through" or "break through" or swamp signals (by image frequency effect).

Durchschlagen der Gleichstromkomponente (über Synchronisierzeichen zum Schirm) d.c. component reaches screen by way of synchronizing pulses.

Durchschlags-festigkeit, Stossimpulse-voltage breakdown strength. -geschwindigkeit propagation speed (of fluid in pipes); rate of insulation breakdown.

Durchschlagspannung breakdown, puncture, rupture, disruptive or flashover potential.

Durchschmelzung burnout, melting, fusion, blowout.

durchschneiden traverse, cut across, cross, intersect.

Durchschnittsverminderung reduction of area (of a test piece pulled apart, resulting in a "waist").

durchschwingen swing through; operating condition in which wave is not interrupted or cut off.

durchseihen strain, filter, percolate.

durchsetzen thread (a film); permeate, traverse, pervade, mix, intersperse, thread (turns by magnetic flux).

durchsichtig transparent, diaphanous, translucent (when only partially transparent), clear, lucid, pellucid, limpid.

Durchsichts-betrachtung rear viewing (of pictures). -bild (cf Durchprojektion) diapositive, lantern slide, glass transparency (picture thrown from rear on frosted glass pane, etc.); translux picture. -dichte transparent density (m.p.). -schirm (Speicherprojektionsröhre) transparent screen (of a storage-type television tube). -sucher direct view finder, eye level finder).

durchstarten eines Flugzeuges "open up" and repeat landing procedure (in ZZ landing method).

durchstechen pierce, stab, cut through.

Durchsteuern operating inside straight portion of characteristic close to point where overload begins; grid excitation sufficient to swing plate current from zero to saturation point (re: amplifier). Load a tube to full capacity (re:transmitter).

Durchstossverfahren descent (or piercing) through clouds, penetration method (in airplane landing).

Durchstrahlung transmission of light or of radiations, penetration by rays, irradiation.

Durchstrahlungs-methode transmission method. -übermiskroskop transmission-type electron or ultramicroscope. -verfahren transmission method (using thin foil, in ultra-microscopy work).

durchstreichen flow, pass, circulate or travel through; cross out, cancel.

durchstreichende Linie trajectory.

Durchziehglas slide.

Düse nozzle, tip (welding), tuyere, twyer.

Düse, Blas- blasting or discharge nozzle, gun. Fang- mixing nozzle.

(steam jet). **Schweiss-** welding tip. **Venturi-** throat or constricted piece of Venturi meter, nozzle-like reducer.

Düsenkopf nozzle tip.

D-Wert quadrantal error due to airplane fuselage, etc.

Dynamik contrast or sound volume ratio between lowest and highest intensities of notes or passages (in music, etc.), practical volume compression and expansion; speech or signal energy volume range. Also: Ratio between undistorted maximum signal strength and noise, noise-signal ratio.

Dynamik-einebnung compression. **-einregelung** adjustment of contrast (by compandor action). **-entzerrer** dynamic expander. **-linie** volume-range or contrast characteristic. **-presser** compressor (in volume-range control). **-regler** volume- or dynamic-range control means, compandor (comprising compressor and expander means), contrast regulator. **-steigerung** (cf. Wuchtsteigerung) dynamic-range expansion. **-verzerrer** dynamic compressor.

dynamischer Lautsprecher, permanent- dynamic loudspeaker with permanent magnet.

Dynamometer, Brems- brake dynamometer, Prony brake.

Dynatronsummer dynatron (AF) oscillator.

E

Ebenbild image, likeness, similarity.
Ebene optical flat (opt.), plane, plain.
Ebene, in einer - mit coplanar with, flush with.
ebene Welle plane wave.
Ebene, Bezugs- datum level, d. plane, reference l., fiducial l. **Blenden-** diaphragm plane. **Brenn-, hintere** back focal plane. **Brenn-, vordere** front focal plane. **Einfalls-** plane of incidence. **Einstell-** focal plane. **Einstellungs-** focusing plane. **Flucht-** vanishing plane. **Halb-** semi-plane, half-p., H-p. **Haupt-** principal plane, p. surface. **Hyper-** hyperplane. **Kristall-** crystal plane, c. face. **Netz-** lattice plane, plane-family of a crystal.
Ebene, Polarisations- plane of polarization. **Drehung der-** polarization error.
Ebene, Reflexions- plane of reflection. **Schmiegungs-** osculating plane. **Spalt-** cleavage plane, c. face. **Symmetrie-** plane of symmetry, crystallographic p. **Teilungs-** plane of division.
Ebenenabstand, Haupt- principal plane distance.
Ebenheit planarity, flatness, smoothness, evenness.
Ebenmass symmetry.
Ebnung, Bildfeld- flattening of image field.
Ebnungslinse, Feld- field flattener.
Echelettegitter echelette grating (opt.).
Echo, stufenförmiges echelon-formation of multiple echoes.
Echo, Hörer- listener echo. **Mehrfach-** multiple echo, reverberation, flutter e. **Schein-** pseudo echo. **Vielfach-** multiple echo, flutter e. **Wellen-** radio (signal or wave) echo, wave reflection (involving one or more hops to ionosphere and back).
Echo-lot sonic altimeter, sound-ranging a., echo depth sounder, fathometer. **-lotung** echo sounding, echo or reflection altitude or distance measurement and ranging; radar or radio locator, sonar work (predicated upon echo principle). **-paare** paired echoes. **-stromdämpfung** echo current attenuation; active return loss. **-unterdrückung** echo suppression, e. killing, damping of reverberations (by sound absorbing material). **-weite** echo area, effective scattering cross-section of radar target. **-werk** swell organ. **-wirkung** echo effect.
echt, dampf- fast to steam. **koch-** fast to boiling. **licht-** fast to light, non-fading.
echtes, absolutes Gehör genuine absolute pitch.
echte Längenmessung direct measurement of length.
Eck, Seiten- lateral summit **Würfel-** corner of a cube, cubic summit.
Ecke corner, angle, edge (of a plane angle), summit (of a crystal), quoin.
Ecke, Mittel- lateral summit.
Eckfrequenz corner frequency, sharp cut-off frequency (filter).
eckig angular, cornered, angled.
edel noble, non-base, rare, precious, rich, vital, inert (gas).
Edel-erde rare earth. **-gas** rare or noble gas, inert gas (in atmos-

phere). **-gas-Glühkathodengleichrichter** tungar (argon-tungsten) rectifier. **-metall** noble metal, precious m., rare m. **-stahl** refined or superior alloy steel.

Effekt, kreismagnetischer gyromagnetic anomaly.

Effekt, Bord- board effect, ship field error (in df work), quadrantal error (due to structural parts, metal, etc.). **Funkel-** low-frequency variation of local emission density. **Richt-** rectification effect; directional e.

Effektbeleuchtung effect lighting, spot or fancy l.

Effektivwert effective value, rms (root-mean-square) value, virtual v.

Effekt-kohle salted carbon, carbon for flame arc. **-lampe** studio light (ordinary hand or table lamp, m.p.). **-szenen** scenics, light effects, etc.

egalisieren equalize, level, regulate (piano).

Ei, Antennen- fish (of trailing antenna).

Eichamt gage or calibration office, standards bureau.

Eichelröhre acorn tube.

eichen gage, measure, standardize, calibrate, test, stamp, adjust (weights), log.

Eich-katzenanker squirrel-cage armature. **-kreis, Arbeits-** working reference circuit. **-mass** gage, standardized measure. **-reizkurven** (typical) visibility curves.

eidesstattliche Erklärung affirmation in lieu of oath, statutory declaration.

eidliche Versicherung affidavit, sworn statement, testimony or declaration.

eigen inherent, self, "eigen", own, individual, characteristic, proper, specific, special; exact, precise, delicate.

Eigen-abstossung der Elektronenwolke repelling action of electron charge, negative space c. **-dichte, Energie-** proper energy density. **-drehimpuls** eigen or characteristic angular momentum. **-energie** "eigen" energy, characteristic e. **-funktion** eigen function, characteristic f. **-geräusch** inherent film noise. **-geschwindigkeit** indicated air speed (of an airplane). **-gewicht** specific gravity, own or dead weight, proper w. **-heit** singularity, particularity, peculiarity. **-kapazität** self-capacitance. **-licht des Auges** eye's own light, self-l. or intrinsic l. of retina (associated with photopsy). **-peilung** board aircraft direction-finding, taking bearings with homing device on board, on two or more stations.

eigenschwingungsfrei aperiodic, deadbeat.

Eigen-spannung residual stress, inherent or internal tension or strain, **-spannung, Druck-** residual compressive stress. **-störung** internal trouble. **-strahler-Übermikroskop** self-emission electron microscope, s. -illuminating e.m. (based on electrons arising on object surface).

eigentümlich peculiar, characteristic, inherent, specific.

Eigen-tumsrecht ownership, proprietary or property right, proprietorship. **-verständigung** inter-phone communication (airplane), inter-vehicular c. **-wärme** specific, body or animal heat. **-wert** "eigen" value, "eigenwert", characteristic, representative, typical or inherent value, characteristic number. **-zeit** proper time. **-zustand** "eigen" state.

Eikonal, Winkel- angle eiconal (opt.).

einachsig uni-axial.
einäugiges Sehen monocular vision, non-stereoscopic vision.
einäschern incinerate, calcine, burn to ash.
ein-atomar, -atomig monatomic.
Einausschalter off-on switch.
Einbandübertragung single-side-band transmission (with vestigial s.-b.).
Einbau inner player (piano).
einbetten imbed, place in matrix.
Einbettungsmaterial matrix or embedding material.
Einbeulungsprüfung bulging test.
Einbiegung curvature, inflection, flexure.
einblasen blow in or into, insufflate, inject.
Einbläser punctiform pits produced by atmospheric pressure in glass.
Einblenden der Mikrophone fade-in and mixing of microphones.
Einblick-linse eyepiece lens. **-rohr** eyepiece tube.
Einbrennzeit heating-up period.
Einbruch (Kurven, etc.) break, breaking in.
Einbruchsmelder burglar alarm.
Einbuchtung bay, bight, indentation, niche.
eindampfen thicken by evaporation, boil down, inspissate.
eindeutig unique, plain, clear, unequivocal, single-valued. **-, doppelt-** one to one.
eindeutige Peilseite absolute direction. **-Richtungsanzeige** unidirectional direction-finding (with sense finding or sensing).
Eindeutigkeitsprinzip (Pauli) exclusion or equivalence principle.
eindicken inspissate, thicken, concentrate, boil down, condense.
eindrehen recess, tap (a screw-thread), indent.
Eindringungsfähigkeit penetrativeness.
Eindruck, Mitten- binaural balance (in sound locating).
Eindrucksgleichheit equality of sensation or impression.
Eindruckverfahren ball test, Brinell hardness t. method.
einebnen level, smooth, flatten (curve or wave).
Einebnung, Dynamik- compression.
Einergang-Kurbel und -welle single-picture crank and shaft (m.p.).
Einfachantenne plain antenna.
einfach-brechend singly refracting, unirefringent. **-frei** univariant, monovariant, having one degree of freedom. **-wirkend** single-acting.
Einfachzacken-schrift mit Abdeckung unilateral variable-area track with single-vane shutter. **-spur** unilateral variable-area track.
einfädeln thread (up) film.
einfädig unifilar.
Einfall, schiefer oder streifender grazing, oblique or glancing incidence. **Glanz-** grazing incidence (at a glancing angle).
Einfallen incidence, snap-in motion; collapsing, implosion (reversed explosion).
einfallendes Licht incident light.
Einfallgeschwindigkeit bombardment velocity.
Einfalls-ebene plane of incidence. **-winkel** angle of incidence. **-winkel, Haupt-** angle of principal incidence.
einfaltige Gruppe one-dimensional group.
Einfang, Neutronen- neutron capture.
einfangen capture, captivate, trap (electrons).
einfarbiges Licht monochrome, monochromatic light, light of one color.
Einfärbung dying, tinting, staining, inking, imbibition.
einfassen border, hem, surround something, enclose, case, bind, trap, trim, set.
Einfassleisten binding strips.
Einflug-richtung direction of approach (in airplane landing).

-schneise approach track. -sektor approach sector. -sender boundary marker beacon.
Einflugzeichen boundary marker signal, come-in s. Haupt- main marker, inner marker or main entrance signal nearest airport boundary (in airplane landing). Vor- outer marker (beacon) signal, fore m. beacon.
Einflussbereich sphere of influence, radius of action.
einförmig uniform, monotonous.
Einfräsung milled slot or recess.
einfrieren freeze up, congeal.
einfügen insert, interpose, intercalate.
einführen thread (up) film, introduce, insert.
Einführungs-isolator lead-in or feed-through insulator, bushing. -isolator mit Vergusskammer pothead insulator. -verlust insertion loss.
Eingabe amendment, petition (in Patent Office practice).
eingängig single-thread (mech.).
Eingangs-bedämpfung input resistance, i. damping. -klemme input terminal. -leitwert input admittance. -pupille entrance pupil. -schlitz entrance slit (spectr.). -wirkleitwert input conductance, i. active admittance.
eingebaute Antenne built-in antenna.
eingefahrene Antenne retracted or drawn-in antenna, reeled-in a.
eingefroren latent, dormant.
eingehen (in eine Gleichung) enter into, occur, or appear in, an equation.
eingekittete Linse cemented lens.
eingeschwungener Zustand steady-state (of current, oscillation, etc.).
eingestrahltes Licht incident light, exciting l.
eingestrichene Oktave one-stroked octave.
Eingiessung pouring-in, infusion, transfusion.
Eingreifen (der Greifer) in-and-out movement (of claws), engagement of c.
Eingrenzung von Fehlern tracking down seat or source of interference or trouble, localize trouble.
Eingriff engagement, mesh, gearing.
Einhaltsbefehl injunction, interdict.
Einhängeblende inset diaphragm.
einhängen des Hörers hang up or replace telephone receiver, clear the line.
Einheit (constructional or structural) unit, integral part; unity.
Einheit, Ladungs-, positive elek. unit positive charge. Massen- unit of mass. Mehrheit von-en array of units, plurality or multiplicity of u. Oberflächen- unit of area. Radiumemanations-curie unit of radon. Raumwinkel- steradian, spheradian. X-, (Roentgenstrahl-) wave-length unit of X-rays, etc., Siegbahn unit, Xu, X unit.
Einheiten und Formeln, Ausschuss für (A E F) Committee on Units and Formulae, Standards Committee.
Einheits-frequenz standard frequency. -funktion von Heaviside unit function of Heaviside. -gitter unit lattice. -ladung unit charge. -masse unit mass. -punkt unit point, principal point. -spin unit spin. -zelle unit cell, elementary c., unit crystal, lattice unit.
einhörig monaural.
einhüllen envelope, enwrap, imbed, shroud, sheath.
Einhüllende envelope, contour, outline.
Einkanaltastung single-channel pulsing (telev. synchronization).
Einkerbung notching, indentation, denting.
einkeilen (in) dent, nick, groove, notch.
einkernig mononuclear, uninuclear,

single-cored.
Einklang syntony, consonance, unison, concord, accord, agreement.
einklinken lock, engage, ratchet.
Einknopfkontrolle uniselector, unicontrol.
einkopieren over-print, double print.
Einkreisempfänger single-circuit receiver, ultraaudion (ultraudion) receiver.
Einkristallfaden single-crystal filament, mono-crystal f.
einlagern intercalate, embed, incorporate, store.
Einlagerung embedding, occlusion, inclusion, insertion, matrixing.
Einlagerungs-rate rate of incorporation (cryst.). -verbindung, komplexe complex intercalation compound.
einlaufen shrink, contract, flow in; run in (as a bearing, etc.).
einläufig unicursal; single-barreled, s.-tracked, s.-channeled.
Einlauf-punkt auf die Rolle point where film is fed onto drum. -seite feed or threading end (m.p.).
einlegen, Berufung gegen eine Entscheidung bei höherer Instanz lodge or take an appeal from a decision to a higher court.
einlegen (Film) thread up the film.
einleiten lead in, initiate, originate, induce, introduce.
einmalig single, solitary, unique. -(er) Vorgang non-recurrent, unique or singular action or phenomenon, single event.
Einordnungsstellung position of grating resulting in a one-order spectrum.
Einpegelung leveling.
einpeilen (Funksender mit Richtempfänger) taking bearings from, or tune in, a radio transmitter, with directional receiver.
einprägen impress, imprint, emboss (sound track on film).
einpressen press-fit, press in.
Einpulsverfahren single-pulse or monopulse method.
Einpunktschreiber single-point recorder.
Einquellenempfänger solodyne receiver.
Einrasten locking by a detent, snap shutting.
Einrede plea, defense. Beschwerde- rejoinder (in an appeal).
Einregelung der Dynamik adjustment of volume range or contrast (by compandor action).
Einregelungszeit building up or waxing period (say, of a control potential).
einreichen file, petition, apply for.
Einrichtung arrangement, device, disposition, equipment, contrivance, outfit, organization, means.
Einriss rent, fissure, tear, flaw.
einrücken engage, throw in gear or mesh.
Einrückhebel trip or engagement lever, starting lever.
Einsattelung crevass or dip (of a resonance or other curve).
Einsatz (acoustic) intonation, onset (of sound, discharge or other action), start, incipiency, initiation, insertion; charge (of a furnace); stake, application, use, what is put to work.
Einsatz der Schwingungen, harter hard start of oscillations. -, weicher gentle or smooth start of oscillations.
Einsatz, Bildkipp- incipience of frame flyback or beam return (telev.). Leuchtfaden- streamer onset. Zeilenkipp- incipient (line) flyback.
Einsatz-blende interchangeable diaphragm. -dichte number of stations (satisfactorily) operable inside a given wave-band. -härtung case hardening, surface cementation. -punkt, Gitterstrom- grid-current point, point of incipient grid-current flow. -spannung (cf Zündspannung) spark or

flashover potential, breakdown potential (in grid-glow tubes), starting potential.

Einsaugemittel absorbent, imbibent, adsorbent.

einsaugend absorbent, absorptive, imbibent.

einschichten imbed, (inter) stratify, interleave, arrange in layers.

Einschlaglupe folding magnifier.

einschliessen enclose, confine, include, seal in, lock, embed, matrix, occlude.

Einschlussröhre sealed tube.

einschmelzen seal-in (a wire or lead in a tube).

einschneiden engrave (sound track on film).

Einschnitt cut-off (spectrum); incision, cut, notch, dent.

Einschnürung squeezing, compression (re:film), bind-up, constriction, recess; waist, reduction in area (metal testing).

Einschnürungs-blende crossover aperture. **-punkt** crossover (in electron-optics).

Einschwing- und Ausschwing Verzerrung transient (non-linear) distortion, build-up and decay d. (in ac., facsimile, etc., associated with hangover, tailing, underthrow and overthrow).

Einschwingvorgang build-up transient oscillation, onset or initiation of impulse.

einseitig offen unilaterally open (-ended).

Einseitenbandübertragung single-side-band transmission (with vestigial s.-b.).

Einsenkung crevass, dip, depression.

einsetzen case-harden (metal).

Einsetzen der Zündung initiation or incipience of striking or firing.

einsickern infiltrate, soak in, ooze in.

Einspannrand, gewellter corrugated suspension means (of loudspeaker cone).

Einsprache acoustic inlet (of a microphone).

einspringen spring in, shrink (of fibers), re-enter (of an angle).

einspringend re-entrant.

Einspruch plea, objection, protest, demurrer, opposition. **-unterstützen** argue in support of an opposition.

einspuriger Tonstreifen single-edged variable-width sound track or record.

Einsteckseiher plug-in inlet strainer.

einstellen set, adjust, position, focus. **-,neutral** adjust or set to neutral.

Einstell-fassung focusing mount (of lens). **-fehler** faulty focusing or spot control, decrease of spot with intensity increase, large diffuse speck on screen in gas-focused picture reproducing tube. **-knopf** focusing knob (of camera). **-lampe, Bild-** framing lamp. **-lupe** focusing magnifier. **-marke** reference mark, gage m., index dot (on dial), measuring m., wander m. (of telescope). **-mikroskop** focusing microscope. **-schraube** set screw, adjusting s., leveling s. **-tafel** focusing board, test chart. **-tubus** micrometer adjustment (electron microscope), adjusting tube.

Einstellung, kritische critical adjustment, delicate a., setting or focusing. **Bildbühnen-** framing, racking (with framing device). **Bild-** centering control, height c., horizontal c., framing (in telev., with rack and pinion for focusing). **Fein-** vernier (dial) adjustment, tuning or setting. **Fern-** long-range focus. **Nah-** short-range focus. **Nullpunkt-** zero adjustment. **Phasen-** phasing, phase adjustment, shift or regulation. **Scharf-** automatic tuning; sharp focusing. **Still-** silent tuning, q.a.v.c. tuning. **Unendliche**-infinity adjustment (of range-finder).

Einstellungs-ebene focusing plane. **-fläche** surface of reference. **-klage** complaint or action for discontinuance, suspension or stay (of infringement, etc.), action for issuance of an injunction, i. suit.

Einstellweite focal range, focusing r.

Einstich puncture, perforation.

einstimmig monophonic, one-voiced, unanimous.

Einstrahl-funkstelle beam station. **-sender** beam transmitter, unidirectional t.

Einstreuung stray effect.

einstufen grade (by quality).

einstweilige Verfügung temporary, provisional or interim injunction, provisional decree.

einsumpfen digest with water, soak wet.

eintasten key on, key in.

eintauchen (cf **Tauch**) dip, plunge, immerse, steep; telescope (a coil).

Eintauch-refraktometer immersion refractometer, dipping r. **-zeit** immersion time.

Einteilchenmodell single-particle model.

einteilen divide, subdivide, distribute, separate, grade, sort, classify, graduate (in terms of degrees or other units), index.

einteilig uni-partite.

Einteilung, Nonius- vernier scale.

Eintrag charge, input.

Eintragende feed or charge end.

Eintragschnecke feed worm.

Eintragung entry, registration.

Eintragungs-bescheinigung certificate of registry. **-zwang** compulsory registration.

Eintritts-blende input diaphragm or capacity disk (re:klystron, resonance cavity). **-feld** area of applicator at skin surface (X-ray work). **-geschwindigkeit** approach or inlet velocity. **-phase** entrance phase (re:filter).

Ein-und Ausrückhebel starting and stopping lever.

ein-und-einachsig orthorhombic (cryst.).

einverstanden sein agree to, acquiesce in.

Einwand demurrer, objection (for instance, of Patent Office Examiner).

einwandfrei satisfactory, unobjectionable, acceptable, tolerable.

Einweggleichrichter half-wave rectifier.

einwegig single-channel, simplex, one-way.

einweichen steep, soak, macerate, digest.

Einwendung plea, defense, objection, demurrer.

einwertige Salze 1 :1 salts.

Einzackenschrift single-edged variable-width sound track or record.

Einzahn-stift striking roller (m.p.). **-scheibe** drive wheel (m.p.).

einzel single, individual; various, unitary, separate.

Einzel-abschnitt-Filter single-section filter. **-bild einer Bildreihe** individual picture or unit frame in a series or sequence. **-darstellung** single representation; separate treatise, monograph. **-heiten** details, particulars, minutiae. **-klang** (sounding of an) individual note. **-linse** unit lens (of focusing field). **-lupe** magnifier unit. **-peilung** taking bearings from one object. **-streuung** single scattering (of electrons). **-vorgang** single or separate process, happening, action, event or reaction.

einziehbar retractile, retractable, withdrawable.

einziehen wind or reel in (an antenna), retract (airplane undercarriage); turn down, decrease (say, diameter of a shaft).

Einziehungsmittel absorbent.

Einzigkeitstheorem uniqueness

,theorem.
Einzugsgebiet cross-sectional area of undistorted portion of field (direction-finding frame).
eirund, halb- semi-oval.
eis-artig ice-like, icy, glacial. **-blau** glacier blue.
Eisen, kadmiertes cadmiated iron. **Band-** hoop, strip or band iron.
Eisen-begleiter elements accompanying iron, i. companions or associates, congeners (if of similar sort). **-bügel** iron staple, yoke. **-drossel** iron-cored choke-coil. **-drossel, Besprechung mit** modulation by voice action with magnetic modulator.
eisenfreie Spule air-cored (choke) coil.
Eisen-goniometer iron-cored goniometer. **-kern, geblätterter** laminated iron core. **-metalle, Nicht-** non-ferrous metals. **-metallurgie** siderurgy. **-panzer** iron shield, i. screen, i. cladding. **-pulverkern** Ferrocart core, compressed iron dust c. **-pulververfahren** magnaflux inspection method. **-rahmen** iron-cored frame (d.f.). **-schluss** path closed, or passing, through iron (for magnetic flux), magnetic shunt, keeper (of permanent magnet). **-späne** iron turnings, borings, filings or chips. **-verlust** iron loss, core l. **-wasserstoffwiderstand** iron-hydrogen resistance. **-weg** iron path, magnetic or ferro-m. circuit.
Eiskalorimeter ice calorimeter.
elastisch gestreut elastically scattered.
elastische Nachwirkung plastic flow persistence, after-flow, elastic after-effect.
Elastizität elasticity, compliance (mechanical filter, ac.).
Elastizitäts-messer elasmometer (for Young's modulus), torsometer. **-modulus** modulus of elasticity (of extension, torsion, compression or volume).
Elektrete electrets.
elektrisch, gleichnamig similarly electrified or charged. **glüh-** thermionic. **ungleichnamig** oppositely electrified or charged.
elektrisch symmetrisch homopolar.
elektrischer Ausfall electric shock.
Elektrizität, Druck- piezo-electricity. **Glas-** vitreous electricity, positive e. **Harz-** resinous electricity, negative e. **Reibungs-** frictional electricity, tribo- e.
Elektrizitätslehre, Bewegungs- electrokinematics, electrokinetics. **Ruhe-** electrostatics.
Elektrode, Abnahme- output electrode. **Beschleunigungs-** accelerator electrode, gun e. (c.-r. tube). **Bildwurf-** target electrode. **Bremsfeld-** reflecting electrode, retarding-field e. (in electron-oscillation, Barkhausen-Kurz or positive-grid tube). **Durchperlungs-** bubbling-type electrode. **Entnahme-** output electrode, catcher e. (klystron). **Gegen-** counter electrode, opposite e., co-operating e.; signal plate, metallic screen (forming foundation of mosaic, in telev. tube). **Nachbeschleunigungs-** second or additional gun or accelerator electrode, after-accelerator, post-accelerator. **Quecksilbertropf-** dropping mercury electrode. **Ring-, abbildende** image focusing ring. **Sammel-** output electrode, collector e., gathering e., catcher (in klystron). **Seiten-** end or wing electrode (in magnetron). **Spitzen-** needle or point electrode. **Steuer-** control(ing) electrode, shield, Wehnelt cylinder or grid (in cathode-ray tube). **Such-** exploring, probe, search or collector electrode (vacuum tube). **Zitter-** vibrating electrode. **Zünd-** ignitor (of ignitron).

Elektroden-abstand electrode spacing, inter-electrode distance or gap. **-fahrsäule** electrode mast or support. **-kapazität** (inter-) electrode capacitance.

elektrodenlos electrodeless, devoid of, or without, electrodes.

Elektrodenschluss electrode short-circuit.

Elektrometer, Faden- filament electrometer. **Goldblatt-** gold-leaf electroscope. **Kipp-** Wilson electrometer (casing tiltable giving gold leaf different positions).

Elektronen, angelagerte trapped electrons. **-, innere** inner electrons (on K.L.M...shells). **-, kernferne** outer (level) electrons, conduction, valence, peripheral, orbital, or planetary electrons. **-, kernnahe** electrons adjacent nucleus, inner (fixed) electrons. **-, klebende** captured electrons, electrons sticking to electro-negative gases, surfaces, etc. **-, kreisende** orbital electrons. **-, positive** positrons, positive electrons. **-, schwere** heavy electrons, barytrons, dynatrons, mesotrons, penetrons, x particles, etc.

Elektronen, Bahn- orbital electrons. **Feld-** field electrons (from cold or auto-electronic emission caused by an electric field). **Gitter-** lattice electrons. **Hüllen-** shell electrons. **Kreisel-** spin electrons. **Leitungs-** conduction electrons (from outer levels). **Rückstoss-** recoil electrons, Compton e. **Superleitungs-** super-conduction electrons. **Unterschale-** sub-shell electrons. **Zerfall-** decay electrons, e. resulting from disintegration (say, mesotrons).

Elektronen-abgabe, glühelektrische thermionic emission, Edison or Richardson effect. **-abgabefläche** electron emission or emitting area. **-abtaststrahl** electron scanning spot or brush (iconoscope), e. pencil, beam or lever. **-auffang** electron capture. **-austrittsarbeit** work function of electrons. **-bahn** electron path, orbit or trajectory, flight of electrons. **-belegung** density of electrons (in beam). **-beugung** electron diffraction. **-bildröhre** electron image tube. **-bremsung** reflection (in Barkhause electron-oscillation tube), deflection (in magnetron); retardation or deceleration of electrons. **-drall** electron spin. **-einfang** electron capture or trapping. **-emission, Auto-** auto-electronic, cold or field emission of electrons. **-energie** electron energy, electron affinity. **-entladung, Glüh-** thermionic discharge. **-fleck** electron spot, scanning s. **-fluss- Diskontinuität** shot effect. **-gas, entartetes** degenerate electron gas. **-gruppierung** phase focusing (in beam tubes). **-hülle** electron shell, electron cloud. **-konzentration** electron focusing. **-laufzeit** transit time of electrons, orbit time of e. **-lawine** avalanche of electrons. **-lehre** electronics. **-linse** electron lens (electromagnetic or electrostatic). **-mikroskop** electron microscope, immersion objective. **-mikroskop-Selbstleuchtverfahren** self-luminous, self-emission or s.-illuminating method (using electrons arising on object surface). **-niveau** electron level. **-optik** electron optics. **-ort** electron position, e. locus. **-pendelung** oscillating of electrons. **-physik** electronics. **-plasma** residual electrons, plasma (region without resultant charge). **-quelle** electron (emission) source, e. emitter. **-raster-Mikroskop** raster microscope, electron-scan m.

-relais electron relay, thermionic r. -rumpf core, kernel, stable electron group. -schalen, beinahe geschlossene nearly closed electron shells. -schleuder electron gun. -schwingungen electron oscillations. -spritze electron gun. -sprungspektrum electron transition (or jump) spectrum. -stauung electron cloud or accumulation (in a virtual cathode).

Elektronenstrahl, gaskonzentrierter gas-focused electron beam. -, geschwindigkeits-gesteuerter velocity-modulated electron beam. -, weisser heterogeneous beam.

Elektronen-strahlabtaster electron scanning pencil, beam, spot. brush or lever. -strahlengang electron ray path. strahlröhre (cf Kathodenstr.) cathode-ray tube; beam tube; thermionic t. (obs), c.-r. tuning indicator t. -strommodulation electron quantity modulation. -tanz oscillatory movements of electrons (due to magnetic or electric forces). -turbine cyclotron and inverted (or Hollmann) forms thereof. -technik electronic art or technology, electronics. -übergangswahrscheinlichkeit transition probability (of electron passing from one level to another level). -überholungsgebiet catchup or overtake region (in which electrons arrange themselves in groups for phase focusing in beam tubes). -überkreuzungsstelle crossover of electrons (where principal rays from first lens cross axis, equivalent to exit pupil of eye). -übermikroskop electronic super microscope, e. ultra microscope. -unterhülle sub-shell (of electrons). -welle electron wave, phase w., de Broglie w. -wolke cloud of electrons. --Eigenabstossung der repelling action of electron charge, negative space charge. -wucht collision force, bombardment f. (of electrons). -zustand, entarteter degenerate electron state.

Elektroniker electronician, electronics engineer.

elektro-optischer Verschluss electro-optic shutter.

Elektrophoresezelle electrophoresis cell.

elektrostatisch geschirmt electrostatically shielded, e. screened. -(er) Lautsprecher electrostatic loudspeaker, condenser l., capacitor l.

Element, Lager- inert cell. -Raster- picture unit, p. element, elementary area. Schalt- circuit element, c. means. Strecken- linear element (math.).

Elementar-ladung elementary charge, electronic c. -wellen Huygens wavelets. -zelle unit cell, elementary c., lattice unit.

Elementenmessung stoichiometry.

Emanation, abgeklungene emanation with diminished or decaying radioactivity. Aktinium- actinon. Radium- radon. Thorium- thoron.

Emanationseinheit, Radium- curie unit of radon or radium emanation (10^{-10} curie-liter = eman.).

Emission, glühelektrische thermionic emission. Autoelektronen- autoelectronic, cold or field emission of electrons.

Emissions-strom, Feld- field emission of electrons, cold e. -vermögen emissivity, emission power.

EMK (elektromotorische Kraft) emf (electromotive force).

EMK von rechteckiger Kurvenform square-shaped or rectangular emf, square-topped emf.

Empfang, Block- block reception, communal or party r. Druck- telescript reception, typed r., teleprinting r. Einquell- solodyne reception. Geradeaus- single-circuit reception, non-heterodyne r., straightahead r. Hör- audible reception, auditory r., aural r., reception by ear, hearing or earphones. **Musa-** multiple-unit steerable antenna reception. **Primär-** single-circuit reception. **Richt-** directional reception. **Rückkopplungs-, mit Hilfsfrequenz** super-regenerative reception. **Schreib-** visual reception, recorder r. **Such-, direkter** single-circuit or straight-ahead (non-heterodyne) reception. **Superheterodyne-, mit selbsterregter Trägerwelle** homodyne reception, zero-beat r. **Tertiär-** three-circuit reception. **Ton-** AF modulated or tone-modulated c.w. reception. **Transponierungs-, Überlagerungs-** heterodyne reception.

Empfänger receiver; collector, pickup (microphone). **Empfänger, verdrahteter** fully wired receiver set.

Empfänger, Allstrom- universal receiver set, a-c/d-c r.s. **Bewegüngs-** velocity microphone, v. pickup; pickup or detector of motion. **Bild-** (cf **Bildempfangstelle**) picture receiver, television or video r., picture reproducing device, p. viewing tube, image reconstructor. **Bildton-** audio-video receiver combination. **Breitband-** broadband or wide-band receiver. **Dreikreis-** three-circuit receiver. **Druck-** pressure (responsive) microphone. **Druckknopf-** push-button or press-button tuned receiver. **Einkreis-** ultra-audion, ultraudion or single-circuit reception. **Einquell-** solodyne receiver. **Fernseh-** television receiver apparatus, video-signal receiver, kinescope (Zworykin), oscillight (Farnsworth). **Gleichstrom-** d.c. electric (radio) set. **Heimfernseh-** home television receiver. **Kontroll-** monitoring receiver. **Peil-** direction-finding receiver. **Peilfunk-** radio compass, direction-finder (avigation). **Pendel-** super-regenerative receiver, Armstrong r. **Primär-** single-circuit receiver. **Richt-, Einpeilen auf Funksender** tuning directional receiver to radio stations (to take bearings). **Schall-** sound collection device, acoustic receiver, sound pickup or microphone. **Schallschnelle-** velocity microphone. **Strahlungs-** radiation responsive or sensitive record or receiving sheet or surface (on which radiations impinge to measure color temperature, etc., such as photographic plate, selenium cell, bolometer). **Superregenerativ-** (Armstrong) super-regenerative receiver. **-und Sender kombiniert** transceiver radio set; microtelephone. **Unterwasserschall-** subaqueous or submarine microphone, pickup device or sound detector, hydrophone. **Wechselstrom-** a.c. electric receiver set. **Zielflug-** homing receiver apparatus. **Zweikreis-** two-circuit receiver set. **Zwischenfrequenz-, mit Kristallsteuerung** crystal-monitored or c. stabilized supersonic or superheterodyne receiver, stenode radiostat.

Empfänger-blechgestell metal chassis **-fläche** radiation-sensitive surface; record sheet. **-reichweite** (distance) range of receiver, distance-getting ability of radio set. **-Senderkombination** microtelephone; transceiver radio set.

Empfangfeldstärke incoming signal level.

Empfänglichkeit susceptibility, receptivity, responsiveness.

Empfangs-antenne, vom Sender entkoppelte balanced receiver aerial (uncoupled from transmitter). **-bild** recorded copy (in facsimile). **-gebiet** service area (in broadcasting). **-spannung, gerichtete** directional incoming or signal potential. -verfahren, Mehrfach- diversity receiving method.

empfinden sense, perceive, feel, apprehend, experience, have consciousness...empfindlich responsive, sensitive or reactive to (say, temperature), be a function of, be dependent on.

empfindlich, blau- blue sensitive. flecken- susceptible to staining or spotting. richtungs- with directional response.

empfindlich machen, licht- photosensitize.

Empfindlichkeit sensitivity, sensitiveness, sensibility, responsiveness; efficiency, yield, output.

Empfindlichkeit, Ablenkungs- deflection sensitivity (reciprocal of deflection factor). Augen- intensity discrimination, contrast sensitivity (embodied in Weber-Fechner law). Hand- hand or body capacitance or capacity effect. Hör- auditory sensitivity, hearing capability. Kerb- stress concentration index. Korrosions- corrodibility, susceptibility to corrosion. Licht- photo-sensitivity. Richtungs- directional response.

Empfindlichkeits-kurve response characteristic. -messung sensitometric measurement (phot.). -verteilung sensitivity distribution, spectral response (of eye). -ziffer sensitivity factor.

Empfindlichmachung sensitization, activation, sensibilization.

Empfindung sensation, feeling, perception, sentience.

Empfindung, akustische acoustic perception. Farb- color sensation, c. perception, chromatic s. Lautstärken- aural sensitivity. Ton- acoustical perception, sound sensation.

empfindungs-gemässe, -gerechte Farbstufe, subjective chromatic scale value. -gerechte Abstufung subjective or natural grading or spacing (in chromaticity or chromatic scale, etc.).

Empfindungs-grenze des Ohres loudness contour of ear, auditory sensation area (between limits of maximum tolerable and minimum perceivable intensities). -lautstärke response of ear to signal strength; sensible loudness of sound. -schwelle, Schall- threshold of audibility or acoustic perception. -zunahme sensation increment.

Emulsions-seite face (of film) bearing emulsion. -steilheit steepness of gradation.

End-alarm, Akt- reel-end signal. -ausschlag full excursion, limiting deflection. -geschwindigkeit terminal velocity. -gewicht fish of a trailing antenna. -impedanz terminating or end impedance, load i. -kapazität top capacity, terminating or end c. -kondensator terminating condenser or capacitor.

endloser Filmstreifen endless film strip.

End-mass, Parallel- standard gage block, end-to-end s. bar. -punkt terminal, terminating point, terminus. -röhre power tube, final t. -röhre, Doppelsteuer- power pentode in which cathode grid is connected with control g. rather than with cathode. -streifen trailer (of a film). -urteil final judgment. -verschluss cable box, terminating box. -verstärkerstufe final amplifier stage, power amplifier s. -wert final, resultant or ultimate value.

energetische Bevorzugung energy preference.

Energie, Ablösungs- excitation energy, work function (of an electron). Anregungs- excitation energy, stimulation energy (say, a potential). Bewegungs- kinetic energy, motional e. **Bindungs-** binding energy, bonding e. **Eigen-** eigen energy, characteristic e. **Elektronen-** electron energy or affinity. **Selbst-** self-energy (of electrons).

Energie-berg energy hill, e. barrier. **-eigendichte** proper energy density. **-erhaltung** energy conservation, preservation of e. **-impulstensor** energy momentum tensor. **-leitung** (energy) feeder lead, feeder downlead, transmission line.

energielos (ohne Energieverbrauch) non-dissipative, wattless.

Energie-messer ergometer. **-niveau** energy level, quantum state.

energiereich full of energy, energy-rich, energetic.

Energie-rückgewinnung durch Bremsen energy recuperation by (regenerative) braking. **-schwelle** threshold of energy. **-stufe** energy level, energy term, quantum state. **-übertragung** energy transfer. **-umwandler** transducer, sink. **-verbrauch** consumption, dissipation, loss or expenditure of energy. **-verbrauch, ohne** non-dissipative, wattless. **-zerstreuung** scatter of energy.

enge Anschmiegung intimate or snug adhesion, adherence, fit, engagement or contact, "hugging."

engjähriges Holz close-grain wood.

entaktivieren render inactive, deactivate, de-energize, desensitize.

entarteter Elektronenzustand degenerate electron state.

entartetes Elektronengas degenerate electron gas.

Entartung degeneracy, degeneration.

Entbindung, Wärme- disengagement of heat, evolution of h., release of h.

Entbrummer hum eliminator, humbucker.

Entbündeln de-bunching (in klystron), unbunching; defocusing.

Entdämpfung gain, regeneration, deattenuation. **-Frequenzkurve** gain-frequency curve.

enteignen expropriate, dispossess.

enteisen de-ice.

entfallen lose (a chance), drop out, fall.

Entfärbung decoloration, bleaching.

Entfärbungsmittel decolorant, bleaching agent.

Entfernung, Kern- inter-nuclear distance. **Sprung-** skip distance, skip zone.

Entfernungs-messer, Basis- base range finder. **-messer, Halbbild-** split-field coincidence range finder. **-schleier** distance fog. **-skala** distance scale.

entfesseltes, bewegliches Mikrophon following microphone.

Entfritter decoherer, tapper.

entgasen extract gas from, outgas, cleanup, degas, getter.

Entgasung, Trocken- dry distillation.

entgegen-arbeiten counteract, buck, work against. **-drehen, dem Wind** head into the wind, crab. **-gesetzte Pole** opposite poles, unlike p.

Entgegenhaltung anticipation, anticipatory disclosure or reference (from prior or earlier art).

entgegen-wirken oppose, counteract. **-wirken, der Resonanz** anti-resonant.

entglasen devitrify.

entgraten debur.

enthärten soften, anneal.

entionisieren de-ionize, unionize.

Entisolierer wire skinner, insulation stripping tool.

Entknurrungswiderstand anti-growl resistance, resistor designed to suppress growling noise due to RF or tickler coil, in regenerative path.

Entkohlung decarburization, decarbonization.

entkoppelte Antenne (vom Sender) balanced antenna.

Entkopplung decoupling, balancing, tuning out.
entkräftigen exhaust, debilitate, weaken, vitiate; invalidate.
entladen discharge; unload (phot.).
Entladeschaltung circuit means causing fast rise of sawtooth potential.
Entladung, aperiodische aperiodic, deadbeat, impulse or non-oscillatory discharge. **-, kontrahierte** contracted discharge. **-, selbständige** self-sustained, spontaneous or unassisted discharge. **-, stille** corona, silent discharge, efflove (in skin therapy). **-, strahlartige** needle-point, corona or streamer discharge, leader-stroke d. **-, überspannte** overvolted discharge. **-, unselbständige** assisted or non-self-sustained discharge.
Entladung, Büschel- brush discharge, corona (ranges between glow and spark discharge). **Dunkel-** dark discharge, dark current. **Fackel-** torch-form of discharge. **Fadenstrahl-** (cf **Fadenstrahl**) thread-ray discharge. **Glühelektronen-** thermionic discharge. **Hochstrombogen-, nichtkondensierte** heavy-current uncondensed discharge. **Kipp-** condenser discharge (in time-base). **Neben-** lateral, stray or secondary discharge. **Schwing-** oscillatory or oscillating discharge. **Spitzen-** point discharge, needle (-gap) d. **Spritz-** initial discharge caused by rapid surge of ions; needle-gap discharge. **Vor-** pre-discharge.
Entladungs-aufbauzeit, Funken- spark-discharge formation time. **-kanal, Vor-** pre-discharge track. **-leuchtstreifen** luminous streamers.
Entladungsröhre, Gas- gas discharge tube or valve. **Kipp-** time-base discharge tube; kipp relay. **Quecksilberhochdruck-** high-pressure mercury discharge tube.
Entladungs-stoss burst, pulse, corona. **-verzug** discharge delay.
entlang streifen brush, skirt or wipe over or along, graze.
entlasteter magnetischer Lautsprecher balanced-armature magnetic loudspeaker.
Entlastung relieving something of weight, traction, stress, etc., say, by counterpoise and the like, unburdening, removal of load. **Spannungs-** stress relief; anti-fatigue or anti-vibration means (for lines). **Zug-** traction relief.
entlüften evacuate, exhaust air, deaerate.
entmagnetisieren de-Gauss, demagnetize, de-energize.
entmischen demulsify, liquate out, unmix, separate, sort (ions in mass spectrometer).
Entmodulierung demodulation, detection, rectification.
Entnahme, Leistungs- energy or power output, delivery, absorption or pickup, tapping, taking or draining of power or energy. **Probe-** sample taking, sampling, taking or selecting at random.
Entnahmeelektrode output electrode, catcher (in beam tube).
entnehmen absorb, pick up, catch or abstract (energy from an electron beam); derive or infer from.
Entorientierung desorientation.
Entquellen (Gele, etc.) shrinkage, cyneresis.
entregen de-energize.
entriegeln unblock, unlatch, trigger, release, trip, unlock (a tube previously cut off, by causing it to break down and discharge).
entrinden strip off bark, bark, decorticate.
Entschädigung indemnity, indemnification, compensation, consideration, damage payment. **-festsetzen** assess or fix damage or indemnity.

-zuerkennen award damages.
Entscheidung decision, sentence, verdict, judgment. **Berufung einlegen gegen eine-** take or lodge an appeal from a decision.
Entscheidungskern crux of a decision.
entschlickern dross.
Entschlüsselung de-coding, ungarbling, deciphering, breaking a (secret) code, decryptographing.
Entspannen einer Feder relaxing, relieving or untensioning of a spring.
Entspannung relaxation, strain, stress or tension relief, untensioning or slackening, untensed state.
entsperren see **öffnen**
Entsprechung equivalent, denotation, analogy.
Entstehung origin, genesis, nascence.
Entstehungs-ort origin, source or seat of generation or production. **-potential** appearance potential, ionization p. **-zustand** nascent state, status nascendi.
Entstörung interference elimination.
entströmen flow, stream, escape or issue from.
Enttrübungsrahmen frame designed to make zero or minimum point sharp (free from night effects, etc.), zero cleaning or sharpening frame.
Entwarnung all-clear signal.
Entweichungsventil escape, drain or outlet valve, vent valve.
entwerfen design, project, plan, trace, sketch, outline, focus or form (an image).
entwickelbar developable.
entwickeln (cf **erzeugen**) develop, evolve, disengage, generate, give off (gases).
entwickelnd, licht- emitting or producing light, photogenic.
Entwickelung, Rahmen- rack or tray development. **Reihen-, Serien-** expansion in a series, power series or seriation (math.). **Umkehr-** reversal processing or development.
Entwickelungs-arbeit development (al) work. **-faktor** Watkins factor, development f., time of appearance. **-rahmen** developing rack. **-steilheit** slope (of an H & D curve). **-vorrichtungen, Film-** film processing equipment.
Entwicklerbottich developing tank.
entzerren correct or eliminate distortion, equalize, compensate.
Entzerrer corrective network, attenuation compensator, attenuator (sound recording and reproduction); sound clarifier. **-zum Anheben und Senken** compensator to accentuate and deaccentuate frequencies or bands, accentuator and de-accentuator, emphasizing and de-emphasizing means.
Entzerrer, Dämpfungs- equalizing or compensation network. **Dynamik-** dynamic expander. **Längs-** series-type attenuation compensator or equalizer. **Quer-** shunt-type attenuation compensator or equalizer.
Entzerrung (Dynamik) expansion (by an expandor). **Ausgleich-** complementary recording (of film). **Phasen-** correction of phase, delay equalizer; echo weighting term.
Entzerrungs-anordnung distortion corrector device or compensator, anti-distortion device. **-bereich** frequency range of equalization. **-drossel** anti-resonant coil. **-filter** filter-type equalizer, network e. **-gerät** rectifying camera (aerial phot.). **-grenze** frequency limit of equalization. **-kette mit Brücken-H-Schaltung** H-type network attenuator (balanced). **-kette mit Brücken-T-Schaltung** T-type network attenuator (unbalanced). **-kreis (zum Hervorheben einiger Frequenzbereiche)** balancing network (to accentuate or underscore certain bands).
entziehen abstract, extract, take

away, withdraw, deprive, rid.

Entziehung, Wasser- dehydration, removal or expulsion of water, drying.

entziffern decipher, decryptograph.

entzünden ignite, kindle, inflame.

Entzündung, Selbst- spontaneous combustion, self-ignition.

Episkop reflecting projector, episcope.

erbreiten broaden.

erbringen produce, adduce.

Erd-alkalimetall alkaline-earth metal. **-antenne** ground or earth antenna, underground or buried antenna. **-anziehung** gravitation. **-anziehungskraft** gravitational force. **-bahn** earth's orbit, terrestrial o. **-bebenmesser** seismometer, seismograph (when recording). **-beschleunigung** acceleration due to gravity. **-drahtnetz** ground mat. **-drahtschleifenmessung** loop test.

Erde, Edel- rare earth.

Erderwiderstand grounder resistance.

Erd-kabel leader cable, buried cable. **-kapazität, Teil-** direct earth capacitance. **-klemme** ground clamp, g. terminal. **-kruste, Ausgleichfläche der** isostatic surface (of earth).

erdmagnetisches Feld geomagnetic field.

Erd-magnetismus terrestrial magnetism. **-schelle** ground clamp. **-schleife** earth or ground circuit.

Erdschluss earth connection, (arcing) ground. **Flammenbogen-** arcing ground. **-fehler** ground fault. **-löschvorrichtung** ground-fault neutralizer.

Erdung, Lichtbogen-, Erdungslichtbogen arcing ground.

Ereignis occurrence, happening, action, event, phenomenon. **-, einmaliges** non-recurrent, unique or singular event or action.

erfahrungsgemäss according to, or in the light of, experience, empirical, usual.

Erfahrungstatsachen empirical, experimental or practical facts or data.

erfassbar, zahlenmässig numerically evaluable.

Erfindergeist inventive genius, creative conception, inventor's spirit.

Erfindungs-gedanke basic or underlying idea of an invention, idea on which an invention is predicated. **-gegenstand** object of invention, disclosure. **-geltungsbereich** scope of invention or patent, command of a p. **-patent** patent for invention, letters p. **-schritt** object of an invention.

erfüllen fulfill, perform, satisfy, comply with, measure up to.

Ergänzung complement, supplement, completion, addendum (addenda, pl.).

Ergänzungs-farbe complementary color. **-stoff** vitamine. **-winkel** complementary angle.

erhaben raised in relief, elevated, embossed; high, grand, salient, outstanding.

erhaben, hohl- concavo-convex. *rund-* convex.

Erhaltung der Energie, Gesetz der law of conservation of energy. **-der Masse** conservation of the mass.

Erhaltungs-gesetz law of conservation. **-satz, Impuls-** theorem of conservation of momentum.

erheben (Anspruch, Klage, etc.) make (a claim), bring or file (a suit or an action). **-, in die n^{te} Potenz** raise to the nth power.

Erhebung elevation, projection, prominence, salience, peak, point, pip (on c.-r. screen tracing).

Erhebungswinkel angle of elevation.

erhitzen heat, warm, fire, incandesce, ignite, anneal.

Erhöhung elevation, prominence, projection, peak, jutting, lobe (as

of a cam), pip (as on a c.-r. screen tracing).
Erhöhungszeichen sharp (mus.).
Erholung recovery, recuperation (from fatigue, stress, etc.).
Erichsen-prüfer Erichsen cupping machine. **-prüfung** Erichsen cupping test.
Erklärung, eidesstattliche affirmation in lieu of oath. **Nichtigkeits-** declaration of nullity.
erlauben allow, permit, authorize.
erlaubt(es) Energieniveau permitted level, p. quantum state, p. energy zone. **-(er) Übergang** permitted transition (according to selection principle).
Erlöschen eines Patentes expiry or expiration of a patent, annulment of a patent.
ermächtigen empower, authorize.
Ermangelung, in...von in default of, in the absence of, failing.
Ermittelung investigation, ascertainment, discovery.
Ermüdung fatigue, sluggishness, inertia, tiredness.
Ermüdungsgrenze fatigue limit.
Erniedrigungszeichen mol (mus.).
Erneuerungsschein certificate of renewal.
erörterbar discussable, arguable, disputable.
erproben test, try out, prove (ordinance)
errechnen calculate, compute, figure (out), reckon.
erregen excite, energize, impulse (by shock), stimulate, activate.
Erreger-lampe (cf Tonlampe) exciter or exciting lamp (m.p.). **-spule, Feld-** field (exciting) coil (of electrodynamic loudspeaker).
Erregung excitation, energization, impulsing. **Fremd-** separate excitation. **Selbst-** self-oscillation, s. -sustained o., spontaneous (unassisted) o., spilling over (of a regenerative receiver, with fringe howl and whistling). **Stoss-** shock, impact or collision excitation. **Summer-** buzzer excitation.
Erregungslampe exciter lamp, exciting l. (operating photo-electric cell).
errichten, eine Senkrechte erect a perpendicular.
Errichtungsprotokoll minutes of proceedings.
Ersatz-anspruch indemnity claim, claim for compensation. **-antenne** artificial antenna, mute a., phantom a. **-bild** equivalent circuit diagram. **-elektron** equivalent or replacing electron. **-fähigkeit** capability or susceptibility of being replaced, equivalency. **-kapazität** equivalent capacity. **-schema** equivalent circuit schema.
Erscheinung phenomenon, action, event, manifestation, appearance. **Nicht-** non-appearance, non-attendance, absence, default, contempt.
Erscheinungsform form of manifestation, phase, state, physical appearance.
erschüttern shake, jar, percuss, chatter, vibrate.
erschütterungsfrei resilient, shock-absorbent, non-vibratile.
erschweren make (more) onerous or difficult, impede, aggravate.
erstarren solidify, freeze, congeal, harden, set.
Erstarrungspunkt freezing, coagulation or solidification point.
Erstattungspflicht liability to make restitution.
Erst-strom primary current. **-wicklung** primary winding.
ersuchen um Gehör file application or petition to be heard.
Erteiler, Lizenz- licenser, grantor, issuer of a license.
Erteilung eines Patentes grant or allowance of letters patent.
Erteilungsakten file wrapper (of a patent application).

Erwärmungs-kraft heating power, calorific p. **-verlust** Joule loss, Joulean heat, thermal or heat l.

Erwartungswert expected or anticipated value.

erweichen soften, plasticize.

erweitern widen, expand, enlarge, dilate, spread.

erweiterte Pupille dilated pupil.

Erwerber acquirer, purchaser, transferee.

Erwerbungskosten acquisition cost, purchase cost or price.

Erwiderung answer, rejoinder, reply, replication (of plaintiff to defendant's answer).

Erz, selbstgehendes self-fluxing ore.

erzeugen (cf **entwickeln**) generate, produce, create, set up, cause, result in.

erzeugend generant. **elektronen-** electronogenic (electron-emission under influence of light, from moles). **farben-** color producing, chromogenic. **fluoreszenz-** fluorogenic, with activator properties. **kälte-** cryogenic, frigorific. **leuchtenergie-** fluorogenic, phosphorogenic, photogenic, having phosphor properties. **licht-** emitting or producing light, photogenic. **phosphoreszenz-** phosphorogenic, causing persistence of fluorescence. **schwingungs-** vibromotive, oscillation-generative.

Erzeugnis, Neben- by-product.

erzwungen(e) Schwingungen forced or constrained vibrations or oscillations. (opposite of free o.).

-(er) Übergang forced or nonspontaneous transition.

Esse stack, chimney.

Estrichgips Keene's cement, flooring plaster.

Etagen-kessel battery boiler. **-ofen** storey furnace. **-rost** step grate. **-ventil** multi-seat valve.

Etagere rack, stand, shelf, support.

Etalonapparat reference standard, calibrated instrument.

eventuell optionally, contingently; if desired or necessary, eventually, ultimately.

Evolventenverzahnung involute tooth gear.

E.W. (Elektrizitätswerk)- Nachrichtenübermittelung power line signaling or intelligence transmission.

Exolierung anodic treatment, anodizing.

Expansionsschalter hydroblast or expansion circuit breaker or switch.

explosiver Laut explosive sound.

Exponentialtrichter exponential horn, logarithmic h. (either of uniform or of multiple-flare type, for loudspeaker). **-, aufgewundener** twisted, coiled or curled exponential horn. **-, gefalteter** folded exponential horn.

Expositionszeit exposure time, exposure scale.

Extinktionskoeffizient total-reflection coefficient, extinction c.

extra matt rough matt.

extremkurze Wellen (cf **Wellenabgrenzung**) ultra-short waves.

Exzenter cam, disk with lobes.

Exzenterbewegung, Herz- cam movement (of Livin).

F

Fabrik-marke trade-mark. **-zeichen anmelden** register a trade-mark.

facettierte Linse bevel-edged lens.

Facettspiegel, Scheinwerfer- segmented or facetted searchlight reflector.

Fach-arbeiter skilled worker, specialist. **-ausbildung** technical education, schooling or training. **-ausdruck** technical term or expression.

Fächer-antenne fan antenna, harp a. or aerial. **-aufblendung** fan fade-in. **-blende** fan fading shutter.

fächerndes Prisma, stark highly dispersive prism.

Fach-genosse professional colleague, p. collaborator. **-literatur** technical literature or press. **-mann** professional man, expert, specialist, man skilled or trained in an art. **-ordnung** classification. **-photograph** professional photographer. **-presse** technical press or literature. **-sprache** technical language, professional terminology. **-welt** technical world.

Fackel-entladung torch form of discharge. **-erscheinung** flicker effect (of a cathode). **-kohle** cannel coal.

Faden, biegsamer ductilized filament; flexible thread or fiber. **-,thorhaltiger** thoriated filament.

Faden, Einkristall- single-crystal filament, mono-c.f. **Glas-** glass fiber or thread; thread-like defect of glass. **Kenn-** colored tracer thread (for color coding). **Leucht-** streamer; glow column (in a tube). **Quarz-** quartz thread, q. filament. **Quecksilber-** mercury column, m. thread. **Strom-** streamer (in discharge); current path, streamline, c. tube.

Faden-dicketaster calipers with jaws to measure thread gage capacitively. **-einsatz, Leucht-** streamer onset. **-elektrometer** filament electrometer. **-galvanometer** string galvanometer. **-glas** spun glass, filigree g. **-kabel (HF-Leitung)** co-axial or concentric cable with silk-thread-supported central conductor. **-kreuz** reticule, cross or hair lines, reticle, graticule, cross-spider hairs. **-kreuzlupe** reticle magnifier, r. glass, r. lens. **-länge, Glimm-** length of neon glow column (in tuning, etc., indicators). **-strahl** fuzzy hose-like pencil of non-uniform cross-section in gas-focussed c. -r. tube, consisting of positive ions and secondary electrons, thread beam. **-strahlen** pencil rays (electrons). **-strahlentladung** thread ray discharge. **-strom** filament current, cathode c. **-umschnürung** serving of thread (on cables) **-zähler** thread counter, linen tester.

fädig, ein- unifilar.

Fading-automatik automatic volume control means (a.v.c.). **-mischhexode** a.v.c. mixer hexode.

Fagott bassoon. **-wasserabguss** bassoon syphon.

Fahne, Rauch- smoke steamer.

Fahnenbildung signal inertia-drag.

Fahr-arm hinged tracing lever (of planimeter). **-aufnahme** running, follow or traveling shot. **-bahn** track, taxiway (airfield),

runway; travel beam.
fahrbares Mischpult dolly or car-type mixer, teawagon console mixer.
fahren run, move, ride, travel.
Fahr-gestell, einziehbares, rückziehbares retractable or retractile undercarriage. **-säule, Elektroden-** electrode mast or support. **-stift** tracing or tracking point (of planimeter). **-strahl** radius vector. **-stuhl, Spulen-, (in Bandbreiteregler)** band-width regulator comprising auxiliary coupling coil parallel to oscillation coil.
Fahrtwindgenerator wind-driven generator.
Fahr-widerstand (Flugzeug) drag. **-zeug** (power) craft (marine, land or air), power-propelled vehicle.
Faktor factor, coefficient, submultiple (of a number).
Faktor, Anregungs- excitation factor. **Anreicherungs-** concentration factor. **Atomform-** atom form factor, atomic structure factor, structure amplitude f., f-value; scattering factor (of X-rays or electrons by gases). **Aufspaltungs-** splitting factor. **Callier-** Callier factor. **Entwickelungs-** Watkins factor (time of appearance) **Filter-** filter factor, screen f. **Form-** shape factor, theoretical stress concentration f. **Füll-** space factor, activity f. or coefficient. **Klirr-** non-linear harmonic distortion factor, klirr, blur or rattling f. **Kugel-** coefficient of sphere (photometry) **Leistungs-** power factor. **Polbedeckungs-** pole arc, pole pitch percentage. **Q-Q** factor, magnification f. **Spur-** tracking error, t. distortion (phonograph). **Streu-** scattering factor, S value (of x-rays). **Trenn-** separating factor. **Übertragungs-** transfer factor, image t. constant. **Verlust-** power factor (e.g., of a cable, condenser, etc.), ratio of watts to volt-amps.; phase angle difference (tan δ). **Verstärkungs-** amplification factor, voltage f. **Widerstand-Reaktanz-** Q factor, magnification f.
fakultativ optional, non-obligatory, non-compulsory, permissive.
fällbar precipitable.
Fall-bär ram, tup, rammer. **-Beschleunigung** gravitational acceleration.
Falle, Ionen- ion trap (in gun of c. -r. tube to attract negative ions). **Quecksilber-** mercury trap (in vacuum pump).
fällen, ein Lot let fall a perpendicular. **-, ein Urteil** render a judgment, decision, sentence, verdict or award, issue a decree.
fallen lassen abandon (a patent, an application).
fallende Charakteristik drooping characteristic, falling c.
Fall-klappenrelais drop indicator relay. **-körper, Zähigkeitsmesser mit** ball-drop type of viscosimeter.
Fällmittel precipitant.
Fall-raum cathode drop or fall space. **-scheibe, -schieberverschluss** drop shutter. **-system-Plattenwechsel** drop system of record changing. **-werk** vertical drop or blow-impact testing machine. **-winkel** angle of inclination.
falsch (e) Ausrichtung misalignment (of image or track). **-(es) Licht** light fog, leakage l., stray l. (in film printing). **-phasig** misphased, dephased, out of phase. **-(e) Strahlung** stray radiation, s. rays.
Falsch-abstimmung mistuning, off-resonance condition. **-anpassung des Widerstandes** mismatching of impedance, mismatched i.
Falschanpassungsfaktor mismatching factor, reflection f., transition f.

Fälschung distortion, vitiation, falsification, adulteration, fraud, forgery, counterfeiting.
Falte pleat, fold, crease, wrinkle, lap.
Falten-filter plaited, folded or fluted filter. **-horn** folded, coiled, curled or twisted horn (of a loudspeaker). **-lautsprecher** folded-horn loudspeaker. **-schlauch** pleated or corrugated hose or tube, accordion tube.
faltig having folds, pleats or creases, wrinkled, puckered,
Faltung (Spektrallinien) faltung (of spectral lines).
Faltversuch folding, doubling or bend-over test.
Falz fold, groove, flute, notch.
Falz-beständigkeit folding endurance (of film). **-fähigkeit** foldableness. **-festigkeit** folding strength. **-membran** non-rigid non-circular (breathing) cone with curved radiating surface. **-widerstand** folding resistance. **-ziegel** interlocking tile.
Familie, Gleichungs- equations grouped in a family.
Fang-anode (cf Auffanganode) collecting or gathering anode. **-blech** baffle sheet (in a magnetron). **-düse** mixing nozzle (steam jet).
fangen catch, capture, captivate, collect, trap, secure.
Fänger, Bild- pickup camera. **Schleifen-** feed and take-up sprocket mechanism (film projection).
Fang-gitter suppressor grid, cathode g. (in power pentode), interceptor g. **-mittel** getter(ing substance) **-platte** target or impactor plate **-röhre, Farnsworth'sche F.** dissector tube **-spiegel** collecting mirror.
Faraday's(cher) Dunkelraum Faraday dark space. **-(ches) Gefäss** Faraday ice pail, F. cylinder. **-(cher) Käfig** Faraday screen, shield or collector.

Farb-angleichung color matching, c. comparison. **-atlas** color scale, c. chart (e.g., of Munsell). **-ausgleich** color balancing (of c. film). **-auszug** color record, chromatic selection. **-auszug, negativer** separation negative. **-deckfähigkeit** concealing, covering or hiding power (of paint); opacity.
Farbe color, dye, paint, pigment, stain, hue, tone, shade, tint. **-, bezogene,** related color. **-, bunte** hue color, chromatic c. **-, unbezogene** unrelated color. **-, unbunte** hueless color, achromatic c., color devoid of hue. **Ergänzungs-** complementary color. **Grund-** primary color; ground or priming c. **Interferenz-** interference color. **Kern-** nuclear stain. **Klang-** timbre, tone color, quality. **Leucht-** luminous paint. **Stamm-** primary color. **Ton-** timbre. **Umdruck-** re-printing ink.
Färbekraft dyeing power, tinctorial p.
Farbemesser, Blut- plethysmograph (tests blood flow), electro-arteriograph.
Farbempfindung color sensation, c. perception.
Farben-abweichung chromatic aberration. **-ausgleich** color balancing (in color film). **-auszugfilter** selective filter, s. screen.
farbenblind color blind.
Farbendruck, Licht- photo-chemical color printing.
farbenempfindliches Zäpfchen color-distinguishing cone (of eye).
Farben-empfindlichkeit des Auges, -empfindlichkeitsverteilung color sensitivity or spectral response of the eye. **-empfindung** color sensation, chromatic s.
farbenerzeugend color producing, chromogenic.
Farben-fehler chromatic defect, color d., chromatic aberration.

-fehlsichtigkeit color-vision deficiency. -filmaufnahme color photograph. korrektion achromatization. -lehre science of color, chromatics. -mass colormeter, colorimeter. -messer chromatometer, colorimeter; c. knife. -messung chromatometry, chromatometrics, colorimetry. -rad rainbow wheel. -reste residual chromatic aberration, uncorrected color.

farbenrichtig orthochromatic (phot.).

Farben-saum color fringe. -schichtfilm dye-coated film. -sehen color perception, color vision. -skala chromaticity scale, color scale, color chart. -spektrum color spectrum, chromatic s. -stufe color gradation, shade, tint. -messer tintometer. -ton tint, hue, color tone. -tüchtigkeit color vision. -unterscheidung color discrimination. -veränderung change of color, discoloration. -vergleich color matching, c. comparing. -wahrnehmung color vision, c. perception. -wiedergabe color reproduction. -zerstreuung chromatism, dispersion, chromatic aberration (opt.).

Farberegler, Klang- tone control, tonalizer.

Färbetabletten varitone tablets.

Farbfehler chromatic aberration.

farbfehlerfrei achromatic.

Farb-film, Natur- natural-color film. -glas colored glass, stained g., c. filter. -gleichung, dreigliedrige three-color equation, tristimulus e.

farbig, flammen- flame or rainbow color. -,mehr- polychromatic, pleochroic (cryst.), multi-colored. -miss- discolored, inharmonious in color. -,regenbogen- rainbow colored, iridescent. -,voll- full or saturated in color.

Farb-kennzeichnung color coding (for identification and wire tracing). -komponente color or image component (in color reproduction). -kopie color print.

farb-korrigiert color-corrected, achromatized. -los colorless, hueless, achromatic.

Farb-lösung staining solution (microscopy). -massystem (IBK) chromaticity scale system, color chart, CIE diagram. -messung colorimetry. -muster color sample -rad inking wheel, i. roller. -raster color embossing. -rasteraufnahme color screen photography.

farbrichtig orthochromatic, isochromatic.

Farb-rolle inking roller, i. wheel. -schaum fringing (in film). -schreiber inker, printer. -schwellenwert liminal value or threshold of intensity of a color. -stufe color gradation, tint, shade. -stufe, empfindungsgemässe subjective chromaticity scale value. -temperatur color temperature. -toleranz color tolerance. -ton (warm, kalt) color hue, tone or tint (warm, cold).

farbton-gleiches, reines Gelb psychologically unique yellow. -richtig orthochromatic.

Farb-treue color fidelity (of color film). -umschlag discoloration, change of color.

Färbung (Schall) timbre, tone or sound color, quality.

Farbunterscheidungsvermogen color discrimination faculty.

Farb-verfahren, Linsenraster- lenticular color printing method. -wiedergabe color reproduction. -zentren F. centers.

Farnsworth'sche Bildfangrohre F. image dissector tube. -scher Vervielfacher F. multiplier tube.

faserförmig fibrous, filiform.

Faserung, Linsen- fibrillation of lens.
Fassondraht sectional, shaped, fashioned or profile wire.
Fassung socket, holder, mount (ing) fitting. **-fur Kristalle** crystal (oscillator or resonator) mount or holder. **-für Objektiv** mount or barrel of lens or objective.
Fassung, Einstell- focusing lens mount. **Kristall-** holder or mount of a crystal. **Linsen-, Objektiv-** objective or lens barrel or mount. **Röhren-** tube socket, valve socket. **Röhren- federnde** cushion socket, anti-microphonic valve holder. **Schneckengang-** helical lens mount.
Faustregel rule of thumb, rough and ready r.
Fazettspiegel, Scheinwerfer- segmented or facetted searchlight reflector.
Feder spring, pen, feather, (mech., corresponds to capacitance in electricity).
Feder anspannen tension, tighten or bend a spring. **-entspannen** relax, relieve or untension a spring.
Feder, Auslösungs- escapement spring, trip or release s. **Blatt-** leaf spring, flat or plate s., reed, blade. **Docken-** poppet spring. **Drehstab-** torsion bar spring. **Druck-** compression spring. **Gegen-** counter (acting) spring, retractile s. **Haupt-** main spring, master s. **Kegel-** volute spring. **Nebenschluss-** shunting spring, off-normal s. **Nut und -** slot and feather (union), slot and key, groove and tongue joint. **Pendel- (des Zerhackers)** vibrator blade, v. reed or spring (of chopper). **Rückfuhr, Ruckzugs-** retractile spring, restoring s. **Schlag-** impact spring. **Schleif-** wiper, slide spring, contact spring. **Schnecken-** coil (ed) spring. **Schrauben-** helical spring, coil s. **Schreib-** stylus, style, recorder pen, inker, scriber.
Feder, Schwing- (des Zerhackers) vibrator spring, reed s. (of chopper). **Sperr-** click spring. **Stromstoss-** impulse spring (of dial). **Zentrierungs-** spider (of loudspeaker coil). **Zug-** tension spring, traction s., retractile s.
Feder-barometer aneroid barometer. **-keil** feather key. **-klinke** spring catch. **-kraft** elasticity. **-kraftantrieb** spring motor, clockwork (motor) drive. **-manometer, Glas-** spoon-type glass manometer. **-motor** spring motor.
federnd springy, elastic, resilient, yielding, cushioned, flexible, supple.
federnd (er) Kontakt spring contact, flexible c., c. spring. **-(e) Leitrolle** spring-loaded idler. **-(e) Röhrenfassung, -(er) Röhrenhalter** cushioned tube holder or socket, anti-microphonic valve holder.
Feder-raster spring catch, s. dentent. **-ring** lock washer. **-satz** spring bank, s. assembly. **-spannung** spring tension. **-uhr, Präzisions-** chronometer.
Federung, Federungswiderstand compliance, mechanical capacitance (reciprocal of stiffness).
Feder-wage spring balance, s. scale. **-weg** spring deflection, s. excursion or elongation. **-werk** spring mechanism. **-zange** spring pliers.
Fehl-abgleichung faulty alignment, misalignment. **-anpassung** mismatching; defective focus. **-anpassungsfaktor** mismatching factor, reflection f., transition f. **-anweisung** incorrect indication or reading, misreading.

-bauerscheinung mosaic structure (of metal crystals).

Fehler (cf **Verzerrung**) error, defect, fault, flaw, failure, distortion, aberration (in lenses), trouble (in circuits).

Fehler, chromatischer chromatic aberration. **-, stroboskopischer** stroboscopic aberration. **Abbildungs-** defect of image, aberration. **Abgleich-** unbalance, balancing error, alignment error. **Beobachtungs-** error of observation, experimental e. **Bild-** picture distortion, image d. **Einstell-** faulty focusing, faulty spot control, decrease of spot with intensity increase, large diffuse speck on screen, in gas-focused picture reproducing tube (telev.) **Farb-, Farben-** chromatic aberration, c. or color defect. **Form-** informality, formal defect. **Geometrie-** geometric distortion. **Geschwindigkeits-** distortion due to synchronous speed differences. **Glass-** glass defects or flaws. striae, cords (when large), bubbles (seeds when small, air-bells when of irregular shape), stones, crystallization bodies. **Gleichgewichts-** imbalance, unbalance. **Gleichlauf-** jitters (distortion caused by non-synchronism in facsimile). **Instrumenten-** variations of bearing due to instruments (df). **Kugelgestalts-** spot-size distortion. **Leitungs-** line failure or fault. **Linsen-** lens defects, l. aberrations. **Norddreh-** northerly turning error. **Öffnungs-** apertural defect, a. effect or error. **Orts-** geometric location error. **Peil-** direction-finding error (e.g., due to night effects), distortion of bearing. **Phasen-** error of phase, phase distortion. **Richtungs-** deviation (of pencil); directional distortion (of bearing). **Teilungs-** faulty pitch (of an image), underlap (of lines). **Tonnen-** barrel distortion, positive distortion. **Trapez-** trapezoidal, keystone or trapezium distortion (due to dissymmetry to earth of deflection plates). **Verschiebe-** lack of synchronism, out-of-frame defect. **Versuchs-** experimental error. **Zerdehnungs- (eines Bildes)** radial distortion of a picture or image (pincushion effect). **Zerdrehungs- (eines Bildes)** rotational or tangential distortion, twist (barrel effect). **Zonen-, Zwischen-** zonal aberration, zonal error.

Fehler-beseitigung curing or elimination of errors or of defects, "trouble shooting." **-dämpfung** balance attenuation, return loss. **-dreieck** triangle of error. **-eingrenzung** tracking down seat or source of interference or trouble, localizing t.

fehlerfrei free from flaws, faults or defects, faultless, unmarred, sound; aplanatic (free from spherical aberration), aberrationless (of a lens). **-, farb-** achromatic. **-, polarisations-** free from polarization error (df).

Fehler-funktion, -gleichung (Gaussian) error equation. **-grenze** limit of error.

fehlerhaft faulty, defective, impaired, having flaws.

Fehlerscheibchen circle of confusion; apertural effect. **-schutzeinrichtung** fault clearing device. **-sucher** aniseikon (electronic crack detector), electromagnetic flaw d. (using flash magnetization), magnaflux inspection means, device used for flaw detection by x-ray or materiology methods.

-tiefenbestimmung stereoscopic radiography for locating defects; depth determination of flaws. -verteilungskurve error distribution curve.

Fehl-griff blunder, mistake, error of judgment, wrong manipulation. -ordnungerscheinung, thermische formation of holes (cryst.).

fehlsichtiges Auge defectively sighted eye, ametropic eye.

Fehl-sichtigkeit, Farben- color vision deficiency. -weisung, Funk- (cf Funk- Beschickung und Funktrübung) direction-finding or directional error, quadrantal error (QE), compass deviation, distortion of bearing. -winkel phase displacement angle, shift angle. -zündung failure to fire or ignite (rectifier, ignitron); misfiring.

Feilhalten (von Waren) delivery on sale (of goods), exposing or exhibiting for sale.

Feilhieb notch.

Feilicht, Feilsel filings, filing dust.

Fein-abstimmung sharp or fine tuning. -bewegung, Zahnrad- slow-motion gear. -blechstrasse sheet rolling mill. -einstellung vernier (dial) adjustment or tuning.

feinfühlig delicately sensitive.

Feingefüge fine structure, microstructure.

feingepulvert finely powdered, pulverized, comminuted or atomized.

Feinheit (des Bildes) detail or (degree of) definition of picture or image. Raster- raster detail, resolution or definition of scanning pattern.

Fein-heitskoeffizient fineness coefficient (of Fabry), coefficient de finesse. -hohenmesser (aneroid) barometer measuring atmospheric pressure at airdrome not reduced to sea level, sensitive altimeter. -korn fine sight; fine grain (phot.).

fein-mahlen grind fine, triturate. -maschig fine-meshed.

Fein-messlehre micrometer caliper or gage. -messchraube micrometer gage screw. -rasterung high-definition scan (of picture). -regulierung sharp tuning (of radio).

feinschuppige magnetische Struktur fine-scaled magnetic structure.

Fein-sicherung glow-tube type, etc., of lightning arrester. -stellschraube micrometer screw, vernier. -struktur fine structure, microstructure (cryst.).

feinverteilt finely divided.

Fein-verschiebung fine displacement. -wage precision balance, microbalance. -zerkleinern fine grinding or crushing, comminution, finely dividing.

feinzerteilt finely divided.

Feinzug finishing block, f. die.

Feld field, compartment, bay, section, panel, pane, frame (film pictures).

Feld, erdmagnetisches geomagnetic field. -, krümmungsfreies flattened field (by field flattening lens). -, totes dead zone (of ribbon microphone), plane of ribbon. -, wölbungsfreies flattened field.

Feld, Auskoppel- output field, coupling f., absorbing f., delivery f. (of a Hollmann electron turbine or inverted cyclotron), catcher f. (of klystron). Bild- printing plane (film printer); frame, picture or image field, picture area. Brems- reflecting field, retarding f. (of Barkhausen electron-oscillation tube). Eintritts- area of applicator (at skin surface, in X-ray work). Gleich- constant or steady (magnetic) field. Gravitations- gravitational field. Kennlinien- family of characteristics or curves. Lampen- bank or panel of lamps, battens (stage lighting). Längs- longitudinal field, paraxial f., f. parallel to axis. **Lösch-** magnetic obliterating field (in magnetic sound recording, telegraphone and magnetophone). **Mittel-** central field (of vision). **Prüf-** test field; proving ground (ordinance). **Quer-** transverse or cross (magnetic) field. **Rückmelde-** revertive signal panel. **Saug-** suction or positive field, field which draws away. **Schärfen-** focal field. **Spiegel-** revertive signal panel, check-back position indicator p. (giving readings of remote instruments). **Stossprüf-** potential impact or impulse testing field. **Stör-** stray field, interference f. **Tasten-** keyboard. **Teilchen-** particle field. **Vergleichs-** matching field (in photometry). **Versuchs-** proving ground, field for making tests or experiments. **Wellen-** wave field.

Feld-ausschnitt, Bild- image area (on scanning disk). **-beleuchtung, hell-** bright ground illumination (micr.) **-bild** field pattern, configuration or map (made in electrolytic tank). **-blende** field stop. **-durchgriff** action of field through aperture lens (el. opt.). **-ebenungslinse** field flattener (opt.). **-elektronen** field electrons (resulting from emission caused by intense electric f.). **-elektronenemission** auto-electronic, cold or field emission (and current caused thereby). **-erregerspule** field (exciting) coil (in electrodynamic loudspeaker).

feldfrei field-free.

Feld-funkentelegraphie military service radio telegraphy, noncivilian wireless. **-generator, Brems-** negative transconductance generator or oscillator (using retarding field circuit), positive grid o. **-krummung, Bild-** curvature of image field. **-krümmungskorrektur** field flattening. **-krümmungsverzerrung** distortion due to curvature or lack of flatness, curvilinear distortion. **-linse** field lens (anterior or front lens in telescope and in microscope). **-messung** plotting field map or configuration of electrostatic field (in electolytic tank), land surveying or measurement. **-profil** field pattern. **-röhre, Zwei-** drift tube working with two fields. **-spule** field coil, magnetizing c. **-stärkenprofil** field pattern. **-stecher** field-glass magnifier. **-strom** field current (emission of electrons due to an electric field). **-tiefe, Schärfen-** depth of focus, d. of definition, d. of field. **-trichtertheorie** field funnel theory. **-verlauf** field map, f. pattern, f. configuration. **-welle** field wave. **-wölbung, Bild-** curvature of image field, lack of flatness of image field. **-zerleger, Bild-** picture scanner, dissector (of Farnsworth), p. exploring means.

Fell, Trommel- drumhead, tympanic membrane, tympanum, eardrum.

Felsöl petroleum.

Fenster projection aperture (film projection); window (in envelope of a cell, iconoscope, etc.); bezel (in casing or panel of radio apparatus).

Fenster, Andruck- pressure plate or gate (m.p). **Belichtungs-** aperture (in aperture plate), photocell window. **Beobachtungs-** bezel, observation window, peephole. **Bild-** picture gate, aperture or trap; range-finder (m.p.). **Bild-, mit zwei Ausschnitten** film gate with two frames. **Druck-** pressure plate or gate (film). **Film-** film gate (of projector), film trap. **Kopier-** printing plane, p. station. **Pendel-** internal pressure plate. **Projektions-** projection aperture. **Rollen-** roller film gate. **Ton-**

sound gate.

Fenster-bilder window transparencies. **-blende** apertural stop or diaphragm. **-druck** gate pressure, gate friction. **-führung** aperture guide. **-kopiermaschine** step printer, intermittent p. **-linse** aperture lens.

fensterlose Roentgenröhre windowless X-ray machine or tube.

Fern-anzeiger telemeter (e.g., metameter of GE Co., using impulse carriers). **-aufnahme** telephotography, long shot. **-besprechgerät** remote-control unit. **-bild** television picture, televised p., telephoto. **-bildgrossaufnahme** (Ferns.) projection video picture (of large size). **-blitzanzeiger** keraunophone. **-drucker** teleprinter, teletyper, stock ticker.

Ferne, Sonnen- aphelion.

Fern-einstellung long-range focus. **-glas, Doppel-** field glass, binocular g.

Fernendunst distant fog.

Fern-hörer, Dosen-watchcase telephone receiver. **Kopf-** headphone, headset, "can". **-kino** telecine, telecast, intermediate-film system (of television). **-linse** telephoto lens. **-meldeleitung** communication or signal circuit or line. **-meldewesen, Flugzeug-** aircraft communication service, avigation signal service. **-messer** telemeter. **-messgerät** telemeter (e.g., metameter of GE Co. using impulse carriers); range-finder. **-messummengeber** telemetric integrator. **-objektiv** telephoto lens. **-punkt** far point (opt.).

Fernrohr, Ablese- reading telescopic tube. **Doppel-** telescope of binocular type. **Doppelblickziel-** double direct- sighting telescope. **Rundblick-** panorama sight. **Ziel-** telescope sight, bombsight, rifle sight, etc.

Fern-rohrbrille telescopic spectacles. **-rohrbüchse** telescope tube, barrel. **-rohrlinse** telescopic magnifier. **-schreiber** telegraph, teletype, teletypewriter, telegraph printer; t. operator.

Fernseh-abtaster televisor, scanner (at sending end). **-abtastung (von Kinofilm)** telecine scan of (motion-picture) film. **-aufnahmewagen** television or electron (pickup) camera truck, video bus, televising car.

Fernseh-bild, plastisches stereocopic or three-dimensional (appearance of) video picture. **-empfänger** television receiver, picture reproducing device, teleview apparatus (e.g., Zworykin's, Farnsworth's oscillight, etc.). **-empfänger, Heim-** home television receiver.

fernsehen televise, teleview.

Fernsehen, direktes direct pickup (and viewing). **Gegen-** two-way television. **Mehrkanal-** multichannel television.

Fernseher (cf **Kathodenstrahlröhre**) television transmitter, video or telev. camera, televisor (e.g., emitron, iconoscope, Farnsworth dissector, etc.).

Fernseh-raster (cf **Raster**) television raster, scanning surface, unmodulated screen pattern. **-röhrenprüfer** monoscope, monotron, phasmajector. **-sender** television transmitter (comprises strictly both video and aural signal transmitters). **-signal** video signal, picture s., telev. s. **-sprecher, -telephon** television telephone, video telephone. **-wandlerorgane** photo-cells and light-relays changing light into current and back (in telev. work), transducer means.

Fernsprech-formfaktor telephone influence factor. **-messtechnik** telephonometry. **-störfaktor** telephone interference factor.

Fern-steuerung distance, telemetric or remote-control action (by

Selsyn type motor). **-thermometer** water temperature gage. **-Übertragung, Wettermessinstrument mit funkentelegraphischer** radio telemeteorographic (or telemetric) instruments, r. sonde. **-wirkung** distance action, remote-control a., telemetric a. **-zeichnung** perspective drawing.

Fertig-erzeugnis finished article, manufactured goods. **-form** finishing mold (glass bottle making).

Fertigung, Fliess- conveyor belt or assembly-line production, large-scale manufacture.

fest (cf **beständig**) solid, firm, unvarying, compact, strong, fast, fixed, stationary, tight, rigid, secure, proof, stable. **-(e) Kopplung** close coupling, strong c.

festbacken cake together (by heat or pressure), clinker.

Festdetektor contact or crystal detector (with non-adjustable cat whisker).

fest-fressen seize, jam, freeze. **-halten, photographisch** record photographically (on record sheet).

Festigkeit strength, solidity, sturdiness, ruggedness, firmness, resistance.

Festigkeit, Dauer- fatigue strength. **Falz-** folding strength. **Feuer-** fire-proofness, refractoriness. **Gestalt-** form stability, non-deformability. **Kerbdauer-** notch fatigue strength. **Knitter-** crumpling strength. **Riss-** crack strength, c. resistance. **Spannungs-** puncture potential, dielectric strength. **Stossdurchschlags-** impact breakdown strength. **Streck-** resistance to stretching or elongation. **Trennungs-** separating strength, static crack strength. **Zerbrechungs-** breaking strength, resistance to breaking strain. **Zerdrückungs-** crush strength, resistance to compression or crushing strain.

Festkondensator fixed or non-variable condenser or capacitor.

fest-machen consolidate, solidify, make concrete or firm; fasten, secure, attach. **-setzen** fix, settle, stipulate, determine, establish.

Festsetzung, Entschädigungs- assessing or fixing damage or indemnity.

Festsitz tight fit, snug f.

feststellbar determinable, ascertainable; lockable (in position), fixable, securable.

feststellen ascertain, determine, establish, state; lock (in position).

Feststellungs-klage declaratory action. **-urteil** declaratory judgment.

festweich semi-solid.

Fett, Vakuum- vacuum grease.

Fett-fleckphotometer Bunsen photometer, grease-spot p. **-schliff** greased (ground-in) joint. **-ton** Fuller's earth.

Feuchtemesser hygrometer, psychrometer.

Feuchtigkeit moisture, moistness, wetness, dampness, humidity.

Feuchtigkeitsmesser hygrometer, psychrometer. **-, registrierender** hygrograph.

feuchtigkeitssicher damp-proof, moisture-proof.

Feuer beacon, (radio) beam station, radio range.

Feuer, Ansteuerungs- line of approach beacon, directional b. **Drehfunk-** rotating radio beacon, omni-directional b. **Funk-, mehrstrahliges** multi-beacon, triple-modulation b. **Navigations-** navigation or maritime beacon, b. or beam station for aerial navigation or avigation.

feuerbeständig resistant to fire or heat, fire-proof, refractory.

Feuerbeständigkeit fire-proofness, refractoriness.

feuerflussig liquid at high temperature, molten, fused.

Feuer-führung firing (schedule).

-punkt focus (opt.); hearth (in mining). -schutztrommel safety magazine (of film projector). -stein flint. --ton fireclay. -versilberung hot-dip silverplating. -verzinkung hot galvanization, pot g. -verzinnung hot-dip tinning.

Figur, stehende stationary pattern or figure.

Figur, Ätz- etch pattern, e. figure. Beugungs- diffraction pattern, d. picture. Fliess- strain pattern. Klang- sound pattern, acoustic p. or figure. Ring- Fresnel or Huygens zone, F. zone plate. Schlag- percussion figure. Staub-dust figure, powder pattern (of Lichtenberg, Bitter, Debye-Scherrer, Hull, etc.). Streck-flow lines, l. of stress, surface bands.

Film, doppelbeschichteter sandwich film, double-coated f. -,gegossener coated film. -,glasklarer pellucid film stock (of maximum transmittance). -,mehrfachbeschichteter double-coated film, sandwich f. -,plastischer stereoscopic film, plastic f. -,unverbrennbarer slow-burning film, safety f. -,verregneter rainy film.

Film, Atmen des...s in-and-out-of-focus effects. Auflaufen des...s take-up (by spool) of film. Einlegen des...s threading of film.

Film, Blank- film base. Doppelschicht- double-coated film, sandwich f. Duppositiv- duplicating positive film, master positive. Farbenschicht- dye-coated film. Flach- non-stereoscopic film, non-plastic f., plain f. Gravar- film with directly engraved track. Hör- (Tonfilmrundfunk) video and audio film, sound telecine or telecasting film. Kino-, Fernsehabtastung von telecine scanning of motion-picture film, scanning of film for television. Klein- substandard film, narrow f. Kultur-, Lehr- educational film, cultural f., instruction or school film. Linsenraster-lenticular film, lenticulated f. Milieu- "milieu" film. Naturfarb-natural-color film. Präge- film with directly embossed track. Raster- lenticulated screen film. Reportage- newsreel, news film, topical f. Roh- blank film, raw stock. Rundfunkton- sound telecine, telecast (broadcasting of video and audio signals). Schmal-substandard film, narrow f. Spezial-, hoher Steilheit high-contrast emulsion film. Stumm-silent film. Trickton- sound cartoon. Umkehr- film resulting from developing by reversal. Vorführungs- display film. Zeichentrick- animated cartoon.

Film-abheber plow (to peel film from teeth. -absatz packing, wear and tear. -abtastung film scanning. -abwickler magazine drum, feeding reel. -architektur setting and composition. -archiv film archives. -atmen in-and-out-of-focus effects. -auflaufspule, -aufwickelspule take-up spool or reel. -bahn film track, f. channel. -bandführung aperture guide. -baukunst setting and composition. -belichtungsstelle film gate. -bewegung, bildweise intermittent film feed or movement. -bewegung, stete continuous film feed or movement. -bildschicht film emulsion, f. coat. -blitz statics (m.p. film). -doppeln film duplicating. -einfadelung, -einführung threading of film. -einschnurung squeezing or compression of film (for making squeeze track). -entwickelungsvorrichtung film processing equipment. -fenster film gate, film trap. -fernsehabtastung telecine scan (of m.p. film). -fernsehsystem intermediate film television system. -fortschaltung film feed, f. motion, f. travel. -führung film feed, film guiding, f. gate. -geber film transmitter, film pickup, f. scanner, tele-cinematographic device. -justierung registration of film. -kameralaufwerk camera film feed mechanism. -kanal film track, film channel. -kern reel, spool. -kittlehre film splicing gage. -messer,-meterzähler footage counter. -methode, Zwischen- intermediate-film (pickup) method of television transmission. -optic, Ton- sound lens, s. optic (m.p.). -projektionsfenster aperture, gate. -registerhaltigkeit registration or stability of film. -regler, film traction regulator (stabilizer). -rolle film take-up spool or reel. -rundfunk, Ton- sound telecine, video and audio broadcast. -schaltung, bildweise intermittent film motion, discontinuous f. feed. -schaltung, stete continuous film feed or movement. -schleife film loop. -schneiden editing. -schwanz trailer run-out (piece of blank film at end). -sensitometer sensitometer, timer. -spannung film tension, f. traction. -spulenhalter film roll holder. -steilheit, hohe high contrast of film. -streifen, endloser endless film strip. -träger film base, f. support. -transport, absatzweiser intermittent film feed. -transport, steter continuous film feed. -transportrolle film (feed) sprocket. -tür film gate. -uhr footage counter. -verarbeitung film processing. -verblitzung dendriform exposure of film (due to static charges). -vertrieb release of film, distribution of f. -vorführung film projection; f. exposure (in gate). -vorlauf lead or precession of sound recording over pictures = 19 frames = 361 millimeters. -vorratsspule film magazine roll, magazine. -wagen, Ton- sound truck, location t., sound van. -wirkung merit or quality of picture. -wölbung film buckling. -zähler footage counter. -zug, Unstetigkeit im flutter in film pull. -zusatzgerät, Ton- sound-head, sound film attachment.

Filter, tonrichtiger true color filter, pure c.f., orthochromatic f.

Filter, Aufteilungs- crossover network (sound production). Band-, von grosser Lochweite broad-band filter. Bewertungs- weighting network. Blau- blue filter, viewing f. Durchlass- band-pass filter, b.-transmission f. Einzelabschnitt- single-section filter. Entzerrungs- filter-type equalizer, network e. Farbenauszug- selective filter, s. screen. Frequenzbereich- band-pass filter. Gitter- lattice filter (used in electric signaling). Glocken- bell jar filter. Kanal- channel filter. Kontrast- filter for selective contrasts, contrast screen filter. Kreuzglied- lattice section filter. Licht- light filter, color f. Mikro- filter for micrographic work. Nadelgeräusch- needle scratch filter. Streifen-, dreifacher tricolor banded filter. Tiefpass- low-pass filter. Trenn- dividing filter, separating f. Verlauf- sky filter (of gradual action). Wellenstrom- ripple filter.

Filterabschluss dch. 1/2 Längsglied mid-series filter termination. -dch. 1/2 Querglied mid-shunt filter termination.

Filter-durchlässigkeit, Glas- filter or glass transmittance or transmittancy. -faktor filter factor, screen f. -federungswiderstand compliance (of mechanical filter). -glied filter section, f. mesh, f.

unit. -lochlage position of transmission range. -massenwiderstand inertance (of mechanical filter). -steilheit sharpness (of cut-off) of selecting network, slope.

Filz-kufe felt pad. -unterlagscheibe felt washer.

Finger, Kontakt- contact finger, wiper.

Finger-anschlag finger stop (of dial switch). -bildner digitorium (for piano practicing). -hut(ionisierungs)kammer thimble ionizing chamber. -regel, Drei- right-hand rule, thumb rule, Fleming's r. -scheibe dial switch, finger disk or wheel.

Fisch-bauchantenne fish-belly antenna (mast antenna in which maximum cross-section and minimum characteristic impedance lie in the middle). -schwanzbrenner bat's wing burner, fishtail b.

Fis-Klappe F sharp key.

Fixier-flüssigkeit fixing liquid, f. liquor. -mittel fixative, fixing agent. -natron sodium hyposulfite, sodium thiosulfate, "hypo". -salz fixing salt.

fixierter Kathodenfleck des Quecksilberlichtbogens anchored spot of mercury arc.

flach flat, without contrast, with low gamma (picture); flat, flat-topped (curve, e.g., in broad tuning).

flach werden level off, flatten out, slope, become less steep, be smoothed.

Flachdruck lithoprinting.

Fläche surface, face, facet, plane, area, zone, plain, level, sheet, flatness.

Fläche, cartesische Cartesian surface. -, deformierte deformed or figured surface. - konstanten Potentials equipotential surface, isopotential s.

Fläche, Abstrahl- radiation surface, radiant s., reflecting s., projector s. (loudsp.). Abstumpfungs- truncated face (cr.). Ausgleich-, der Erdkruste isostatic surface (of earth). Äquipotential- isopotential or equi-p. surface or plane. Bild- image area, picture field; image plane (opt.). Brenn- caustic surface (geometry). Empfänger- radiation sensitive surface; record sheet. Gleit- glide plane, slip p. (cryst.), sliding surface, slide. Grenz- interface, boundary layer, surface of contact. Grund- basal surface, base, area. Ketten- catenoid. Knoten- nodal surface, n. plane (opt.). Kristall- crystal face, c. plane. Leit- deflection vane (in ldspk. horn, to reduce directional property), labyrinth, baffle. Lichtverteilungs- isophote, isolux diagram, luminosity or isophotic surface or curve. Neben- secondary face. Niveau- level, surface, equipotential s., isopotential s. Potential-, konstante equipotential or isopotential surface. Prismen- prismatic surface, p. face. Rückstrahlbrenn- catacaustic (surface). Schärfen- surface of sharp or distinct vision. Schiebungs- slip plane (cryst.). Spalt- cleavage plane, c. face (cryst.). Stirn- end face, face, front surface. Strahl(ungs)- emitting surface, emitter area (electrons), radiator surface, radiant surface. Translations- translation plane, slip p. Trennungs- cleavage plane, parting p., surface of separation. Wellen-, Fresnel'sche Fresnel zone (of half-period elements).

Flächen-aberration-, -abweichung zonal aberration (opt.). -antenne sheet antenna. -anziehung surface attraction, adhesion, adsorption.

flächenartige Strombahn (cf flächenhaft) current sheet, laminar

or areal path of current.
Flächen-ausmessung planimetering. -element (cf Bildpunkt) elementary area, unit a. -erhaltungsgesetz principle of conservation of area.
flächenförmig areal, laminar, of sheet-like nature or form.
Flächenglimmlampe glow-lamp or neon l. with plate-shaped cathode, plate neon l.
flächenhaft areal, laminar, involving two-dimensionality, of sheet or plane form. -(e) Berührung surface or areal contact (as in osculation). -(e) Spule laminar or strip-like coil, non-filamentary c.
Flächen-helle, -helligkeit luminous brightness of a surface, intrinsic brilliance. -inhalt surface area. -integral area integral (of a surface). -kathode plate-shaped cathode, large-surface c. (as distinguished from crater-shaped, punctiform orfilamentary c.). -lampe, -leuchte bank or pillar of lamps or luminous sources, battens. -messer planimeter, integraph (for curves). -potential, Grenz- interface potential.
flächenreich polyhedral, with many faces or sides.
Flächen-satz theorem of conservation of areas. -stück area element, unit a., elementary a.
flächentreue Projektion equal-area projection (in mapping).
Flächenwinkel plane angle. -winkel, Grenz- interfacial angle (dihedral between two crystal faces). -zeichnung, Spalt- cleavage plane marking, crystal m.
flächenzentriert face-centered (cryst.).
flacher Frequenzgang flat frequency characteristic (free from peaks and dips).
Flach-film plain film, non-plastic f., non-stereoscopic f. -gewinde square (screw) thread. -heit (eines Bildes) flat or non-contrasty quality or condition of a picture or image.
flächig ...faced, areal, laminar.
flächig, drei- trihedral, three-faced. viel- polyhedral, many-faced. vier- tetrahedral, four-faced. zwei- dihedral, two-faced.
Flächner, Halb- hemihedron. Voll- holohedron.
Flach-spule flat coil, pancake c. -streicher flat squeegee, strickler.
Flackereffekt flicker effect.
flackern flicker, flare, flash.
Flackerrelais flashing relay.
Flageolettöne harmonics.
Flammbläser blasting magnet, blowing m.
Flamme, leuchtende luminous flame. -, nichtleuchtende roaring flame. Lock- pilot flame. Lötrohr- blowpipe flame.
flammenbeständig flame-proof, f. -resistant
Flammenbogen flame arc, electric a. -erdschluss arcing ground.
flammenfarbig flame-like, flame-colored.
Flammen-mikrophon flame microphone, f. transmitter. -spektrum flame spectrum.
flammicht watered (of fabrics), veined or grained (of wood).
Flammpunkt flashpoint, ignition p., kindling p.
Flanken-spiel backlash (of screw thread). -steilheit (cf Filtersteilheit) width of transition interval (between transmission and attenuation bands); steepness of sides or slopes of a curve, or of upstroke and downstroke of an impulse. -streuung side leakage, s. stray, end s.
Flansch, Kühl- radiator vane, cooling vane.

Flasche, Kugel- balloon flask, spherical flask.

Flaschenzugbügelhaken block shackle.

Flattereffekt flutter effect (called wowow when changes per second are less than 6, flutter 6-30, gargle 30-200, and whiskers when over 200 ps.)

flattern flutter, flicker, wave.

Flatter-russ lampblack. **-widerhall** multiple reflection (between two parallel surfaces), flutter echo.

flau flat, without contrast (c. often diminished by white tint permeating whole picture); limy or blooming condition, fuzzy (of picture); flattened or flat (characteristic or response).

flau werden flatten out.

flechten plait, twist, braid, interweave.

Fleck, Kathoden-, fixierter anchored cathode spot (of an Hg arc). **Laue-**Laue spot. **Licht-** spot (in scanning); hot spot (of film). **Reflexions-** flare ghost, f. spot.

flecken spot, stain, speckle, mottle, patch.

fleckenempfindlich susceptible to staining or spotting.

Fleckentheorie patch theory (of metals).

Fleck-schärfe (sharpness of) spot focus. **-Verformung, -verzerrung** spot distortion.

flicken repair, patch, mend.

Flickerpeilung Robinson flicker direction-finding method, make and break, loop reversing or switching method.

fliegen, gegen den Wind fly up wind. **-mit dem Wind** fly with tail wind.

fliegende Kathode floating cathode.

Flieger-aufnahme aerial picture, airplane p. or photo. **-horizont** artificial horizon. **-visier** anti-aircraft sight.

Fliehkraft centrifugal force, flywheel f. **-regler** centrifugal governor.

Fliess-behinderung inhibition of plastic deformation. **-bereich** plastic range. **-eigenschaften** rheological properties.

Fliessen, plastisches viscous flow, streamline f., laminar f., steady f. **-,Wissenschaft der Verformung** unter rheology.

Fliess-feder fountain pen. **-Fertigung** production on conveyor belt or assembly line. **-figur** strain figure. **-grenze** flow limit, yield point. **-körper** fluid or liquid body or substance, non-solid. **-verzug** yield or flow distortion or deformation. **-vorgang** flow process.

Flimmer, Zwischenlinien-, Zeilen- inter-line flicker, shimmer, weave (telev.).

flimmerfrei flickerless. **-(e) Verschlussblende** non-flicker shutter. **-(e) Wiedergabe** flickerfree reproduction.

Flimmerfrequenz fusion frequency, no-flicker f., critical flicker f.

flimmern flicker, glisten, glitter, sparkle, scintillate.

Flimmer-peilung loop reversing or switch method, Robinson flicker d. f. method. **-photometer** flicker photometer.

flittern glisten, glitter, sparkle, scintillate.

flocken flake, form flakes or flocks, flocculate.

Flodel purfling (mus. instrument).

Flöte flute, flageolet. **-mit Stimmzug** flute with tuning slide.

Flöte, Harmonie- organ accordion. **Liebes-** flute d'amore. **Papageno-** pandean pipe. **Piccolo-, Pickel-** piccolo flute. **Stimm-** tuning or pitch pipe. **Stock-** stick flute. **Terz-** third flute. **Trommel-** fife.

Fluchtebene vanishing plane.

fluchtgerecht truly aligned or flush (with).

Fluchtlinientafeln alignment charts.

flüchtig volatile, fugitive, fleeting,

transient, cursory, superficial, evanescent.
Flüchtigkeit volatility, fugacity, evanescence, fleetingness.
Fluchtpunkt vanishing point.
fluchtrecht flush, aligned, in true alignment.
Flug, Gehörziel- aural indication or reading of homing device, aural method. **Schleifen-** ("U") turn, loop. **Sichtziel-** visual homing flight.
Flug-bahn path of flight, trajectory **-bahnbild** trajectory diagram.
Flügel wing, vane, flap, lobe, blade.
Flügel, Abdeck- masking or shutter blade or wing. **Beruhigungs-** intermediate blade, anti-flicker blade, balancing blade. **Radiometer-** radiometer vane. **Zwischen-** intermediate blade, anti-flicker b., balancing b.
Flügel-horn bugle, vocal horn. **-messer** vane-type fluid meter. **-mutter** winged nut, butterfly n., thumb n.
flügeln wing.
Flügel-rad screw wheel, worm w., s. propeller. **-schraube** wing or thumb screw or nut.
Flug-fernmeldewesen avigation or aerial navigation signal service. **-funkdienst** aeronautical radio service. **-geschwindigkeit** rate of flying speed, traveling velocity (electrons, etc.).
Flughafen airport, airdrome, landing field, flying f. **Verkehrs-** commercial airport, civil a. **-befeuerung** airport beacon service. **-grenze** airport boundary, a. border line.
Flug-leiter, Boden- ground control operator. **-leiter, Peil-** ground direction-finding operator. **-ort** (flying) position (of craft). **-probleme** avigation, aeronatical or flying problems. **-sicherung** avigation or aerial navigation safety, air communication safety. **-strecke** airway, avigation route. **-wesen** aviation, aerial navigation, avigation, aeronautics.
Flugzeug, anlaufendes approaching airplane. **-, schwanzloses** tailless airplane. **Nurflügel-** tailless airplane.
Flugzeug-abhörapparat airplane sound locator or detector. **-aufnahme** aerial photograph, airplane p. **-aufsatzpunkt** ground contact point, landing point. **-befehlsübertragung** inter-aircraft voice or command communication. **-führer** pilot; navigator, avigator. **-führung** avigation or piloting of an airplane. **-gleichgewicht** balance or trim of an airplane. **-längsachse** fore-aft line of airplane, major axis. **-leitung** aircraft guiding or guidance, avigation, piloting. **-navigation** aircraft guidance, piloting of aircraft, avigation. **-propeller, Versetz-** variable-pitch (feathered) airscrew or propeller. **-schieben** crabbing of an airplane. **-stützpunkt** airplane base, base airport. **-versetzung** drift, lateral displacement or shift (in course) of an airplane, deviation from course, yaw.
Flugziel flying course, airplane heading or head (towards destination), home.
Fluoreszenzauslöschung fluorescence quenching, evanescence, f. extinction (e.g., by poison).
fluoreszenzerzeugend fluorogenic.
Fluoreszenz-messer fluorometer, phosphoroscope (measures decay). **-strich** fluorescent screen tracing or line. **-unterdrückung** quenching of fluorescence, suppression or killing of f.
fluoreszieren fluoresce, scintillate (when excited by α rays), luminesce.
Fluoritlinsensystem semi-apochromatic lens system.
Fluss flow, flux, circulation, stream,

current, drift (of electrons).
Fluss, Streu- leakage flux, stray flux. **Windungs-** turn flux, flux turns, magnetic linkage.
flüssig liquid, fluid, non-solid, aquiform. **-machen** liquefy. **-werdend** liquescent.
flüssig, feuer- liquid at high temperature, molten, fused. **streng-** viscous, semi-fluid, difficultly fusible.
Flüssigkeits-dämpfung liquid or fluid damping or dash-pot action. **-dichtemesser** hydrometer, liquid density meter, areometer. **-linse** (oil or water) immersion objective. **-prisma** liquid prism.
Fluss -mittel fluxing material, flux. **-richtung** low-resistance or forward direction (of a dry or oxide rectifier, e.g.) **-spat** fluorspar, fluorite. **-verkettung** flux linkage.
Flutlicht floodlight.
Fokustiefe depth of focus, d. of field.
Folge sequence, series, succession, consecution, future, consequence, result, conclusion, sequel, compliance.
Folge, Bild- series or sequence of optical or pictorial actions. **Klang-** sequence or series of acoustic actions. **Linsen-** lens combination, system of lenses. **Reihen-** sequence, consecution, succession, series, order **Ton-** series or sequence of acoustic or sound actions.
Folge-kern, Nach- product nucleus. **-kontakt** make-before-break contact. **-produkt** metabolons, products of successive disintegration of radio-active parent material. **-punkt** consequent points or poles (magn.) **-reaktion** consequent reaction, secondary r. **-rohr** piloted thyratron (fires second in a trigger circuit). **-rung** deduction, inference, conclusion, induction, corollary. **-schalterkamm** sequence switch cam
Folie, Träger- supporting foil.
Foliendurchmesser record (disk) diameter.
Forcierkrankheit strain disease (metals).
Förderband, Bandförderer band or belt conveyor, c. belt.
Fördermechanismus, Platten- plate feed mechanism (of electron microscope).
fördern convey, transport, feed, haul; boost, promote, favor, further, advance, help, hasten.
Form, Fertig- finishing mold (glass, bottle making). **Mutter-** master mold (cer.), parental form. **Press-** pressure mold. **Vorgathering** mold (glass manuf.). **Zwischen-** passage type.
Formänderung deformation, distortion, strain, warp.
Formänderungsfähigkeit (plastic) deformability, rheologic or flow property.
Formant harmonic determining timbre or tone color, formant. **Hall-** resonance formant. **Haupt-** main or basic formant. **Unter-** sub-formant
Format, Bild- picture or frame size or format.
formbar plastic, capable of being plastically shaped or fashioned, plastically deformable or formable, moldable, fictile.
Formbeständigkeit stability of shape, non-deformability.
Formelbild structural formula.
Formeln, Ausschuss für Einheiten und (AEF) Committee on Units and Standards, Standards Committee. **Raum-** spatial formulae.
formengleich conformal.
Form-faktor aspect ratio (picture). **-faktor, Fernsprech-** telephone influence factor. **-fehler** formal defect, informality.

-festigkeit stability of shape, non-deformability. -gebung fashioning, shaping, molding, imparting shape by outside action (e.g., pressure). -gebung, spanabhebende (und spanlose) shaping or fashioning by machining with (and without) removal of material by cutting tools (in the form of chips, etc.)

form-gerecht, -getreu true to form or shape, undistorted, having fidelity, faithful (e.g., in amplitude or phase of signals).

Formierung forming (of a valve or cell); activation, sensitization.

Formling molded article or blank.

formlos amorphous, formless, non-crystalline.

Form-mangel abhelfen oder heilen remedy or cure formality defect. -maschine, Ruttel- jolt molding machine, jar ramming m.

formschlüssig form-closed or -locking.

Form-stich shaping pass (in rolling). -stück, Ton-, mehrzügiges multiple tile or clay conduit.

Formulierung, Strich- chemical formula using dashes or lines to represent bonds.

Formziffer shape factor; theoretical stress concentration factor.

Forschungs-anstalt research institution. -gebiet research field or domain. -geist inquisitiveness, inquiring or searching spirit.

Fortdiffundierung diffusing away.

Fortepedal forte pedal, loud p.

fortführen carry away, convey; continue, prosecute (a case).

fortheben, sich cancel out or eliminate (a factor).

fortlaufend sequential, continuous, uninterrupted, non-intermittent. -(e) Reihe unbroken, uninterrupted or continuous sequence or series.

Fortpflanzungs-geschwindigkeit velocity of propagation. -konstante propagation constant, transfer c. -mass (hyperbolical) line angle.

Fort-rückung translation, advance, feed, shift, removal, moving away. -schaltrelais stepping or impulsing switch or relay. -schaltung, Film- film feed, travel or movement. -schaltwerk stepping mechanism.

fortschreitende Welle progressive wave, advancing w.

Fortschreitungs-bewegung translational movement, stepping, propagational m. -geschwindigkeit speed of progression, s. of travel.

Fortschritt advance, progress, improvement, advancement (in an industrial art, etc.)

Foto... see Photo...

Fourier'sche Intergralzerlegung Fourier integral seriation.

Fourierreihe, doppelte double Fourier series.

Frage-bogen questionnaire, list of queries. -recht right to cross-examine. -stellung interrogatory.

Fransen fuzz, fringe. Beugungs- diffraction fringes.

fräsen cut, mill, face.

frei, drehungs- irrotational, rotation-free. farbfehler- achromatic. funken- non-sparking, non-arcing, arcless. kohle- carbon-free, free from carbonaceous admixture. polarisationsfehler- free from polarization error (in df, etc.).

...frei -less, -free, non-, an-, free from, devoid of.

frei(e) Elektronen (Leitungselektronen) free electrons, conduction e. -(er) Platz hole (nuclear theory). -(e) Schwingung free oscillation. -(e) Wegl änge, mittlere mean free (length of) path.

Frei-antenne outdoor antenna. -fallbogenlampe gravity-feed arc-lamp. -gabe unblocking, release, disengagement, opening, clearing, liberation.

Freiheitsgrad degree of freedom. **-zahl** variance (of a system).

Freilauf-bremse free-wheeling brake, coaster b. **-kupplung** free-wheel clutch, slip coupling.

Frei-lichtaufnahme outdoor or exterior shooting or shot. **-luftdurchführung** open-air wall-duct, leadin out-door bushing (insulator). **-luftisolator** outdoor insulator. **-schwinger** free radiator (loudspeaker).

freitragend self-supporting, s.-supported.

Fremdbestandteil foreign, extraneous or external matter, material or substance.

fremderregt (er) Magnet electromagnet, non-permanent m. **-(er) Schwingröhre, -(er) Sender** master-excited oscillator or transmitter.

Fremd-erregung separate excitation. **-gas** foreign gas.

fremdgesteuerte Steuerstufe crystal-stabilized master stage. **-unselbständige Kippmethode** distant-operated non-self-running time-base method.

Fremd-körper foreign or extraneous body or substance. **-licht-Lichtmodulator** light relay of Kerr cell or Karolus c. type (to re-convert currents into brightness variations with constant, independent ray source). **-peilung** ground direction-finding (bearing or directional data supplied to airplane from ground stations). **-schichtverfahren** self-luminous or s.-emissive method (using electrons on surface of object, in electron microscope). **-tone** alien frequencies or tones (in sound reproduction).

Frequenz, mittlere medium frequency, mid frequency. **-,technische** commercial frequency, industrial f.

Frequenz, Abschneide- cut-off frequency, critical f. **Abtast-** scan frequency (= lines x frame frequency per second). **Beruhigungs-** no-flicker frequency, fusion or critical f. (m.p.). **Bild-** picture frequency, frame f., video f., visual f. **Differenz-** difference frequency. **Durchdringungs-** penetration frequency, critical f. **Durchgangs-** pass wave, p. frequency. **Ecken-** corner frequency, sharp cut-off frequency (of a filter, etc.). **Einheits-** standard frequency. **Grenz-** cut-off frequency, limiting f. **Gruppen-** group frequency (number of wave trains per second), spark f. **Hilfs-, Rückkopplung mit** super-regeneration. **Kreis-** pulsatance, angular velocity, cyclic frequency, frequency in radians (ω). **Leitkreis-** transit-time frequency (magnetron). **Mittel-** medium-wave frequency, mid freq. **Mutter-** master frequency. **Null-** frequency at which phase shift is zero. **Pendel-** electron oscillation frequency, quench or bias f. furnished from auxiliary oscillator circuit (in super-regeneration). **Raster-, Rasterwechsel-** field frequency (in interlaced scanning). **Resonanz-** resonance frequency, natural f. **Schwankungs-** frequency flutter. **Schwellen-** threshold or critical frequency, photo-electric threshold. **Seiten-** side frequency (one of sum or difference frequencies). **Spiegel-** image frequency. **Sprech-** voice frequency, speech f. **Steuer-** pilot or synchronizing frequency. **Summen-** summation frequency, sum f. **Teil-** component frequency. **Trenn-**

cut-off frequency. **Überhör-** supersonic, ultrasonic or ultra- audio frequency. **Umlauf-** rotational frequency (electrons). **Unterhör-** subaudio or infrasonic frequency. **Unterresonanz-** submultiple resonance frequency, subsynchronous f. **Verschmelzungs-** critical flicker frequency, fusion f., no-flicker f. **Wach-** watch frequency, distress f., f. of international automatic alarm signal. **Wellenzug-** group frequency, wave-train f., spark f. **Welligkeits-** ripple frequency. **Wiederholungs-** repetition frequency, f. of recurrence. **Winkel-** angular frequency, radian f., pulsatance. **Zeichen-** signal frequency. **Zeilen (wechsel)-** line frequency, strip f., horizontal scan f. **Zwischen-** intermediate-beat or transfer frequency (in superheterodyning).

Frequenz-abbau frequency division, f. submultiplication. **-abgrenzung s. Wellenabgrenzung.-abhängigkeit** frequency dependence, having a f. effect, condition of being affected by, or a function of, frequency. **-abstand** interference guard band, tolerance frequency. **-abweichung** frequency deviation, drift or swing, lilt (slow fluctuation). **-amplitude** amplitude proportional to frequency. **-bereichfilter** band-pass filter. **-bereichuberstreichung durch Kondensator** frequency range or band swept, scanned or covered by a condenser or capacitor. **-Entdämpfungskurve** frequency-gain curve.

Frequenzgang frequency characteristic or response curve, influence of frequency, f. effect. **-, flacher** flat frequency characteristic (without peaks or dips). **-der Amplitude, starker** marked dependence of amplitude upon frequency. **-normung** standardization of frequency response curve.

Frequenzgemisch frequency spectrum.

frequenzgerader Kondensator straight line frequency condenser.

Frequenzgerät, Silben- syllable vodas (Bell Co.).

frequenzgesteuerte Senderöhre frequency-stabilized transmitter tube.

Frequenz-hub frequency fluctuation, swing or variation (between two limits); f. deviation (above and below the assigned center or resting f., in f. modulation), amount by which instantaneous carrier f. differs from r.f. **-kennlinie** frequency-response characteristic. **-kurve, Entdämpfungs** gain-frequency curve.

frequenzmässig, hoch- for high frequency, so far as high frequency is concerned, relative to hf.

Frequenz-messer frequency meter, wave m., ondometer, sonometer. **-messer, Zungen** vibrating reed frequency meter. **-modulation** frequency modulation. **-reihe** series of frequencies. **-schnitt** cross-over of frequencies (in sound film production), point where wave-bands handled by loudspeakers intersect. **-schwankung** lilt (slow fluctuations of frequency), drift of f., f. flutter, f. departure, swinging (momentary variation of received f.). **-schwankungsschwächer** discriminator (converts drift into direct potential difference). **-selektion** frequency selection. **-sieblochbreit** transmitted band, band-width of a filter, spacing between cut-off points. **-skala** frequency spectrum, f. scale. **-sperre, Spiegel-** image frequency stopper. **-steuerung (von *Senderohrschwingungen*)** frequency stabilization (of transmitter tube oscillations).

Frequenz-teiler (cf Frequenzweiche) frequency divider, f. submultiplier. **-übersetzung** frequency transformation.

frequenzunabhängig free from frequency

effect, independent of f., f.-independent.

Frequenz-unterdrückung, Spiegel- image frequency stopper, image frequency rejection. -unterscheidung frequency discrimination. -verdreifacher frequency tripler. -verschiebung frequency distortion, tone d.-verwerfung slow frequency drift (in transmitter or receiver apparatus occurring after tuning). -verzerrung attenuation-frequency distortion, amplitude-f.d., attenuation d.,f. distortion. -wandlung heterodyne or superhet. action; frequency change,f. transformation. -weiche dividing filter or network, separator (sync and video signals,in telev.) comprising limiter tube and network. -wiedergabe frequency response characteristic.

Fresnel's Mitführungskoeffizient Fresnel coefficient of drag.

fressen seize, bind, freeze, attack, corrode.

freundlich, licht- photophilic, photophilous, favoring highlights over shadows.

Freyagerät German version of radar or radio locator.

Frisch-arbeit fining process (metals). -aufnahme new recording, re-recording or new pickup of film (sound track film). -wirkung oxidizing or purifying reaction.

Frist grace, limitation, respite, term, (fixed) time allowance or extension. moratorium.

fristgerecht in time, within time limit.

Frist-gesuch dilatory plea. -gewährung grant of respite. -verlängerung time extension, respite, grace.

fritten frit, sinter, concrete.

Frittporzellan soft porcelain, frit p.

Front, Wellen-, geneigte tilted . wave front or wave head.

frontale Halbkugellinse hemispheric front lens.

Frontwelle onde de choc, impact wave, bow wave (which precedes a projectile).

Frosch cam, dog, adjustable stop, arm, frog (of violin bow). -perspektive worm's eye view.

frostbeständig resistant to frost, freeze-proof.

Frostpunkt freeze point.

Frühzündung premature ignition, pre-ignition.

F S P see Funkseitenpeilung

Fuge joint, junction, slit, groove, crack.

Fuge, Dehnungs- expansion joint. Schweiss- welding seam or line, weld, shut. Stoss- joint, junction.

fühlbare Abstimmung tuning condenser sensibly detained or braked on passing carrier, "lazy tuning".

Fühler-kondensator pickup condenser, scanning c. -lehre thickness gage.

Fühlhebel tactile, feeler, contactor or probing lever, scanner, pickup means; indicator of micrometric caliper.

fühlig, fein- delicately sensitive.

Fühlnadel selecting needle or pin, pecker. Kugelspitzen- scanning or pickup needle with spherical point (sound record).

Fühlratsche micrometer friction thimble.

führen lead, conduct, guide, pilot, steer, channel, carry (current), support, hold.

Führerstand (Flugzeug) pilot's cockpit, p. compartment (airplane).

Führung, seitliche lateral guidance (of airplane).

Führung, Fenster- aperture guide, gate.

Film- film feed, f. guiding, f. gate. **Nach- (des Rahmens)** re-setting or re-adjustment (of frame). **Nieder-** downlead. **Rück-** return path, return circuit, r. wire, r. lead. **Schmelz-** smelting practice, conduct of heat, smelting schedule. **Seiten-(von Film)** lateral guidance, side guiding (of film). **Wellen-** wave guide (either conducting or dielectric).

Führungs-band control strip (of a printer). **-kanal** film track, f. channel. **-loch** (guide) perforation (for engagement of sprocket). **-rille** track or groove (of phonograph record). **rolle** sprocket wheel (of film feed). **-schlitten** slide (carriage).

Fülle, Klang- sound volume.

Füll-faktor space factor, activity coefficient. **-flüssigkeit** immersion liquid (microscope). **-körper** filling body, packing material. **-kurve** single-loop oscillogram. **-mittel** filler, filling, stuffing or loading material or compound. **-perspektive** plenary perspective **-strich** filling mark, gage m. or line.

fünf-dimensional in terms of five-dimensional world. **-eckig** pentagonal.

Fünferalphabet five-unit code.

Fünfflach pentahedron.

fünfgliedriger Mischer five channel mixer.

Fünf-polrohre pentode (either power pentode or screen-grid p.) **-strömealphabet** five-unit code. **-stufenschwacher** five-step weakener (phot.)

Funk, Bild- radio picture, wired picture, facsimile transmission.

Funkazimuth, rechtweisender oder geographischer true bearing. **-, missweisender oder magnetischer** magnetic bearing.

Funkbake radio beacon, radio range (sometimes marker beacon).

Funkbake, Kurs- radio range or beacon. **Landungs-** landing beacon. **Richtungs- (zum Anfliegen)** runway localizing beacon, terminal marker beacon (used in approach procedure).

Funk-beamter radio station official, board radio operator. **-beschicker, automatischer oder mechanischer** automatic (radio) quadrantal error compensator, cam compensator. **-beschickung (cf Funktrübung)** goniometric error or direction-finding deviation and its correction, clearing or compensation; shift of minimum or zero point; (local) calibration of radio compass.

Funkbeschickungs-aufnahme calibration of direction-finder. **-drahtschleife QE** (quadrantal error) compensating or clearing loop. **-kurve** error curve. **-kurvenscheibe** compensator cam, zero-clearing c.

Funk-besteck radio fix. **-bild** photo-radio, radio picture, wired p., radiophotogram. **-drosselspule** air-gap reactance coil.

Funkeleffekt. sparkling, scintillation, glitter; low-frequency effect of local emission density (of electrons).

funkeln sparkle, scintillate, glitter.

Funken, Gleit- slide spark, creepage s. **Öffnungs-** break spark, sparking at break. **Schliessungs-** make spark. **Unterbrechungs-** break spark, wipe s., circuit-opening s.

Funkenentladungs-Aufbauzeit formation time of spark discharge.

funkenfrei sparkless, non-sparking, non-arcing.

Funken-geber spark coil, Ruhmkorff c. **-kanal** spark track. **-linien** spark lines. **-löscher** spark quench, blowout or extinguisher, s. killer. **-mikrometer** micrometric spark-gap or spark-discharger, spintharometer. **-sender, Ton-**

singing or musical (quenched) spark transmitter.

Funkenstrecke, Kugel- sphere (spark) gap, ball gap, Lösch- quenched spark gap. Nadel- needle (spark) gap. Reihen- multiple spark gap. Schutz- spill gap. Ton- quenched gap. Überspannungs- surge arrester, s. gap, s. absorber.

Funkensystem, Takt- timed spark system (wireless transmission).

Funkentstörer static eliminator.

Funken-Überschlag spark breakdown, flashover, sparkover. -verzögerung spark lag. -zähler spark counter.

Funk-fehlweisung radio df error, directional e., quandrantal e. (QE), compass deviation, distortion of bearing. -feuer radio beacon, radio range; marker beacon.

Funkfeuer, mehrstrahliges multi (-ray) beacon, multiple-or triple-modulation beacon. Dreh- rotating radio beacon or range. Haupt- main route or airway radio beacon. Kreis- rotary beacon, omnidirectional beacon, circular navigational beacon (sending non-directional signals in all directions). Landungs- landing beacon (for instrument or blind landing). Richt- directional beacon station or transmitter s. Strecken- (air) route beacon.

Funk-feuerkennung call signal or distinctive signal of a beacon station. -freund radio amateur, r, fan, ham, -geber radio transmitter; spark coil (obs.). -horchdienst radio intercept service. -mutung radio prospecting, r. metal locating. -ortung radio position finding, obtaining r. fix.

Funkpeil-scheibe bearing plate of radio df. -schwund radio fadeout or fading. -sektor approach sector. -station, -stelle radio direction-finder station or post (on ground or on craft). -stationmittellinie center line of approach sector.

Funk-platte spark plate (condenser of automobile radio). -querverbindung intercommunication channel. -rufzeichen radio call, code signal, "signature" (of a station). -schatten radio shadow, dead spot, pocket. -schneise radio beacon course, equisignal track, corridor of approach. -schwund fading, fadeout (of radio signals).

Funkseitenpeilung, beschickte od. verbesserte true or corrected radio bearing. -, unverbesserte oder abgelesene uncorrected or read radio bearing.

Funksender, Einpeilen eines Richtempfängers auf einen tuning in a transmitter station with directional receiver, to take bearings. Leit- radio range, r. beam station, r. beacon.

Funkspektrum see Wellenabgrenzung

Funkspruch, geschlüsselter radio code message, coded r.m. -spruchweg radio channel.

Funkstelle, Boden- aeronautical ground radio station, g. direction-finder s. Einstrahl- beam station. Gross- high-power radio station, long-distance or long-range s. Land- ground radio station; shore or coast s. Leit- net control station. Luft- aircraft radio signal station.

Funktechnik radio engineering, r. art; r. communication or signaling art. -techniker radio technician, r. engineer.

funktelegraphische Verbindung radio telegraphic connection, r. signal communication (channel).

Funktelephon radiophone.

Funktion, konjugierte conjugate function. Ausgangs- trial function. Bessel- erster Art

Bessel function of the first kind. **Eigen-** eigen function, characteristic f. **Einheits-** unit function (of Heaviside). **Hyperbel-** hyperbolic function. **Kraft-** power function. **Kugel-** spherical function. **Orts-** position function. **Winkel-** trigonometric function, angular f., function of angles.

Funk-trübung uncleared zero, blurring or lack of precision of minimum or zero (in radio d.f. or bearing). **-verkehrsbezirk** radio controlled aerial navigation (or avigation) district or zone. **-wechselsprechen** alternate two-way radiophone communication. **-wesen** radio technology, r. signaling art.

Furche furrow, channel, wrinkle, ruling or line (of diffraction grating), groove or track (with undulations, of phonograph or gramophone disk).

Fuss pinch, press, squash, stem or foot (of a lamp, tube or valve). **-,umgestülpter** re-entrant squash or press. **Strahlerzeuger-** gun press (of c.-r. tube).

Fusskreis root line, dedendum l.

fusslose Röhre loctal tube.

Fussplatte base plate (camera).

Fusspunkt foot (math.), nadir (point opposite the zenith, in astronomy). **-,Gegen-** antipode.

Fusspunkt-kurve pedal (or podal) locus curve. **-linie** podal line. **-widerstand** terminal impedance or base-loading (of antenna).

Futter, Mantel- shell lining.

Fütterungsstoff lining material.

G

Gabe, Impuls-, Stromstoss- pulsing, impulsing, emission or transmission of impulses. **Impuls-, rückwärtige** revertive pulsing.

gabelartig forked, furcated, bifurcate.

Gabel-klammer forked clamp. **-umschalter** hook switch. **-zinken, Stimm-** tuning fork prong or tine.

Galgen, Mikrophon- microphone boom.

Galle, Glas- glass gall, sandiver.

Galtonpfeife Galton pipe.

galvanisch galvanic, conductive.

galvanische Kopplung resistance coupling.

Galvanometer, kompensiertes rheograph (of Abraham) in which inertia and damping are compensated. **Bändchen-** twisted strip or band galvanometer. **Faden-** string galvanometer, vibrating g. **Saiten-** Einthoven g. **Stoss-** ballistic galvanometer.

Galvanostegie galvanoplastics, electroplating, electrodeposition.

Gamma, Kopie- negative gamma.

Gang motion, working state, gear, course, passage, pass, travel, stroke; trend, drift; heat run; thread or spire of screw; action (of camera); function, dependence, variation, response (characteristic).

Gang, toter lost motion, backlash, play. **-zügiger** intimate or positive threading of gear (without play or backlash).

Gang, Chargen- heat run. **Dämpfungs-** frequency response of attenuation, attenuation-frequency characteristic. **Frequenz-** frequency characteristic, frequency-response curve, effect of frequency. **Frequenz-, starker, der Amplitude,** marked dependence of amplitude on frequency. **Gefälls-** gradient variation. **Gleich-** unison, synchronism. **Hin-** forward motion or working stroke, travel, pass or journey, scansion (telev.) **Hin- und Her-** reciprocating motion, rocking movement. **Orts-** local variation. **Rück-** return stroke, retrace or flyback (of pencil); decline. **Selbst-** automatic feed, operation or action. **Strahlen-** path of rays, ray tracing, geometric configuration of rays. **Tages-, örtlicher** local diurnal variation. **Temperatur-** temperature dependence. **Vorwärts-** direct action (of camera)

gangbar salable, marketable; practicable.

Ganghöhe, Windungs- pitch of turns.

gängig ....-thread(ed), pitched.

gängig, links- left-handed, counterclockwise. **rechts-** clockwise, right-handed. **scharf-** (screw cut) with triangular thread.

Gangunterschied path or phase difference (of waves, rays, etc.)

ganz whole, entire, total, complete. **-(es) Vielfaches** integral multiple. **-(e) Zahl** integer, whole or integral number.

Ganz-heit totality, whole, aggregation. **-metallwand** all-metal (projection) screen. **-polkegel** polehode cone, body cone. **-ton** whole tone.

ganzzahlige Ladung integral spin, charge.

Ganzzahligkeit integralness, property of being an integer.

Gar-brand finishing burn (ceramics). **-brenne** fired to maturity (ceramics).
Gare refined state, finished s.
Garnierung bushing, lining.
Garnitur fittings, trimmings, mounting.
Gasabgang gas issue, withdrawal or offtake, g. delivery.
gasartig gasiform, gaseous.
Gasaufnahme occlusion or absorption of gas (in metals). **-aufzehrung** gettering, absorption of gases. **-ausbruch** gas eruption, (violent) liberation of g., escape of g. **-austreibung** gas cleanup or expulsion, outgasing, degasing, gettering.
gasbehaftete Oberfläche gas contaminated surface, s. with adsorbed g. (skin).
Gas-bindung gettering. **-blasenbildung** cavitation, occlusion of gases (in metals, etc.). **-dichtemesser** dasymeter. **-entladungsrohr** gas discharge tube or valve. **-entwicklung** evolution, disengagement or giving off of gas.
gas-förmig gasiform, gaseous. **-gefüllte Röhre** gas-content tube, gaseous t., gassy t.
Gas-konzentrierung gas focusing, ionic f. (in c.-r. tube). **-schalter, Druck-** (cross) airblast switch. **-schmelzschweissung** gas torch autogenous welding. **-strecke** gaseous path or gap (in a tube). **-verschluss** gas seal, bell and hopper, cup and cone arrangement (furnace). **-wage** gas specific gravity balance. **-zelle** gas-filled photo-emission cell.
gattieren classify, sort.
Gattung kind, genus.
gaufrieren goffer, emboss.
Gaze, Draht- wire gauze, mesh or cloth.
gebend s. **erzeugend**
Geber, Bild- pickup camera. **Film-** film transmitter, pickup or scanner, telecinematographic or film scanning device. **Funk-** radio transmitter, spark coil (obs.) **Funken-** spark coil, Ruhmkorff c. **Impuls-** impulsing device, impulse transmitter; tonal generator, tonic transmitter. **Kontakt-** contact maker, timetapper or marker. **Kristalltakt-** crystal monitor, c. stabilizer. **Takt-** impulsing means, impulser, time beater, t. tapper, metronome.
Gebiet, Druck- pressure or compression area or region. **Einzugs-** cross-sectional area of undistorted portion of field (re: df frame). **Empfangs-** service area (broadcasting). **Interferenz-** mush area (chain broadcasting). **Orts-** local area. **Reiss-** discontinuity or breakoff region (in oscillations). **Stör-** mush area (in chain broadcasting), disturbed area or zone.
Gebilde structure, system, organization, form, image.
Gebilde, Gewebe-, winzige ultramicrons (of eye tissue). **Leiter-** conductor structure, c. system, array. **Leucht-** luminous pattern (lines, bands, or continuous spectra). **Luftleiter-** aerial extension, a. structure, a. network. **Raum-** space diagram; solid 3-dimensional structure. **Sieb-** selective system, filter means, network.
Geblase, Knallgas- oxyhydrogen blowpipe.
geblättert laminated.
Gebrauch custom, usage, practice, use.
gebrauchsfertig ready for use, readied.
Gebrauchs-kopie usable or working copy (for display). **-last** working load, service l. **-muster** (German) petty patent, utility-model patent. **-vorschrift** directions for use.
Gebühr fee, royalty, tax, dues,

duty, payment. **Bekanntmachungs-** lay-out fee. **Zeugen-** witness fee. **Zuschlags-** extra fee, supplementary fee or dues, extra tax.

gebundene Ladung bound charge. **-Warme** latent heat.

gedacht imaginary, fictitious, conceived, assumed.

gedackte Pfeife stopped pipe.

gedämpft(e) Anzeige deadbeat, aperiodic or non-ballistic reading or indication. **-(er) Raum, mittel-** moderately live room.

Gedanke, Erfindungs- basic idea underlying an invention.

gedeckte Lichter covered lights. **-Pfeife** stopped pipe.

gedrängt, gedrungen compact, compendious, of restricted dimensions.

Gefahr danger, peril, risk, hazard, chance, jeopardy, menace.

Gefäll, Grenz- limiting slope, l. gradient. **Potential-** potential gradient, p. barrier, threshold of p.

Gefällsgang gradient variation.

gefalteter Exponentialtrichter folded exponential horn (of a loudspeaker).

Gefäss container, vessel, tank, jar, pot, box. **-, Faraday'sches** Faraday ice pail, F. cylinder.

Gefäss-barometer cistern barometer. **-heberbarometer** combination cistern and syphon barometer.

gefasste Linse mounted lens.

Gefässwirkungsgrad bulb efficiency (rectifier).

Geflecht plaited or woven work, network, reticulation.

gefrierbar coagulable, freezable, congealable.

Gefrierpunkt-lehre cryoscopics. **-messer** cryometer. **-messer und Druckmesser** manocryometer.

Gefüge structure, texture, bed, stratum. **Fein-** fine structure, micro-s. **Grob-** macro-structure. **Klein-** fine structure, micro-s. **Streifen-** laminated structure, banded s.

Gefühl, Raum- space feeling, spatial feeling (in vision).

gegebenenfalls if necessary, under certain circumstances, in an emergency, optionally.

gegen den Wind fliegen fly up wind.

gegenarbeiten buck, oppose, counteract, work in opposition.

Gegen-beleuchtung back (ground) illumination. **-beweis, direkter** rebutting testimony or evidence. **-bild** counterpart, antitype, replica. **-druck** counter-pressure, back-p. reaction. **-druck, Strahlungs-** radiation reaction.

gegeneinander schalten connect in opposition, c. differentially.

Gegenelektrode counter-electrode, opposite e., cooperating e.; signal plate or metallic screen (forming foundation of mosaic in telev. tube).

gegenelektromotorische Kraft counter electromotive force (c emf), back emf.

Gegen-feder counter(-acting) spring, retractile s. **-feldmethode** retarding potential method. **-fernsehen** two-way television. **-fusspunkt** antipode. **-gewicht** counterpoise, counterweight, counterbalance. **-gewichtantenne** screened aerial. **-keil** counter-wedge. **-klage** cross-claim, counter-claim, **-kontakt(pol)** co-operating contact or terminal, opposite contact. **-kopplung** negative or stabilized feedback, reverse f., anti-regeneration, degeneration. **-kraft** reaction, counter(-acting) or opposing force. **-kufe** counter pad. **-kurs** course, head or flight away from port, outward c., opposite c. **-lager** abutment, any relatively immovable point or surface

sustaining pressure.
gegenläufig oppositely directed or oriented, of contrary sense.
Gegen-läufigkeit opposition, o. of phases, etc. **-modulation** push-pull modulation, m. in opposition. **-mutter** lock, clamp, check, jam or binding nut. **-partei** adversary, adverse party, "the other side" **-phasigkeit** phase opposition, anti-phase condition. **-polung** opposition of polarity, opposite p., reversal of p. **-probe** contrasting sample, c. test, check test. **-satz, im... zu** contradistinct from, in contradistinction to. **-schaltung** differential connection, connection in opposition. **-sehen, Vorbeisehen beim** not to look straight at each other (in two-way video telephony). **-seher** 2-way video or television apparatus.
gegen-seitig mutual, reciprocal, one another (when more than two) each other (when two), contralateral (on the opposite side). **--mischbar** consolute. **-sinnig** in or of opposite or contrary sense, direction or orientation (e.g., of rotation); irrational, devoid of sense, non-sensical, illogical. **-sinnig in Reihe** series opposing.
Gegensonne parhelion, mock sun.
Gegenstand object, (sometimes:) image of cathode formed at crossover by first lens, in c.r. tubes, and acting as object for second lens, subject-matter.
Gegenstand, körperlicher solid or three-dimensional object. **Abbildungs-** imaging object. **Erfindungs-** object of invention, disclosure. **Übertragungs-** object to be televised or transmitted by video signals, teleview o., televised o.
Gegenstands-glas object glass **-weite** distance from electron source (aperture or crossover) to lens, object d.
Gegen-stoff antisubstance, antibody, antidote. **-strahl** reflected ray, reflection. **-strom** counter-current, inverse c., reverse c. **-stromrelais** reverse-current relay. **-stück** counterpart, match.
Gegentakt-arbeiten push-pull operation, work in phase opposition. **-audion** push-pull grid detector. **-bremsfeldaudion** electron-oscillation detector or rectifier (negative anode associated with leak resistance and condenser) in push-pull circuit scheme. **-kippgerät** push-pull time-base. **-mikrophon** push-pull microphone. **-noiseless** split-wave noiseless recording. **-schrift** push-pull track.
Gegen-teil opposite, contrary, antithesis, converse, contrast. **-versuch** control or check test or experiment. **-werk** equivalent, counter, value. **-wind** contrary wind, head w.
gegenwirkend counteracting, reacting, antagonistic.
Gegen-wirkung, akustische acoustic reactance. **-zelle** counter electromotive force (battery) cell.
gegliederte Bildteile, verschieden picture portions of different gradation, composition or organization.
Gegner adversary, objector, opposer, opponent, "the other side."
gegossener Film coated film.
gegründet auf based upon, predicated on, grow out of, result from, conditioned by.
Geh-Steh-Apparat start-stop type of telegraph printer.
Gehalt, Kern- nuclear concentration, content of nuclei.

Gehäuse case, box, shell, housing, cabinet, chamber (camera). **Horer-** receiver case, r. shell. **Trieb-** gear casing, g. box, g. housing.

Geheim-kamera detective camera, concealed c. **-telephonie** garbled, scrambled or secret telephony (based upon speech band inversion, use of scrambler circuits, speech inverters, etc.)

geheizte Röhre, indirekt heater type or indirectly heated tube.

Gehör...auditory, acoustic, audio, aural, hearing.

Gehör, echtes, absolutes genuine absolute pitch. **um-ersuchen** file application or petition to be heard, petition for a hearing.

Gehör-empfindungsskala auditory or aural sensation scale. **-maskierung** auditory making. **-schädigung** hearing impairment. **-schärfe** acruity of hearing. **-schärfemessung** audiometry. **-zielflug** aural homing.

Gehrung mitre, bevel.

Geige violin. **Stroh-** violin in which bridge vibrates a diaphragm attached to a horn.

Geigen-harz rosin, colophony. **-lack** violin varnish.

Geist, Erfinder- inventive genius, inventor's spirit. **-Forschungs-** researching or inquiring spirit, inquisitiveness.

Geister, Gitter- spectral (grate) ghosts.

Geisterbild ghost image.

Gekrätz- Metall- waste metal.

gelagert, drehbar- rotationally supported, journaled, fulcrumed, pivoted.

Gelände, Aufnahme- location.

Gelände-arbeit field work (geology, etc.) **-winkelmesser** angle of site instrument, clinometer.

gelatinebeschichteter Blankfilm gelatine-coated film base

Gelb-giesser brass founder. **-glut** yellow heat.

Gelbildung gel formation.

Geldstrafe, eine - ausklagen sue for a penalty.

Gelenk joint, link, articulation, hinge. **Kardan-** universal joint, cardan j., hall and socket j. **Knie-** knee joint, knuckle j., elbow j. **Kreuz-** universal joint, Hooke's j. **Kugel-, Universal-** ball and socket joint, universal j., cardan j.

gelenkig flexible, pliable, supple, articulate, jointed, linked.

Gelenk-kette link chain. **-kondensator** swing-out condenser. **-kupplung** Hooke's joint. **-viereck** linked or articulated quadrilateral or quadrangle.

gellend shrill, yelling, tingling.

gelten be valid or in force, be alive, apply to, hold good for.

geltend machen, einen Anspruch advance or make a claim.

Geltungsbereich einer Erfindung scope or command of an invention.

Gemagerät German type of radar.

gemasert speckled, streaked, veined, grained.

gemeinsam common and joint, shared.

gemeinschaftlich mutual.

gemeinschaftlich und einzeln jointly and severally.

gemeinschaftlich betrieben ganged (of condensers), geared or locked together, operated jointly or in interlocked relationship.

Gemeinschafts-antenne party antenna, block a., communal a., centralized or shared a. **-sender** chain broadcast station.

Gemisch, Klang- sound spectrum.

Gemischkomponenten spectral components (sound, etc.)

gemischtpaariges Kabel composite cable.

gemischte Stimme mutation stops (organ)

genau exact, correct, precise, close, faithful (of reproduction).

Genauigkeit, fünfstellige five-figure accuracy.

Genehmigung approval, approbation, permission, allowance, assent, consent, sanction, concession.

General-nenner common denominator. **-unkosten** overhead or general expense. **-vollmacht** general power of attorney.

Generator, Ablenkungs- deflection generator, sawtooth g., time-base generator. **Bremsfeld-** negative transconductance oscillator or generator, positive-grid or retarding-field o. **Fahrtwind-** fan-driven generator, wind-driven g. (airplane). **Impuls-** impulse generator, surge g., "lightning" g. **Lochscheiben-** light chopper. **Prüf-** signal generator, all-wave oscillator, test o. **Quadratwellen-** square or rectangular wave generator. **Resotank-** micro-wave generator (electron-oscillation tube confined in a cavity, e.g., c. resonator magnetron). **Sägezahn** sawtooth generator, ratchet g. **Steuer-** drive oscillator, master o. **Ton-** musical generator, tonal g. **Wanderwellen-** surge generator, impulse g., "lightning" g.

Geometriefehler geometric distortion.

Geometrisierung der Wellenmechanik geometric derivation of wave equation.

geordnet, regellos with random orientation.

geordneter Zustand ordered state or arrangement (of a structure), perfect configuration (of units in crystal lattice).

gepackt, dicht close-packed.

gepanzerte, wenig streuende Spule iron-core deflecting yoke (with minimized spot and pattern distortion).

gepeilter Sender tuned-in beacon or transmitter (to take bearings).

gepufferte Lösung buffered solution.

Gerade straight-line characteristic or "curve," (Schottky) line.

gerad(e) Beleuchtung direct illumination(micr.) **-(er) Kondensator, frequenz-** straight-line frequency condenser or capacitor. **-(er) Kondensator, kapazitäts-** straight-line capacity condenser. **-(er) Kondensator, wellen-** straight-line wave condenser. **-(e) Oberharmonische** even harmonic.

gerad-linig (recti)linear, in direct proportion. **-sichtiges Prisma** direct-vision prism. **-wertig** of even valence.

gerasterte Schicht, -(r) Schirm mosaic screen, photosensitized m.

Gerät implements, gear, apparatus, outfit, equipment, instrumentation, tools. **Lichtton-** sound film head or attachment.

Geräteglas apparatus glass (stock).

Geräusch s.a. Akustik, Klang, Schall.

Geräusch (Radio, Film, etc.) und akustische Eigenarten: Noise (in radio, film, etc.) and acoustic peculiarities,

Apparatgeräusch system noise (in recording).

Bildgeräusch frame noise.

Bildstrichgeräusch frame line noise (form of motor boating).

blecherner Klang tinny sound.

Bratpfannengeräusch frying noise, zoop.

Brodeln boiling.

Brummton, Netz- mains hum, hum.
dumpf all-bottom, boomy, dull, tubby.
Eigengeräusch inherent film noise.
gellend shrill, tingling, yelling.
Geschwindigkeitsschwankungs-Geräusch -flutter, wowows (when intensity pulsation or variation is up to 6 cycles a second); flutter (6-30); gargle (30-200), and whiskers (over 200).
gleichmässiges Geräusch smooth noise, uniform n.
gleitender Ton glissando, sliding note.
Grundgeräusch (back) ground noise.
hallend reverberant (multiple echoes).
Heiserkeit pops (at splices), raucousness, hoarseness.
helle Klangfarbe strident or shrill timbre.
heulen howl.
Heulton warble tone, hf warble note, "multi-tones."
hoher Ton tweeter, treble, high-pitched or high-frequency note or sound.
hohler Ton boomy sound, dull s. (higher pitches cut off).
Kellerton tunnel or boom effect.
Klappern chatter, rattle.
Klebestellengeräusch blooping, splice bumps, dull sound due to blooping patches.
klimpernder Klang harsh or strident sound (with strong harmonics or high partials), shrill, tinkling, chinkling s.
Klingen sounding, ringing (of a tube), microphone noise.
klirren clink, clatter.
knacken click, crack, crackle, crepitate.
knallen detonate, explode, report, crack, pop, pistol shot noise.
knarren bloop, jar, crack, rattle creak, squeak.
knattern crack.
knirschen grind, crunch.
knistern sizzle, grate, crackle, rustle.
knurren growl, snarl, rumble.
kochen boil.
Kohlekorngeräusch carbon noise (microphone).
Kornrauschen film grain noise.
krachen crack, crash, crackle, rustle.
krächzender Ton all-top sound or voice.
kratzen scratch, scrape, rasp, bumps (film splices).
kreischend strident, all-top (of voice).
lautgetreu of acoustic fidelity, orthophonic.
Lautsprechergeräusch rattle of loudspeaker.
lebendig live, realistic.
Mikrophon-geräusch microphonic noise, valve n. **-schmoren** mike stew.
moiréartiges Geräusch moiré pattern of noise.
nachhallend reverberant, echoing, resonant.
Nadelgeräusch needle scratch n. chatter, record noise.
Nebengeräusch extraneous noise.
Netz(ton)brummen mains hum.
offen hollow (of sound).
Perforationsgeräusch sprocket hole modulation or noise.
pfeifen whistle, ring, squeal, howl, sing.
Plattengeräusch record noise, surface n., needle scratch. **-schlag** disk wobble.
plötzliches Geräusch impulse excitation noise.
prallen rattle.
prellen(am Kontakt) chirp, thump, clink, chatter (caused at

contacts and keys).
quieken squeal, squeak.
raspeln rasp.
rasseln rattle, clatter.
Rauhigkeit hissing, raucousness, hoarseness.
Raumgeräusch set noise.
Rauschen (von Röhren) tube or valve noise (thermal agitation of electrons), hissing, rustle.
Rauschen, Schrammen- scratch noise. **Widerstands-** circuit noise (due to Brownian movement).
reiner Ton pure note or tone.
sausen rush, whiz, whistle, hum.
schallen sound, resound, ring.
Schallplattengeräusch record noise, surface n.
schallhart acoustically rigid or hard, non-absorbent.
schallweich sound absorbent.
scharf shrill, piercing, strident.
scharren scrape, scratch.
Schlag disk wobble (phonograph).
Schlagen(Kontakt) key click, thump, or chatter.
Schmutzgeräusch dirt noise.
Schmoren frying, mike stew.
schnarren, burr, buzz, jar.
Schrammenrauschen scratch noise.
Schrotgeräusch shot noise (due to Schottky effect).
schwirren whiz, whir, buzz, hum.
Selbsttönen singing or squealing (of a tube).
Staubrauschen dust noise.
Strichgeräusch frame-line noise (sort of motor boating).
summen hum, buzz.
surren motor boating.
tiefe Töne low AF or bass notes; bassy condition (in sound reproduction, low AF overemphasized).
trillern warble, trill, quaver.
überschreien overmodulate.
unreiner Ton impure tone, ragged tone or note.
Wärmerauschen thermal noise (due to thermal motion of electrons).
Widerstandsrausch circuit noise (due to Brownian movement).
Wobbeln warble, wobble,
Zischen fuzzy noise (of film), hiss, sizz, sizzle, fizz, high harmonic alien tones, in film reproduction.
Zwitschern birdies, canaries.

Geräusch, Bild- frame noise. **Eigen-** inherent film noise. **Lautsprecher-** rattle of loudspeaker. **Mikrophon-** microphone or valve noise. **Neben-** extraneous noise. **Perforations-** sprocket noise (m.p.). **Raum-** set noise. **Wärme-** thermal noise (sort of Brownian movement of electrons in input circuit).
Geräuschabstand signal-noise ratio.
geräuscharm of low noise, noiseless, silent, silenced.
Geräusch-bekämpfung noise abatement, suppression or attenuation. **-beseitigung (an Tonklebestellen)** blooping elimination. **-beseitigung, automatische** silent tuning, noise suppression, noise gate. **-filter, Nadel-** needle scratch filter. **-kasten (Erzeugung besonderer Geräusche)** equivalent acoustics (for generation and imitation of noises and sound effects). **-losigkeit** noiselessness, silence, quietness, calm(ness), absence of noise. **-messer** noise meter, sound-level meter (e.g., measuring in terms of decibels), noise measuring set, acustimeter. **-messer (in elektrischen Kreisen)** psophometer. **-muster** noise pattern. **-spiegel** noise level. **-unterdrückungsstromkreis** noise reduction, squelching, suppression or silencing circuit. **-verdeckung** overriding of noise.
gerecht, flucht- truly aligned or

flush (with). form- true to form or shape.

gerechte Farbstufe, empfindungs- subjective chromatic scale value.

Gerechtigkeit justice, equity, equitableness.

gereckte Probe stretched specimen.

gereifelt grooved, milled, knurled, fluted.

Gericht, Bezirks- district court.

gerichtet, einseitig- unidirectional(ized).

gerichtet(e) Empfangsspannung directional signal or incoming potential. -(er) Lautsprecher directional loudspeaker. -(es) Licht parallel light (rays).

gerichtlich belangen bring legal or judiciary action or proceedings against.

Gerichts-barkeit erster Instanz original jurisdiction. -befehl writ or warrant. -beisitzer assessor, associate judge. -bezirk court circuit, court district. -hof, zuständiger court of competent jurisdiction. -stand competency of a court, jurisdiction, venue. -verhandlung hearing, trial.

gerieft, geriffelt grooved, milled, knurled, fluted.

geringhaltig low standard, worthless, of low content.

Gerippe, Kristall- crystal skeleton, skeletal c., crystallite.

gerissene Kurve dotted-line curve, broken-line c. or graph.

Gerüst structure, skeleton, frame, rack, scaffold. Trocken- drying frame or rack.

Gerüstkran gantry crane.

gesamt total, entire, aggregate, complete, whole.

Gesamt-ansicht total view, general v., assembly v. -heit totality, entirety, assembly, generality, whole. -spin total spin. -verhalten general behavior. -wirkungsgrad overall efficiency, total e., commercial e.

Geschädigter aggrieved, injured party.

gescheckt cross-hatched.

Geschehnis action, event, occurrence, happening, phenomenon.

geschichtet, zwischen- inter-stratified.

geschichtete Kathode coated (oxide) cathode, photo- c. with light-sensitive layer.

Geschicklichkeit skill, cleverness, art, dexterity, expertness.

geschlagene Saite percussed string, p. chord.

geschlitzter Stöpsel split plug.

geschlossen, in sich closed upon itself, self-contained.

geschlüsselter Funkspruch coded radio message.

geschmeidig pliable, flexible, supple, soft, ductile, malleable.

Geschoss-bugwelle nose wave of a projectile. -geschwindigkeit projectile velocity.

Geschütz-aufsatz optical instruments for artillery. -zielender gunlayer

Geschwindigkeit (in Mechanik) velocity (being, in mechanics, the analog of current in electricity).

Geschwindigkeit, ungeordnete velocity of agitation. Abtast- pickup velocity, p. rate (phonograph). Anfluss- velocity of approach, afflux v. (of fluid). Apparat zur Bestimmung der- velocity selector. Austritts- exit, efflux, muzzle or emergence velocity. Boden- ground speed (of airplane). Durchschlags- propagation speed (of fluids in pipes); rate of insulation breakdown. Eigen- air speed (of airplane). Einfall- bombardment velocity. Eintritts- approach or inlet velocity. Flug- rate of flying speed, traveling velocity (electrons, etc.). Fortpflanzungs- speed of propagation or transmission. Fortschreitungs- speed of progression or travel, forward speed. Grund- ground speed.

Gruppen-envelope velocity, group v. Kriech- rate of creep (of metal). Masseteilchen- particle velocity. Mundungs- muzzle velocity. Phasen- phase speed, p. velocity, wave v. Rückkipp- flyback speed, time-base unlock s. Teilchen- particle velocity (ac.) Umfangs- circumferential speed, peripheral s. Wanderungs- rate of migration, crawl or creep, m. velocity.

Geschwindigkeits-abtastung variable-speed scan. -amplitude volume current (ac.). -fehler distortion due to synchronous speed differences. -feld velocity gradient, v. field pattern.

geschwindigkeitsgesteuerter Elektronenstrahl velocity-modulated electron beam or pencil (klystron, etc.).

Geschwindigkeits-gitter accelerator grid. -messer speedometer, tachometer, tachygraph, (when recording). -messer, Wind- -anemometer. -potential velocity potential.-raum momentum space, velocity s. -schwankungen wowows (when intensity pulsation is up to 6 cycles per second); flutter (when 6-30 cps.); gargle (30-200) and whiskers (over 200). -steuerung velocity modulation. -verteilung velocity distribution, d. in energy. -wechselgetriebe change-speed gear.

gesehnte Wicklung short-pitched winding, chord w.

Gesenk die, block forging d., swage. -schmiede drop forge.

Gesetz, Inkrafttreten eines -(es) commencement of an act or law.

Gesetz, Ähnlichkeits- similarity principle. Brechungs-, Snellius'sches Snell's law of refraction. Erhaltungs- law of conservation. Flächenerhaltungs- principle of conservation of areas. Impulserhaltungs- principle of conservation of momentum. -,Lambert'sches emission law, Lambert l. Teilungs- law of partition. Zufalls- law of probability, l. of chances.

Gesetz-blatt, Reichs- imperial law journal, i. gazette. -buch code, statute book.

Gesetzesauslegung law interpretation.

gesetzlich lawful, legal, licit, legitimate, statutory.

gesetzliches Hindernis statutory bar.

gesetzmässig according to law or statutes, conformable to (natural) law, in a regular or legal way.

Gesetzmässigkeit lawfulness, legality, legitimacy, procedure according to statutes.

gesicherte Mutter lock nut, jam n.

Gesichts-eindruck visual impression. -feld field of view, f. of vision. -feldblende field stop. -sinn vision, faculty of seeing. -strahl visual ray. -wert face value, facial brightness (of a screen performer). -winkel visual angle, optic a., facial a.; camera or viewing angle.

Gespensterstrom ghost current.

Gesperr, Anker- anchor escapement. Malteserkreuz- cross wheel,(cam and) star w. of Maltese cross, Geneva stop or intermittent movement.

Gespinst spun fabric, yarn goods.

Gesprächspausen intervals of no speech.

gespritzt squirted, extruded; sprayed, atomized. -(er) Glühfaden squirted filament. -(es) Metall die-cast metal.

gestaffelt staggered, echeloned, graded.

Gestalt, Kugel- spherical shape, bulbous or ball form.

Gestaltfestigkeit form stability,

non-deformability.
gestaltlos amorphous, non-crystalline, formless.
Gestalts-fehler, Kugel- spot size distortion. -veränderung change of form, change of shape, deformation.
Gestaltung formation, configuration, form, shaping, fashioning, organization, state. Doppel- dimorphism. Neu- re-organization, re-arrangement, modification, re-formation.
Gestell frame, stand, rack, bay, shelf, mount. Verstärker- amplifier rack, repeater bay. Wässerungs- washing rack (phot.).
gesteuerter Oszillator, leitungs- line-controlled oscillator.
gestochene Schärfe microscopic sharpness (phot.).
gestreckt straight, stretched, linear.
gestreut, elastisch und unelastisch elastically and unelastically scattered. kern- deflected by nucleus.
Gesuch application, petition. -steller applicant, petitioner.
geteilter Ring split ring, segmented
getreu, amplituden- of equal amplitude, with a. fidelity. form- true to form, undistorted, orthoscopic (opt.) klang- orthophonic, of high fidelity.
getreue Wiedergabe faithful reproduction, fidelity in r., good definition of image (with fine detail formed by a lens, telev. or facsimile receiver or a c.-r. oscilloscope).
Getriebe, Ausgleich- differential gear. Kegelräder- bevel gearing, mitre-wheel g. Planeten- planetary gearing, epicyclic gear train. Reduktions- reduction gear, stepdown gear. Schraubenzahn- helical gear, spiral g., worm. Umlauf- planetary gear, sun and planet g. Winkelverzahnungs- double helical gear.
Getriebelehre kinematics.
Getter-abschiessen set off or cause vaporization of getter. -pille getter tab, g. patch, g. pill. -spiegel getter patch, getter film.
gewährbar allowable, grantable.
gewähren (ein Patent) allow a patent, grant a p., issue a p.
Gewebe tissue, texture, textile, fabric, web, cloth, gauze, netting. Draht- wire gauze, w. cloth, w. mesh. Hart- indurated fabric, impregnated f.
Gewebegebilde, winzige ultramicrons (of eye tissue).
gewellter Einspannrand corrugated suspension means (of cone.)
gewerblich industrial, concerning trade or profession.
Gewicht, Atom- atomic weight. Gegen- counterpoise, counterweight, counterbalance. Mol- molar or gram-molecule weight.
gewichtsanalytisch gravimetrical.
Gewichtsprozent percent by weight. -verhältnis proportion by weight.
Gewinde thread, winding, coil. -schneiden tapping, thread cutting.
Gewinde, Aussen- male screw thread, outside s.t. Flach- square thread. Innen-, Mutter- female or internal screw thread. Rechts- right-hand screw thread.
Gewinde-lehre thread gage. -steigung pitch of a screw thread. -stift headless screw, grub s. -taster thread caliper.
gewölbt arched, arcuate, domed, convex. -(e) Lamellen, kugelförmig dished laminae.
gezackte Linie jagged line, serrated l., notched l., dentated l., dented l.
gezupfte Saite plucked string or chord.
Gier, Elektronen- electron affinity.

Gierungsmesser yaw meter.
giessen coat (a film).
Giesser, Gelb- brass founder.
Giesserei, Kunden- jobbing foundry.
Giessform casting mold.
Giltigkeit validity, lawfulness, legality, life (of a patent).
Gipfel, Wellen- peak, crest or hump of a wave.
Gipfel-hohe ceiling. **-spannung** peak voltage, crest v.
Gips plaster of Paris, gypsum, calcium sulfate.
Gisklappe G sharp key.
Gitter, automatisch vorgespanntes automatic- or self-biased grid. **-,freies** floating grid, free g. **-,weitmaschiges** open-meshed grid, wide-meshed g.
Gitter, Anodenschutz- screen grid, plate shield. **Beschleunigungs-** accelerator grid (of c.r. tube); screen g. (of tetrode). **Beugungs-** diffraction grid. **Brems-** suppressor grid, cathode g. (in pentode tube). **Dichte (regulier)-** (charge) density grid, density control g. **Echelette-** echelette grating (opt.). **Einheits-** unit lattice. **Fang-** suppressor grid, cathode g. (in power pentode), interceptor g. **Geschwindigkeits-** accelerator grid. **Grund-** fundamental lattice. **Haupt-** parent lattice. **Hohl-** concave grating. **Kathodenstrahlröhren-** grid or shield, Wehnelt cylinder, of a cathode-ray tube. **Konkav-** concave grating. **Kreisbogen-** arcuate grid. **Kreuz-** cross grating. **Kristall-** crystal grating, surface lattice. **Maisch-** stirrer, rake (brewing). **Plan-** plane grating. **Raum-** space-charge grid; space lattice, "raumgitter". **Reflexions-** diffraction grating with parallel reflecting surfaces. **Saug-** space-charge grid. **Schicht-** stratified lattice. **Schirm-, Schutz-** screen grid; shield g. (in thyratron). **Stau-** suppressor grid, baffle g. **Steg-** grid in which all wires are parallel to axis and surround cathode cage-fashion. **Strich-** ruled grating. **Strom-** atomic lattice. **Stufen-** echelon grating. **Zug-** positive grid, space-charge g.
Gitter-ableitung grid leak (resistance). **-absorption, Grund-** fundamental lattice absorption. **-antenne** parasol-type antenna. **-aufstellung, Rowland'sche** Rowland mounting of concave grating. **-aufweiterung** expansion or increase in lattice spacing. **-aussteuerung** grid (voltage) swing, grid sweep. **-beeinflussung, -besprechung** grid control, grid modulation. **-block** grid condenser; staircase-like structure of echelon grating. **-blockierung** negative biasing of grid, g. cut off. **-durchgriff** grid transparency, g. controllance, g. penetration factor. **-ebenenabstand** inter-lattice-plane distance. **-einheit** lattice unit, unit cell, elementary cell. **-einsatzpunkt** grid current point, point of incipient grid current flow. **-elektronen** lattice electrons. **-filter** lattice filter. **-furchen** rulings or ruling grooves (of diffraction grating). **-geister** spectral grate ghosts. **-gleichrichtung** grid-current rectification or detection (in audion valve). **-haltedraft** grid stay wire, g. supporting w. **-holm** grid stay or supporting member. **-interferometer** grating interferometer. **-ion, Zwischen-** interstitial ion. **-kathodenstrecke** grid-cathode path. **-konstante** grating constant, g. space; lattice c. **-loch**, lattice vacancy, l. hole. **-masche** grid mesh. **-mast** lattice pole, girder p. **-mikrophon** grille-type microphone. **-mittenabstand** distance between centers of successive rulings (in diffraction grating); grating constant.

-modulation grid modulation, g. control (Brit.) -nebenschluss grid leak (resistance). -platz lattice place, l. point. -punkte lattice points. -rohr, Doppel- bigrid valve, twin-grid tube, four-electrode t. (pliodynatron, tetrode, etc.). saum grid skirt (in telev. tube). -schwingung (Kristall) lattice vibration (cryst.) -spektrograph, Vakuum- vacuum grating spectrograph. -steg grid stay; solid portions forming g. meshes; grid strip used in lieu of wire. -steuerung grid (voltage) sweep, g. swing, g. excursion, g. excitation, g. control (in ignitron, thyratron, etc.); g. modulation. -störungen lattice distortions, l. dislocations (cryst.). -strebe grid stay, g. supporting means. -strich grating line, ruling or ruling groove of a grating.

Gitterstrom-aufnahme drawing or taking of current by grid. -aussteuerung grid current swing or sweep. -einsatzpunkt grid current point, point of incipient grid current flow or where g. begins to draw current. -gleichrichtung grid current detection, cumulative grid detection. -punkt grid current point, point of incipient current flow at g.

Gitter-verschiebung lattice distortion, l. dislocation. -vervielfältiger mesh multiplier. -vorspannung, automatische automatic or self-biasing of grid. -wickeldrahte grid coil wires.

Glanz, Halbmetall- submetallic lustre.

Glanz-einfall grazing incidence. -garn glace or glazed cotton, lustrous or mercerized yarn. -messer gloss meter, glossimeter. -winkel glancing angle.

Glas s.a. Linse, Objektiv

Glas, berichtigendes correcting lens. -,spannungsloses strainless glass, strain-free g. -,ungefasstes rimless lens.

Glas, Aufsteck- slip-on lens segment. Brillen- spectacle glass. Durchzieh- slide (micr.) Farb- colored glass, color filter, stained g. Hart- pyrex glass. Lese-, viereckiges oblong reading glass. Milch- frosted glass, opal g., light-differing g. Muschel- globular lens. Objekt- slide, mount (micr.). Rauch- smoked glass, tinted gl. Sammel- converging lens; preparation tube, specimen t. Solin- crown glass. Spiegel- plate glass. Trüb- opal glass, frosted g., light-diffusing g. Überfang- flashed opal glass.

glas-ähnlich, -artig glasslike, glassy, vitreous.

Glas-bildner vitrifier. -blase glass bubble, seed (when small), air bell (when of irregular shape). -faden glass thread, thread-like defect of glass. -federmanometer spoon-type glass manometer. -fehler (cf Schlieren) glass defects or flaws; glass striae or chords (when heavy), bubbles (seed when small); airbell (when of irregular shape), stones; crystallization bodies, cloudiness, strain. -flächenspiegelung glare or reflection of glass surfaces, specular r. -galle glass gall, sandiver.

glasig glassy, vitreous.

glasklarer Film pellucid film stock (having maximum transmittance).

Glas-kolben glass bulb. -plattenstaffeln (Michelson) echelon spectroscope. -satz glass batch, composition or charge. -schaum glass gall, sandiver. -scheitel lens vertex. -schlieren schlieren, striae or chords (when large), reams (when broken by rolling into fine bands). -schliffverbindung ground-glass joint.

-schmutz glass gall. -schneider, Roll- wheel-type glass cutter. -schweiss glass gall. -spannung strain or internal tension of glass. -stöpsel glass stopper. -trübung glass cloudiness.

glastechnisch glass technological.

Glas-Überfangschicht glass lining. -vergütung refining or treating of optical glass (e.g., by producing low-reflectance coating, dimming, etc., as by Bausch and Lomb magnesium-fluoride anti-reflection film on glass).

Glasur, halbmatte half-matt glaze.

glätten smooth (ripples, fluctuations or pulsations of currents); make level, even or uniform, stabilize.

Glattheitsprüfer smoothness meter or tester, glossimeter (measures ratio of reflection to total light), surface analyzer.

Glättung smoothing (ripples of current, potential, etc.). Spannungs- voltage stabilization, v. smoothing.

Glättungs-kondensator smoothing condenser or capacitor. -röhre glow-tube stabilizer.

Glauben, in gutem- in good faith, bona fide.

Glaubwürdigkeit credibility, trustworthiness, credit.

gleichachsig co-axial, equi-axial.

Gleichbeanspruchung static loading, constant l.

gleich-bedeutend equivalent, tantamount, synonymous. -belastete Phasen balanced phases.

Gleichen, Kurs- loxodromes, rhumb lines. Missweisungs- isogenic lines, isogonals.

gleichentfernt equidistant, equally spaced (apart).

Gleichfeld constant or steady field

gleichförmig uniform, steady, smooth, even, homogeneous.

Gleichgang unison, synchronism.

gleich-gerichtet equidirectional, unidirectional, equi-axial, of like orientation. -gestaltet isomorphic, isomorphous

Gleichgewicht balance, equilibrium, poise, trim (of airplane), stability. -, radioaktives radioactive equilibrium (being either transient or secular). Brücken-bridge balance. Druck-, osmotisches isosmotic equilibrium.

Gleichgewichts-fehler unbalance, imbalance, lack of balance or equilibrium. -lehre statics. -potential equilibrium potential (of mosaic). -verhältnis equilibrium ratio. -zustand state of equilibrium or balance, stable or stabilized state.

Gleichheit state of balance or equality, likeness, sameness, uniformity, constancy. Druck-, osmotische isosmotic equilibrium. Phasen- in-phase state, coincidence of phases. Spiegel- mirror symmetry,

Gleichheits-photometer, Lummer Brodhun L.-B. contrast photometer, total-reflection p. -zeichen equality sign.

Gleich-klang unison, consonance, resonance. -lauf, lokal geregelter local synchronizing (e.g., by a tuning fork).

gleichlaufende Kondensatoren gang(ed) condensers, synchronized capacitors.

Gleichlauf-fehler, Auflaufen der compounding or integration of errors or differences in synchronism. -impuls synchronizing impulse, corrective i. -signal, -zeichen synchronizing pulse, sync signal; tripping s.

Gleichluftstrom direct flow of air, d.c. air f.

gleich-machen equalize, balance, stabilize, make even or uniform, steady. -mässig uniform, steady, constant, smooth. -namig like, similar (of poles). **-phasig** co-phasal, in phase, (equi-) phased. **-richten** rectify, detect, demodulate.

Gleichrichter, linearer linear rectifier, detector or demodulator. **-,quadratischer** square-law rectifier or detector.

Gleichrichter, Doppel(weg)- full-wave rectifier. **Einweg-,Halbweg-** half-wave rectifier. **Hochvakuum-** kenotron, electron-type rectifier. **Kontakt-** contact rectifier, crystal r., natural r. **Pendel-** vibrating rectifier, tuned-reed r. **Schwingkontakt-** vibrating-reed rectifier. **Sperr-** barrier-film rectifier, blocking-layer r., metal r. electronic r., cuprox r. **Trocken-** dry rectifier, metal r., copper r., cuprous-oxide r., selenium-oxide r. **Verstärker-** amplifying detector. **Vollweg-** full-wave rectifier. **Zweifachdioden-** duodiode rectifier.

Gleichrichter-pille oxide or metal rectifier of reduced size (e.g., Westector). **-rückzündung** back-firing, arcing-back (of Hg rectifier), backlash (imperfect rectification). **-voltmesser** rectifying voltmeter. **-zelle** rectifier cell, barrier-layer c.

Gleichrichtung beider Halbwellen full-wave rectification. **-einer Halbwelle** half-wave rectification **-,unvollständige** backlash

Gleichrichtung, Anoden- plate-current rectification or detection. **Gitter-** grid-current rectification or detection.

Gleichrichtung, Volumen- volume rectification.

gleich-schenklig isosceles. **-sinnig in Reihe** series aiding.

Gleichstrom, Luft- current of breath (in sound recording).

Gleichstromempfänger d.c. electric radio set.

gleichstrom-vormagnetisierte Spule d.c. controlled saturable or three-legged reactor or coil. **-überlagert** superposed on d.c., superposed with d.c.

Gleichtakt-aufnahme in-phase recording (on film). **-sprossen-Doppelschnürspur** variable-density double squeeze track. **-sprossenspur** single variable-density track. **-sprossenschnürspur** variable-density squeeze track. **-verstärker** in-phase amplifier.

Gleichteilung equipartition.

gleich-temperiert isothermal, uniformly heated. **-tönend** unisonant, consonant.

Gleichung, eine - nach n auflösen solve an equation with respect to n.

Gleichung, quadratische quadratic equation. **Ausgangs-** starting equation, initial e. **Bedingungs-** equation of condition. **Bestimmungs-** defining equation. **Farb-, dreigliedrige** three-color equation, 3-stimulus e. **Kontinuitäts-** equation of continuity. **Summen-** summation equation. **Wiederhall-** Eyring formula. **Zustands-** equation of state.

Gleichungs-ansatz laying down a formula. **-determinante für Leitung** determinant for line equations. **-familie** equations grouped in families.

Gleichverteilungssatz law or principle of (energy) equipartition (of Maxwell-Boltzmann).

gleich-vielfach equimultiple. **-warm** isothermal, uniformly heated.

Gleich-wellenbetrieb chain or network broadcast (from a key station, on one wave), operation of stations on one w. **-wertigkeit** equivalence.

gleich-winklig isogonal, equiangular, equi-angled. **-zeitig** simultaneous, synchronous, isochronous, contemporaneous, contemporary, co-incidental.

Gleisweg rail track.

Gleit-bahn chute (of film), slide, slideway, shoot. **-drahtbrücke** slide wire bridge. **-element** glide element.

Gleiten, bildsames plastic deformation.

gleitender Ton sliding note, glissando.

Gleit-fläche glide plane, slip p. (cryst.); sliding surface, slide. **-flug, Seiten-** side slip (airplane landing). **-funken** slide spark, creepage s. **-Kontakt** sliding contact (with self-cleaning or self-wiping action). **-lager** plain bearing, slide b. **-lineal** slide rule, s. bar. **-linien** slide or slip bands or lines. **-modulus** modulus of elasticity (in shear), torsion m., rigidity m. **-richtung** slip direction (cryst.). **-schuh** pressure pad, p. guide. **-sitz, Schlicht-** plain sliding fit.

Gleitung slip (cryst.). **Zwillings-** twin slipping.

Gleitungskoeffizient slip coefficient.

Gleitweg glide path, landing curve, l. beam. **-, äquipotentieller** isopotential glide path.

Gleit-wegbake glide path beacon. **-winkel** glide angle.

Glied member, term (math.); limb, joint, link; section or mesh (filter), element.

Glied, L,T,H, und π (oder II)- L,T,H, and pi type section or mesh (of a filter, attenuator or network). **Ableitungs-** shunt element. **Brücken-**lattice section. **Dämpfungs-** attenuation network or mesh. **Kettenleiter-** network mesh. **Kreuz-** lattice section (of filter or network). **Längs-** line arm, series element. **Mittel-** intermediate member (of a series). **Quer-** cross arm, shunt element. **Regel-** regulating attenuator. **Schwächungs-** attenuator. **Sperr-** suppressor. **Spin-** spin term. **Übergangs-** transition type, transition member. **Verbindungs-** connecting link. **Wechselwirkungs-** inter-action term. **Zwischen-** intermediate member, connecting m., link, intermediate.

gliedern arrange, organize, classify, divide methodically (say, contents of a book, etc.); articulate, joint. **verschieden gegliederte Bildteile** picture portions of different nature, make-up or organization.

Gliederung organization, arrangement; jointing, articulation.

gliedrig, vier- four-membered, tetragonal, quadrinomial (math.) **-(e) Farbgleichung, vier-** four-color equation.

Glimm-entladung glow-discharge (spontaneous or unassisted, in glow-tube), brush d. (when occurring at sharp points), corona. **-fadenlänge** length of (neon) glow column (in tuning indicator).

Glimmlampe gaseous-conduction tube, glow-tube (e.g., neon tube), flashing lamp (in sound recording). **Flächen-** neon tube with flat plate-shaped cathode, plate neon tube or lamp. **Polsuch-** polarity-finder glow-tube.

Glimmlampen-gerät, Universal- universal neon-tube tester. **-kippschaltung** neon lamp time base circuit scheme. **-schallaufnahme** glow-lamp sound recording (Aeolight).

Glimmlicht-anzeiger flashograph-type indicator, Tune-A-Light tuning indicator. **-gebiet, Vor-** sustaining voltage range, prephotoglow region (arises at a voltage slightly below striking voltage).

Glimmröhre, Abstimm- tuning indicator glow-tube (e.g. flashograph). **Amplituden-** glow-tube amplitude indicator, resonance or tuning indicator (e.g., flashograph).

Resonanz- resonance indicator glow-tube (flashograph).

Glimm-säule light column, glow c. (of tuning indicator). -spannung glow point or potential, breakdown p. (indicates incipient glow-discharge in grid-glow tube); anode p. (at which glow-discharge begins, in photocell), stopping p. (lies slightly below glow p., in photo-glow or grid-glow t., and is the value to which p. must be dropped to stop glow-d. once it has started). -(strecken)spannungsteiler glow-tube potentiometer. -summer glow-tube type buzzer oscillator (produces relaxation waves), neon time base. -verlust corona loss. -zelle photo-emissive gas-filled cell (operating at a potential slightly below critical discharge potential), photo-glow tube.

Glocke globe, bell, jar, glass; receiver (of air pump).

Glocken-filter bell-jar filter -klöppel bell hammer, bell striker. -magnet bell-shaped magnet. -spiel carillon (set of bells), orchestra bells. -trichter bell funnel.

glühen heat, incandesce, anneal, glow, fire, ignite.

glühelektrisch thermionic.

Glüh-elektronenabgabe thermionic emission, Edison or Richardson effect. -elektronenentladung thermionic discharge. -faden, -biegsamer ductilized filament. -faden, gespritzter squirted filament. -fadenpyrometer disappearing-filament pyrometer, optical p., monochromatic p. -kathodengleichrichter mit Edelgas tungar (argon-tungsten) type rectifier. -körper incandescent body or mantle, Welsbach mantle. -rückstand residue on ignition. -span mill iron, hammer scale.

Glutmesser pyrometer.

gnomonische Karte gnomonic chart (straight-course chart).

Gold-bergkeil Goldberg wedge, neutral(grey) w. -blatt gold leaf, g. foil. -blattelektrometer gold-leaf electroscope or electrometer. -glimmer yellow mica.

goldhaltig auriferous.

Goldquarz auriferous quartz.

Goniometer, Eisen- iron-cored goniometer. Reflexions- reflecting goniometer.

Grabstichel graver, burin.

Grad, vom ersten linear.

Grad, Aussteuerungs- modulation percentage, percentage modulation; carrier amplitude (in tele-transmission). Blähungs- expanding property, swelling p., imbibition power. Durchlass- (von Wasser) transmission factor (of water), radiation through water. Winkel- degree of angle, radian (unit of circular measure of angle).

Gradation contrast.

Gradationskurve, Steilheit der gamma value or slope of response line.

Grad-bogen graduated arc, bow, protractor. -netz graticule, reticule, reticle, map grid.

Gramm, Atom- gramatom.

Grammophon, Licht- sound-track film player (without pictorial actions).

Granalien shot, granulated metal.

Graphitbelag Aquadag coating.

Grätzschaltung Grätz (full-wave) rectifier circuit organization (four electrolytic r. per phase in a bridge scheme).

Grau-keil neutral (grey) wedge, Goldberg w. -keilstufe intermediate shading value (of wedge). -schleier grey fog (phot.). -strahlung grey radiation. -stufe intermediate tonal or shading value (between white and black, of picture).

Gravarfilm engraved sound film, embossed groove recording.

Gravitationsfeld gravitational field.

Greifer claw (of in-an-out movement or feeder mechanism); pull-down (of projection printer). Bild- threading mechanism, in-and-out claw. Justier- pilot claw (m.p.).

Greifer-maschine step-type printer. -mechanismus pull-down claw mechanism. -spitze dowel pin (film feed). -stift film feeder pin, pilot p., registration p.

Greifloch finger hole.

Grenz-bedingung limiting condition, critical c., boundary c. -durchmesser limiting diameter (of dielectric line).

Grenze limit, boundary, bound, end, limitation, confines, demarcation. Grenze, Empfindungs-, des Ohres loudness contour of ear, auditory sensation area. Entzerrungs- frequency limit of equalization. Ermüdungs- fatigue limit. Loch- transmission range of band-pass between cut-off points. Scharfenfeld- camera lines. Schmerz- threshold of feeling.

Grenz-fall limiting case, critical case. -fläche interface, boundary layer, surface of contact. -flächenpotential interface potential. -flächenspannung interfacial tension. -flächenwinkel interfacial angle (dihedral between two crystal faces). -frequenz cut-off frequency, limiting f. -gefälle limiting slope, l. gradient. -lastwechselzahl limiting number of load alternations. -linie boundary line, demarcation l. separation l., limiting l., junction l. -schicht interface, boundary layer. -strahl border line ray, infra-Roentgen r., grenz r. -wellenlänge threshold wave length. -wert limiting value, critical v., saturation v. -winkel critical angle, limiting a. (of refractometer).

Gries, Stör- low noise in picture background due to thermal effect, Brownian movement, mechanical properties of tubes and low-frequency fluctuations of local emission density, i.e., shot effect, etc.

Griff, Durch- grid transparency, penetration factor, transgrid action, gain reciprocal 1: µ Kordel- milled knob, knurled k. Rück- inverse grid transparency.

Griffbrett finger board (of string instrument).

Griffel style, stylus, (slate) pencil. -stopfen stopper with thumb piece.

grobfädig coarse-threaded, c. filamentous.

Grob-gefüge macro-structure, coarse or gross s. -höhenmesser (cf. Fein-) barometer measuring actual atmospheric pressure reduced to sealevel.

grobkristallinisch coarsely crystalline.

Grob-sicherung arrester of spark-gap type. -struktur reduced (or reduction of) resolution. -strukturanalyse gross, coarse or macro-structure analysis.

Grossaufnahme (big) close-up (m.p.). Fernbild- large viewing screen projection (for public entertainment).

Grösse quantity, magnitude, entity, size, dimension, format (of pictures, etc.). -, abgeleitete subsidiary quantity, derived q. -, konjugierte, komplexe conjugate, complex quantity. Wirkungs- action quantity, a. magnitude.

Grössenlehre mathematics; geometry.

grössenmässig dimensional(ly), as regards size or quantity.

Grössenordnung order of magnitude.

Gross-flächenlautsprecher power loudspeaker, (public) address l., announce, call or paging l. system. -funkstelle high-power radio station, long-distance signal station. -kreis (peilung) great-circle (used in bearing calculation).

-kreisrichtung great-circle direction or course. -lautsprecher power loudspeaker, (public) address loudspeaker (system). -projektion large picture projection (with magnascope). -schallübertragungsanlage (public) address loudspeaker system, announce, call or paging l. system. -struktur coarse structure, gross s., macro s.

Grösstwert maximum (value), maximal or crest v.

Grossumfangkopie "hi-range" or wide-range print.

Grübchen-bildung, Gruben- pitting, formation of pits.

Grund, Ätz- resist, protective coating. Ausschliessungs- reason for exclusion.

Grund-bande fundamental band. -farbe primary, ground or priming color. -fläche basal surface, base; area. -flüssigkeit suspending liquid. -form primary, basic or fundamental type or form.-geräusch (back)ground noise.- -stromkreis (back) ground noise reduction circuit, squelching, suppression or silencing circuit.-geschwindigkeit ground speed. -gitterabsorption fundamental lattice absorption. -gleichungen, Maxwell'sche Maxwell electromagnetic equations. -harmonische fundamental frequency. -helligkeit background, mean or average shading component, d.c. component. -körner fundamental substance, parent s. -kurs true course, head-on c. -linie, Anflug-center line of approach sector. -masse ground mass. -material matrix, elementary material, base m., ground m.-nachbildung basic network.-niveau background level (or average illumination, telev.). -platte mounting plate, bed p., base, chassis. -punkt cardinal point (of a lens). -richtungswinkel base angle (gunnery). -riss ground plan, floor p. -sätzliches fundamentals, rudiments, basic or elementary facts. -scheitelbrechwert primary vertex refraction. -schicht primary or fundamental layer. -schleier inherent fog, ground f. (of emulsion resulting in background noise). -schwingung fundamental or first harmonic oscillation or vibration. -stimmen foundation stops (organ). -teilchen smallest particle, fundamental p., atom. -ton pitch, keynote, fundamental or primary tine or frequency; back-ground (noise). -zug leading or salient feature or characteristic, basic trait. -zustand ground state (of an atom). -zustand, Zurückfallen in return or re-transition to lowermost energy level.

Grünauszug green record.

grünempfindlich green sensitive.

Gruppe, angeregte exciton (in dielectric breakdown).

Gruppen-erscheinungen group phenomena (cryst.). -frequenz group frequency, number of wavetrains per second, (wave) train frequency, sparking frequency. -geschwindigkeit envelope velocity, group v.; group retardation. -schaltung series-parallel connection. -wähler group selector (aut. telephony).

Gruppierung, Elektronen- phase-focusing (in beam tubes), bunching.

Guckloch peephole, bezel (in radio set panel).

Guitarre guitar.

Gültigkeit eines Patentes anfechten challenge validity of a patent, plead or argue invalidity of a p.

Gummi, Hart- ebonite, hard rubber, vulcanite.

Gummi-andrückrollen rubber-tired presser or pad rollers. **-chromverfahren** bichromated gum process. **-dichtung** rubber gasket, r. packing. **-drucklack** gum printing varnish. **-harz** gum resin. **-propf** rubber stopper. **-puffer** rubber grommet. **-stopfen** rubber stopper. **-überzug(auf Rollenlaufbahn)** rubber-tread(of pad roller).

günstigst optimal, most favorable, most propitious.

Gurtband belt strap, b. band.

Gürtel belt, girdle, band, zone. **Klima-** climatic belt.

Guss, Hart- chilled casting. **Spritz-** (pressure) diecasting, feeder casting, die molded casting.

Guss-aluminium cast aluminum. **-blase** flaw or defect in a casting. **-haut** casting scale, c. skin. **-masse** pourable compound, dope.

Gutachten expert opinion, judgment, decision, verdict.

Güte quality, merit, grade, "goodness"; legibility or definition (in telegraphy, television, etc.) **Bild-** (figure of) merit or quality of pict. **Übertragungs-** transmission performance (rating).

Güte-wert, -zahl figure of merit, quality (Q) factor (of coil, etc.)

gutschliessend tight-fitting, tightly sealed.

G. W. (Gruppenwähler) group selector (in automatic telephony).

H Brückenschaltung, Entzerrungskette mit balanced H-type attenuator network.

H Strahlen H rays (consisting of H particles, positive hydrogen p. or protons).

H & D Kurve Hurter & Driffield curve (phot.).

haarartig hair-like, capillary.

Haardraht capillary wire, Wollaston w.

haar-faserig filamentous. -förmig capillary, hair-shaped.

Haar-riss craze, micro flaw, (hairline) crack, capillary fissure. -rohr, Heber- capillary syphon, c. tube syringe. -rohranziehung capillary attraction.

haarscharf very (or microscopically) sharp, precise or correct.

Haarzellen hair cells, capillary c.

Habannröhre type of split-anode magnetron; dynatron oscillator (though the latter strictly is predicated on secondary emission).

hachieren hatch.

Hacker chopper.

Hafenbefeuerung, Flug- airport beacon service.

haftbar liable, responsible, answerable.

haften adhere, stick, cling.

Haft-spannung adhesive stress. -vermögen adhesive power, adhesion.

Hahn (stop) cock, tap, faucet, cutoff (vacuum pump).

Hahn, Kreuz- four-way cock. Licht- electromagnetic mirror vibrator (with oil damping, for variable-area recording). Schraubquetsch- screw pinchcock. Tropf- dripping cock. Vakuum- vacuum tap. Viel-wege- multi-way stop cock. Vier-wege- four-way cock. Wechsel- change cock.

Hahn-fett tap grease, joint g., vacuum g. -kegel, -küken plug of a cock. -stopfen cock stopper.

Häkchen little hook, clasp.

Haken hook, catch, fastening. Band- bridle wire, tie hook. Sperr- pawl, catch, latch, trip, dog, click, detent, ratchet. Wider- barb.

Hakenharfe hook harp.

Halb-achse semi axis. -achsenverhältnis ellipticity coefficient. -art sub-species. -bildentfernungsmesser split-field coincidence range finder. -blende half stop, scale. -dunkel twilight, semi-darkness.

halbdurchlässig semi-permeable, semi-transparent.

Halbebene semi-plane, half-p., H-plane.

halbeirund semi-oval.

Halb-erzeugnis, -fabrikat semi-finished product, intermediate p. -flächner hemihedron.

halbgeschränkter Riemen quarter-turned belt.

Halbheit imperfection, half-measure, half-way step, "half-cooked" condition.

halbieren halve, bisect, disect, cut in half, part.

Halbierende, Winkel- bisector (of angle).

Halbierungslinie bisector, bisectrix (line).

halbklassische Bildkraftberechnung semi-classical image-force calculation.

Halb-kokung semi-coking, partial carbonization. -kreis semi-circle, hemicycle. -kugel hemisphere. -kugellinse semi-circular lens, hemispherical l. **-längsglied, Abschluss eines Filters durch** mid-series termination of a filter. **-leiter** semi-conductor, poor c.

halbmatt half matt.

Halb-mattglasur half-matt or semi-matt glazing. **-metallglanz** sub-metallic lustre. **-öffnungswinkel** semi-apertural angle. **-periode** semi-period, half-period, half-cycle. **-prisma** hemi-prism (cryst.) **-querglied, Abschluss eines Filters durch** mid-shunt termination of a filter. **-raster, Ineinanderpassen der** interlocking or meshing of fractional scans (in interlaced scanning). **-raum** half-space.

halbrund half-round, semi-circular, hemi-spherical.

Halbschatten half-tone, half-shade, partial shadow, penumbra. **-apparat** half-shade apparatus, half-shadow analyzer. **-grenze** penumbra boundary. **-kompensation** penumbral compensation.

Halbton semi-tone, semi-note; sharp (mus.) **-bild** half-tone picture, mezzo tinto (in black and white, with grey intermediate shading values).

Halbtotale close-medium shot, medium shot.

halb-tragbar semi-portable. **-versenkt** semi-sunk, semi-recessed. **-versilberter Spiegel** beam splitter (densitometer).

Halb-vokale liquid consonants, semi-vowels (liquidae). **-weggleichrichter** half-wave rectifier.

halbwegs fairly, substantially, half-way.

Halb-weltschicht half-value layer. **-wellenpotential** half-wave potential.

Halbwert-breite half-width, w. at half maximum intensity (opt., spectral line and band). **-dicke** half-value thickness or layer. **-druck** half-value pressure. **-punkt** half-maximum point. **-zeit** half-value life, half-l. period.

Halbwinkel half angle.

halbzahliger Spin half-integral spin, half-odd s.

Halb-zähligkeit half integer. **-zeit** period of half life or change, semi-period, time-to-half value, half-life period. **-zelle** half cell.

Halleffekt, Raum- liveness.

hallend reverberant, with multiple echoes.

Hallformanten resonance formants.

Halligkeit liveness (realistic acoustic condition).

Hallraum acoustically live room, reverberating chamber or enclosure. **-zahl** chamber coefficient (ac.).

Halo-bildung halation (television, etc.). **-chemie** chemistry of salts, halochemistry.

Hals, Schwanen- goose neck. **Stöpsel-** neck, ring, collar or sleeve of a plug. **Trichter-** throat of a loud-speaker horn, neck.

Hals-bandkurve collar curve. **-lager** journal bearing.

haltbar durable, strong, stable, fast, permanent (of colors).

Haltbarkeit, Dauer- fatigue endurance.

haltbarmachen stabilize, preserve.

Haltedraht, Gitter- grid stay wire, g. supporting w.

halten hold, support, keep, retain, stop, restrain.

Haltepunkt break.

Halter holder, support, handle, keeper. **Distanz-** spacing means, spacer, separator. **Röhren-, federnder** cushioned tube holder or socket, anti-microphonic valve holder. **Wicklungs-** winding form, coil f., skeleton f., former.

Halterelais guard relay, holder r.

Halterung, Kristall- crystal holder, c. mounting.

Halterungssystem holder, mounting or supporting system.

Halte-spule (eines Relais) restraining coil, holding-on c. (of a relay). **-strom** retaining current, holding c. **-vorrichtung** chassis (of loudspeaker). **-zeit** critical range, c. interval, thermal retardation.

haltlos unstable, unsteady unsupported, unattached, untenable.

Hammer, Pendel- Charpy's impact testing machine.

hämmerbar malleable.

Hammerkopf (für Bass, Mittellage und Diskant) piano hammer (for bass, tenor and treble).

Hand, an - von by reference to, by the aid of. **auf der - liegen** (it is) obvious, evident or patent, goes without saying.

Handbuch handbook, manual, compendium.

handelsgängig commercial, marketable

Handelsmarke trade-mark.

Hand-empfindlichkeit hand (or body) capacity effect. **-feilenklobe** hand vise. **-griff** handle, grasp, grip, knob, button. **-kamera, Berufs-** professional hand camera. **-kapazität** body capacity effect, hand c.e. **-koffer** carrying case, suitcase.

Handlung plot, story, treatment, theme, scenario, shooting script (m.p.).

hängen bleiben "take hold" (in silent tuning).

hantieren work, handle, manipulate, manage, operate.

Harfe, Doppelpedal- double-action harp. **Haken-** hook harp. **Klavier-** keyed harp, piano h.

Harfenantenne harp antenna, fan a.

Harmonieflöte organ accordion.

hart (es) Bild crisp or harsh picture, high-contrast p., contrasty p., hard p. **-(e) Komponente der Höhenstrahlung** penetrating or hard component of cosmic rays. **-(e) Röhre** hard tube, h. valve, high-vacuum t. **-(e) Teilchen** primaries.

härtbar hardenable, temperable, indurable, capable of being hardened or cemented.

Härte, Auslenk- deflection hardness (force required to deflect phonograph needle point 100 μ). **Schall-** sound or acoustic stiffness, rigidity, ratio pressure amplitude: velocity amplitude as function of frequency.

Härte-bad, Fixier- fixing and hardening bath. **-bildner** salt causing hardness (in water). **-messer, Strahlen-** penetrometer (for X-rays), qualimeter.

härten harden, temper, indurate.

Härteprüfer hardness tester (Brinell, Rockwell sclerometer and scleroscope).

hartgeschlossenes Rohr rigidly terminated tube (ac.).

Hart-gewebe indurated or impregnated fabric. **-glas** pyrex. **-gummi** hard rubber, ebonite, vulcanite. **-guss** chilled casting. **-lagerhärtung** induction hardening.

Härtling (hard) dross.

Hartlötung hard soldering, brazing.

Härtung, Ausscheidungs- precipitation hardening. **Brennstrahl-** flame hardening. **Einsatz-** case hardening. **Oberflächen-** case hardening, surface cementing.

Härtungs-biegeprobe bending test in tempered state. **-kohle** hardening carbon, cementing c.

Hartwerden (einer Roentgenröhre) hardening of an X-ray tube (due to clean-up effect and gain of vacuum).

Harz, Dammar- Dammar resin or gum.

harzhaltig resinous, resinic, resiniferous.

Haspel reel, spool, windlass.

Haufen cluster, aggregation,

accumulation, piles, agglomeration, clump. Molekül- molecular clusters.

Häufigkeit frequency, incidence, abundance.

Häufigkeitsverhältnis abundance ratio.

Haufwerk, Kristall- aggregation of crystallites.

Haupt-achse prime axis, major a., vertical axis (of a rhombic crystal). **-azimuth** principal azimuth. **-ebene** principal plane, p. surface. **-ebenenabstand** distance between principal planes or points. **-einfallwinkel** angle of principal incidence. **-einflugzeichen** inner marker signal, main entrance s. (nearest airport boundary, in airplane landing), main marker. **-formanten** main formants, basic f. **-funkfeuer** main route or airway radio beacon. **-gitter** parent lattice. **-linie** principal line (spectr.). **-nenner** common denominator. **-patent** parent patent, main or original patent. **-peilstelle** main control direction-finder station (of an air communication safety district comprising several ground df stations). **-punktbrechwert** principal point refraction. **-punkte** principal points, unit p. **-quantenzahl** principal quantum number. **-regler** main governor or regulator (of a turbine). **-rohr** main oscillation generator (in independent drive system). **-rolle** leading star. **-rollendarsteller** principal, star, star performer.

hauptsächlich main, chief, principal.

Haupt-sammellinse second lens (in electron gun, c.-r. tube). **-satz** fundamental principle, f. law, f. theorem, axiom (math.). **-schnitt, erster (I.)** primary plane, tangential p., meridional p. (opt.). **-schnitt, zweiter (II.)** secondary plane, sagittal p. (opt.). **-sender** main oscillator, m. transmitter. **-signal** inner marker signal, main entrance s., main marker (in airplane landing). **-streckenfeuer** main or trunk route or airway beacon. **-taste** master key. **-uhr** master clock. **-valenzkette** primary valence chain. **-verhandlung** main oral hearings, m. proceedings. **-zahl** cardinal number. **-zipfel** major lobe (of space pattern). **-zug** main object (of an invention), outstanding or principal characteristic feature or trait.

Haustelephon domestic telephone, private home telephone, house 'phone, telephone extension, intercommunication t.

Haut skin, hide, film, membrane. **Guss-** casting scale, c. skin. **Licht-** cathode glow. **Oxyd-** skin or film of oxide.

Haut-abstand (focal) skin distance. **-widerstand** dermal or skin resistance. **-wirkung** skin effect.

Heaviside'sche Einheitsfunktion H. unit function.

Hebdrehwähler vertical and rotary selector.

Hebedaumen cam, tappet, finger, disk with lobes.

Hebel, drehbar gelagerter pivoted lever, fulcrumed l. (when with two arms). **-, hin- und hergehender** rocking lever, oscillating l.

Hebel, Aufzieh- lever ratchet. **Auslöse-** trip lever, detent lever. **Einrück-** starting lever, tripping lever, engaging l. **Einrück- und Ausrück-** starting and stopping lever. **Fühl-** tactile lever, feeler, scanning l., pickup l., probing l., indicator (of micrometric caliper). **Schwung-** balance beam, rocker, rocking lever. **Sperr-** pawl, latch, arresting lever, lock l.

Hebel-arm, Strahl- pencil leverage, deflectibility of p. **-detektor**

lever-adjusted cat whisker crystal detector. -schalter, Zweipol- double-lever switch. -wage beam scale, lever s. -wirkung leverage, lever action, purchase.

Hebepunkt fulcrum.

Heber, Kontrast- (dynamic-range) compandor (comprising compressor and expander, to improve signal-noise ratio).

Heber-barometer syphon barometer. -haarrohr capillary syphon. -pumpe syphon pump. -rohr syphon tube, syphon. -schreiber syphon recorder.

Hebeschritt, Wähler- vertical step of selector (automatic telephony).

Heck stern.

Heft handle, hilt; number or issue of a magazine, stitched booklet.

Heilgerät therapeutic apparatus (e.g. using short waves, rays, etc.).

Heimfernsehempfänger home television receiver.

heimlauf backstroke, return s. or motion; retrace or flyback (of pencil or beam, in telev. tube, often gated by blanking pulse).

Heiserkeit raucousness, hoarseness.

heissbrüchig hot-short.

Heissdampf superheated steam.

heissgrädig difficultly fusible.

Heiss-kühlung cooling by evaporation. -leiter hot conductor (with negative temperature coefficient), third class conductor.

Heiz-kathoden-Roentgenröhre hot-cathode X-ray tube, Coolidge tube. -körper heater (e.g., for indirectly heated cathode), heating element; radiator. -kraft heating power, calorific p. -lötstelle hot junction (of a thermocouple). -mantel heating jacket. -unruhe fluctuation in heating current, ripples. -wendel heater spiral, heater coil.

hell bright, brilliant, light, pale, clear, transparent, limpid; strident, shrill (ac.).

Hellanpassung bright adaptation, light a., photopia.

Hell-Dunkel-Intervall light-dark range. -linien light-dark-lines.

helle Klangfarbe strident timbre, shrill t. -Stelle (im Film) clear portion (in film).

Helle, Flächen- brightness of a surface, intrinsic brilliance.

Hellfeldbeleuchtung bright ground illumination (micr.).

Helligkeit (als physikalische Grösse) photometric brightness, luminance, radiance (in candles per unit area). -(als psychologische Grösse) visual sensation. -eines Objektivs speed of a lens or objective.

Helligkeit, mittlere average background illumination (telev.).

Helligkeit, Flächen- surface brightness, intrinsic brilliance. Grund- background, mean or average shading component, d.c. component. Objektiv- speed of a lens or objective. Raum- volume brightness. Ton- pitch (sound sensation).

Helligkeits-angleichung brightness matching. -schwebung brightness flutter. -signal, Bild- video signal, picture shading s. -steuerung brilliance modulation, brightness control, intensity m. -stufe intermediate brightness stage, i. grey value (between black and white), shading or tonal v. -umfang brightness range, contrast, key (of a picture). -verteilung brightness distribution (of highlights and of shadows). -wechsel je Sekunde frame frequency per second, number of illuminations of a picture point p.s. -werte brightness values. density v., shading v.

Hell-rotglut bright red heat. -schreiber Hell-type recorder. -sektoren shutter openings. -steuerung modulation to light condition.

Hemieder hemihedron, hemihedral form, h. crystal.

hemmen stop, check, brake, impede, obstruct, retard, inhibit, clog, lock.

Hemm-rad brake wheel, escapement w. **-schuh** brake shoe.

Hemmung escapement, retarding action, brake a., stopping a., check, hindrance, restraint, inhibition (in reactions).

Hemmungs-grund disability (law). **-körper** restraining substance, decelerator.

Henkel handle, lug, ear.

Herabsetzung stepping down (of potential, gearing, etc.), decrease (speed, by step-down gear); submultiplication.

heranlotsen guide to (airplanes, etc.).

herausarbeiten work out of (say, a solid piece, by cutting, punching, etc.), blank out.

Heraus-führung lead-in (wire or conductor), lead-out wire. **-gabe** restitution, return, delivery, surrender, issuance, edition (of a publication).

heraus-hören apperceive (say, a certain note in a spectrum). **-pipettieren** pipet out, remove with pipette. **-spritzen** spurt or spout out, gusher, gush. **-treten** protrude, emerge, step out of.

herrschende Verhältnisse prevailing circumstances.

herrühren originate, proceed or come from.

herstellen make, produce, manufacture, close (a circuit, a connection).

Hertz-Einheiten cycles per second, cps. **-'sche Welle** Hertzian wave, H. radiation, electric wave.

herunter-gehen let down (of an airplane). **-klappen** let down in hinges, hinge down, drop (as a table leaf). **-regeln, -steuern** regulate (a quantity) in a downward direction, regulate down.

hervor-bringen produce, create, set up, develop. **-heben** make conspicuous or prominent, underscore, highlight; accentuate, emphasize, underscore (certain frequencies or bands). **-ragen** project, stand out, be salient.

Hervorrufungszeit developing time.

Herz-bahn cardioid-shaped path (of electrons). **-charakteristik** cardioid characteristic or space pattern. **-exzenterbewegung** cam movement of Livin. **-spannungskurve** electro-cardiogram.

heulen howl.

Heul-signal siren. **-ton** sound with large number of modes of vibration, warble frequency tone, high-frequency warble tone ("multitones").

Hexodenkappe top cap of hexode tube.

H-Glied H-type section, filter mesh or network m.

Hiebton Aeolian note.

Hilfs- auxiliary, subsidiary, subordinate, ancillary, accessory, standby.

Hilfsfrequenz pilot frequency auxiliary f.; local f. oscillation (in superheterodyne). **-rückkopplung** super-regeneration (working with quench voltage).

hilfs-funke pilot spark. **-landeplatz** emergency or auxiliary landing place or field. **-mittel** auxiliaries, accessories, supplemental or expedient means, makeshift, adjuvants. **-netz** auxiliary (control) grid. **-peilstelle** auxiliary, associate or affiliated ground direction-finding station (subject to main control station). **-phasenkreis** starting-winding circuit (of split-phase induction motor). **-sender** auxiliary transmitter, local oscillator (in heterodyne scheme). **-stoff** adjuvant (substance), auxiliary or accessory material or agent. -träger auxiliary carrier, sub-carrier. -voramt sub-control station.

Himmelfernrohr astronomic telescope, astrographic t.

Himmels-gegend point of compass, quarter of the heavens. -mechanismus astronomical dynamics, celestial mechanism or mechanics.

Hinaufpendeln (gradual) rise in resonance (amplifier).

Hindernis bar, barrier, difficulty, obstacle; estoppel (law). ,gesetzliches statutory bar, legal impediment estoppel. Luftfahrt- avigational (air navigation) obstacle.

Hinderungsgrund sein be a bar to (law).

hineinziehen draw in, drag in, incorporate, involve.

Hinführung und Rückführung lead and return (path or circuit wires); scansion and fly-back (telev.).

Hin-gang, -lauf forward move, f. working stroke, travel, pass or journey; scansion, scanning motion, (active-line) sweep (telev.).

Hinlauf-Rücklauf-Verhältnis sweep-flyback ratio.

Hinleiter outgoing lead, o. wire, o. conductor.

hinten in the rear, behind, posteriorly.

Hinter-ansicht rear elevation, r. view. -beleuchtung Rembrandt illumination, half-back i.

hinterdrehen back-off, relieve eccentrically.

hintere Brennweite back focal length.

hintereinander sequentially, consecutively, serially, one after the other.

Hintereinanderschaltung series connection, seriation, cascading, concatenation or tandem connection (of motors).

Hinter-grundeffekt background effect. -grundprojektion back (ground) projection.

hinterlegen deposit, file, record, register.

Hinterlinse back lens.

hinterschnitten undercut.

Hinter-setzer false front. -wandzelle photo-voltaic barrier-layer cell with posterior metallic layer.

hinüberreissen carry over (in a distillation).

Hin- und Herbewegung shuttle, rocking, oscillating or reciprocating motion.

Hinuntertasten carrier caused to drop to low value (zero) by synchronizing signal.

Hirschleder buckskin.

Hitzdraht-anemometer, -sonde hot-wire anemometer. -strommesser thermal ammeter, hot-wire a.

hitze-beständig stable or resistant to heat, heat-proof, thermostable, refractory. -empfindlich sensitive to heat, susceptible to thermal action.

hitzen, über- superheat, overheat.

Hitzmesser pyrometer.

Hoch-antenne open outdoor antenna (installed at, or reaching, a point high above ground). -ätzung relief etching, r. engraving (metal bears outline of pattern in resist). -aufnahme vertical or upright picture or shot. -belastbarkeit high-load or heavy-load or current-carrying capacity.

hochblau bright blue, azure.

Hochdruck high pressure; relief print. -kondensator pressure-type capacitor (in nitrogen tank). -rohr, Quecksilber- high-pressure mercury discharge tube.

Hoch-frequenzbereich treble range, treble band (mus.). -frequenzheilgerät short-wave therapeutic apparatus. -frequenzkamera high-

speed camera.
hochfrequenzmässig for high frequency, so far as high frequency is concerned, relative to high frequency.
Hochfrequenztelephonie, leitungsgerichtete carrier-current or wire-guided radio telephony.
hochgekohlt high-carbon, with high carbon content (of metal).
Hoch-glanz high polish, h. lustre, brilliance. **-heim'sche Verspiegelung** special mirror plating or "silvering" (of I.G. Farben). **-intensitätskohle** high-intensity carbon (arc).
hochkantig on edge, edgewise.
Hoch-lautsprecher tweeter loudspeaker, treble l. **-leistungslautsprecher** high-power loudspeaker, (public) address l., announcer, call or paging l.
hoch-lichtstarkes Objektiv high-speed or ultra-speed objective or lens. **-rund** convex. **-schmelzend** high-melting, difficultly meltable.
Hoch-schule, technische techologic-al institute, polytechnical university, technical college. **-spannungskondensator** high-potential (fixed) condenser or capacitor.
hochspannungssicher safe against high-potential breakdown, having high puncture strength, proof against high p. and shock.
Höchst-besetzungszahl maximum number of electrons in shell. **-leistung** maximum power, peak p., crest p.; extreme record performance.
Hochstrombogen heavy-current arc, large-c. arc. **-entladung, nicht kondensierte** heavy-current uncondensed discharge.
Höchstwertanzeiger crest indicator, peak i.
Hoch-tastung working impulse transmitted on voltage substantially above normal; transmitter for impulse-keying operated at voltage far above normal. **-tonkonus, -tonlautsprecher** tweeter or treble (cone) loudspeaker.
Hochvakuum-gleichrichter kenotron, electron-type rectifier. **-röhre** high vacuum tube, hard t.
hochzeiliges Bild high-definition (telev.) picture.
hochziehen pull out (of a dive, airplane).
Höcker hump (of resonance curve); bump, protruberance; pip (of c.-r. screen tracing).
Hof, Licht- halo, halation.
hoffreie Platte, licht- non-halation plate.
Höhe, Breiten- meridian altitude. **Objektiv-** height of objective or lens. **Sprung-** pedestal height or level; amplitude of return or flyback (of beam or pencil, in telev.). **Windungs-** pitch of turns.
Höhen high (audio) frequencies, h. pitches. **Absenkung in den-** de-accentuation or attenuation of higher (audio) frequencies, using treble de-emphasizer means.
Höhen-einstellung height control (telev.). **-lader** supercharger (with high blower ratio). **-lage** height, altitude, position in vertical sense. **-messer** altitude meter, altimeter (for instance, echo or reflection type sound ranging altimeter), statoscope, terrain clearance indicator (of W.E. Co.). **-messer, Fein-** (aneroid) barometer measuring atmospheric pressure at airdrome not reduced to sealevel, sensitive altimeter. **-messer, Grob-** barometer measuring actual atmospheric pressure reduced to sealevel. **-schritt** vertical step (in automatic telephone equipment). **-strahlen** cosmic rays. **-strahlenschauer, -strahlenstösse** bursts

or showers of cosmic rays, c. ray track (traced by odoscope). -strahlung, harte Komponente der penetrating or hard component of cosmic rays.

höhensymmetrisch symmetric about vertical axis.

Höhen-verstellung adjustment in elevation. -wert altitude value (in vision). -winkel angle of elevation, azimuthal a., angular height, vertical visual a.

Höhepunkt culminating point, summit, top, crest, maximum, critical point.

hoher Ton tweeter, treble, high-pitched or h.-frequency note or sound.

höhere Instanz higher court, superior c., c. of appeals.

hohl all bottom, boomy, dull (of sound or voice); hollow, concave (of mirror).

hohl(er) Ton boomy sound, dull s. (high AF frequencies cut off).-(e) Wiedergabe reproduction with high acoustic frequencies cut off or de-accentuated.

Höhle, Augen- socket of eye, orbit, orbital cavity. Ohr- ear cavity.

hohlerhaben concavo-convex.-

Hohl-gitter concave grating. -heit cavity, emptiness.

höhlig honeycombed, having cavities or pits.

Hohl-kathode concave cathode, hollow c. -kehlenhalbmesser fillet radius. -metallrohr, Wellenausbreitung in wave propagation in wave guide or metal pipe (dielectric or conducting). -prisma hollow prism.

Hohlraum reverberation chamber (ac.); (blackbody) cavity. Resonanz- resonance cavity, r. chamber, rhumbatron. -bildung cavitation (in supersonic waves). -kabel, Papier- airspace paper-core cable, dry-space c., dry-core c. -resonator cavity resonator, chamber r., rhumbatron. -strahlung black-body radiation, cavity r.

Hohl-rohrleitung hollow pipe line, h. wave-guide (either dielectric or conducting); concentric line, co-axial l., pipe l. -schicht airspace. -schraube female screw. -sog cavitation. -spiegel concave (mirror) reflector.

Höhlungsresonanz cavity resonance (of rhumbatron and klystron).

Holm, Flugzeug- spar of an airplane wing. Gitter- grid stay, g. supporting member. Tragflächen- wing spar.

Holoeder holohedral form of crystal.

Holoedrie holohedrism.

Holz, engjähriges close-grain wood. Knet- plastic wood. Kohl- charcoal (of wood).

Holzfaser wood fiber, ligneous fiber, woodpulp. -stoff lignin, cellulose, ligno-cellulose.

Holz-mehl wood meal, w. powder, w. flour. -pech wood pitch. -pfeife wood pipe (of organ). -schliff mechanical woodpulp. -schnitt wood stock, w. block, engraving, w. cut. -stoff woodpulp, cellulose. -wolle excelsior, wood-wool.

holzzerstörend lignicidal.

homochrom of uniform color, homochromatic, homochromous.

Honigwabenspule honeycomb coil.

hörbar audible, aurally perceptible, auditory, capable of being heard or acoustically perceived or sensed. -machen render audible, audibilize.

Hörbarkeits-lautstärkeregler audibility network. -schwelle threshold of audibility or of auditory or acoustic perception (represented by audiogram).

Horch-dienst, Funk- radio intercept service. -gerät radio or sound locator, listening or detecting device (e.g., to spot airplanes by sound, based on echo principle), range- and distance-determining device, intercept receiver, asdic (submarine detector). **-ortung** radio intercept service, sound location, s. ranging.

Hör-empfang auditory reception, aural r., r. by ear, hearing or earphones. **-empfindlichkeit** hearing ability, auditory sensitivity.

Hören, einohriges monaural listening or reception. **-, zweiohriges** binaural or plastic hearing or audition, two-channel listening. **Richtungs-** aural determination of direction (of airplanes), airplane spotting. **Rück-** side tone.

Hörer, Schau- aural and visual audience, video-audio a., persons receiving audible and visual impressions simultaneously. **Sprech-** microtelephone, telephone handset, transceiver, walkie-talkie set.

Hörer-echo listener echo. **-gehäuse** receiver case, r. shell.

Hörfilm (Tonfilm zur Rundfunkübertragung) sound film broadcast, s. telecine or telecasting.

Hörfrequenz audio, audible, tonal or acoustic frequency, A.F. **Über-** supersonic, super-audible or ultra-audio frequency. **Unter-** sub-audio frequency.

hörig, doppel- binaural. **ein-** monaural.

Horizont, künstlicher Flieger- artificial horizon.

Horizontal-ablenkung horizontal or line scan, sweep or deflection. **-wechsel** horizontal or line synchronizing pulse or cycle.

Hör-kopf sound (pickup) head. **-muschel** earpiece, earcup.

Horn, Exponential-, aufgewundenes twisted or curled exponential or logarithmic horn (of loudspeaker). **Exponential-, gefaltetes** folded exponential loudspeaker horn.

Horn, Ruf- signal horn. **Signal-** bugle. **Wald-** French horn, concert h.

hornig horny, horn-like, corneous, callous.

Hör-peilung auditory direction-finding. **-samkeit** acoustic properties, acoustics. **-schärfe** acuity of hearing. **-schwelle** audibility threshold. **-sprechschalter** talk-listen switch (in inter-communication systems).

Hosenrohr syphon pipe, Y-pipe.

Hub stroke, throw, travel, excursion; percentage modulation. **Frequenz-** frequency swing or deviation (above and below the assigned resting frequency, in f.-modulation), amount by which instantaneous carrier f. differs from resting f. **Rück-** return stroke, back s.

Hubmagnet vertical stepping magnet (automatic telephony).

Hügel, Potential- potential barrier.

Hülle cover (ing), case, casing, wrapper, envelope, sheath, cloak, shell, hull (ship), contour, shroud. **Atom-** atom shell. **Dampf-** steam jacket; vaporous envelope or sheath. **Elektronen-** electron shell, e. shroud. **Leucht-** luminous sheath or envelope, aureole, photosphere (of sun), l. shroud. **Schutz-** protective cover, p. coat, p. sheath.

Hüllen-elektronen shell electrons. **-kurve** envelope. **-rohr** encasing tube or pipe, jacket.

Hülse bush, sleeve, socket, case, collar, barrel.

Hülsenschliff female part of ground joint.

Hupe reed horn.

Hüpfer, Hüpferschalter contactor (in multiple-unit control system).

Hütchenkondensator ceramic hood-shaped fixed condenser or capacitor.

Hüttenarbeit smelting, founding or foundry work.

Huygens'sche Elementarwellen Huygens wavelets.

H W Z (Halbwertzeit) half-value life,

half-life period.
hydrophob non-hygroscopic, non-absorbent, water- or moisture repellent.
Hyperebene hyperplane.
hyperfeine Struktur hyperfine structure.
Hyperfunktion hyperbolic function.
Hysteresemesser hysteresis meter.
-, schreibender hysteresigraph.
HZ (Hertz) cycle per second, Hertz unit (of frequency).

*

IBK-System C I E (Commission Internationale d'Eclairage) chromaticity scale.

Iconometer iconometer, wire-frame view-finder.

Ignitron-Stromtor-Kombination mutator.

Im allgemeinen as a general rule, generally speaking, by and large.

Im Grunde genommen basically or fundamentally speaking, as a matter of principle.

Im übrigen as to the rest.

Immersionslinse immersion lens, i. objective (with oil or water, of microscope); double aperture lens.

Impedanz s. a. Scheinwiderstand.

Impedanz, Abschluss- terminating, end or load impedance or impedor. **Bewegungs-** motional impedance. **Blindkomponente der -** reactive impedance component. **End-** terminating impedance, end i., load i. **Streu-** leakage impedance. **Stromkreis mit-** impedor. **Wirkkomponente der -** active, dissipative, effective or resistance component of impedance.

Impedanzglied, Reihen- series impedance element or impedor.

Impfen inoculate, vaccinate, inject.

Imprägniermittel impregnating agent, impregnant.

Impuls momentum (in dynamics); impulse, pulse (telev.), surge (when of unidirectional polarity).

Impuls, Austast- blanking pulse (to gate retrace of pencil, in telev. tube). **Bahn-** orbital moment. **Bildsynchronisier-** frame synchronizing impulse, low sync. i. **Dreh-(Spin)** angular momentum (spin). **Gleichlauf-** synchronizing impulse; correcting i. **Kipp-** tripping impulse. **Öffungs-** break impulse.

Impuls, Takt- synchronizing impulse, sync signal. **Zeilensynchronisier-** line synchronizing impulse, high sync. **i. Zeilenfolge-** line sequence impulse, line set impulse, partial scan impulse (in interlaced scanning).

Impuls-abtrenner, Synchronisier- synchronizing separator, amplitude s., clipper. **-erhaltungssatz** theorem of conservation of momentum. **-gabe** impulsing, pulsing, i. transmission or emission. **-geber** impulse generator, surge g., impulsing device, i. transmitter; tonal generator, tonic transmitter. **-generator(cf -geber)** impulse generator, surge g., lightning g. **-messer** peak or crest (voltage) meter.

Impulsmodulierte Wellen pulse-modulated waves (as in radar).

Impuls-ölschalter oil-blast circuit breaker. **-peilverfahren** impulse direction-finding method. **-platte** signal plate (iconoscope). **-prüfung** flash or impulse test (of insulation). **-quantenzahl** spin quantum number. **-summe** sum of momenta. **-tensor, Energie-** energy momentum tensor. **-transporttheorie** momentum transportation theory (of turbulent flow). **-verfahren, Doppel-** double impulse method.

Inaktivierung inactivation, deactivation; racemization (opt.). **-, optische** racemization (of a quartz crystal).

Inanspruchnahme claim, right of taking advantage of.

Inbetriebsetzen setting in motion, rendering operative, starting, initiating.

Indeterministische Auffassung indeterministic conception or interpretation.

Indexstrich index mark, i. line, gage m.

indirekt geheizte Röhre heater type tube, indirectly heated tube. **-(er) Dampf** latent steam.

Induktion, Wechsel- mutual inductance.

induktionsfrei(es) Kabel screened (conductor) cable. **-(er) Widerstand** non-inductive resistance, effective r., ohmic r.

Induktionsspule, variable, variometer.

Induktive Kopplung inductive coupling, transformer c.

Induktivität, punktformige lumped inductance, concentrated i. (or electrical inertia). **-, verteilte** distributed inductance, continuous loading. **Streu-** stray inductance, leakage i.

Industriegerate commercial apparatus, industrial a. (as distinguished from precision and laboratory equipment).

ineinandergreifen interlock, interengage, intermesh (as partial scans).

inertmachen de-activate, devitalize, de-energize.

Influenz electrostatic induction.

Inhalt content (s), area, volume, substance, capacity. **Bild-** picture area, p. content or subject-matter, "information". **Flächen-** (surface) area.

Inhaltsangabe table of contents, index.

Inhibierung stay of proceedings, estoppel.

Inhomogenität non-homogeneousness, lack of homogeneity, heterogeneousness.

Inklinationswinkel magnetic inclination, m. dip.

Inkrafttreten commencement (of an act or law), coming into force.

Innen-antenne indoor antenna or aerial. **-aufnahme** indoor shot, studio s.; oscillogram directly recorded on photographic emulsion inside vacuous space. **-gewinde** internal thread, female t. **-grenzlehre** plug gage. **-photographieoszillograph** internal photography oscillograph. **-raumdurchführung** indoor bush(ing). **-taster** inside calipers.

innere Quantenzahl inner quantum number. **-Reibung** solid viscosity, internal friction.

innermolekulare Kette inter-molecular chain.

innewohnend inherent, intrinsic, innate.

Insel island.

Inselartige Ablagerung, Inselbildung "island" formation.

Insgesamt altogether, substantially, on the whole.

Instand-haltung servicing, maintenance work. **-setzung** repairing, reconditioning, restoration, rehabilitation.

Instanz, Gericht erster court below, lower court. **Gericht höchster-** supreme court. **Gericht zweiter oder höherer -** court above, higher c., c. of appeal. **Gerichtsbarkeit erster -** original jurisdiction.

Instrument, Blech- brass instrument (mus.). **Dreheisen-** moving-iron instrument (of plunger, vane or repulsion type). **Mess-, gedämpftes** deadbeat measuring instrument, aperiodic m.i. **Mess-, ungedämpftes** ballistic measuring instrument. **Schreib-** graphic or recording instrument.

Instrumentausrüstung instrumentation.

Instrumentenfehler variations of bearing due to instruments (df), measuring instrument error.
Integral, Austausch- exchange integral. **Raum-** space integral.
Integralrechnung integral calculus.
Integrierbar integrable.
Intergrieren über integrate over.
Intensitäts-ausfall fading. **-messgerät** level recorder. **-steuerung** intensity, brightness or brillance modulation (of c.-r. tube). **-verfahren** variable-density method (of sound recording).
Interferenz-bild interference figure. **-farbe** interference color. **-gebiet** mush area (in chain broadcasting).
Interferenzialrefraktor interferometer.
Interferenz-streifen interference fringes (opt.) **-ton** beat tone, interference note.
Interferometer, Gitter- grating interferometer.
Interkombinationen inter-system combinations.
Intermediär intermediary, intermediate.
Intermittenzschwingung relaxation oscillation, sawtooth o., ratchet o., timebase o.
Intervall-faktor interval factor. **-regel** interval rule (of Landré).
intervenieren intervene, interpose, interfere.
Intonation voicing.
intonieren tune (a piano), intone.
in Tritt kommen fall in step, synchronize, phase-focusing (in beam tubes).
Inverter inverter (thyratron).
Invertzeitrelais inverse-time relay.
Ionen, Locker- defective ions, "Locker" i. **Sprüh-** spray ions. **Zwischengitter-** interstitial ions. **Zwitter-** dual, hybrid or amphoteric ions.
Ionen-beweglichkeit ionic mobility. **-falle** ion trap (in electron gun of c.-r. tube). **-komponente, Wasserstoff-** pH value, hydrogen ion concentration. **-rückkopplung** ion feedback. **-rumpf** ion core. **-rundlauf** circulation of ions. **-spaltung** ionic cleavage, ionization. **-stärkemesser** ionometer. **-steuerrohr** thyratron. **-windvoltmesser** ionic wind voltmeter. **-wolke** ionic atmosphere.
Ionisationswärme heat of ionization, h. of dissociation.
Ionisierungs-kammer, Fingerhut- thimble ionization chamber. **-spannung** ionization potential, ionizing p., p. sufficient to detach electron from its atomic bond. **-spielaufbau** Townsend structure or build-up. **-stufe** stage of ionization.
Iris-auf- und-Abblendung irising in and out. **-blende** iris diaphragm.
irisieren irisate, iridize, iridesce.
irreführen confuse, mislead, misdirect.
Irrung correction.
Isoklinen isoclinics (lines of equal darkness, in photoelasticity).
Isolations-messer insulation tester, megger. **-prüfung, Stoss-** impact, impulse or flash test. **-widerstand** insulation resistance, insulance (reciprocal of leakance).
Isolator, unvollkommener imperfect dielectric, leaky insulation. **Abspann-** terminal insulator, strain i. **Abstand-, Distanz-** stand-off insulator, spacing i. **Durchführungs-, Einführungs-** feed-through insulator, lead-in i., wall duct, bushing. **Freiluft-** outdoor insulator. **Kreuzungs-** transposition insulator. **Pilz-** mushroom, petticoat or umbrella insulator (long leakage path).

Stütz- pin insulator, rod i.
Wanddurchlass- wall lead-in or bushing insulator.
Isolier-masse insulating compound, i. paste, moldable or fictile i. substance, "dope." -masse, gepresste molded insulation. -röhre buffer tube, isolating valve.
Isotopieverschiebung isotope effect, i. shift.

J

Jacarandaholz rosewood.
Jahresgebühr annual fee or tax.
Jahreszeitliche Anderungen seasonal variations.
Jalousiebretter louvres.
Japan-lack Japan lacquer, J. varnish., -wachs Japan wax.
jeweils under given or prevailing conditions; at the time, in each case.
Jodieren iodizing (of a negative).
jungfräuliche Kurve virgin curve.
Justler-greifer pilot claw (m.p.) -stift pilot pin, registration p. (film printing.)
Justierung, Film- registration of film.
Jutebeilauf jute filler (cable making.)

K Absorptionskante, K. wert K edge, K absorption limit.

K wert octantal component of quadrantal error (QE) due to fuselage.

K and L Schale-Abschirmungskoeffizient K and L shell absorption coefficient (screening function.)

Kabel, gemischtpaariges composite cable. **-, induktionsfreies** screened (conductor) cable. **-,konzentrisches** co-axial line, concentric l. or cable, pipe l. **-,mehradriges** multi-wire cable. **Erd-** leader cable, pilot cable, buried c. (for airplane landing.) **Faden-** co-axial or concentric cable with silk-thread-supported conductor. **Leit-** leader cable, pilot cable (for airplane landing.) **Papierhohlraum-** airspace paper-core cable, dry-core c. **Photozellen-** low-capacity shielded cable between photo-cell and amplifier, video c., concentric or co-axial c.

Kabel-schuh cable eye, c. socket, connector lug, sweated thimble. **-verlegung** cable laying, burrying of a c. **-wachs** cable wax.

Kabine, Kino-, Vorführungs- projection room, p. box; p. booth (in back projection.)

kadmiertes Eisen cadmiated iron.

Käfig, Auffänger- collector electrode (in Farnsworth dissector.). **Faraday'scher** Faraday shield, F. screen, F. wire cylinder. **Rollen-** roller cage.

Käfig-antenne cage antenna. **-spule** canned coil, shield c., cage c. **-transformer** shielded transformer.

Kaliber bore, pass, caliber, diameter; groove. **-dorn, Toleranz-** tolerance plug gage.

kalkartig calcareous, limy.

Kalk-schleier chalk fog. **-spat** calcite.

Kalorimeter, Eis- ice calorimeter. **Kondensations-** steam calorimeter.

Kalotte cup-shaped, cap-shaped or hemispherical part, calotte.

Kalottenmembran hemispheric or cap-shaped (metallic loudspeaker) diaphragm.

kalt(er) Farbton cold color hue.

kalt(e) Kathode cold cathode (as in photo-tube and BH type gaseous rectifier).

Kaltaushärten cold tempering.

kalt-bläsig refractory (of metal.) **-brüchig** cold short-brittle.

Kälteeinheit frigorie.

kälteerzeugend frigorific, cryogenic.

Kälte-punkt solidifying point (of oil). **-regler** cryostat.

Kalt-lack cold varnish. **-schlag** cold heading die. **-sprödigkeit** cold brittleness. **-stauchmatrize** cold upsetting die. **-verformung** cold working, c. shaping.

Kamera, Atelier- studio camera. **Berufshand-** professional hand camera. **Geheim-** detective camera, concealed c. **Hochfrequenz-** high-speed camera. **Laufboden-** base-board extensible camera. **Loch-** pin hole camera. **Mikrospektral-** microspectroscopic camera. **Nebel-** cloud chamber. **Rundblick-** panoramic camera. **Spreiz(en)-** strut camera, extension c. **Ton-** sound recording camera. **Trick-** trick camera. **Wilson-** fog, cloud

or expansion chamber (of Wilson).
Kamera-auszug camera extension. -balgen camera bellows. -fahrgestell camera truck, dolly. -laufwerk feed mechanism. -mann, Atelier- studio cameraman. -mann, Ton- recordist.
Kamm, Folgeschalter- sequence switch cam.
Kämmen der Turbinendichtungen rubbing of labyrinth seals.
Kammer chamber, room, space, camera, compartment. Aufzeichnungs- recording chamber (of electron microscope). Fingerhut(ionisierungs)- thimble ionizing chamber. Klein- small (air) chamber for ionizing work. Leistungs- energy chamber (of klystron). Luft- air pressure space (of a loudspeaker.) Nachhall- reverberation room, r. space or enclosure. Wilson- Wilson fog, expansion or cloud chamber.
Kammer-gericht Supreme Court (of Prussia.) -lautsprecher, Druck- pressure chamber loudspeaker, pneumatic l. (without diaphragm.) -ton "A" of tuning fork.
Kamm-lager (concealed) collar thrust bearing. -rad cog wheel. -zwecke tack.
Kampfstoffe (chemical) warfare substances.
Kanadabalsam Canada balsam.
Kanal, Film-, Führungs- film track, f. channel (in film camera.) Funken- spark track. Sende- transmission channel, signaling c. Teil- sub-channel. Übertragungs- transmission channel. Vorentladungs pre-discharge track.
Kanalfilter channel filter.
kanalig, mehr- multi-channel.
Kanal-strahl canal ray, positive r., diacathode r., anode r. -tastung, Ein- single-channel pulsing (telev. synchronization.)
Kanonenbronze gun metal, g. bronze.
Kanonier, Richt- gunlayer.
Kante, ablaufende trailing edge. Leit- leading edge, l. tip (of airplane wing). Schmal- narrow edge, n. side.
Kanten-schärfe resolution on border (of a picture.) -schema Deslandres diagram (of spectral band systems.) -system zero-line system, edge s.
kantenweise edgewise, on edge.
kantenwirkung edge effect, fringe e.
Kapazität s. a. Kondensator
Kapazität, akustische acoustic capacitance, a. compliance (reciprocal of a. stiffness.) -,mechanische compliance. -,punktförmige concentrated capacitance, lumped c. -,verteilte distributed capacitance, self-c.
Kapazität, Beruhigungs--smoothing capacity. Dach- top (loading) capacity. Eigen- self-capacitance. Elektroden- (inter-) electrode capacitance. Ersatz- equivalent capacity. Kipp- time-base condenser. Röhren- (inter-)electrode capacitance. Sättigungs- saturation capacity. Schalt- circuit capacitance. Selbst- self-capacitance, distributed capacity (of a circuit, etc., due to proximity effects.) Streu- stray capacitance, distributed c., spurious capacity. Teilerd- direct earth capacitance. Verdunkelungs- shading condenser. Verlust- imperfect capacity, leaky c.
kapazitätsarm of low capacity or capacitance.
Kapazitätsbrücke capacitance bridge, farad b., Wien b.
kapazitätsgerader Kondensator straight-line capacity condenser.
Kapazitäts-messer faradmeter. -probe capacity test. -reaktanz capacitive reactance, negative r.
kapazitiver Blindwiderstand capacitive reactance, negative reactance, condensance.
Kapotte, Abblende- dimming cap.

Kappe cap, hood, top, dome, crown, can. Aufsteck- slip-on cap. Hexoden- top cap of hexode tube. Lichtschutz- light-hood (phot.) Ohren- ear muff.
Kapsel capsule, case, box (printing), cap, chill, mold (foundry), sagger (cer.). Mikrophon- microphone button (housing), capsule. Sprech- condenser microphone.
Kapsel-blitz flashlight capsule. -guss casting in chills, chilled work. -ton sagger clay, capsule c.
karburiert, öl- oil carburetted.
Kardan-gelenk, -kupplung universal joint.
Kardinalpunkte cardinal points (opt.)
Kardioide, Wechsel- switched cardioid (space pattern).
Kardioidenbildung cardioid formation (of space pattern or directional diagram.)
kariert cross-hatched.
Karneol carnelian.
Karolus-Kerr Mehrfachplattenzelle Karolus multi-plate Kerr cell.
Karte, gnomonische gnomonic chart (straight-course chart.) Merkator- Mercator chart. Strecken- route (airway) map.
Kartei card index, c. catalog.
Kartenaufnahme, photographische photographic mapping.
Karten-kurs map course. -netz graticule. -projektion, Merkator- Mercator or cylindrical projection (orthomorphic p.) -winkelmesser map protractor.
kartesische Fläche Cartesian surface.
Karton, Bristol- bristol board.
Kartothek card index, c. catalog.
kaschieren place on a support, base or other surface; coat with paper.
Kaskaden-schaltung cascade connection, cascading, concatenation or tandem connection (of motors.) -verstärker cascade amplifier, repeating a.
Kassette spool box, darkslide (phot.); cassette (X-ray work); sagger, coffer, casket. -,leise silenced magazine. Anlege- clamp-on darkslide. Tageslicht- daylight loading magazine (m.p.).
Kassetten-rahmen darkslide carrier. -schieber shutter of darkslide.
Kastagnetten castanets.
Kastanienholz chestnut wood.
Kasten, Geräusch- equivalent acoustics (for special sound effects and noise imitations.) Potential- potential well, p. hole. -drachen box kite.
Katheten perpendicular and base adjoining hypothenuse (in a right-angled triangle.)
Kathode cathode, filament, emitter (of electrons.) -,direkt geheizte direct-heated cathode. -,geschichtete (oxide-) coated cathode; photo-c. with light-sensitive layer. -,indirekt geheizte indirectly heated cathode (in heater-type tube), unipotential c., iso-p. c. -,kalte cold cathode (as in photo-tube and in BH type gaseous rectifier). -,körperliche physical cathode, material c.
Kathode, Anti- target, anti-cathode (of X ray tube). Flächen- plate-shaped cathode (of neon tube as distinguished from point or crater.) Hohl- concave cathode, hollow c. Oxyd- oxide-coated filament, dull emitter. Paste- oxide-coated (pasted) cathode. Photo-, zusammenhängende mirror type or continuous photo-cathode, photoelectric c. Punkt- point-source cathode, crater c. Quecksilber- mercury-pool cathode. Raster (photo)- mosaic, rastered photocathode screen. Rein- bright emitter (often thoriated c.) Schicht- cathode bearing a layer

or coat (of oxide, photo-sensitive substance, etc.). **Siedekühlung-** vapor-cooled cathode.

Kathoden-anheizzeit cathode heating time (till stable temperature is reached), thermal time constant. **-ansatzpunkt** spot (of arc) formed on cathode. **-dunkelraum,** Crookes cathode dark space, Crookes' dark s., Hittorf's dark s. **-fackelerscheinung** cathode flicker effect. **-fallableiter** valve arrester. **-fleck des Hg-Lichtbogens, fixierter** anchored cathode spot of mercury arc. **-licht(haut)** cathode glow. **-linie** cathode line. **-neuaktivierung** cathode re-activation, c. rejuvenation. **-siedekühlung** vapor cooling for cathode. **-spratzen** flicker or spitting of cathode. **-sprung** cathode jump of potential.

Kathodenstrahl-abtaster electron-beam scanner. **-bündel** cathode-ray beam or pencil; gun. **-intensitätskontrolle** cathode-ray beam intensity modulation (by shield, grid or Wehnelt cylinder). **-rohre** thermionic tubes (obs.), beam tube; c.-r. tube, oscilloscope, iconoscope, emittron, and other tubes operating with beam and screen, Lenard t. **-rohre, gasgefullte** gas-filled c.-r. tube, soft t. working with a c.-r. (for ionic or gas focusing).

Kathoden-verstärker cathode follower. **-zerstaubung** cathode disintegration, c. spatter, splutter or sputter.

Kathodo-lumineszenzlampe cathode-glow tube. **-phon** cathodophone, diaphragmless microphone, glow-discharge m.

Kation cation (positive ion moving toward cathode, negative ion called anion).

Katzenauge bull's eye (small thick lens).

kaubar masticable.

Kauf nehmen, in accept something (at the expense of), be tolerable or tolerated.

kauflich commercial (grade), merchantable, purchasable.

Kaustik caustic (opt.).

Kaution stellen give bail, bond or security.

Kavitation cavitation.

Kegel, Anflug- sector of approach. **Ganzpol-** polehode cone, body cone. **Licht-** cone of light, luminous c. **Rastpol-** herpolnode, space cone. **Reibungs-** pole of friction. **Seh-** angle of vision, cone of v. (camera work.) **Strahlen-** cone of rays. **Stumpf-** truncated cone, frustum.

Kegel-antenne, Doppel- double-cone antenna, cage a. **-druckprobe** cone (indentation) test. **-feder** volute spring. **-lehre** taper gage. **-mantel** cone-shaped shell. **-projektion** conical projection (of a map). **-rad** bevel wheel, cone w., mitre w. **-radgetriebe** bevel gearing, mitre-wheel g. **-schnitt** conic section. **-stumpf** truncated cone, frusto c., frustum. **-ventil** conical valve, ball v.

Kehle throat, larynx, channel, flute.

kehlen groove, slot, flute, channel, key.

Kehl-kopfmikrophon laryngophone, throat microphone. **-nahtschweissung** fillet weld.

Kehr-doppelwendel reversed double loop. **-wendel** return spiral, reversed s., helix, coil or loop. **-wert** reciprocal value, inverse v. **-wert, Zahigkeits-** fluidity (reciprocal of viscosity).

Keil wedge, key, cotter, jib, liner, shim. **Dreiecks-** triangular wedge. **Goldberg-** Goldberg wedge, neutral (grey) w. **Streifen-** photometric wedge. **Stufen-** step tablet, stepped (photometric) absorption wedge (X-ray, etc., work).

Keil-flach spheroid. **-nute** keyway. **-schwärzungsmesser** circular wedge densitometer.
Keim nucleus or seed (cryst.), germ. **Kristall-** crystal nucleus, seed c. **Reif-** digestion nucleus.
Keim-kristall crystal nucleus, nucleus crystal. **-ling** nucleus (cryst.), germ, embryo. **-wirkung** nuclear action, action of nuclei (cryst.) **-zenter** grain center.
Keller-armatur boiler fittings. **-ton** tunnel effect, boom e.
Kenndämpfung iterative attenuation constant.
Kenndämpfungswiderstand iterative impedance; characteristic i. (for uniform line).
Kennedy-Heaviside Schicht Kennedy-Heaviside layer.
Kenn-faden (colored) tracer-thread (for c.coding). **-grösse** characteristic quantity, c. magnitude. **-grösse, Übertragungs-** transfer characteristic. **-leitwert** indicial admittance.
Kennlinie, Arbeits- operating or working characteristic or curve, performance c. **Dynamik-** volume (range) characteristic. **Frequenz-** frequency-response characteristic. **Richt-** directional diagram or space pattern (df); rectification characteristic. **Schwing-** oscillating characteristic, resonance c.,c. of an oscillatory system. **Touren-Belastungs-** speed-load characteristic. **Übernahme-** transfer characteristic.
Kennlinien-feld family of characteristics or curves. **-kippung** characteristic straightening. **-knick, oberer** upper bend or knee of characteristic. **-kick, unterer** lower bend or knee of characteristic.
Kennsignal tuning note (characteristic audio signal for tuning),c. sign or signal (of Morse code or other identification), signature of a station, theme.
kenntlich machen characterize, identify, make conspicuous, identifiable or distinguisable.
Kennung, Funkfeuer- call or distinctive signal (of a beacon station.)
Kennungssignal distinctive, identifying or code signal, signature.
Kenn-widerstand characteristic impedance, image i., indicial i. **-winkelmass** iterative phase constant. **-zahl** characteristic factor, office code. **-zeichen** charact. or distinguishing mark or feature, sign, identification, symptom; object (of invention or patent).
kennzeichnen characterize, distinguish, code.
Kenn-zeichnung, Farb- color coding. **-ziffer** characteristic.
Kerbe, scharfe sharp notch, nick or groove; slot, slit. **Rund-** round notch, Charpy notch (with root radius.)
Kerb-dauerfestigkeit notch fatigue strength. **-empfindlichkeit** stress concentration index. **-nute** keyway. **-schlagzähigkeit** notch impact tenacity.
kerbverzahnt splined (externally or internally.)
Kerb-wirkungszahl fatigue stress concentration factor. **-zähigkeit** notch tenacity.
Kern nucleus, core, grain, kernel, pip, stone, pith, heart, crux (of a matter.) **-,geblätterter** laminated core. **Eisenpulver-, gepresster** compressed iron-dust core, Ferrocart c. **Film-** reel, spool. **Kristall (isations)-** center or nucleus of cristallization. **Lamellen-** laminated core. **Luft-** air core. **Masse-** compressed iron-powder or dust core, molded c., Ferrocart c. **Nachfolge-** product nucleus. **Spulen-** bob or core of a spool (m.p.) **Staub-** duct or molded core, Ferrocart c. **Verunreinigungs-** impurity nucleus.

Wasserstoff- H particle, hydrogen particle or nucleus, proton, deuteron (when of heavy hydrogen or deuterium.) **Zwischen-** compound core, "intermediate".

Kern-abstand inter-nuclear distance. **-anregung** nuclear excitation. **-bildung** germination, nucleation. **-distanz** internuclear distance. **-durchdringung (Potentialschwelle)** nuclear penetration function. **-farbe** nuclear stain.

kernferne Elektronen planetary, peripheral, orbital or outer (-level) electrons (farthest from nucleus), valence e., conduction e.

Kerngehalt nuclear concentration, content of nuclei.

kerngestreut deflected by nucleus.

Kernguss cored work (foundry).

kernig, ein- uninuclear, mononuclear. **mehr-, viel-** multinuclear, polynuclear, having a plurality of nuclei or cores, polycyclic.

Kern- körper(chen) nucleolus. **-ladungszahl** charge on nucleus, nuclear charge number, atomic number. **-magneton (K.M.)** nuclear magneton. **-moment, magnetisches** nuclear (magnetic) moment.

kernnahe Elektronen inner(fixed) electrons, e. adjacent nucleus.

Kern-physik nuclear physics. **-potentialschwelle, Durchdringung der** nuclear penetration function. **-schatten** umbra. **-schliff** male part of ground-in joint. **-spin** nuclear spin. **-spule, Masse-** dust-core coil, compressed-iron-core coil. **-teilung** nuclear division. **-theorie** nuclear theory. **-umänderungsprozess** nuclear disintegration process. **-umwandlung** nuclear transmutation. **-verlust** core loss, iron l. **-wechselwirkung** nuclear inter-action. **-zerplatzen** nuclear fission or disintegration.

Kerr Mehrplattenzelle, Karolus- Karolus multi-plate Kerr cell.

Kerze, römische Roman candle, magnesium torch, smoke pot.

Kerzenstärkencharakteristik candle-power characteristic.

Kessel boiler, vessel, tank, pot. **Druckwind-** compressed-air tank, blast pressure tank. **Etagen-** battery boiler.

Kesselsteinbildung boiler scale formation, incrustation.

Kette filter, network; channel, chain, train, series, bondage. **-, elektroakustische** channel (in sound recording or reproducing), electro-acoustic transducer. **-, innermolekulare** intra-molecular chain. **Abstimm-** tuning arrangement comprising a plurality of oscillation or tuning circuits. **Drossel-** low-pass filter. **Entzerrer-** corrective network, equalizing n. **Gelenk-** link chain. **Neben-** side chain, subordinate c. **Sieb-** ladder-type filter, band-pass f. or network. **Sperr-** low-pass filter. **Sternglied-** T network. **Transport-** conveyor chain. **Verstärker-** amplifier cascade. **Verzögerer-** delay network. **Widerstands-** voltage divider, potentiometer.

Ketten-bruchmethode method of infinite continued fractions. **-durchhang** slack or sag of a chain. **-fläche** catenoid. **-flaschenzug** chain tackle block. **-glied** network or filter mesh or section.

Kettenleiter iterative network, recurrent structure. **-erster Art** II (π) network. **-, mehrgliedriger** multi-mesh network. **-, mit 1/2 Längsglied endender** mid-series terminated network. **-, mit 1/2 Querglied endender** mid-shunt terminated network. **-, vielgliedriger** multi-mesh network, multi-section n. **-, zweigliedriger** two-section network.

Ketten-leitermasche mesh, section of a network. **-linie** catenary. **-molekül** chain molecule. **-rad** sprocket wheel. **-rollenscheibe** chain sheave. **-schaltung** relay chain circuit. **-widerstand** iterative impedance.

K-H Schicht Kennedy-Heaviside layer.

Kiefernholz pinewood.

Kieselzinkerz willemite.

Kimm (sea) horizon. **-tiefe** dip of horizon.

Kino motion-picture theater, film playhouse, movie.

Kino, Fern- telecine, telecast, intermediate-film television system. **Koffer-** portable projector. **Spielzeug-** home movie.

Kino-film-Fernsehabtastung telecine scan of film, pickup of pictures from moving-picture f. **-kabine** projection booth. **-positiv** moving picture positive. **-skala** station name projection scale.

Kipp... time-base, relaxation, "kipp," trip, triggering.

Kipp-drossel cut-off choke (regulates battery charging). **-einsatz, Bild-** incipient frame flyback. **-einsatz, Zeilen-** incipient line fly-back. **-elektrometer** Wilson electrometer (gold leaf given different positions by tilting case).

Kippen retrace, reversal or return of pencil (in scanning); trip over, trigger; tip, dump, tilt.

Kippentladung condenser discharge (in time-base).

Kippentladungsröhre time-base discharge tube.

Kipper, Bild- frame time-base. **Zeilen-** line time-base.

Kipp-erscheinung jump phenomenon, swing p. (in arts other than television). **-frequenz** relaxation frequency, time-base f. **-gerät** time-base, deflector means, relaxation oscillator, trigger device. **-gerät, Gegentakt-** push-pull time-base. **-hebelschalter** toggle switch (with projecting lever). **-impuls** tripping impulse, **-kapazität** time-base condenser. **-kreis** time-base circuit, trigger c., relaxation c. **-moment** maximum torque (mech.) **-oszillator, Sperr-** blocking oscillator. **-periode** time-base period. **-schaltung, Glimmlampen-** neon-lamp time-base circuit scheme. **-schwinger** time-base oscillator.

Kippschwingmethode, fremdgesteuerte, unselbständige distant-operated (non-self-running) time-base method. **-, selbständige, mitgenommene** self-running controlled time-base method.

Kippschwingung relaxation, sawtooth or ratchet oscillation, time-base o., kipp o. **-selbständige -** self-running time-base. **Bild-** frame or picture time-base oscillation or impulse. **Zeilen-** line time-base impulse or oscillation.

Kipp-schwingungsdauer relaxation period. **-sender, Sperr-** blocking oscillator. **-spannungsrücklauf** flyback of sweep potential, return of time-base p. to zero. **-spiegel** oscillating mirror, vibrating m.

Kippung, Kennlinien- characteristic straightening.

Kippvorgang time-base action, ratchet a.

Kirsche, Schmelz- igniting pellet.

Kirschrotglut cherry-red heat, cherry redness.

Kissen, Luft- air cushion, dashpot.

kissenförmige Verzeichnung pincushion distortion, pillow d. (telev.).

Kittlehre, Film- film splicing gage.

klagbar werden bring action or suit (against).

Klage, Einstellungs- action for discontinuance (of an infringement, etc.) **Feststellungs-** declaratory action. **Nichtigkeits-** action or suit for cancellation, annulment or invalidation. **Patent-** patent suit, patent action, p. litigation. **Patentverletzungs-** action or suit for infringement of patent rights.

Schadenersatz- action for damages. Unterschiebungs- suit for passing off goods as though of another person.

Klage-abweisung dismissal of claim, non-suit. -beantwortung reply, response, answer, defense (plea). -grund cause of action.

klagen bring action, sue, file a suit.

Klagepatent patent in issue, in suit, at bar.

Kläger plaintiff, complainant. Berufungs- appellor. Wider- cross or counter-claim plaintiff.

Klageschrift plaint, bill of charges, statement of claim, bill of particulars, plaintiff's declaration.

Klammer clamp, clasp, clip; bracket (round or square, used in mathematical formulae, etc.). (Round brackets are strictly known as parentheses in typography). Gabel- forked clamp. Verbindungs- brace, clip connector.

Klang s.a. Ton, Akustik, Geräusch und Schall.

Klang, blecherner tinny sound. -,zusammengesetzter complex (musical) sound. Doppel- echoed sound (hearing sound twice, i.e., by electric transmission and by transmission through air). Einzel- (sounding of an) individual note or tone. Gleich- unison, consonance, resonance.

Klang-bild sound pattern, acoustic p. -blende tone-shading means, tone switch, t. control, tonalizer. -drossel tone-control choke (coil). -einsatz acoustic intonation, onset of a sound. -farbe timbre, tone color, quality.--,helle strident timbre, shrill sound. -färbemittel tone-shading means, tone or timbre control m., tonalizer. -farberegler tone control, tonalizer. -figur sound pattern, acoustic figure. -fülle sound volume. -gemisch sound spectrum.

klanggetreu of high fidelity, orthophonic (of reproduction of sound).

Klang-intensitätsbereich dynamic range of sound. -reiniger acoustic clarifier (on ldpsk. baffle).

Klappe (clack) valve, flap, trap, lid, damper, stop, key (of wind instrument), pallet (of harmonium).

klappen strike, clapper; clap, flap, fold, move in various directions, often about pivots or hinges, to open or close, as the lid of a chest or the cover of a box. auf-, hoch- fold, raise or lift to open or erect, on hinges, as the cover of a talking machine. weg- swing or fold out of the way. zu- drop, move, swing or let down (for shutting) on hinges.

Klappen-horn key horn. -mechanismus key work, k. action.

klappern chatter, rattle.

Klappensignal see Klatsche.

Klapp-kamera folding camera, collapsible c. -kassette hinged darkslide. -sucher, umlegbarer reversible folding finder. -variometer hinged-coil variometer.

klar clear, distinct, pellucid, lucid; intelligible, articulate. -bestimmt uniquely determined, definitely or clearly defined.

Klär-bad clearing bath. -bassin, -bottich settling or clearing tank, tub, vat or reservoir.

Klarinette clarinet, clarionet.

klarmachen standby, make ready, alert.

Klärmittel, Klärungsmittel clarifying or clearing agent, clarifier, fining substance.

Klartonverfahren noiseless film method.

Klatsche beat or clappers (to mark sound and picture tracks for synchronization).

Klatschenbild beat picture.

Klaue claw, paw, pawl, catch, clutch, grasp.

Klauen-kupplung claw clutch, c. coupling.

Klaviatur keyboard, bank of keys.
Klavierharfe keyed harp, piano harp.
Klebäther collodion.
Klebband friction tape, adhesive or sticky t.
Klebemittel adhesive, agglutinant, cement.
Kleben sticking (of a keeper), freezing, adhering, clinging; cementing, gumming, glueing. **-der Elektronen** capturing or sticking of electrons (to electro-negative gases, surfaces, etc.) **Film-** blooping, patching or splicing (of film).
Klebe-presse blooper, splicer (for making blooping patches in film). **-stelle, Ton-** bloop, blooping patch or splice (on sound film). **-stellengeräusch** splice bump, dull sound due to blooping patch, wowows, blooping. **-tisch** splicing table, s. bench.
klebrig sticky, adhesive, pasty, viscid, glutinous.
Klein-bildphotographie microphotography, photomicrography. **-film** narrow film, sub-standard f. **-gefüge** micro-structure, fine structure. **-kammer** small (air) chamber, thimble (for ionizing work). **-lampe** miniature lamp. **-lader** trickle charger. **-lebewesen** micro-organism. **-lichtbildkunst** photomicrography, microphotographic art.
kleinlückig of fine porosity, close-grained.
Kleinröhre miniature tube (of acorn, midget, etc. type).
kleinsten Quadrate, Methode der method of least squares.
Kleinstmaschine miniature machine.
Kleinumfangkopien "lo-range" prints.
Klemme terminal, binding post, connector, clamp.
Klemmeffekt pinch effect (in needle reproduction of sound).
Klemme, Eingangs input terminal. **Erd-** ground clamp. **Verbindungs-** connector means, binding post, terminal.
Klemmenwiderstand terminal impedance, t. resistance.
Klemmschraube clamp, lock or thumb screw; binding post, terminal.
Klick, Tast- key thump, k. click.
Klima-gürtel climatic belt. **-technik** air conditioning.
Klimatisierung air conditioning.
klimpernd harsh or strident (of a tone or sound containing strong harmonics and high partials), grating, shrill, tingling.
Klingeldraht bell wire, annunciator w.
Klingen sounding, ring (of a tube).
Klingfähigkeit ringing noise, microphonic n. (of a tube).
Klinke, Batterie- battery jack. **Doppelsperr-** double dogs, pawls or detents. **Rast-** detent, latch, pawl. **Sperr-** pawl, catch, latch, click, detent, ratchet.
Klinken-scheibe notched disk, ratchet disk. **-stecker** jack, plug switch.
Klinkwerk trip gear, release g.
Klirrdämpfung correction of non-linear harmonic distortion.
Klirren scratch (of a loudspeaker), clink, clatter.
Klirrfaktor non-linear harmonic distortion factor, klirr f., blur or rattle f.
klopfen beat, knock, tap.
Klopffeind anti-knock agent (engine).
Klöppel, Glocken- bell hammer, b. striker. **-stange** clapper rod, c. stick.
klumpig clotted, clotty, lumpy.
Klystron klystron (ultra-high-frequency oscillator with two rhumbatrons acting as buncher and catcher).
K.M. (Kernmagneton) nuclear magneton.
knacken crack, crackle, click, crepitate.
Knagge cam, catch, peg, stop, lug, bracket.
Knall detonation, explosion, report, crack, pop, pistol shot noise.
Knall-gas, Wasserstoff- detonating

gas mixture (of H and O), oxyhydrogen gas. **-gasgebläse** oxyhydrogen blowpipe. **-pyrometer** explosion pyrometer. **-satz** detonating composition. **-schutzgerät** pistol shot silencer, crack s., noise killer (radio). **-welle** shock wave (ac.)

knapp close, tight, narrow; scanty, meager, brief, concise.

Knarre ratchet.

knarren crackle, rattle, crack (of loudspeaker), creak, squeak.

Knebel-bolzen toggle bolt. **-griffschraube** tommy screw.

kneifen nip, squeeze, pinch.

kneten knead, pug (cer.)

Knet-holz plastic wood. **-legierung** plastic alloy, wrought a., malleable a.

Knick break (in curves); crack, flaw, kink, bend or knee (e.g., upper and lower, of tube characteristic).

knicken crack, break, buckle, collapse.

Knick-festigkeit flexibility (of a film), breaking strength, buckling strength (of metal). **-punkt, -stelle** kink or knee, break (of a curve). **-widerstandsfähigkeit** folding endurance (of film), breaking or buckling strength.

Knie knee, bend (of a curve); elbow, knuckle.

knieförmig knee-shaped, geniculate.

Kniegelenk knee joint, elbow j.

Kniff (cf. **Trick**) artifice, trick, device; crease, pinch.

knistern crackle, crepitate, rustle (silk).

Knitterfestigkeit crumpling strength.

Knöllchen nodule, little knob or node.

Knolle lump, knob, node, nodule, knot.

knollig knobby, knotty, bulbous.

Knopf button, knob. **Druck-** push button, press button. **Einstell-** focusing knob (camera). **Trieb-** milled head, pinion h. (micr.).

Knopf-empfänger, Druck- pushbutton-tuned receiver, receiver with automatic tuning. **-karte** instruction chart. **-röhre** shoebutton tube, acorn t.

Knoten node, nodule, knot, knob; plot. **-abstand** internodal distance. **-ebene, -fläche** nodal plane, n. surface (opt.). **-linie** neutral line, common l., nodal l. **-punkt, -stelle** nodal point (opt.); junction p. **-strahl** beam or pencil of non-uniform section, in gas-focused c.-r. tubes, due to poisons and secondarv electrons. **-verbindung** transfer joint, temporary bridge.

Knüppelisolator club-type insulator, pin i.

knurren growl, snarl, rumble.

Kochbecher beaker.

koch-beständig resistant, stable or fast to boiling (dyes, etc.) **-echt** fast to boiling (dyes).

kochend(er) Akkumulator agitated or gassing accumulater (just disconnected from charger). **-(es) Geräusch** boiling noise.

Kochpunktmesser ebullioscope.

Koeffizient s.a. **Faktor** und **Beiwert**.

Koeffizient, Ausnutzungs- utilization coefficient. **Extinktions-** total-absorption coefficient, extinction c. **Gleitungs-** slip coefficient. **Mitführungs-** Fresnel coefficient of drag. **Schwächungs-** absorption coefficient. **Vereinigungs-** combination coefficient.

Koerzitivkraft coercive force, coercivity (sometimes: retentivitv). **-messer** coercimeter.

Koffer, Speisungs- current feed box. **-apparat** box equipment, trunk unit (for sound and picture pickup). **-kino** portable projector equipment.

Kohärenzlänge der Wellenzüge coherence length of wave-trains.

Kohäsionsfestigkeit cohesion strength.

Kohle coal, charcoal, carbon, carbonaceous material. -,salzgetränkte mineralized carbon, impregnated c. Docht- cored carbon. Effekt- salted carbon, c. for flame-arc. Härtungs- hardening carbon, cementing c. Hochintensitäts- high-intensity carbon (for arcs). Rein- solid carbon.

kohlefrei carbon-free, non-carbonaceous.

Kohlen-korn carbon granule, c. grain. -korngeräusch carbon noise (of microphone). -mikrophon, atmendes breathing of microphone (slow changes of resistance and response of carbon m.)

kohlen char, carbonize, carburize.

Kohlennachschub carbon feed.

Kohlepuppe (Trockenelement) carbon rod (of dry cell).

Kohlholz charcoal.

kohlschwarz coalblack, jet-black.

Kokosfaser coco fiber, coir (fiber from husk of coconut).

Kolben piston, plunger, ram, bulb, iron or bit (for soldering). Mess- graduated flask, measuring f., gage f. Röhren- bulb or vessel of a tube (radio). Schliff- ground-in glass stopper. -membran piston diaphragm. -stange piston rod.

Kollektiv des Okulars field lens of eyepiece.

Kollektor commutator, revolving contact-maker, distributer, rotary switch.

Kollimationslinie collimating mark (opt.).

Kollisionsverfahren interference action or procedure (law).

Koma-abbildung,-bild comatic image. -reste höherer Ordnung independent coma of higher order.

Kombinations-aufnahme combination exposure, composite shot, mask method of shooting. -blende Goerz effect shutter. -register combination stop (organ). -stimme mutation stop (organ). -töne combination (difference or summation) tones.

Komikbilder film cartoons, funnies.

Kommando-gerät data computer, ballistic director (gunnery). -stelle order point, order place, o. station (bridge); impulsing or control-action transmitter (remote control). -übertragung zwiwchen Flugzeugen inter-aircraft command or voice communication or signalling.

kommutieren reverse (polarity, connections, etc.)

kompakt compact, solid, non-apertured.

Komparator, selbstrechnender computing recording comparator.

Komparserieraum extras room.

Kompass, Kreisel- gyro-compass Ortungs-,Peil- navigation compass. Tochter- auxiliary compass, repeater c. Trocken- dry compass.

Kompass-büchse compass bowl, c. kettle. -kurs compass course, steered c. -rose compass card, c. scale. -strich point of compass.

Kompendium Goerz effect set (to obtain different effects, fading, etc. in motion picture work).

Komplementwinkel complementary angle.

komplexe konjugierte Grösse conjugate complex quantity.

Komponente, ausserordentliche extraordinary component (of rays). -,ordentliche ordinary component (of rays). Blind- wattless, idle, quadrature, reactive or reactance component (of impedance). Wirk- active, wattful, energy or inphase component.

Komprimieren der Laufzeit transit-time compression.

Kondensanz capacitive reactance, condensance.

Kondensationskalorimeter steam calorimeter.

Kondensator s. a. Kapazität

Kondensator, frequenzgerader straight-line frequency condenser or capacitor. (Note: Capacitor nowadays is preferable). -,gleichlaufender ganged condense synchronized condenser. -,kapazitätsgerader straight-line capacity condenser. -,quadratischer square-law condenser. -,unvollkommener leaky condenser. -,wellengerader straight-line wave (length) condenser. Abgleich- aligning condenser, trimming c. Blätter- foil capacitor, tubular c., Mansbridge (paper) c. Block- blocking condenser, stopping c., insulating c., fixed c. End- terminating condenser. Fest- capacitor, fixed or invariable condenser. Fühler- pickup or scanning condenser (used in sound reproduction). Gelenk- swing-out condenser, book c. Glättungs- smoothing condenser. Hochdruck- pressure-type capacitor (in nitrogen high-p. tank). Hütchen- ceramic hood-shaped capacitor. Mehrfach-, Mehrgang- multiple condenser, ganged c., synchronized c., tandem c. Mittellinien- logarithmic condenser (relation between capacity and angle obeys l. law). Nachstimmungs- aligning condenser, trimming c.; vernier c. Quer- shunt condenser, by-pass c., bridging c. Quetsch- compression type condenser. Schluss- terminating condenser, short-circuiting c. Sperr- stopping condenser, insulating c., blocking c. Trocken- dry electrolytic condenser. Überbrückungs- bridging condenser, by-pass c. Verkürzungs- shortening condenser, padding c. Vervielfältigungs- condenser-type d.c. multiplier. Wickelblock- tubular condenser, paper capacitor, mica c. (made of metallized or foil-coated insulation strip). Zeitkreis- time-base condenser.

Kondensator-abschlusswände end shields (forming "tub" of ganged condenser). -belagverlust surface leakage loss of a condenser. -klemme condenser type bush or terminal. -leitung high-pass filter. -mikrophon condenser microphone, electrostic m. -rückstand residual charge of condenser. -sirene condenser siren (produces pulsating current). -wanne tub-like container with end plates or shields of dielectric or sheet-metal to accommodate a condenser; ganged c., multiple c.

kondensierte, nicht-, Hochstromentladung uncondensed heavy-current discharge.

Kondensorspule condenser coil (of electron microscope).

Konduktanz, Röhren- transconductance (of a tube).

Kondukte wind trunk (of an organ).

konforme Abbildung conformable representation.

Königsholz king's wood.

Konizität conicity, conicalness, coning, taper.

konjugierte Funktionen conjugate functions. -komplexe Grösse conjugate complex quantity. -Schnittweite conjugate intercept.

Konkavgitter concave grating.

konphas in-phase, in p. coincidence, co-phasal.

Konsistenzmesser viscosimeter.

konsolidieren consolidate, set, concrete, solidify.

Konsonanten, stimmhafte sonant, voiced or vocal consonants. -, stimmlose non-vocal, surd or breathed consonants. -deutlichkeit consonant articulation.

konstant constant, invariable, non-varying, fixed, stable, steady, stabilized.

Konstante constant, parameter. Ansprech- sensitivity or response (of microphone). Dämpfungs- attenuation constant (real part of transfer or propagation c.) Dielektrizitäts- dielectric constant, specific inductive capacity, permittivity. Fortpflanzungs- propagation constant, transfer c. Gitter- grating constant, g. space; lattice c. Kraft- force constant. Richt- rectification constant. Verlust- attenuation constant. Verschiebungs- displacement constant specific inductive capacity (in vacuo). **Wellendämpfungs-** wave attenuation constant. **Zerfalls-** disintegration constant, transformation c., radio-active constant, decay coefficient.

Konstanthaltung stabilization.

konstruktiv constructional, practical, concrete, constructive.

Kontakt, ausgefressener pitted contact, worn c. **-, federnder** spring contact, flexible c., c. spring.

Kontakt, Abreiss- arcing contact (of a switch). **Arbeits-** marking contact, make or operating contact (closure occurring upon energization of relay). **Folge-** make-before-break contact; dummy c. **Gleit-** sliding contact (self-wiping and s.-cleaning). **Leerlauf-** spare or vacant terminal or contact. **Ruhe-** spacing contact, rest c., non-operating c. (closure occurs upon de-energization of relay); contact of rear part of key lever. **Schiebe-** adjustable slide, slide contact, cursor. **Schlepp-** make-before-break contact, trailing c. **Wackel-** variable contact, loose or defective c. **Wechsel-** make-and-break contact. **Wisch-** self-cleaning or -wiping contact.

Kontakt-abdruck contact print (phot.) **-arm** wiper. **-bürste** contact brush, wiper. **-finger** contact finger, wiper. **-geber** contact maker, time tapper, time marker. **-gleichrichter** contact detector, crystal d. or rectifier, natural d. **-hebel** contact lever. **-kleben** sticking of contacts. **-mittel** catalyst, contact agent, c. substance. **-prellen** contact chatter, c. click, c. thump, c. chirp. **-satz** bank of contacts. **-schieber** adjusting slider, slide contact, cursor. **-schiene** contact bar. **-schmoren** scorching, freezing or melting together of contacts. **-substanz** contact substance, catalyzer, catalyst. **-wirkung** catalysis, contact effect. **-zunge** spring contact, reed, blade, ditton. **-zusammenschmoren** scorching, freezing or melting together of contacts.

Kontingentierung quota system.

Kontinuitätsgleichung equation of continuity.

Kontinuum continuous spectrum.

Kontra-Alt contralto. **-Bass** -contra-bass (brass wind instr.) **-Fagott** contra bassoon.

kontrahierte Entladung contracted discharge.

Kontraktionskoeffizient Poisson's ratio (lateral contraction: longitudinal extension), rho ratio.

kontraktlich contractual, by contract or agreement.

Kontrast contrast, gamma (γ) **-empfindlichkeit, -empfindung** contrast sensibility or sensitivity, intensity discrimination. **-filter** filter for selective contrasts, contrast screen, c. filter. **-hebung** compandor action, dynamic-range action means (comprising compressor and expander) to improve signal-noise ratio. **-photometer mit Würfel** cube-type contrast photometer, contrast p. with cubical cavity.

kontrastreiches Bild contrast picture, high-gamma p.

Kontraststeigerung Callier effect, print contrast.

Kontrollbild monitoring picture (telev.).

Kontrolle control, check-up, supervision, monitoring, regulation, governing.

Kontrolle, Aussteuerungs-, mit Neonröhre neon tube volume indicator. **Einknopf-** uni-selector, uni-control operation. **Modulationsgrad-** modulation meter.

Kontroll-empfänger monitoring receiver. **-lehre** master gage. **-sender** monitor transmitter, check-up t.; service test oscillator.

Konturenversteilerung improvement in resolution and black-white contrast, insuring higher definition (telev.); expansion.

Konus, weitgeöffneter wide-open cone or pencil. Hochton- tweeter or treble cone loudspeaker. Riffel- corrugated cone. Tiefen- low-frequency cone loudspeaker, woofer. -antenne, Doppel- cage antenna. -membran diffusion cone, conical diaphragm; woofer.

Konventionalstrafe (monetary) penalty, contractual fine.

Konvergenz, Winkel- half convergency error.

Konvergenzwinkel convergence angle. -unterschied binocular parallax difference.

Konzentration, Anfangs- pre-focus. Elektronenoptik- focussing by electron lens. Gas- gas focusing (in c.r. tube), ionic f. Nach- re-focusing. Selbst- gas focusing, self-focusing. Vor- (cf Vorsammellinse) preliminary focusing, pre-focus, first focus action (in c.r. tube).

Konzentrieren focusing, concentration, fasciculation.

konzentrisches Kabel co-axial cable or line, concentric c., pipe l.

Konzert-flügel concert grand piano. -piano upright grand piano.

Koordinaten-bild pattern of co-ordination. -papier cross-section paper, ruled p., co-ordinate p. -system, Polar- polar co-ordinates. -system, rechtwinkliges rectangular co-ordinate system.

Koordinationszahl co-ordination number.

Kopf, verlorener deadhead, crop end, wastehead. Banden- band head, b. edge (spectr.) Düsen- nozzle tip. Einspann- fixing head, clamping h. Hör- sound (pickup) head. Lawinen- head of surge. Magnetisierungs- magnetic recording head (for sound, on wire or tape). Revolver- lens turret (camera). Ring- annular recording head. Schicht- striation head. Schreib-recorder head. Teil- index head. Verschluss- cable terminal.

Kopf-fernhörer headphones, headset, "can". -höranschluss telephone jacks. -hörerbügel harness, head band, strap. -welle onde de choc, head or front wave, bow w., impact w., shell w., ballistic w.

Kopie, Ansichts- release print. Arbeits- studio copy, s. print. Farb- color print. Gebrauchs- working copy, usable copy (for display). Grossumfang- high-range ("hi-range") or wide-range copy or print. Kleinumfang- "lo-range" print. Lavender- lavender print. Mehrfach- multiple print. Montage- first answer print. Muster- master copy, m. print. Photo- photostatic copy, photostat, photographic c., photoprint. Schnitt- edited print. Theater- release print. Ton- sound film used for reproduction. Verleih- release print, distributing p. Zwischen- lavender print, l. copy.

Kopieranstalt printing shop.

Kopieren, Durchlauf- continuous film printing. Verkleinerungs- reduction printing.

Kopier-fenster printing plane, p. station. -gamma negative gamma (γ).

Kopiermaschine printer. -, automatische automatic printer. -, optische projection printer, p. printing machine. Durchlauf- continuous film printer. Fenster- intermittent step printer. Schlitz- slit printer, contact printer.

Kopier-schwärzung printing density. -stelle printing drum. -werk printing studio, p. shop or plant.

Koppel, Oktav- octave coupler (organ).

Koppelnavigation navigation by dead

reckoning. **-ort** fix, dead-reckoning position (avigation).

Kopplung, direkte direct coupling (by a resistance, capacitance or inductance common to both circuits), impedance coupling, conductive c. **-,feste** close coupling, tight c., strong c. **-,galvanische** resistance coupling, galvanic c. **-,induktive** inductive coupling, transformer c. **-,kapazitive** capacitive coupling, electric c., electrostatic c. **-,lose** loose coupling, slack c., weak c. **-,veränderliche** vario-coupler **-,wilde** stray, spurious or undesired coupling.

Kopplung, Gegen- negative or stabilized feedback, reverse f., anti-regenerative coupling, degeneration. **Mit-** positive regeneration, p. feedback; spurious or undesired coupling or feedback. **Rück-** regeneration, reactive coupling, feedback c., reaction. **Sperrkreis-** parallel resonance coupling. **Spinimpuls-** spin orbit coupling. **Streu-** stray coupling, capricious c., spurious c.

Kopplungs-block invariable or fixed coupling capacitor. **-faktor** coupling factor, c. coefficient. **-röhre** coupling tube. **-spule, veränderliche** vario-coupler. **-verlauf** coupling coefficient, c. factor. **-widerstand** transfer impedance, coupling resistance.

Korb, Lautsprecher- loudspeaker chassis frame. **Schutz-** arcing ring, guard r. (of insulator).

Korb-bodenspule spiderweb coil, basket c. **-flasche** carboy, demijohn. **-wicklung** basket winding, low-capacitance w.

Kordel-griff milled knob, knurled k. **-mutter** milled nut, knurled n.

kordieren mill or knurl (an edge).

Korn front sight, muzzle s.; grain, bead, granule.

Korn. Raster- silver globule (of iconoscope mosaic), silver plug (in barrier-grid m.) **Schleier-** fog grain (of film). **Sucher-** backsight (of view-finder).

Korn-anhäufungen grain aggregations. **-bildung** granulation, crystallization.

Körner center point, punch mark, gage p. **-rauschen** carbon granule noise (microphone).

Kornet cornet.

Korngrenzen grain boundaries (cryst.)

Körnigkeit graininess, granularity, granulation.

Körnigkeitsmesser granulometer.

Korn-rauschen film-grain noise. **-zerfall** grain disintegration.

Körper, optisch inaktiver racemate, racemized material (quartz), racemic substance.

Körper, Fliess- fluid or liquid body, non-solid (substance). **Fremd-** foreign, external or extraneous matter or substance, undesired admixture, impurity, contamination. **Füll-** filling body, packing b. **Glüh-** incandescent body, incandescing b., incandescent mantle (Welsbach).

Körper(chen), Kern- nucleolus.

Körper, Lager- bearing box, housing or body. **Rotations-** body or solid of rotation or revolution. **Schicht-** stratified structure, s. body. **Stamm-** parent substance, p. body.

Körperbeschaffenheit constitution.

Körperchen corpuscle, particle.

körperlich solid, three-dimensional, bodily, corporeal, physical, material (e.g., of a cathode), molar.

körperlose Spule structure in which whole mass of vibrating system is contained in oil.

Körper-schaft body corporate, corporation. **-schall** sound conducted through solids. **-strahlung, Schwarz-** black-body radiation,

cavity r. -teilchen particle, corpuscle.
Korpuskeln, Licht- photons.
Korrektionsglied correction factor, corrective term.
Korrektur-abzug proof sheet. -scheibe compensating cam (df).
Korrespondenzprinzip (Bohr's) correspondence principle.
korrigiert, farb- color corrected, achromatized.
Korrosions-bildner corrosive, corrodent, corroding agent, rusting agent. -dauerfestigkeit corrosion fatigue strength. -empfindlichkeit corrodibility, susceptibility to corrosion. -verhinderung corrosion inhibition.
kosmische Ultrastrahlen cosmic rays.
kostenpflichtig abweisen dismiss (a suit or legal action) with costs.
krachen crack, crash, crackle, rustle.
Krachtöter (automatic) silent tuning; noise suppressor, noise gate (in film work, etc.)
krächzender Ton all-top sound or voice.
Kraft force, power, energy (force in mechanics = voltage or emf in electricity). -,auflösende resolving power. -,atombindende atomic combining power, valence. -,lebendige-kinetic force, vis viva, momentum.
kraft meines Amtes by virtue of (power vested in) my office.
Kraft, rücktreibende repelling force, deflecting f. (acting on electron pencil). -,schwingungserzeugende vibromotive force. -,zwischenmolekulare intermolecular force.
Kraft, Abstossungs- repulsion, repulsive force, repellency. Abstrebe- centrifugal force. Anstrebe- centripetal force. Antriebs- driving power, propelling force or p., propulsion. Anziehungs- attraction, attractive force or power, adhesive f. Bewegungs- motive or motor force or power, motional f., kinetic f. Bild- image potential, i. force. Brech- refractive power, refractivity. Direktions- directive force, directing f., versorial f. Erwärmungs- heating power, calorific p. Färbe- tinctorial power, dyeing p. Feder- elasticity. Flieh- centrifugal force. Gegen- reaction, counter-acting force, opposing f., reactive f., antagonistic f. Massen- inertance (of mechanical filter). Prall- resiliency, elasticity; ricocheting force. Rand- force acting at an edge, marginal force. Richt- directive force, directing f., versorial f. Rückstell- resiliency, restoring force, retractile f., elastic f., restitution force or pressure. Schieb- thrust, thrusting power, shear. Schnell- elasticity. Schwimm- buoyancy, floating capability. Schwing- centrifugal force; vibratory power, vibrating p. Spann- tension, extensibility, expandibility, elasticity. Spring- springiness, power of recoil. Stoss- percussive power, motive p. Teil- partial force, componental f. Thermo- thermoelectromotive force or power, Seebeck effect. Umfangs- circumferential force, tangential f., peripheral f. Verbindungs- combining power, bonding strength, affinity. Wurf- projectile force. Zieh- drawing power, tractive power, traction, attraction, pulling, dragging or hauling power or force. Zug- traction, tractive force, tension, attraction.
Kraft-audion power grid-current detector. -äusserung manifestation or effect of force. -berechnung, Bild-, halbklassische semi-classical

image force calculation. -fahrzeug motor vehicle, power craft. -funktion power function. -konstante force constant. -linie line of force, tube of force. -linienschräglage slope of force lines. -loserklärung declaration of invalidity or cancellation. -mass unit of force. -messer dynamometer. -messer, Koerzitiv- coercimeter.

kraftschlüssig (cf zwangsläufig) tensionally or positively connected (by positively acting means), force-locking.

Kraft-schwingröhre power oscillator. -stossmesser ballistic pendulum. -stufe power stage, final s. (of an amplifier). -wechsel energy exchange.

Krählrechen rake, rabble.

Kran, Gerüst- gantry crane.

Kranz, Rad- wheel rim, tire. Spiegel- row, array, rim or set of mirrors, mirror drum (telev. scanning). Zahn- gear rim, tooth rim.

Kranz-abtaster, Linsen- lens drum scanner. -brenner ring burner, crown b.

krarupieren krarupize, load (a line) continuously.

kratzen scrape, scratch, rasp, grate.

Kratzer, sichtbare visible scratches.

kräuseln curl, crisp, crimp, mill (coins).

Kräuselung rippling (of a d.c.).

Kreis (s.a. Stromkreis). circuit, circle, orbit, circular path.

Kreis, nichtschwingungsfähiger aperiodic or non-oscillatory circuit. -, widerstandloser resistanceless circuit. -, umschreibender circumscribing circle. Achtel- octant. Arbeitseich- working reference circuit. Bei- epicycle. Brems- negative anode circuit. Dunst- atmosphere. Fuss- root line, dedendum l. Halb- hemicycle, semicircle. Kipp- trigger circuit, time-base c., relaxation c. Krümmungs- curvature circle (lens). Kurs- direction or bearing circle. Licht- circle of light-halo. Luft- atmosphere. Mittags- meridian. Nutz- active circuit, useful c., signal c. Parallelresonanz- parallel resonance circuit, rejector c. Roll(bahn)- cycloidal path (of electrons). Saug- acceptor circuit, impedance or absorption wave trap. Schwungrad- parallel resonance circuit, flywheel c. Sechstel- sextant. Selektiv- selective circuit, selector c. Serienresonanz- series resonance circuit, acceptor c. Sieb- selective circuit, filter c. Spannungs- potential circuit (of ammeter). Sperr- rejector circuit, wave trap. Stoss- impulse circuit, impulsing c. Teil- pitch circle, p. line, divided c., graduated c., dial, g. circular plate. Übertrager- intermediate circuit, transformer c. Um- periphery, circumference; surroundings, environs, ambiency. Ureich- master reference system. Viertel- quadrant. Zerrungs- annulating network, suppressing n. Zerstreuungs- circle of confusion, blur circle.

Kreis-bahn circular path, orbit. -bahnbewegung motion in a circle or orbit, orbital movement, circular m. -bewegung circular motion, gyration. -bewegung, Rollbahn- cycloidal path motion. -blende iris diaphragm. -bogen arc of a circle; circular arch. -bogengitter arcuate grid.

kreischende Stimme all-top voice (or sound).

Kreisdiagramm circle diagram, circular loci. Doppel- tangent circle pattern, figure-8 diagram or (space) pattern (df).

Kreisel top, gyro gyroscope, gyrostat.

Kreiselektron spinning electron.

Kreisel-kompass gyro-compass. **-molekul** top molecule, spinning m. **-neigungsmesser** bank-and-turn indicator. **-pumpe** centrifugal pump, turbine p. **-verdichter** rotary compressor, centrifugal c.

kreisend circulating, rotary, gyratory, eddying, vortical (i.e., of vortex motion).

Kreis-frequenz pulsatance (ω), angular velocity, cyclic frequency, f. in radians. **-funkfeuer** rotating or omnidirectional beacon, circular navigational beacon (sending non-directional signals in all directions). **-funktion** trigonometrical function, angular f. **-grad** degree of a circle or arc, d. of angular measure, radian. **-keilschwärzungsmesser** circular wedge densitometer. **-lauf** cycle, circulation (of fluid), circular course, circuit. **-lochscheibe** film scanning disk with apertures circularly rather than spirally arranged.

kreismagnetischer Effekt gyromagnetic anomaly.

Kreis-messung cyclometry. **-parameter** circular parameter. **-prozess** cyclic process, cycle. **-scheibe mit veränderlichen Sektoröffnungen** sector disk (shutter) with variable apertures. **-spannung** "circle" voltage. **-spule** tuning coil. **-strome** circular currents, c. flowing in a circular path. **-welle** circular electric wave.

Kreuz sharp (mus.).

Kreuz, Abstimmungs- tuning indicator (magic eye, etc.). **Achsen-** coordinate system. **Faden-** reticule, reticle, graticule, cross line, hair l., cross spider hairs. **Leucht-** reticle image (reflector sight). **Malteser-** cam and star wheel of Maltese cross, Geneva intermittent movement, cross-wheel. **Strich-** cross lines, c. ruling, reticle.

kreuzen cross, intersect, traverse, transpose.

kreuzförmig cross-shaped, cruciform.

Kreuz-gelenk universal joint, Hooke's j. **-gelenkkupplung** cross joint. **-gesperre, Malteser-** Geneva stop. **-gitter** cross grating, surface lattice. **-glied** lattice section (of network or filter). **-hahn** four-way cock. **-libelle** cross level, **-modulation** cross modulation. **-peilung** fix, bearing taken from two or more ground direction-finders, cross bearing. **-rahmenantenne** crossed coil antenna.

kreuzsaitig cross strung, overstrung.

Kreuz-schaltung s.H- & T-Brückenschaltung. **-schlitten** compound slides. **-teilungsmast** cantilever type of tower (with double-tapered mast).

Kreuzungs-isolator transposition insulator. **-punkt, Über-** crossover (of electrons in c.-r. tube).

Kreuzverhör cross examination, cross questioning.

kribbeln swarming (of film).

kriechen crawl, sneal, creep (of materials), leak (of current).

Kriech-galvanometer quantometer, fluxmeter. **-geschwindigkeit** rate of creep (in metal). **-strom** creep current, sneak c., surface leakage c.

Kristall, einachsiges uniaxial crystal. **-,langstrahliges** needle-like elongated crystal. **-,linksdrehendes** levogyrate crystal. **-,optisch inaktives Quarz-** racemic quartz crystal, racemized c., racemate. **-,zweiachsiges** biaxial crystal.

Kristall, After- pseudomorph. **Biegungs-** "bender" crystal. **Doppelplatten-** biquartz crystal, bimorph c., bimorph cell. **Drehungs, Drillings-** "twister" crystal. **Ein-** monocrystal, single c. **Holoeder-** holohedral form of crystal. **Keim-** crystal nucleus, nucleus c. **Mehr-** polycrystal. **Misch-** mix crystal, mixed c., crystalline solid solution.

Quarz-, fertiges piezoid, finished blank. Quarz-, optisch inaktives racemic quartz crystal, racemized c., racemate. Schwing- crystal oscillator. Seignettesalz- Rochelle (salt) crystal. Spaltungs- cleavage crystal. Sphäro- spherical crystal, sphero c. Tannenbaum- dendrite. Zweielement- bimorph crystal (either of bender or twister type). Zwillings- (cf Doppelplatten-) twin crystal.

Kristall-achse crystal axis, crystallographic a. (x = electric a.; y = mechanical a.; z = optical axis). -achsenmesser conoscope (determines optical axis). -achsenrichtung crystallographic axis orientation. -anschuss crop of crystals, crystallographic growth, crystallization, shoot into crystals.

kristallartig crystal-like, crystalline.

Kristall-ausscheidung separation of crystals. -bau crystal structure, crystalline form. -baufehler crystal (structural) defects. -beschauer inspectoscope. -biegungsschwingung crystal flexural vibration. -bildung crystal formation, crystallization, granulation (of sugar). -dickenschwingung thickness vibration of a crystal. -drehungsschwingung, -drillungsschwingung torsional vibration. -ebene crystal plane, c. face. -ecke crystal angle. -faden, Ein- single-crystal filament, monocrystal f. -fassung (piezoelectric) crystal holder or mount. -fläche crystal face, c. plane. -gerippe crystal skeleton, skeletal c., crystallite. -gitter crystal grating (used for diffraction); crystal lattice. -gitterschwingung crystal lattice vibration. -halterung crystal holder, c. mount. -haufwerk crystallite aggregation. -inaktivierung, optische racemization.

Kristallisations-kerne centers of crystallization, nuclei of c. -wachstum crystal growth, germination.

Kristall-kanten crystal edges. -keim crystal nucleus, seed c. -kern nucleus of crystallization, center of c. -kernbildung crystallization, nucleation, germination (a germ being required in induced crystallization). -körperchen, rundes globulite, spherulite. -längsschwingung longitudinal vibration of a crystal. -lautsprecher piezo-electric loudspeaker. -lehre crystallography. -linse crystalline lens. -mikrophon piezo-electric microphone, crystal m. -querschwingungen transverse vibrations (including flexural v.), contour vibrations or oscillations. -schnitt crystal cut (x cut or Curie cut, also known as the zero angle cut, face-perpendicular cut and normal cut; y cut, 30-degree cut or face-parallel cut; T cut = 35 degrees to z axis). -schwingung crystal vibration. -splitter crystal grain or chip. -steuerung crystal stabilizing, crystal control. -taktgeber (beim Sender) crystal monitor, c. control means, c. stabilizer. -winkelmessung crystal goniometry.

kritisch critical, delicate.

Kröpfung crank.

Krücke rake, rabble, crowbar.

Krümmer, Rohr- pipe bend, elbow or knee. -radius radius of curvature, r. of gyration.

Krümmung, Bildfeld- curvature of image field.

krümmungsfreies Feld flattened field (by flattening lens).

Krümmungskorrektur field flattening.

Krummwerden warping, buckling, crooking, bending.

Kübel vat, tub, pail.

Kubieren raise to third power, involution process (math.)
Kubikwurzelziehen extraction of cube root, process of evolution.
kubisch-raumzentrierte Metalle body-centered cubic metals, cubic body-centered m. (cryst.)
Kubizierung calculate volume of; raising to third power, involution process (math.)
Kufe pressure pad, p. guide (m.p.) **Filz-** felt pad. **Gegen-** counter pad.
Kugel ball, sphere, globe, bullet, shot, bulb, bead.
Kugel, After- spheroid. **Halb-** hemisphere. **Ulbricht'sche** Ulbricht sphere.
kugelähnlich spheroidal, ball-like.
Kugel-blasen, Stahl- steel shot blast. **-blitz** flashbag; globular lightning, ball l.
Kügelchen small bulb, globule, spherule, globulite, pellet, small ball or shot.
Kugel-densitometer, integrierendes integrating spherical densitometer. **-druckprobe** ball pressure test, indentation or Brinell ball impression t. **-einsprache** spherical mouthpiece. **-faktor** coefficient of sphere (photometry). **-flasche** balloon flask, spherical f.
kugelförmig spherical, globular, ball-shaped, orbicular. **nicht-** aspherical, non-s. **-gewölbte Lamellen** dished laminae.
Kugel-funkenstrecke ball-gap, sphere gap, spark-gap with ball electrodes. **-funktion** spherical function. **-gelenk** ball-and-socket joint, universal j. **-gestalt** spherical form, ball shape, bulbous s. **-gestaltsfehler** spot size distortion. **-körperchen** globulite, spherulite (cryst.) **-kupplung** ball-and-socket joint, universal j. **-linse** spherical lens. **-mikrometer** spherical micrometer. **-mikroskop** ball stage microscope. **-photometer** sphere or globe photometer (of Ulbricht). **-rücksprung** ball rebound, ricocheting. **-schale** partial sphere, cup-shaped, or cap-shaped or hemispherical part, spherical shell, calotte. **-schliff** ground-in ball-and socket joint. **-spiegel** spherical mirror, s. reflector. **-spitzenfühlnadel** scanning or pick-up needle with spherical point (sound engraved film).
kugelsymmetrisch spherically symmetric, sphero-s.
Kugel-tisch ball stage (micr.). **-ventil** ball valve. **-welle** spherical wave.
Kühler cooler, refrigerator, condenser; radiator (automobile). **Rückfluss-** reflux condenser, return-flow c. **-mantel** cooling jacket, condensing j.
Kühl-mantel cooling jacket, condensing j. **-rippe** cooling or radiating, fin, vane or flange. **-schlange** cooling coil, spiral condenser pipe or tube.
Kühlung, Heiss- cooling by evaporation.
Kühlvorrichtung cooling, refrigeration or chilling device.
Küken plug of a cock.
Külbchen ball, piece, lump, parison (of glass).
Kulissenaufblendung side-curtain fade-in.
Kulturfilm cultural film, educational f.
Kümmerform degenerate form.
Kundengiesserei jobbing foundry.
Kündigung (giving of) notice, denunciation (of a treaty), termination.
Kundschaft clientele, customers, clients, following, goodwill.
Kunst-ausdruck technical term, art term. **-griff** artifice, trick. **-holz** artificial wood, imitation w. **-leder** leatherette, art leather, artificial l., l. imitation. **-leitung** artificial line, cable or

circuit, balancing or equalizing network.
künstlich artificial, synthetic. **-(e)Antenne** artificial antenna, mute a., dummy a., phantom a. **-(e)Radioaktivität** induced or artificial radio-activity. **-(er)Zug** forced draft.
Kunst-produkt, -stoff synthetic product, artificial p., synthetic, plastic. **-trick, Schleif-** optical grinding artifice.
Küpe vat, boiler, copper.
Kupfermantelrelais copper-jacketed relay.
Kuppe cusp, crest, peak, meniscus (of a liquid colum).
kuppeln couple, clutch, interlock, gang (e.g., condensers), join, unite.
Kupplung, Freilauf- free-wheel clutch, slip coupling. **Gelenk-** Hooke's joint, universal j. **Kardan-** universal joint. **Klauen-** claw coupling, c. clutch. **Kreuzgelenk-** cross joint. **Kugel-** ball-and-socket joint, universal j. **Leerlauf-** slip coupling, free-wheeling clutch. **Rutsch-, Schlupf-** slip clutch, s. coupling.
Kurbel und Welle, Einergang- single-picture crank and shaft (m.p.)
Kurbel, Tret- foot pedal, treadle.
Kurbel-antenne reel antenna, trailing-wire a. **-aufnahme** hand-drive shot or shooting. **-kröpfung** crank (of a shaft). **-scheibe** crank disk, cam. **-schleife** crank guide. **-zentrum** crank center.
Kurbler crank turner.
Kurs, ablaufender course or direction away from airport, outward course. **-, missweisender** magnetic heading, m. course, uncorrected bearing. **-, rechtvoraus** head-on course (fore-aft axis of craft directed to objective). **-, rechtweisender** true (north) course. **-mit Vorhaltewinkel** course heading directly into wind, crabbing.
Kurs, Grund- true course, head-on course. **Karten-** map course. **Kompass-** compass course, steered c. **Soll-** prescribed course, charted c. **Steuer-** steered course, compass c. **Vorhaltewinkel-** course heading directly into the wind, crabbing. **Wind-** course headed directly to objective, corrected c. heading in wind (same as map c. in absence of wind). **Wind-, rechtweisender** true heading. **Ziel-, ablaufender** course away from transmitter or beacon. **Ziel-, anlaufender** target course towards beacon. **Ziel-, missweisender** magnetic course to steer. **Zwischen-** compound course.
Kurs-absetzen laying off or plotting position lines or bearings, shaping course for. **-abweichung** (temporary) deviation from course, yaw. **-fehler** course error. **-funkbake** radio range, r. beacon. **-gleiche** loxodrome, rhumb line. **-kreisel** directional gyro. **-linie, achterliche** astern course, direct line away from transmitter. **-linie, vorliche** ahead course (towards beacon), "go-to" indication. **-peilung** course bearing (taking bearing by directional signals, to keep craft on course). **-richtung** head, course bearing. **-scheibe** bearing plate. **-stativ** class microscope. **-steuerung, automatische** automatic or mechanical piloting. **-weiser** radio range, r. beacon. **-weisung, rechtweisende** true course to steer, radio guidance or pilotage (from ground). **-weisung, missweisende** magnetic course to steer by radio guidance or pilotage (from ground). **-winkel** azimuth of target (in gunnery); magnetic a.
Kurve(s.a.Kennlinie, Charakteristik) curve, graph, diagram, characteristic. **-dritter Ordnung** cubic curve. **-, flache** flat curve, flat-topped curve(e. g., in broad tuning), flat characteristic. **-, flaue** flattened curve (not rising steeply), flat characteristic. **-gerissene** dotted-line or broken-line

curve or graph. -, jungfräuliche virgin curve. -, zweispitzige double-peaked curve, double-hump c. Abkling- decay curve. Auslauf- coasting curve, deceleration c. Begrenzungs- envelope, contour line. Blatt- folium. Durchlass- selectivity curve (of df receiver, etc.) Eichreiz- (typical) visibility curve. Empfindlichkeits- response curve, r. characteristic. Füll- single-loop oscillogram. Funkbeschickungs- error curve. H & D - Hurter & Driffield curve. Halsband- collar curve. Hüll- envelope. Längungs- elongation curve, stretch c. Leit- directrix, pitch or contour of cam compensator(df). Ohren- auditory sensation curve. Peil- directional diagram, space pattern. Resonanz-, mit zwei Höckern double-hump resonance characteristic or curve. Schlangen- serpentine. Schwärzungs- characteristic film or H & D (Hurter & Driffield) curve. Sinus- sine curve, curve of sines. Übertragungs- transmission (frequency) characteristic. Umrandungs- envelope. Verfestigungs- plastic stress-strain curve. Wachstum- growth curve. Wende- turn in landing (by ZZ method), U turn. Zapfen- spectral response of eye cones. Zerfall- decay curve.

Kurven-einebnung flattening of a curve. -form, rechteckige rectangular curve shape (say, of an emf). -lichtmethode variable-area recording method (of sound). -schaar family, group or system of curves or graphs. -scheibe, Funkbeschickungs- compensator or zero cleaning cam. -schreiber oscillograph, oscilloscope (when no graphic record is made), curve tracer. -schwanzstück toe region of film characteristic. -zeichnen mapping a curve, plotting a graph (say, with stylus). -zug, abklingender decay train.

kurz-brennweitiges Objektiv short-focus lens or objective. -brüchig short-brittle.

Kurzflammlampe hard light.

kurz-gefasst brief, concise, abbreviated, compendious, of reduced length or style. -halsig short-necked. -lebiges Produkt short-life, short-lived, periodic or transient product.

Kurzschluss short-circuit, "short," bridge, by-pass, short-circuit path. -, akustischer acoustic feedback or short-circuit. -, vollständiger dead short-circuit.

kurzsichtig near-sighted, short-s., myopic.

Kurzschluss-brücke, -bügel short-circuiting bridge. -läufer squirrel-cage rotor (of induction motor).

Kurz-wellenwandler short-wave converter. -zeichen symbol.

Küvette trough, vessel; bulb. Absorptions- absorption cell.

K Wert octantal component of quadrantal error (or QE) due to fuselage.

L

L Glied L-type section, filter s., mesh of network.
L Regler L-section attenuator.
Labialpfeife flue pipe (of organ).
labil unstable, non-stable, labile.
Labium lip (mus.).
Labyrinthdichtung labyrinth seal, l. gland (of turbine).
Lack, Tauch- dipping varnish.
Lackmuspapier litmus paper.
La Cour'sches Rad phonic wheel.
Lade, Wind- wind chest, sound box (of organ). **-pumpe** charging pump, compression p.
Lader, Klein- trickle charger.
Ladeschaltung circuit means causing slow rise of sawtooth voltage.
Ladestrom charging current (of dielectric).
Ladung, freie free charge. **-, ganzzahlige** integral spin, i. charge. **-, gebundene** bound charge, induced c. **-, punktförmige** lumped loading. **-, ungebundene** free charge. **Einheits-** unit charge.
Ladungs-ausgleich charge equalization. **-austauschoperator** charge exchange operator. **-bild** charge pattern, c. image (of iconoscope). **-dichte** charge density. **-dichtesteuerung** charge density modulation. **-einheiten, positive elektrische** unit positive charges. **-sinn** nature of a charge, sign. **-speicher-Bildabtaster, einseitiger** single-sided mosaic-type of storage iconoscope. **-speicherung** charge storage, c. accumulation. **-träger** charge carrier. **-trägertrennung** sorting or separation of ions and electrons (in a mass spectrometer). **-vermögen, elektrisches** electrostatic capacity. **-zahl, Kern-** nuclear charge number. **-zusammenballungen** bunching of charges, b. of electrons.
Lage layer, stratum, bed, coat, coating, state, situation, position, location, site; bearing (df).
Lägel barrel, keg.
lagenrichtig positionally or geometrically correct.
Lagensicherung locking in position.
Lager bed, layer, stratum; bearing, bush. **Mit einem einzigen - versehene Drehspule** unipivot moving coil. **-mit Weissmetallfutter** babbitted bearing.
Lager, Gegen- abutment. **Gleit-** slide bearing, plain b. **Hals-** journal bearing. **Kamm-** (concealed) collar thrust bearing. **Spur-** thrust bearing. **Spurzapfen-** end thrust bearing. **Steh-** pillow block bearing, pedestal b. **Stein-** jeweled bearing. **Weissmetallfutter-** babbitted bearing.
Lager-bock pedestal of a bearing. **-büchse, -buchse** bearing bush, bushing of a b. **-element** inert cell. **-härtung, Hart-** induction hardening. **-körper** bearing box, b. house. **-prozess** ageing, storage process.
Lagerung, Schneiden- knife edge supporting or suspension. **Spitzen-** pivot jewel, jewel bearing, point suspension or support. **Stift-** pin suspension, p. supporting.
Lagerzapfen journal pin.
Lambertgesetz cosine emission law, Lambert l.
Lamelle, Pumpen- vane (in Gaede vacuum pump).
Lammellen, kugelförmig gewölbte

dished laminae.

Lamellen-kern laminated core. **-magnet** compound magnet, laminated m. **-spiegel** strip-like mirror (in screw scanner).

laminieren laminate; draw (in spinning).

Lammfell lambskin.

Lampe, punktförmige lamp with point-shaped or crater light source (e.g., tungsten arclamp, Pointolite). **Aufnahme-** studio light. **Bienenkorb-** beehive neon lamp. **Bildeinstell-** framing lamp. **Blink-** flash lamp. **Blitz-** flash (light) lamp. **Bogen-, sprechende** speaking arc, Simon or Duddell. **Effekt-** studio light. **Einstell-** framing lamp. **Erreger-** exciter lamp, recorder or reproducer l. **Flächen-** bank of lamps, pillar of l. (studio, etc.), battens (stage). **Flächenglimm-** flat-plate neon tube. **Glimm-** glow tube (e.g., neon tube or lamp), gaseous-conduction t., flashing lamp (for sound recording). **Kathodolumineszenz-** cathode glow-tube. **Klein-** miniature lamp. **Kurzflammen-** hard light. **Langflammen-** soft light. **Leuchtstoff-** fluorescent, luminescent or luminous lamp or light, tube using phosphor substance. **Mehrfaden-** multiple-filament lamp. **Melde-** pilot lamp, telltale l., alarm l., indicator l. **Mikronernst-** Nernst microscope lamp, **Polsuchglimm-** polarity finder glow-tube. **Punkt-** point-type neon tube (e.g., tungsten arc lamp, Pointolite). **Pust-** blow lamp. **Quecksilberdampf-, mit Steuerelektrode** mercury-vapor tube of thyratron or ignitron type with control grid. **Reduktor-** adapter-transformer lamp. **Reserve-** spare lamp. **Säulen-** pillar (lamp), battens. **Schlitz-** slit lamp (micr.). **Signal-** signal lamp, telltale l. **Sofitten-** (overhead) scoop lamp, scaffold l. **Spiegelbogen-** reflected (or reflector) arc- lamp. **Ton-** exciter lamp (operating photocell). **Wolframmbogen-** pointolite.

Lampen-aggregat bank of lamps, pillar of l. **-arbeit** blast-lamp work (glass making). **-arm** lamp bracket. **-feld** bank of lamps, panel of l. **-fuss** lamp squash, l. press. **-gehäuse** lamp chamber, l. housing. **-stempel** lamp squash, l. press. **-überlastung** overrunning of a lamp (operated under overvolted condition).

Landé Intervallregel Landé interval rule.

Landeachse correct bearing or line for approach (and landing, of an airplane). **Blind-** blind or instrument landing line or path.

Lande-bahn landing runway. **-dienst, Blind-** beacon service (to assist in blind or instrument landing of airplanes). **-funkfeuer** landing radio beacon. **-platz** landing field, l. strip, airport, airfield. **-platz, Hilfs-** emergency or auxiliary landing field or airport, landing strip.

Landesgericht national court.

Landfunkstelle ground radio station, shore or coast s.

Landgericht district court or county c., general court of justice.

Landung, Nebel- blind landing, instrument l.

Landungsfunk-bake, -feuer landing beacon (for instrument landing).

langbrennweitige Linse long-focus lens or objective.

Länge, Ost oder West east or west longitude.

langelförmig of extended oval shape.

Längen-ausdehnung linear expansion. **-bruch** longitudinal fracture. **-durchschnitt** longitudinal section. **-mass** linear measure. **-messung, echte und unechte** direct and indirect measurement of length.

längentreu isometric.

Langflammenlampe soft light.
lang-gestreckt extended, elongated, oblong. -gliedrig long-linked, long-membered. -halsig long-necked. -lebig long-lived. -lebiges Produkt long-life product.
länglichrund elliptical, oval, lenticular.
Langloch oblong hole, slot. -hebel slotted lever.
Längsachse major axis, longitudinal a., fore-aft line (ships, airplane, etc.).
langsam abfallendes Relais slow-release relay. -ansprechendes Relais slow-response relay, slow-operating r. -lösend slowly releasing or disengaging; s. dissolving.
Langsambinder slow-setting cement.
Längsarm line-arm, series a. (of a network).
Langschliff long-fibered (mechanical) pulp.
Längs-durchschnitt longitudinal section. -entzerrer series-type attenuation equalizer or compensator. -feld longitudinal field, f. parallel to axis, paraxial f. -glied(einer Kette) series element, s. mesh (of network). -glied, in 1/2 - endendes Filter mid-series terminated filter. -peilung taking bearings from d.f. station situated more or less in direction of travel or course. -schärfe definition in line direction.
längsschiffs fore and aft.
Längsschwingung longitudinal vibration.
langstrahliges Kristall elongated, needle-like crystal.
Langstreckendienst long-range or long-distance avigation or navigation service.
Längsstreifen longitudinal stripes, l. striae, l. streaks.
Längs- und Querverschiebung axial and cross adjustment (electron microscope).
Längs-widerstand (Filter) line resistance, series r. (of a filter). -zweig line arm, series arm.
Längung extension, elongation.
Längungskurve elongation curve, stretching c.
lang-wellig long-wave, of long wavelength. -wierig protracted, wearisome, tedious, time-consuming.
Lappen flap, tab, lobe, flange, lug; rag. -schraube thumb screw.
lappig lobed, lobate; flabby, flaccid.
Lärm-abwehr, -bekämpfung combating of noise, noise abatement.
Lasche bond, cover plate, fish plate, splice strip (rail track).
Last, Gebrauchs- working load, service l.
Lastwechsel cyclic stress, s. application cycle. -zahl alternating stress number, n. of load alternations.
Lasurstein lapis lazuli.
Laternen-bilderkasten lantern slide box. -bildermasken lantern masks. -einfassleisten lantern slide binding strips.
lau tepid, lukewarm, mild, balmy.
Laueflecke oder -punkte Laue spots.
Lauf course, path, run, race, current; barrel (gun). Heim- return or back stroke or motion, flyback (of pencil or cathode spot). Hin- forward move, f. working stroke, travel, pass or journey, active-line scansion (of spot or pencil, in telev.). Kreis- cycle, circulation (of a fluid), circular course, circuit. Leer- no-load operation, idling. Rück- return or retrace, flyback (of pencil or cathode spot).
Lauf-bahn einer Rolle mit Gummiüberzug rubber tread of pad roller. -bild moving picture, animated p., one of a sequence or series of pictures or pictorial actions. -bodenkamera base-board extensible

camera. **-einschaltung** throwing in (of a gear), starting to run (of a mechanism, etc.).

laufend(en) Band, am—fabrizieren conveyor-belt manufacture, assembly line or large-scale m. **-(es) Maass** linear measure. **-(e) Nummer** serial number, consecutive n. **-veränderlich** running variable.

Läufer rotor; cursor contact, sliding c.

Läufer, Kurzschluss- squirrel-cage rotor (of induction motor).

Läufermühle edge mill.

Lauffläche tread, bearing surface, journal. **-gewicht** sliding weight. **-glasur** flow glaze.

läufig, links- moving to left, m. counter-clockwise.

Lauf-richtungskoeffizient directional effect (in film processing). **-ring** race, raceway, ball-race (of a bearing). **-schaufel** runner blade, bucket (of turbine). **-umschaltung** reversal of motion. **-werk** feed mechanism, movement, working gear. **-werk, Ton-** sound-film feed mechanism. **-winkel** transit angle. **-zahl** variable number, running variable. **-zapfen** journal, neck (of rolling mill). **-zeit** transmission time, transit t., propagation t. **-zeit, Elektronen-** electron transit time, orbit time of electrons. **-zeit-einfluss** (electron) transit time effect. **-zeitensteuerung** velocity modulation.

Laufzeit-geräte (cf. **Lautzeitröhre**) devices predicated on transit time wherein electrons are bunched in proper phase to result in space-charge concentrations and formation of groups. **-glieder** means involving transit time effects or phase delay. **-komprimierung** transit time compression. **-korrektur** echo or delay weighting term (telephony), delay equalizing. **-röhre** drift tube, beam t., t. involving transit time or phase effects. **-schwingungen** electron oscillations. **-störungen** delay distortions. **-verzerrungen** transit-time distortion, phase d.

Laugenflüssigkeit liquor.

launisch capricious, erratic, fickle.

lauschen listen in.

Lauschmikrophon pickup microphone, p. transmitter detectophone (connected to head-phones).

Laut (articulated) sound. **-, explosiver** explosive sound. **Selbst-** vowel. **Zisch-** sibilant, friccative.

Laute lute.

lauter pure, clear, genuine, unalloyed, sheer.

läutern purge, refine, purify, clear, clean, cleanse, clarify, rectify, wash.

lautgetreu of high fidelity, orthophonic.

Lautheit loudness.

lautlos silent, quiet, noiseless.

Lautschriftband sound recording steel tape.

Lautsprecher, elektrostatischer capacitor or condenser loudspeaker. **-, fremderregter dynamischer** electrodynamic or moving-coil loudspeaker. **-, gerichteter** directional loudspeaker. **-, induktordynamischer** moving-iron loudspeaker. **-, permanent-dynamischer** permanent-magnet(pm) or moving-coil dynamic loudspeaker. **-, trichterloser** direct-radiator loudspeaker (without horn).

Lautsprecher, Abhör- monitoring loudspeaker. **Druckkammer-** pneumatic loudspeaker (without diaphragm), pressure-chamber l. **Exponentialtrichter-** exponential or logarithmic loudspeaker (with uniform or multiple flare in horn). **Falten-** folded, curled, coiled or twisted horn loudspeaker. **Freischwinger-** direct-radiator loudspeaker, free-r. l. **Gross-** power loudspeaker.

Grossflächen-, Hochleistungs- power loudspeaker, (public) address l., announce l., paging l., call l. Hoch(ton)- tweeter loudspeaker, treble l. Konusmembran- cone loudspeaker. Kristall- piezo-electric loudspeaker. Luftkammer- air-chamber loudspeaker (c. coupling diaphragm and horn). Pilz- omni-directional (360°) exponential horn loudspeaker (mostly mounted on pole). Pressluft- pneumatic loudspeaker, compressed-air l. (without diaphragm). Riffel- Riffel loudspeaker (with corrugatiins and substantially horizontal radiation). Riffelfalten- folded and corrugated horn loudspeaker. Tieftonkonus- low-frequency cone loudspeaker, woofer. Vierpol- balanced-armature loudspeaker (with permanent magnet).

Lautsprecher-abstrahlfläche projector surface, radiating s. (of a loudspeaker). -anlage, Betriebs- works address or public a. loudspeaker system (in factories, etc.). -antrieb motor element. -haltevorrichtung chassis of a loudspeaker. -horn, aufgewundenes twisted loudspeaker horn, coiled h. -horn, gefaltetes folded loudspeaker horn.

Lautsprecher-korb chassis frame of loudspeaker. -leitfläche deflecting vane, labyrinth. -leitwand deflecting vane; baffle (to prevent acoustic feedback). -luftkammer air space, a. chamber. -spinne spider of loudspeaker (inside or outside voice coil, to keep coil centered). -tonführung labyrinth of loudspeaker.

Lautsprechertrichter, aufgewundener twisted loudspeaker horn, coiled h. -, gefalteter folded loudspeaker horn. -hals throat of loudspeaker horn. -mundöffnung mouth opening of horn, largest flare. -mundstück throat (narrow inlet) of horn.

Lautsprecher-vorhof air chamber of loudspeaker (between horn and cone). -wagen sound truck, loudspeaker t.

Lautstärke, Empfindungs- response of ear to signal stimulus, sensible loudness of sound. Stör- noise level.

Lautstärkemesser, Neonröhre- neon tube volume indicator.

Lautstärken-bereich der Stimme human voice intensity or dynamic range. -empfindung aural sensitivity. -stufe loudness unit, sensation unit.

Lautstärke-regler, logarithmischer volume control with resistance varying according to a logarithmic law. -schwankungen fadings. -umfang volume range, dynamic r.

Lavenderkopie lavender copy, l. print.

Lawine, Elektronen- electron avalanche.

Lawinenkopf head of a surge.

lebender Kalk quicklime.

lebendige Kraft kinetic force, vis viva, momentum.

Lebendigkeit liveness, realistic properties (in architectural acoustics).

Lebensdauer life period, (working or useful) life, life time (say, of a quantum).

Lebewesen, Klein- micro-organism.

lebhaft lively, brisk, active, vivacious; bright, vivid, gay (of colors).

lebig, lang- long-lived.

Lecherleitung Lecher-wire line.

Leckstrom leakage current, leakance c.

Lederharz (India) rubber. -stulpen leather cuffs.

leer empty, blank, void, vacant, hollow.

Leere vacuum, emptiness, vacancy.

Leergewicht tare, weight allowance for container, weight empty.

Leerlauf no-load (operation), idling.

-impedanz open-circuit impedance, no-load i. **-kontakt** spare or vacant terminal or contact, dumb or dummy c. **-kupplung** slip coupling, free-wheeling clutch. **-nocke** delayed pulse trip cam. **-widerstand** open-circuit impedance.

Leer-scheibe idler (wheel). **-stelle** hole, vacancy (in lattice). **-stellenmechanismus** vacancy mechanism (metals). **-versuch** blank experiment, no-load test.

Legierung, sparstoffarme alloy low in critical or scarce metals. **Automaten-** machining alloy stock, alloy readily workable in automatic machine tools. **Knet-** plastic alloy, malleable a., wrought a.

Lehm-guss loam casting. **-stein** unburnt brick, adobe.

Lehrdorn inner cylinder or plug gage.

Lehre, Augen- ophthalmology. **Bewegungs-** mechanics (re:machines); kinetics (re:motion of bodies and forces acting thereon); kinematics (re:motion in the abstract); dynamics (force action on bodies). **Bewegungselektrizitäts-** electrokinematics, electrokinetics. **Dreiecks-** trigonometry. **Elektronen-** electronics. **Farben-** science of color, chromatics. **Filmkitt-** film splicing gage. **Fühler-** thickness gage. **Getriebe-** kinematics. **Gefrierpunkts-** cryoscopics. **Gewinde-** thread gage. **Gleichgewichts-** statics. **Grössen-** mathematics, geometry. **Innengrenz-** plug gage. **Kegel-** taper gage. **Kontroll-** master gage. **Kristall-** crystallography. **Licht-** optics, photology (or physiological o.). **Loch-** internal caliper, plug gage. **Natur-** natural science, n. philosophy. **Normal-** normal gage, master g. **Phasen-** solution theory, doctrine of phases. **Rachen-** gap gage. **Ruheelektrizitäts-** electrostatics. **Schraub-** micrometric gage, micrometer. **Schub-** sliding gage, s. caliper. **Seh-** optics. **Uhrschraub-** dial micrometer. **Wärmekraft-** thermodynamics. **Wasserkraft-** hydrodynamics. **Wurf-** ballistics. **Zahlen-** arithmology, numerology.

lehrenhaltig true to gage.

Lehr-film educational film, instruction f., school f. **-satz** abstract theorem, a.doctrine. **-schau** educational exhibition.

leicht-flüchtig readily volatile. **-flüssig** easily fusible, e. meltable, e. liquefiable; mobile. **-löslich** easily soluble.

Leichtmetall light metal (of low sp gr).

leicht-schmelzbar easily fusible, e. meltable. **-siedend** low-boiling, of low ebullition (point).

Leim glue, size.

leimartig gluey, gelatinous, glutinous.

Leine line, chord, cord.

leise Kassette silenced magazine.

Leiseabstimmung silent tuning.

Leiste ledge, border, ridge, molding, strip. **Einfass-** binding strip.

Leistung power, work, performance, energy. **Abschalt-** rupturing capacity, circuit-breaking c., circuit-opening c. (of a switch). **Blind-** reactive power, r. volt-ampere, var. **Schall-** response (of an acoustic system). **Schein-** apparent power. **Spitzen-** record (achievement, performance or accomplishment). **Verlust-** (power) dissipation, power loss. **Verzerrungs-** distortion power (in d. volt-amps.). **Watt-** wattage, power in terms of watts. **Wirk-** (cf **Blind-**) active power, actual p., true p. (in watts).

Leistungs-aufwand power input, p. expenditure, energy dissipation. **-bedarf** power absorption, p. dissipation, p. requirements. **-bedarfspegel** power reference level. **-entnahme** energy output, power o., delivery, absorption or pickup of energy or power; taking, tapping or draining of power or energy. **-fähigkeit** performance, efficiency, serviceability, power, ability, resolving power (of a microscope). **-faktor** power factor. **-faktorzeiger** power factor indicator or meter. **-kammer** energy chamber (of a klystron).

leistungslos non-dissipative, wattless.

Leistungs-messer power meter, ergometer. **-reserve** margin of power, standby p. **-umsatz** energy exchange, transduction. **-verstärker** power amplifier.

Leitbahn transit path. **-bewegung** transit path motion. **-welle (Magnetron)** transit-time or electron path oscillation (of magnetron).

leiten conduct, guide, lead, govern, steer, pilot, route (a message).

Leiter, Aufnahme- first cameraman, chief c. **Aussen-** "outer", outside conductor or wire. **Halb-** semi conductor, poor c. **Heiss-** third-class conductor, hot c. (with negative resistance). **Hin-** outgoing lead, o. wire, o. conductor. **Ketten-** iterative network, recurrent structure. **Null-** neutral wire. **Rück-** return wire, r. lead, r. conductor. **Ton-** musical scale (natural, diatonic or just scale, and tempered s.). **Voll-** compact or unstranded conductor.

Leiter-gebilde conductor structure, c. system. **-gebilde, Luft-** aerial extension, a. structure, a. network, array of antennae. **-stück** element of conductor, conductor element. **-system** conductor system, c. structure.

Leitfähigkeit, dielektrische leakance, dielectric conductance. **-,elektrische** conductance. **-,magnetische** magnetic conductance, permeability, permeance. **-,spezifische** conductivity. **Super-** supra conductance, super c. **Temperatur-** temperature diffusivity, thermal conductivity. **Über-** supra conductivity, super c.

Leit-fähigkeitsgefäss conductivity cell. **-fläche** deflecting vane (mounted in horn to reduce directional properties of a loudspeaker), labyrinth, baffle. **-funksender** radio range, r. beacon. **-funkstelle** net control station. **-kabel** leader cable, pilot c. (d.f.). **-kante** leading edge, l. tip (of airplane wing). **-kreisfrequenz** transit-time frequency (of a magnetron). **-kurve** directrix; pitch or contour (of a cam compensator). **-linie** beacon course, beam, equi-signal line, e.-s. course or zone; directrix. **-richtung** course on beam or equisignal line. **-rolle** guide roller, idler, pad roller (without sprockets). **-rolle, federnde** springloaded idler. **-satz** rule, guiding principle, basic p. **-stelle** ground direction-finder station (sending directional or position data to craft). **-strahl** guide-ray, localizer beam, radio beacon course, landing beam, radius vector (math.). **-strahl, Strecken-** long-distance or route navigation (or avigation) beam. **-strahlsender** beacon, radio range or beam station, aerophare. **-strahlwinkel** angle of beam. **-strom** conduction current; pilot c. (in wired radio). **-system, Licht-** condenser relay system (in film printing).

Leitung conduction, conductance, transmission, electric circuit, wire, (supply) line, cable, lead conduit, guide, piping, mains. **-,angepasste** matched line. **-,konzentrische** co-axial line or cable, concentric c., pipe l.

-,kurzgeschlossene short-circuited line. -,offene (am Ende) open-ended line. -,verkabelte cable line, stranded-wire l. -,zusammengesetzte compound circuit.

Leitung, Antennenring- multiple receiver connection to an antenna. Ausgleichs- balancing network, b. line. Energie- energy feeder lead, e. leader, downlead, transmission line. Fern- long-distance line, toll circuit. Hohlrohr- conductive (hollow or pipe) line, wave-guide. Kondensator- high-pass filter. Kunst- artificial line, a. cable, a. circuit, balancing or equalizing network. Lecher- Lecher-wire circuit or line. Quer- shunt line, cross l.; leakage conductance, leakance. Regler- pilot wire. Rohr- pipe line, co-axial cable, concentric l. Rückwarts- high-resistance direction (of rectifier). Simultan- earthed phantom, composite circuit. Speise- (aerial) feeder lead, down-lead, energy feeder cable or wire. Umweg- by-pass lead. Vorrats- spare circuit, standby c. Weg- directions for tracing or tracking trouble or for "trouble shooting". Wellen- ultra high-frequency feeder line; wave guide (either dielectric or conducting).

Leitungs-dampfung line loss, l. attenuation. -elektronen conduction electrons (from outer levels of atoms). -fehler line failure, line fault.

leitungs-gerichtete Hochfrequenztelephonie wired or carrier current RF telephony, t. along wires by RF. -gesteuerter Oszillator line-controlled oscillator.

Leitungs-induktivitat lead inductance, line i. -messer conductometer. -nachbildung balancing network, artificial b. line. -rundfunk wire broadcast, AF rediffusion. -schema diagram of connections, circuit d., hook-up. -schleife loop, metallic circuit. -strom-Modulation conduction-current modulation. -stück stretch of line, portion of l., line element. -taubheit (im äusseren Ohr) conduction loss or impairment of hearing, conduction deafness (in outer ear). -übertragung wired radio, w. wireless, carrier telephony or telegraphy, wire broadcast. -verstärker telephone (circuit) repeater, communication line r. -wasser tap water, city mains w. -wechselstrom conduction a-c.

Leitvermögen, richtungsabhängiges asymmetric conductance. -,spezifisches specific conductance, conductivity.

Leitwand baffle, deflecting vane (of loudspeaker).

Leitwert admittance, conductance, susceptance (being, respectively, the reciprocals of impedance, resistance and reactance, the latter either inductive or capacitive). -,spezifischer conductivity. Blind- susceptance. Eingangs- input admittance. Eingangswirk- input active admittance, input conductance. Kenn- indicial admittance. Rückwirkungs- reactive admittance, susceptance (either inductive or capacitive). Schein- admittance. Ubergangs- transconductance, transfer characteristic. Wirk- conductance.

Leit-zahl, Wärme- thermal conductivity. -zone equi-signal or glide zone or path.

Lenardröhre c.-r. tube of Lenard (used in radiology and early experiments on phosphors).

Lese-band (ore) picking belt conveyor. -glas, viereckiges oblong reading glass.

Lettern-gut, -metall type metal.

lettig clayey, clayish.

Leuchtdauer illuminated period, light p., period or duration of glow, luminescence or luminosity; phosphorescence p.

Leuchtdichte brightness (in terms of stilb, sb, or apostilb, asb, or other units), intrinsic brilliance (of a luminous surface). **Bildwand-** screen brightness, s. luminous density. **Dreifarben-** tricolor RD (reflection density) value. **Schwarz-** brightness of black.

Leuchte lamp equipment (m.p.). **Flächen-** bank or pillar of lamps, battens. **Natrium-** sodium lamp.

Leuchtelektronen emitting electrons, optical e.

leuchten emit light, shine, glow, luminesce.

Leuchten, Nach- afterglow, phosphorescence, persistence of light or of vision (of an image, on retina).

leuchtende Flamme luminous flame.

Leuchtenergie erzeugend photogenic, fluorogenic, phosphorogenic (as by manganese, in zinc sulfide).

Leuchter, Selbst- self-luminous substance, fluorescent s., phosphor.

Leuchtfaden luminous or glow column; streamer. **-einsatz** streamer onset.

Leucht-farbe luminous paint. **-fleckdurchmesser** spot diameter. **-gebilde** luminous pattern (lines, bands, or continuous spectra). **-hülle** luminous sheath, l. envelope; photosphere (of sun).

Leuchtkraft illuminating power, luminosity. **-, gemessene** luminance, photometric brightness. **-bestimmung** photometry. **-verminderer, Nach-** poison, killer (of phosphorescence). **-verstärker** phosphorogen.

Leucht-kreuz reticle image (reflector sight). **-masse, -material** phosphor, luminescent or phosphorescent substance or material. **-mittel** illuminant. **-punkt** (luminous) spot. **-rakete** flare. **-resonator** glow-tube resonance indicator. **-röhre** luminous discharge tube; fluorescent t. **-schirm** fluorescent screen, luminescent s., phosphor s.; target. **-schirmsubstanz** phosphor, fluorescent substance (for c.-r. tube screens). **-spur** tracer trajectory. **-strahlquelle, einfarbige** monochromatic illuminator, monochromator. **-stofflampe** fluorescent lamp, luminous or luminescent l. or tube, t. containing phosphor. **-streifen (bei Entladung)** luminous streamer (in discharge). **-verfahren, Selbst-** self-luminous or s. -emissive method (using electrons on object surface). **-Zeiger** dial light resonance indicator. **-zifferblatt** luminous dial.

L Glied L-type section, filter mesh or network m.

Libelle level, spirits or water l. **Kreuz-** cross level.

licht(e) light, pale, bright, luminous, clear, inside (of diameter). **-(e)Weite** inside diameter, i. width, lumen, clear aperture (of lens), useful or effective d.

Licht, auffallendes incident light. **-, ausfallendes** emergent light, transmitted l. **-, direktes** specular light. **-, durchfallendes** transcident light, transmitted l. **-, einfallendes** incident light. **-, einfarbiges** monochromatic light, homogeneous l. **-, eingestrahltes** incident light, exciting l. **-, falsches** light fog, leakage l., stray light (film printing). **-, gerichtetes** parallel light rays. **-, spektral zerlegtes** spectroscopically dispersed or separated light. **-, unzerlegtes** undispersed light.

Licht, Blitz- flashlight, flash bulb, magnesium light, lime l. **Flut-** floodlight. **Gegen-** backlight, counter light (m.p.). **Kathoden-** cathode glow. **Neben-** light-shot (film). **Nord-** aurora borealis. **Ober-** overhead scoop light, scaffold l., skylight. Ruhe- unmodulated light, no-modulation l., average or steady illumination, no-sound lighting (in sound recording). See- marine phosphorescence. Sofitten- overhead scoop, scaffold light. Spitzen- highlights (of a picture). Streu- stray light. Sucher- searchlight. Süd- aurora australis. Ton- exciter lamp (operating a cell). Wechsel- brightness variations (striking photo-cell), light intensity v.

Licht-abstufung gradation of light. -anzeiger, Glimm- flashograph, Tune-A-Light tuning device. -ausbreitung auf Film distribution of exposure on film.

Lichtbild photo-image, photograph. -,unvergängliches imperishable picture or photograph. -entzerrung rectification of aerial photographs. -kunst, Klein- microphotography, photomicrography.

Licht-blende light stop, diaphragm, apertured partition; light shield (in sensitometer). -blitz pulse of light, flash of l., scintillation. -bogen, Erdungs- arcing ground.

lichtbogenbeständig arc-proof, a. resistant.

Lichtbogen-erdung arcing ground. -löschkammer arc quench chamber.

lichtbrechend optically refracting, light r., refractive or refringent.

Licht-brechungsvermögen (optical) refractive power, refractivity. -bündelvignettierung vignetting of cone of light. -chaos chaotic light (of eye or retina), photopsy.

lichtchemisch photo-chemical, actinic.

Lichtdämpfer silk, light softener (m.p. studio).

lichtdicht (sealed) light-tight or l.-proof.

Lichtdruck photo-mechanical printing; photographic p.; radiometer-vane effect.

lichtdurchlässig light transmissive or transmitting, diaphanous, pervious or permeable to light rays, transparent, translucent.

Lichtdurchlässigkeit light transmittance, (specific) light transmittancy.

licht-echt fast to light, non-fading. -elektrisch (cf photo-) photoelectric; electronogenic. -elektrische, -empfindliche Zelle photo (-electric) cell, light microphone. -empfindlich machen photo-sensitize.

Lichtempfindlich-keit photosensitivity; actinism. -keitskurve des Auges response curve of eye.

lichtentwickelnd emitting light, producing l., photogenic, with luminophor properties.

Lichter highlights (in pictures), lights, transparent spots or areas (in photography, printing, etc.). -,gedeckte covered lights.

lichterzeugend emitting light, producing light, photogenic, with luminophor powers.

Licht-farbendruck photo-chemical color printing. -filter light filter, color f. -fleck hot spot (on film); spot (of light). -fleckverzerrung spot distortion.

lichtfreundlich favoring high lights over shadows, photophilic, photophilous.

Lichtfreundlichkeit light transmittancy, l. transmissivity.

lichtgebend giving or emissive of light, luminous, photogenic, with luminophor properties.

Licht-gewinn light increment, l. gain. -grammophon sound-track film player (without pictorial accompaniment). -hahn electromagnetic mirror vibrator (often) with oil damping (for film recording by variable-area method). -haut cathode glow. -hof halation, halo. -hof, Reflexions- halation by reflection.

lichthoffreie Platte non-halation plate, plate free from halo.

Lichthof-schutzmittel anti-halo

means, anti-halation substance. -störung halo disturbance.

Licht-kegel cone of light, luminous cone. **-korpuskeln** photons. **-kreis** circle of light, halo. **-lampe, Ton-** exciter lamp. **-lehre** optics, photology (or physiological o.). **-leitsystem** condenser relay system (film printing). **-loch** pupil (of eye), opening for light. **-messung** photometry, optometry (re:eye). **-mikroskop** optical microscope, light m., light-optical m.

Lichtmodulator light relay, any device adapted to reconvert current into light or brightness variations, e.g., television receiver lamps, Kerr cells, modulated neon lamps, c.-r. tubes, etc. **-, selbstleuchtender** light relay of the glow-tube, neon arc lamp, sodium lamp, etc., type. **Fremdlicht-** light relay of the Kerr cell, Karolus c., etc. types adapted to reconvert currents into brightness variations by the use of a constant independent ray source.

Licht-mühle radiometer. **-netzantenne** light-socket antenna, l.-circuit a. **-optik** light optics, physical o., geometrical o. of light rays. **-pause** photographic tracing, blue print. **-phonograph** sound-track film player (without pictorial accompaniment). **-quant** light quantum, photon. **-quelle** source of light, illuminant. **-quelle, -punktförmige** punctiform light source, pointolite, tungsten-arc lamp. **-quellenbild** light source image. **-rakete** flare. **-reflexionsvermögen** light reflectance, l. reflextivity. **-regler** light relay, l. modulator (Kerr cell, etc.), light control means. **-reiz** luminous stimulus.

Lichtrelais light relay (Kerr cell, Karolus cell, modulator, recorder lamp, etc.); device converting current into light variations. **-, elektrooptisches** electro-optical cell, Faraday c., light-relay depending on F. effect. **Überschall-** supersonic light-relay.

Licht-säule light column (of glow tube). **-schacht** light tunnel (of sensitometer). **-scheue** photophobia. **-schleuse, Vierband-** four-band or f.-ribbon light valve.

lichtschluckend light absorbing, optically absorptive, optically absorbing.

Licht-schranke light barrier. **-schutzkappe** light hood (phot.).

lichtschwaches Bild low-luminosity picture, p. of low brightness.

Licht-schwächung light absorption, l. loss, l. diffusion, l. dimming, l. intensity drop, cutting down or reduction of l. **-sinnprüfer** visual photometer. **-sirene** light chopper, c. wheel. **-spalt** light slit, l. aperture, l. opening. **-spielhaus** moving picture theater, film playhouse, movie. **-spritze (Ewest'sche)** recorder lamp with constricted neon arc (system Ewest). **-spur** light track, l. pattern, l. tracing.

Lichtstärke luminous intensity, light i.; power, speed, F number (of a lens). **-, gemessene** luminance, photometric brightness. **Objektiv-** speed or power (of a lens, etc.), F number.

Lichtstärkemesser photometer.

lichtstarkes Objektiv high-speed or high-power objective, fast o. **Objektiv, hoch-** ultra-rapid lens, ultra-high-speed l.

Licht-stärkeverteilung light intensity distribution (on photo-cathode). **-staub** luminous dust (of dark visual field). **-steuerung** light modulation; l. scanning. **-steuerungseinrichtung** light valve, light relay. **-steuerzelle, Debye-Sears** supersonic light valve, D.-S. l. modulator.

Lichtstrahl, wandernder flying spot (for spotlight scanning). **-(es), maximaler Ausschlag des** clash point of light valve (in sound recording). -abtaster spotlight scanner, scanning brush, s. pencil.

Licht-streuung light scatter, diffusion. -strom luminous flux, light f. (in lumen units); lighting or lamp current. -stromdichte luminous flux density (in lumens). -strommesser lumen meter (of Blondel, etc.), photometer, lumeter. -stufe der Kopiermaschine printer step. -summe total light emitted, l. sum.

Lichtton photographic sound recording. -ansatz photographic sound-film head. -aufnahme photographic sound film recording. -gerät sound-film head or attachment. -punkte elementary shading values, e. density v. -spalt recording slit. -streifen veränderlicher Breite squeeze track. -zusatz sound-film head or attachment.

Lichtumfang range of light oscillations.

lichtundurchlässig light-tight, opaque to l., impervious to l., non-diaphanous.

Licht-ventil light valve, l. relay (e.g., selenophone, Kerr cell, neon lamp, etc.). -verteiler beam splitter (m.p.). -verteilungsfläche luminosity curve, isophotic c., isophote, isolux diagram. -wagen light-truck. -wechsel variation of exposure. -weg, Satz vom ausgezeichneten law of extreme path. -werbung light advertising, propaganda using illumination or light effects. -wissenschaft optics, photology (or physiological o.). -zähler counter tube. -zelle photo- (electric) cell. -zerhacker light (beam) chopper, episcotister. -zerlegung dispersion of light, decomposition of light (into primary colors).

lichtzerstreuend light diffusive, dispersive or scattering.

Liderung packing.

Liebesflöte flute d'amore.

Ligand attached atom or group.

Lindenholz lime.

lindern alleviate, soften, relieve, mitigate.

Lineal straight edge, rule.

Linearantenna plain or straight (one-wire) antenna or aerial.

linear(er) Gleichrichter oder Detektor linear rectifier or detector. -(e) Vergrösserung magnification in diameter.

Linearisierung linearizing action, correction of non-linear distortions (in an amplifier), say, by reverse feedback.

Linearskala slide-rule dial.

Linie, abgewickelte evolute. -, ausgehackte jagged line, serrated l., notched l., dented l. -, ausgezogene solid line (in graphs, etc.). -, durchstreichende trajectory. -, falsche grating ghost. -, gezackte jagged line, serrated l., notched l., dented l. -, in eine —bringen align, rectilinearize, straighten out in a line. -, in gerader —mit in alignment, or flush, with. -, strichpunktierte dash-dot line. -, verbotene forbidden line. -, zusammengesetzte multiplet (in complex spectrum).

Linie, Anflug- center of approach sector. Anschlag- striking line (piano). Aussen- contour line, boundary l. Begleit- satellite line. Bild- focal line, image l. Blick- line of sight, l. of vision. Brenn- focal line, image l. Gleit- slide or slip band or line. Grenz- boundary line, demarcation l. Grund- center line of approach sector. Halbierungs- bisector (line), bisectrix. Haupt- primary line. Kenn- characteristic curve or graph, indicial diagram. Knoten- neutral line, common l., nodal l. Kollimations- collimating mark. Kraft- tube of force, line of f. Mantel- directrix, line along a shell or cylinder. Mittags- meridian. Mittel- median line, center l., axis, equator. Neben- secondary line, branch l.; satellite (spectr.). Nordlicht- green auroral line. Rad- cycloid (stroboscopic aberration), epicycloid. Richt- guide line, guiding l. Richtkenn- rectification characteristic; directional diagram, space pattern. Schau- graph, curve. Scheide- boundary line, separation l. Schmelz- fusion curve. Schrauben- helix, helical line. Seh- line of sight, l. of vision, l. of collimation. Sinus- sinuous line, sinusoidal l., sinuoid. Spektral- spectrum line, spectral l. Stand- position line. Strom- stream line, flow l., force l. Translations- slip band. Trennungs- dividing line, boundary l., partition l., line of separation (of a solid moving through liquid). Umgrenzungs- contour, line of demarcation or separation, boundary line, peripherical l., circumferential l., circumscription l. Verschiebungs- line of displacement. Visier- line of sight. Wellen- wavy line, undulatory l., sinuous l. Winkel- diagonal. Wurf- line of projection, curve of p., trajectory. Zeuge- generating line, generatrix. Zug- tractrix.

Linien-bilder, aufgelöste resolved line patterns. -breite breadth or width of line; spot diameter. -gleicher Spannung isoclines (indicating stress lines in glass and other materials). -komplex multiplet. -raster ruled plate, line screen or grating; raster.

linienreich rich in lines, with abundance of l.

Linien-skala variable-density sound track. -spektrum line spectrum. -steuerung velocity modulation. -summe line integral. -umkehr line reversal. -verbreiterung broadening or spread of line or strip (telev.). -zug line, trace (of a graph).

linig, krumm- curvilinear.

linksdrehend levo-gyrate, levo-rotatory, counter-clockwise rotating.

Links-drehung levorotation, left-handed polarization. -händigkeit left-handedness. -kurve left-hand (or port) curve or loop (in airplane landing, etc.).

links-laufende Wellen sinistro-propagating traveling waves or surges. -läufig moving to the left, counterclockwise.

Links-quarz left-handed quartz, levogyrate crystal. -schraube left-handed screw.

Linse (s.a. Objektiv, Lupe, Glas) lens, objective glass; bob (of a pendulum).

Linse, eingekittete cemented lens. -, facettierte bevel-edge lens. -, fehlerfreie perfect, aberrationless lens. -, gefasste mounted lens. -, gekittete cemented lens. -, krümmungsreduzierende field-flattening lens. -, kurzbrennweitige short-focus lens. -, langbrennweitige long-focus lens. -, lichtstarke high-power lens, h.-speed l. -, magnetische, mit Eisenpanzer ironclad magnetic lens, iron-core deflector unit (of c.-r. tube). -, scharfzeichnende achromatic lens. -, ungefasste rimless lens. -, ungekittete uncemented lens, broken-contact l.— veränderlicher Brennweite und v. Bildgrösse zooming lens for variable focus and variable magnification, vario-objective, -, verkittete cemented lens. -, weichzeichnende soft-focus lens. -, weitwinklige wide-angle lens. -, zusammengesprengte uncemented lens (lens units closely fitted together, as in Cornu device), broken-contact l.

Linse, Augen- eye lens. Beleuchtungs- condensor lens, illuminating l., bull's eye l. Beschleunigungs-, mit 2 Lochelektroden double-aperture (electron optic) accelerator lens. Brenn- burning glass, convex lens. Butzenscheiben- bull's eye lens. Doppel- double lens, doublet. Doppelschicht- double-layer lens. Durchgriff-(cf. Lochblenden-) aperture lens (through which field or anode may act or draw electrons). Einblick- eyepiece lens. Einzel- unit lens. Elektronen- electron lens (either electromagnetic or electrostatic). Feld- anterior lens (in telescope and microscope), field l. Feldebnungs- field flattener. Fenster- aperture lens. Fern- telephoto lens. Flüssigkeits- (oil or water) immersion lens. Halbkugel- semi-circular lens. Haupt- second lens (in electron gun). Hinter- back lens. Immersions- (oil or water) immersion lens, double aperture lens (electron-opt.). Kugel- spherical lens. Lochblenden- (cf. Durchgriffs-) simple aperture lens (electronic). Lochscheiben- aperture lens (in scanning disk). Muschel- globular lens. Punktal- toric lens. Sammel- condensor lens, convergent l., positive l. Schau- viewing lens. Scheitel- zenithal lens, vertical l. Stufen- echelon lens. Trocken- dry lens, d. objective. Verzögerungs- cutoff lens, stopping l. (el.opt.). Vorder- front lens, field l. Vorsammel- cathode lens, first focusing l. (like Wehnelt cylinder, aperture disk, etc., in c.-r. tubes). Vorsatz- auxiliary lens, supplementary l., l. attached in front, magnascope (to enlarge projector image). Zerr- anamorphic lens, anamorphosing l., distorting l. Zerstreuungs- dispersive lens, divergent l., negative l. Zusatz- supplementary lens. Zwiebel- biconvex lens. Zylinder- cylindrical lens.

Linsen-abweichung aberration of a lens. -achse optical axis of a lens.

linsenähnlich lenticular, lenticulated, lens-shaped.

Linsen-deckglas anterior surface of a lens. -einschiebebrett slidable lens panel. -faserung, Augen- fibrillation of lens (of eye). -fassung lens barrel, l. mount. -fassung, Vignettierwirkung der vignetting effect or trimming of light by lens mount. -fehler lens defects, l. aberrations. -folge lens combination, system of lenses. -grundpunkt cardinal point (of a lens). -kranzabtaster lens drum scanner. -prüfer lensometer. -rasterfarbverfahren lenticular or lenticulated lens color printing method. -rasterfilm lenticular film, lenticulated f. -satz lens combination, system or assembly of lenses. -scheibe lens disk (scanner). -scheinwerfer spotlight. -scheitel vertex of lens, apex of l. -system, Fluorit- semi-apochromatic lens system, fluorite l.s. -trommel lens drum (scanner). -trübung clouding of a lens, opacity of a l. -vergütung s. Glasvergütung. -weite, lichte clear aperture of a lens (useful or effective diameter). -zusammenstellung lens combination, lens assembly.

Lippenpfeife flue pipe, flute p.

Lissajousfiguren Lissajous figures, L. curves. -apparat kaleidophone (of Wheatstone).

Literatur, Fach- technical literature, t. press.

Litze strand, stranded wire, braid, lace, cord, string.

Litzen-draht stranded wire, litzwire. -spule litzwire coil.

Lizenz, Zwangs- compulsory license.

Lizenz-erteiler licenser, grantor, issuer of a license. -gebühr royalty, license fee. -nehmer licensee, grantee, concessioner.

ln natural, Naperian or hyperbolic

logarithm = $\log_e$ or $\log_e$.

Loch eye, pore, pocket, opening, aperture, perforation, puncture; hole (of nucleus). **-auftreiben** enlarge, expand, broach or ream a hole. **-, ausgetuchtes** bushed hole.

Loch, Blenden- anode aperture (cathode-ray tube). **Führungs-** guide perforation (for sprocket engagement). **Gitter-** vacancy or hole in lattice. **Greif-** finger hole (in dial switch). **Licht-** opening for light admission, pupil (eye). **Schau-** peep hole; inspection hole, bezel.

Loch-blende apertured partition, anode aperture, diaphragm or shield (of c.-r. tube), stopping aperture. **-blendenlinse (cf. Durchgrifflinse)** apertured disk lens; simple (electron-optic) aperture lens. **-breite (eines Frequenzsiebes)** transmitted band, transmission range, bandwidth, band pass (of a filter), spacing between cutoff points. **-breite, Bandfilter von grosser** broad-band filter. **-elektroden, Beschleunigungslinse aus zwei** double-aperture accelerator lens (in c.-r. tube).

löcherig punctured, perforated, apertured, porous, pitted.

Löchertheorie hole theory (re: nucleus).

lochfrassähnliche Zerstörung pitting, destruction in form of pits, honeycombing.

Loch-grenze transmission range of band pass (range between cutoff points), channel width or limit, cut-off frequency. **-kamera** pinhole camera. **-lage (eines Filters)** position of transmission range or of band pass (of filter). **-lehre** internal caliper gage, plug g. **-niete** hollow rivet, tubular r. **-probe** drift test.

Lochscheibe, konzentrierende focusing aperture lens or apertured disk (mounted anteriorly of cathode, in c.-r. tubes). **Kreis-** film scanner (with holes arranged circularly rather than spirally). **Mehrfach-** multi-spiral scanning disk. **Spiral-** Nipkow disk, spiral d. (for scanning). **Vierfach-** quadruple spiral scanning disk.

Lochscheiben-anode apertured disk anode, ring a. **-generator** light chopper. **-linse** aperture disk lens.

Loch-spirale scanning (hole) spiral; spiral row of apertures (e.g., in Nipkow disk). **-trommel** scanning drum (of Jenkins).

Lochung perforation, punched holes (in film), punching.

Lochweite band-width, band-pass, band between cutoff points; inside diameter. **Bandfilter grosser-** broad band filter. **Wellenberg-** peak separation, inter-peak distance, inter-crest d.

Locker-ionen defective ions, "Locker" ions. **-stellen** loose spots, l. places (cryst.).

Lockflamme pilot flame.

Löffelbohrer spoon bit.

Logarithmen, dekadische common logarithms.

Logarithmus der Belichtung log exposure.

lösbar detachable; soluble, dissoluble, dissolvable.

losblättern detach in leaves or scales, exfoliate, defoliate.

löschen (cf auslöschen) turn off (repeater); extinguish, quench, slake (lime), obliterate, clear, cancel, blot, erase, wash out (a magnetic record).

Löscher, Funken- spark quench, blowout or extinguisher, s. killer. **Röhren-** tube quench.

Lösch-feld, magnetisches obliterating field, erasing f. (of magnetophone, etc.). **-funkenstrecke** quenched spark gap. **-kalk** quicklime, slaked l. **-kammer, Lichtbogen-** arc quench chamber. **-kohle**

quenched charcoal. -magnet obliterating magnet (in magnetic sound recording equipment), wash-out m. -spannung stopping potential (at which glow discharge is stopped; s.p. is lower than glow p., in photo-cells and g.-tubes), extinction p., critical p. (at which tube goes out), cut-off p. (of thyratron). -taste cancel key, clearing k., release k. -wirkung quench effect. -zündspannungsdifferenz extinction-striking potential difference.

lösen solve, dissolve; loosen, untie, detach, disengage, disconnect.

Löser solving agent, solubilizer.

Löslichkeits-druck solubility pressure. -produkt solubility product.

löslich machen solubilize.

los-löten unsolder. -trennen separate, sever, tear apart, detach.

Lösung solution; discharge (of a gun); separation, detachment, opening (of a connection), severance. -, ungepufferte unbuffered solution. Abschwächungs- reducer solution. Zehntel (normal)- tenth normal solution, decinormal s.

Lösungs-druck solution pressure. -magnet releasing magnet, trip m.

Lot fällen let fall a perpendicular.

Lot, Behm- fathometer, echo depth sounder, sonic altimeter. Messingschlag- brass solder. Mittel- median perpendicular.

Löt-kolben soldering bit. -rohr blowpipe. -rohrflamme blowpipe flame.

lotsen, heran- guide to, pilot (a craft by radio signals).

Lötstelle soldered joint, shut; junction of a thermo-couple. Heiz- hot junction of a thermocouple.

Lötstreifen tag strip, terminal s., connection s.

Lotung sounding (determination of depth or altitude), probing (for depth).

Lötung soldering, brazing (by hard solder); agglutination; adhesion. Doppel- double seals (in Dewar vessel, etc.).

Lotung, Echo- echo sounding, reflection s. (for depth and altitude measurement).

Lötung, Hart- hard soldering, brazing. Selbst- autogenous soldering.

Loxodrome oblique line, loxodromic line or course, great-circle bearing.

L-Regler L-section attenuator.

Lücke gap, void, space; deficiency. auf - stehen positioned so as to fill gaps (e.g., in a parallel row), staggering. -, positiv geladene positive hole (cryst.).

Lücke, Bildsynchronisierungs- synchronizing gap (between traversal of frame and beginning of next), underlap or interstice period in which sync signal is introduced. Mischungs- miscibility gap.

Lücken-bildung unsaturated linkage. -satz vacancy principle. -synchronisierungsverfahren gap, underlap, interstice or interval synchronizing method (sync signal introduced between end of line traversal and beginning of next line).

lückig porous, honeycombed, pitted.

Luft-abschluss exclusion of air, absence of a., hermetic seal, air-tight s. -alarm air raid alarm. -aufnahme air absorption, a. occlusion; aerial photograph, airplane picture. -auftrieb air buoyancy. -bild aerial image, i. formed in space by optical system. -bildaufnahme aerial photograph, airplane picture. -bildkarte aerial mosaic; photographic map.

luftdicht air-tight, hermetical.

Luftdraht (cf Antenne) aerial (wire), antenna. -gebilde aerial structure, radiating system, array of antennae.

-verkürzungskondensator shortening condenser. -verlängerungsspule antenna load coil, aerial loading inductance.

Luft-druck air pressure, atmospheric p., pneumatic pressure. -durchlässigkeit air permeability, a. perviousness, -einschluss air cavity, a. bubble, a. occlusion.

luftelektrische Störungen atmospherics, statics.

lüften ventilate, air, aerate; raise or lift (clear of something).

luftentzündlich inflammable in contact with air, pyrophoric.

Lüfter ventilator, fan.

Luftfahrt, Verkehrs- commercial aviation, c. avigation. -hindernis air navigation obstacle, avigational o. -navigation avigation, aerial navigation.

Luft-funkstelle aircraft radio signal station. -gebilde aerial structure, radiating system. -gleichstrom current of breath (in sound recording). -gütemesser eudiometer. -hahn air cock. -kammer air pressure space (loudspeaker). -kammerlautsprecher air-chamber loudspeaker (c. coupling diaphragm with horn). -kern air core. -kissen air cushion, dashpot. -kreis atmosphere.

luftleerer Raum vacuum, vacuous space, exhausted s.

Luft-leerspannungssicherung vacuum arrester. -leitergebilde aerial extension, a. structure. -mast ("Schnorchel") breathing mast, underwater air intake (supplying air, for Diesel engines, to submerged submarines). -meer atmosphere. -plattenspektroskop Fabry-Pérot interferometer, étalon (when not variable). -rahmen air-cored frame (d.f.). -reibungsverlust windage loss. -sauger aspirator. -säure carbonic acid. -schall sound conducted or transmitted through air. -schleier air fogging (chemical action of air). -schraube, Versetz- variable-pitch (feathered) propeller or airscrew. -schraubensteigungsmesser propeller pitch indicator. -schutzleiter air warden. -spule air-core coil, loop or frame (d.f.). -strahl air jet. -strom, Gleich- direct flow of air, d.c. air-flow. -strompendelung air jet pendulation. -strömungsmesser anemometer. -verdichter air condenser, a. compressor. -verdünnung air rarefaction, vacuum. -verkehrskontrolle airways traffic control, avigation regulation. -welle sky wave, space w., indirect w. -wichte specific gravity of air. -widerstand aerodynamic drag (on airplane). -winkel angle of the wind. -zutritt access of air, admission of a.

Lumièreplatte autochrome plate.

Lumineszenzlampe, Kathoden- cathode-glow tube.

Luminophore phosphors, luminophors, luminescent substances.

Lummer-Brodhun'sches Photometer oder Würfel L.-B. contrast (or total-reflection) photometer, cube p. (with cubical cavity).

Lunkerung shrinkage, contraction, piping, blowhole formation, cavitation (in metals).

Lupe, stark vergrössernde high-power magnifier. Ablese- reading lens. Einstell- focusing magnifier (lens). Einzel- magnifier unit, individual m. Fadenkreuz- reticle magnifying lens. Feldstecher- field glass magnifier. Fernseh- telescopic magnifier. Präparier- dissecting magnifier. Stativ- stand magnifier.

Lupen-aufnahme, Kamera für Zeit- camera for slow-motion picture projection. -ständer lens stand. -verfahren, Zeit- high-speed shooting, low-speed projection method.

Luvwinkel drift angle, a. of lead.

l.W. (lichte Weite) inside diameter, i. width, lumen.

Lyra lyre.

M

Maass s. Mass

Mach'sche Streifen Mach bands.

Mächtigkeit size, thickness, magnitude, power, strength.

Madenschraube headless screw.

Magazinbalg reservoir bellows.

magern make lean or poor, shorten, reduce plasticity (of ceramics).

Magnesiaglimmer magnesia mica (biotite).

Magnesiumlicht magnesium light.

Magnet, fremderregter non-permanent magnet, electro- m. Abnahme- pickup magnet. Aufnahme- recording magnet (for steel-band or wire recording of sound), pickup magnet. Auslösch- obliterating magnet, wash-out m. (to wipe away sound track or recording on wire or tape, for re-use). Einrück- trigger magnet, trip m., starting m. Glocken- bell-shaped magnet. Hub- vertical magnet, stepping m., lifting magnet. Lamellen- laminated magnet, compound m. Lösch- obliterating magnet, wash-out m. (sound recorded on wire or tape). Lösungs- release magnet, trip m. Topf- ironclad magnet, pot-shaped m. Wähler- bank of (stationary) contacts, selector magnet (aut. telephony).

Magnetfeld,Quer- transverse magnetic field.

Magnetfeldröhre magnetron (or permatron) working with a constant magnetic field (strictly, a m. operates with a variable m.f.)

magnetisch(e)Belegung magnetic induction, m. charge, seat of m. induction or m. field. -(er)Effekt, kreis- gyromagnetic anomaly. -(e)Leitfähigkeit permeance, magnetic conductance, permeability. -(er)Schirm magnetic shield, m. screen, can (of a coil). -(e)Spannung magnetic potential (line integral of m. intensity). -(er)Tonaufzeichner magnetic sound recorder, magnetophone, Poulsen telegraphone, Blattnerphone (modern form of P.t.)

Magnetisierungskopf magnetic sound recording head (on wire or tape).

magneto-optische Drehung magneto-optic rotation, magnetic r., Faraday effect (rotation of plane of polarization).

Magneton,Kern- (K.M.) nuclear magneton.

Magnetorotationsspektrum magnetic rotation spectrum.

Magnetron, Zweischlitz- two-split magnetron, two-segment m.

Magnetron-cycloidhöhe cycloidal height of a magnetron. -leitbahnwelle transit-time or electron path oscillation. -leitkreisfrequenz transit-time frequency of a magnetron. -schwingungen magnetron oscillations (3 kinds: electronic, negative- resistance and rotating field o.)

Magnet-schale magnetic shell. -schenkel magnet limb, m. leg. -spannschraube magnetic chuck. -spule solenoid, field coil. -summer magnetic interrupter, buzzer type of audio-frequency oscillator. -tonverfahren magnetic sound recording method, telegraphone s.r.m.

Mahl-feinheit fineness of grind, f. of grain, f of comminution. -gut material to be ground or milled.

Maischgitter stirrer, rake.

Makroachse macro-axis (of crystal).

Mal mark, spot, sign, token; stigma.

mal times (x).
Malerfirnis painter's varnish.
Malteserkreuz Maltese cross, cross wheel, (cam and) star wheel in Geneva intermittent movement. -gesperre Geneva stop.
Mängel (in einer Anmeldung) defects (in a patent application).
mangels in default of, in the absence of, because of lack of.
manipulieren manipulate, handle.
Manko shortage, deficiency, deficit.
Mannigfaltigkeit manifoldness, multiplicity, variety, diversity.
Mann-loch manhole. -schaft men, force, crew, gang.
Manometer, Glasfeder- spoon-type glass manometer. McLeod- McLeod (vacuum) gage.
Manschette flap, collar, cuff.
Mantel casing, jacket, case, shell, sheath, surface (math.), nappe, sheet, envelope, cover. Heiz- heating jacket. Kegel- cone-shaped shell. Wasser- water jacket.
Mantel-fläche generated surface, shell s. -futter shell lining. -gefäss, Vakuum- thermos or Dewar vessel exhaust jacket. -linie directrix. -rohr tubular shell, -transformator ironclad transformer, shell t. -welle wave on outer surface of coaxial cable, "shell" wave.
Mantisse mantissa (fractional part of a logarithm).
Manual keyboard, manual.
Manuskript scenario script (m.p.).
Marien-bad water bath. -glas selenite, mica.
Mark marrow, pith, pulp, core.
Marke, Einstell- reference work, bench m., gage m., measuring m., wander m. (of telescope, to establish "stereoscopic contact"). Fabrik- trade-mark. Zeit- time scale, t. marking. Ziel- sight reticle or graticule.
Markenplatte index plate (of telescopic sight).
Markieren mark, define, outline.
Markierung(Szenenwechsel) nothing on film edge. Unterscheidungs- distinctive marking, telltale m.
Markierungs-sender marker beacon, marker. -vorrichtung marker, puncher, notcher (m.p.).
Masche, Gitter- grid mesh. Kettenleiter- mesh or section of a network.
maschig meshed, netted, reticulated. grob-, weit- wide-meshed, coarse-m. (e.g., of a grid).
Maschinensender alternator transmitter.
Maser speckle, spot, mark. -holz burwood.
maserig speckled, streaked, grained, veined.
Maske mask, mat (in film recording). Laternenbilder- lantern mask. Vignettier- vignetting mask.
maskieren mask, blank, blanket, conceal, camouflage, cover, dissimulate, eclipse.
Maskierung, Gehör- auditory masking (loss of sensitivity of ear).
Mass measure, scale, dimension, proportion, degree, extent. auf -drehen turn true (to template or gage). auf genaues - bringen finish or true up (to template or gage). -,laufendes linear measure.
Mass, Aussen- over-all dimension, outside d. Belichtungs- exposure level (of Decoutes). Dämpfungs- unit of attenuation, image attenuation constant. Eben- symmetry. Eich- standardized gage, s. measure, calibration unit. End- standardized gage block, end-to-end standard bar. Farben- colorimeter. Kraft- unit of force. Längen- linear measure. Parallelend- standardized gage block, end-to-end standard bar. Phasen- phase constant, wave-length

c. **Richt-** standard, gage. **Strich-** line standard. **Vierpolwinkel-** image phase constant. **Winkel-** phase constant, wavelength c. (denotes phase shift between input and output potential, and is the imaginary component of image transfer constant): set square. **Zeit-** time scale, t. marking.

Mass-abweichung (cf. Toleranz) off-size condition, dimensional discrepancy or difference. **-analyse** volumetric analysis.

Masse mass (in mechanics; analogous to inductance in electricity and inertance in acoustics).

Masse, aktive active paste. **Einheits-** unit mass. **Grund-** ground mass. **Isolier-, gepresste** molded insulation. **Leucht-** phosphor, luminescent or fluorescent substance. **Ruhe- (der Teilchen)** rest mass (of particles).

Massekern dust core, molded c. **-spule** dust-core coil, compressed iron core c. Ferrocart c.

Massen-absorptionskoeffizient mass absorption coefficient. **-anziehung** gravitation. **-dichte** mass density. **-einheit** unit of mass. **-gesteine** unstratified rocks. **-güter** bulk goods; large-scale manufactured goods. **-kopie** release print, quantity production of prints. **-kräfte** mass, inertia (in mechanics, mechanical filters, etc.; analogous to inertance in acoustics and inductance in electricity). **-moment** moment of inertia, rotational i. **-platte** pasted plate (of battery). **-plet** electric mass. **-punkt** quantum, point-mass (in quantum theory), rotator; center of mass, c. of inertia, centroid. **-schwankungskoeffizient** mass absorption coefficient. **-spektrograph** mass spectrograph, m. spectrometer. **-spektrum** mass spectrum. **-strahler** mass radiator. **-widerstand** inertance (ac.). **-zahlen** mass numbers. **-zentrum** center of mass, c. of inertia, centroid.

Masseteilchen-Geschwindigkeit particle velocity.

Mass-faktor dimensional constant, d. factor. **-gabe, nach** in accordance with, in proportion w.

massgebend determinative, determinant, decisive, authoritative, competent, conclusive.

Massgefass measuring vessel, graduated v.

masshaltig true to size, measure or dimension (inside tolerance range, found by limiting gage).

-mässig suffix signifying "in the manner of", "so far as...is concerned"; according to, in terms of. For instance, hochfrequenzmässig: so far as RF is concerned, or, for RF.

massig massive, massy, molar.

mässig, gesetz- in accordance with law or statute, conformable to (natural) law, in a regular or legal way. **-, grössen-** dimensionally, quantitatively, numerically; pertaining to, or in relation to, quanta. **-, hochfrequenz-** for RF, so far as RF is concerned, relative to RF. **-, quanten-** pertaining or in relation to quanta. **-, zahlen-** numerically.

mässigen moderate, temper, mitigate.

massiv solid, strong, molar.

Mass-kunde metrology. **-nahme** mode of action, (precautionary) step, expedient, technique. **-nahme, Schutz-** preventive, protective, safety, or precautionary step, measure or means, safeguard. **-regel** cf. **-nahme**. **-röhre** measuring tube, graduated t., burette.

Masstab scale, size, measure, yardstick, proportion, gage, standard. **Parallelend-** end-to-end measuring bar, standard gage block. **Vergleichs-** standard of comparison or reference, comparative "yardstick". **Zeit-** time scale.

Mass-stabtransformator gage (group) transformer. **-stock** gage,

yardstick, linear measure. -system, Farb- (CIE) chromaticity scale system, CIE diagram. -zahl measure or numerical criterion.
Mast, Gitter- lattice pole, girder p. Kreuzteilungs- cantilever type of tower, double tapered mast. Luft- ("Schnorchel") breathing mast, underwater air intake (supplying Diesel engine air when submarine is submerged).
Mast-armatur pole hardware, fittings. -befeuerung mast beacon.
Material, entgegenstehendes anticipatory references (found in prior art and literature). -, zertrummerbares fissionable or disintegrable substance. -prüfer, elektronischer aniseikon. -prüfung testing of materials. -wanderung flow of material, creep of m.
Materie-welle matter wave, de Broglie (elementary particle, electron, proton) wave, electron w., phase w. -zertrümmerbarkeit disintegrability or fissionability of matter.
Matrize metal master, master negative, mother (in phonograph record manufacture), matrix die. Kaltstauch- cold upsetting die. Press- negative stamper. Stanz- cutting die-plate, punching die.
matt dead, faint, feeble, weak, dull, matt (photographic paper).
mattgeschliffen ground, frosted, delustered, matted.
Mattglas ground glass, frosted g.
mattieren tarnish, deaden, dull, deluster, matt, frost, impart a matt surface or look to, grind (glass).
Mattierungsmessung film polish measurement, glossimetric m.
Mattlack matt varnish.
Mattolein matoleine (retouching varnish).
mattrot dull red.
Mattscheibe ground or frosted glass (pane or plate). Visier- focusing screen.
Maul-trommel Jew's harp. -weite interpolar distance, air-gap (of horseshoe magnet).
m.a.W. in other words, that is to say.
Maxima, Neben- secondary or lateral radiations, s. lobes or ears (in space pattern).
maximale Aussteuerung des Lichtstrahles clash point of light valve (in sound recording).
Maximalpeilung maximum signal (strength) method (df).
Maximumwertzeiger peak indicator, crest i.
Maxwell'sche Grundgleichungen Maxwell electromagnetic equations.
Maxwellverteilung Maxwell (statistical) distribution.
mC milli-curie.
McLeod Manometer McLeod vacuum gage.
mechanischer Filter, Federungswiderstand compliance of a mechanical filter. -Filter, Massenwiderstand inertance of a mechanical filter.
Medien media, mediums.
Meerfarbe sea green.
mehlartig like meal or flour, farinaceous.
Mehlgips earthy gypsum.
mehradriges Kabel multi-wire cable.
Mehraufwand extra or additional expenditure or means (in equipment, etc.) Röhren- tube complement, extra outlay for tubes,
Mehr-deutigkeit ambiguity, equivocation. -drehung multirotation.
mehrfach manifold, multiple, repeated.
Mehrfachbelichtung superimposing, multiple exposure.
mehrfachbeschichteter Film sandwich film.
Mehrfach-bild multiple image (in telev.) -echo multiple echo (reverberation), flutter e. -empfangsverfahren diversity reception method (to minimize fading). -kondensator multiple condenser, gang c., tandem c., synchronized c., gang capacitor. -kopien multiple prints.

-nachrichtensystem multiplex signal system. -platten Karolus-Kerrzelle Karolus type of multiple-plate Kerr cell. -röhre multi-valve, dual v., multipurpose tube, multi-unit t. -rückkoplung multiple regeneration. -spirallochscheibe multi-spiral scanning disk. -strahler bank or pillar of lamps, battens. -telegraph multiplex telegraph. -telephonie multiple telephony. -verstärker multistage amplifier. -zackenschrift multilateral sound track. -zeichen multiple signal, echo s.

mehr-fädige Lampe multiple-filament lamp. -farbig polychromatic, pleochroic, multi-colored.

Mehr-gangkondensator ganged condenser or capacitor, synchronized c., multiple c., gang capacitor. -gitterröhre multi-grid tube (tetrode, pentode, etc.)

mehrgliedrig multi-mesh, m. -section (e.g., of a network).

Mehrheit plurality, multiplicity, majority. -von Einheiten array of units, (plurality or multiplicity of u.

Mehrkanalfernsehen multi-channel television.

mehrkanalig multi-channel.

Mehrkantprisma polyhedral prism.

mehr-kernig with more than one nucleus or core, polynuclear. -köpfige Besatzung multiman crew. -köpfiges Stationspersonal multi-manned station.

Mehr-kristall polycrystal. -lagenspule multi-layer coil, banked c., pile c., honeycomb c. with banked winding. -platten Karolus-Kerrzelle Karolus multiple-plate Kerr cell. -punktschreiber multi-point recorder.

mehr-stimmig polyphonic (of an organ). -strahliges Funkfeuer multi(-ray) beacon, multiple- or triple - modulation b. -teilig multipartite. -wegig having multi-channel or m. -path properties. -wegige Übertragung multi-path transmission.

Mehr-welligkeit multi-wave property. -zahl plurality, multiplicity; majority.

mehrzügiges Tonformstück multiple tile or clay conduit.

Meilerkohle charcoal.

Meissel chissel, pledget, stylus or style (for sound recording), engraver tool.

Meister, Bild- director of photography, picture chief, p. director. Ton- monitoring operator (working at mixer desk), recordist, sound engineer, director of s.

Meister-negativ metal master (in making phonograph records), master negative. -positiv mother record, master positive.

Melde-feld, Rück- revertive signal panel. -lampe alarm lamp, pilot l., telltale l.

Melder (s.a. Anzeiger) indicator

Melder, Abstimmungs- tuning indicator. Einbruchs- burglar alarm. Rück- revertive signal means, position-repeating m., check-back p. indicator (in telemetric arrangements). Stations- station indicator, tuning i. Wellenbereich- wave-band indicator.

Membrane, als Ganzes schwingende piston diaphragm, diaphragm moving like a p. or en bloc. -, atmende flexible, non-reflecting diaphragm. Falz- non-rigid non-circular breathing cone (with curved radiating surfaces). Kalotten- hemispherical diaphragm, cap-shaped d. Kolben- piston diaphragm. Konus- conical diaphragm, cone d., diffusion cone. Nawi- cone

diaphragm with curve profile, curvilinear c., para-curve. Riffel- pleated diaphragm (of loudspeaker), d. with pleated corrugations. Schirm- screened diaphragm (of microphone). Sektor- (non-rigid) sector diaphragm, sectorial cone.
Membraneinspannung suspension of diaphragm.
membranloses Mikrophon diaphragmless microphone, cathodophone.
Membranschwingamplitude excursion amplitude of diaphragm.
Mengeeinheit unit of quantity.
Mengen-bestimmung quantitative determination, q. analysis. -verhältnis quantitative relationship, proportion (of ingredients), q. composition.
Meniskuskante meniscus edge.
menschliche Stimme, Lautstärkeverhältnis human voice intensity range.
Mensur measure (string instruments), scale (organ); measuring vessel; measuring, mensuration.
Merkatorkarte Mercator chart.
Merkatorkartenprojektion Mercator projection, cylindrical p., orthomorphic p.
merkbar perceptible, noticeable, appreciable.
Merk-mal characteristic feature, c. object (of an invention), c. mark, sign, indicator, index, criterion. -punkt index point, fiducial p., reference p., point de repère (in Baudot apparatus), bench mark.
Mesonen-bahnende mesotron track end. -zerfall meson decay.
Mesotrontheorie meson theory, t. of mesotrons (also known as heavy electrons, barytrons, dynatrons, penetrons, x particles, etc.)
Messbereich measuring range, measurement r., indication or reading r.
Messer (s.a. Anzeiger, Melder, Messgerät, Messung, Meter, Probe, Prüfung, Zähler) measurer, meter, measuring instrument or device; knife, cutter, blade. Abklingzeitfluorometer. Abschleif- grinding blade, g. knife. Abtrift- drift meter. Augen- ophthalmometer, optometer. Augenabstand- interpupillary distance gage. Augenbrechungs- skiascope, retinoscope. Ausdehnungs- dilatometer, extensometer. Basisentfernungs- base range finder. Beben- seismometer, seismograph. Beleuchtungs(stärke)- illumination meter, i. photometer, illuminometer. Belichtungs- exposure meter (in sound recording, etc.), brightness m., turbidity m. Beschleunigungs- accelerometer. Blutfarbe- plethysmograph (tests blood flow), electroarteriograph. Brand- pyrometer. Brechungs- refractometer. Brennweiten- focometer. Dampfdichte- dasymeter. Dampf(druck)- steam gage, manometer, tensimeter. Dampfdruck-, osmotischer osmometer. Dämpfungsloss or transmission (efficiency) measuring device, decremeter (measures logarithmic decrement or damping of circuits). Deckkraft- cryptometer (measures concealing power of paint). Dehnbarkeits-, Dehnungs- ductilimeter, dilatometer, extensometer. Dichte-, photographischer (photo-) densitometer. Dicken- thickness gage, calipers. Dosis- dosimeter, dosage m., intensimeter. Druck- piezometer, liquid manometer, fluid m., pressure gage. Druck-, osmotischer osmometer. Druck- und Gefrierpunkt- manocryometer. Dunst- atmometer. Durchlässigkeits- transmissometer (for semi-transparent media). Elastizitäts- elasmometer (measuring Young's modulus), torsometer. Energie- ergometer. Entfernungs- range finder, distance meter. Erdbeben- seismometer, seismograph (when

recording). Fallkörperzähigkeits- ball-drop viscosimeter. Farben- chromatometer, colorimeter. Farbenstufen- tintometer. Feinhöhen- (aneroid) barometer measuring atmospheric pressure at airdrome (not reduced to sea level), sensitive altimeter. Fern- telemeter. Fern-, mit Summenangeber telemetric integrator. Feuchte- hygrometer, psychrometer. Feuchtigkeits, registrierender hygrograph. Film- film footage meter or counter. Flächen(aus)- planimeter; integraph (for curves). Flügel- vane-type fluid-flow meter. Fluoreszenz- fluorometer. Fluss- fluxmeter (for magnetic measurements). Flüssigkeitsdichte- liquid density meter, areometer, hydrometer (measuring sp.g.) Frequenz-(cf Wellenmesser) frequency meter, onometer, sonometer. Gasdichte- dasymeter. Gefrierpunkt- cryometer. Gehörschärfe- audiometer. Geländewinkel- angle-of-site instrument, clinometer. Geräusch- sound level meter, noise m., n. measuring set, psophometer (for telephone noises), acustimeter. Geschwindigkeits- speedometer, tachometer. Gierungs- yawmeter. Glanz- gloss meter Glätte- glossimeter; surface analyzer. Gleichrichtervolt- rectifying voltmeter. Glut- pyrometer. Grobhöhen- (aneroid) barometer or altimeter measuring actual atmospheric pressure reduced to sea level. Halbbildentfernungs- split-field coincidence range-finder. Herzschlag - cardiac tachometer. Hitz- pyrometer. Hitzdraht- hot-wire anemometer (and other measuring instruments using a hot w.) Höhen- altimeter, altitude meter (of echo or reflection type), sound ranging altimeter, statoscope. Hörschärfe- audiometer. Hysterese- hysteresigraph. Impuls- peak meter, crest m. (for potentials). Ionenstärke- ionometer. Ionenwindvolt- ionic wind voltmeter. Isolations- insulation tester, megger. Kapazitäts- faradmeter. Kartenwinkel- map protractor. Kochpunkt- ebullioscope. Koerzitivkraft- coercimeter. Konsistenz- viscosimeter. Körnigkeits- granulometer. Kraft- dynamometer, force meter. Kraftstoss- ballistic pendulum. Kreiselneigungs- bank-and-turn indicator (aviation). Kreiskeilschwärzungs- circular wedge densitometer. Kristallachsen- conoscope (determines optical axis of crystals). Lautstärke- volume indicator. Leistungs- power meter, ergometer. Leitungs- conductometer. Leuchtkraft- photometer. Lichtstärke- photometer, lumen meter, lumeter. Lichtstrom- lumen meter (of Blondel), photometer, lumeter; wattmeter for lighting circuits. Luftdruck- barometer, manometer. Luftdruck-, registrierender barograph. Luftfeuchtigkeits- hygrometer, psychrometer. Luftgüte- eudiometer. Luftstrom- anemometer. Mattierungs- film polish meter, glossimeter. Neigungs(winkel)- clinometer. Neonröhrenlautstärke- neon tube volume indicator. Oberflächenspannungs- surface tensiometer. Offnungswinkel- apertometer. Pegel- transmission measuring set. Permeabilitäts- permeameter. Politur- glossimeter. Pulsschlag- cardiotachometer. Pupillenweite- korometron. Quotienten- quotient meter, ratio m. Röhrenspannungs-, Röhrenvolt- vacuum tube voltmeter. Roentgendosis- X-ray dosimeter. Roentgenstrahlenhärte- penetrometer, qualimeter. Roentgenstrahlen- Roentgen meter, iontoquantimeter (Duane's), ionometer. Rückstrahlungs- reflectometer. Sauerstoffeudiometer. Saug- vacuometer. Schalenwindstärke- cup-type anemometer. Schallstärke- phonometer, sound level meter. Scheitelbrechwert- apex refractometer. Schlüpfungs- slip meter. Schnittbildentfernungs- split-field range-finder. Schwärzungs- densitometer. Schwere- barometer, gradiometer (measures gradient of earth's gravity field). Schwere-, registrierender barograph. Schwingaudionwellen- autodyne wave or frequency meter. Sicht (barkeits)- visibility meter (of Lukiesh-Taylor, etc.) Sonnenstrahlungswärme- solarimeter, pyroheliometer, actinometer (for solar flux density measurements). Spiegelneigungs- Abney level (to measure vertical angles). Steigleistungs- rate of climb meter ("variometer" measuring both rise and descent). Steilheits- derivator, tangent meter (for curves). Strahlen- radiometer (for heat rays), actinometer (for actinic r.) Strahlenbrechungs- refractometer. Strahlenhärte- penetrometer (for x-rays). Strom-, registrierender recording ammeter. Summerwellen- buzzer wave meter. Temperatur- bolometer (of high precision). Temperatur-, thermoelektrischer thermel. Tiefen- fathometer (e.g., supersonic and sonic types), depth finder. Ton- tonvariator, pitchpipe; frequency record. Tonschwankungs- flutter meter. Trägerfrequenzfern- telemeter, metameter (of GE Co., using carrier waves). Tropfen- stactometer, stalagmometer. Trübungs- turbidimeter, opacimeter, nephelometer. Übersteuerungs- overload indicator. Umfangsgeschwindigkeits- tachometer (e.g., stroboscopic using strobotron). Vakuum- vacuum gage, ionization g., vacuometer. Verbrauchs- supply meter. Wanderwellen- klydonograph, surge indicator. Wärme- calorimeter, thermometer. Wärme-, thermoelektrischer thermel. Wassersiede- hypsometer. Wellen- (s.a. Frequenzmesser) wave meter, frequency meter, Fleming's cymometer, ondometer. Widerstands- ohmmeter, megger. Windgeschwindigkeits- anemometer, wind meter. Winkel- protractor, goniometer, angle gage (for eye), anglemeter (for measuring external angles), theodolite (for surveying). Wirkverbrauchs- active power meter. Zähigkeits- viscosimeter, viscometer. Zeit- chronometer, time piece, clock, watch. Zeit-, registrierender chronoscope, chronograph. Zungenfrequenz- reed frequency meter.

Messer-schalter knife (blade) switch. -zeiger knife-edge pointer.

Mess-filmeinrichtung time recorder on film. -form parison (bottle making). -gefäss measuring vessel, graduated v., measure.

Messgerät, Fern- telemeter, metameter (of GE Co., using carrier impulses). Intensitäts- level recorder. Mikrozeit- microchronograph. Wetter-, mit funkentelegraphischer Fernübertragung radio telemeteorographic instrument. Zeit- chronographoscope.

Messing-blasinstrument brass wind instrument. -schlaglot brass solder. -späne brass shavings, b. turnings.

Messinstrument, gedämpftes aperiodic or deadbeat measuring instrument. -, ungedämpftes ballistic measuring instrument. Wetter-, mit funkentelegraphischer Fernübertragung radio-telemeteorographic instrument, r. sonde.

Mess-kolben graduated flask,

measuring f., gage f. -kunde metrology. -lehre, Fein- micrometer caliper, m. gage. -methodik measuring methodology. -okular micrometric eyepiece. -pegel measuring level, expected l. -schallplatte frequency record, phonograph r. bearing sliding or constant notes, parlophone record (producing warble note). -technik, Fernsprech- telephonometry. -trommel graduated drum. -umformer measuring transducer.

Messung, quantitative quantitative measurement, q. analysis. -, stationäre deadbeat measurement. Abnahme- acceptance test, works test. Elementen- stoichiometric measurement, stoichiometry. Entdämpfungs-Frequenz- gain-frequency measurement or test. Farb- colorimetry, chromatometrics, chromatometry. Feld- land surveying, l. measuring; mapping or plotting of field configuration or field pattern (in an electrolytic tank). Gehörschärfe- audiometry. Kreis- cyclometry. Kristall- crystallometry. Kristallwinkel- crystal goniometry. Längen-, echte direct measurement of length.-, unechte indirect measurement of length. Leuchtkraft-, Licht- photometry, optometry (re: eye). Mattierungs- film polishing measurement, glossimetric m. Strecken- end-to-end measurement, straight-away m.

Mess-wandler measuring transformer, instrument t. -zahl numerical value, measured v., measuring v. -zweige measuring arms.

Metall, gespritztes die-cast metal. Edel- noble metal, precious m., rare m. Erdalkali- alkaline-earth metal. Misch- mix metal, alloy m., "Misch" m. Spiegel- specular metal.

Metall-azid metallic azide. -beschreibung metallography. -blatt (thin) sheet of metal, m. foil, lamina, lamination. -gekrätz waste metal. -glas enamel. -kassette metal darkslide. -kunde metallography. -pickelbildung (corrosion) pitting. -probe assay, test for metal; metal testing. -schliff polished section, micro-section; metal filings. -späne metal shavings, turnings, borings or chips. -spritzmethode metal spray method (with Schoop gun). -tuch wire cloth. -versetzung alloying, alloy. -vorrat metal reservoir, pool of Hg (in a mercury-vapor arc rectifier, etc.)

Metastellung meta position.

Meter, Spektralphoto- spectrophotometer. Vakuum- vacuometer, vacuum gage.

Meter-zahl, ausgefahrene reeled-out yardage, paid-out y. (airplane antenna). -zähler (Film) film footage counter.

Methodik methodics, methodology. Mess- measuring methodology.

Mikro-densogramm microdensitometer record. -filter filter for micrographic work.

Mikrometer, Funken- micrometric spark-discharge gap. Kugel- spherical micrometer. Objekt- stage micrometer. Okularnetz- micrometer with reticle eyepiece. Über- ultra-micrometer.

Mikrometerschraube micrometric screw, micrometer.

Mikro-nernstlampe Nernst lamp of a microscope. -nutsche micro-funnel, m. suction filter.

Mikrophon, bewegliches, entfesseltes following microphone. -, lichtempfindliches light-sensitized cell, l. microphone. -, membranloses diaphragmless microphone, cathodophone, glow-discharge m. -, nichtrichtungsempfindliches astatic microphone. Atmen des -s breathing (slow changes in resistance and response) of a carbon microphone. Band-, Bändchen- ribbon microphone, band m., velocity m. Besprechungs- sound pickup microphone. Bewegungs- velocity microphone, pressure- gradient m. Brust- breast-plate microphone or transmitter. Doppelkohle- push-pull carbon microphone. Einblendungs- fade-in and mixing microphone. Flammen- flame microphone, f. transmitter. Gegenkontakt- push-pull microphone. Gitter- grille-type microphone. Kehlkopf- laryngophone, throat microphone. Kohle-carbon microphone. Kondensator- condenser microphone, electrostatic m. Kristall- piezo-electric microphone, crystal m. Lausch- pickup microphone, p. transmitter, detectophone. Post- solid-back transmitter, carbon-back microphone. Raum- non-directional (polydirectional) microphone. Richt- directional microphone (unidirectional, bidirectional, etc.), non-astatic m. Schall- acoustic microphone. Schüttel- granule microphone. Stab- carbon-stick microphone. Tauchspulen- moving-coil microphone.

Mikrophon-ansprechkonstante sensitivity constant of microphone. -antenne microphone boom, m. outrigger. -atmen breathing of microphone (due to slow changes in resistance and response, in carbon m.) -einblendung fade-in and mixing of microphones. -einsprache acoustic inlet of microphone. -galgen microphone boom. -geräusch microphonic noise (carbon noise), valve noise. -schmoren mike stew (noise). -verstärker microphone amplifier, m. pre-a.

mikrophotographisch photomicrographic.

Mikro-photozellen globules of mosaic screen (in iconoscope). -raffer camera for stop-motion or time-lapse motion micrography.

Mikroskop, bildaufrichtendes image-erecting microscope. Durchstrahlungsüber- transmission-type ultramicroscope. Einstell- focusing microscope. Elektronen- electron microscope. Elektronenraster- electron scan microscope, raster m. Hornhaut- corneal microscope. Kugel- ball-stage microscope. Licht- optical microscope, light m., light-optical m. Raster- electron-scan microscope, raster or screen-scan m. Schatten- shadow microscope. Über- electron ultra-microscope, e. super-microscope.

Mikroskop-aufbau,-aufhängung microscope mounting. -bügel saddle of microscope, bracket of m.

Mikroskopie microscopic optics.

Mikroskopiker microscopist.

Mikroskop-revolver revolving nose-piece. -schleuse airlock (chamber) of electron microscope. -spektralkamera microspectroscopic camera. -träger microscope stage. -verschiebung,Längs- und Quer- axial and cross adjustment (in electron microscope).

Mikro-telephon microtelephone transmitter, transceiver, combination handset. -wage micro-balance. -wellen micro or hyperfrequency waves (1 to 100 cm). -zeitmessgerät microchronograph.

Milchglas opal glass, frosted g., light-diffusing g.

mild mild, soft, gentle.

Milderungsgründe extenuating circumstances, mitigating c.

Milieufilm "milieu" film.

Milopam German trade-name for polyvenyl chloride.

Minderer, Druck- pressure reducer, relief valve.

Mindest-voltgeschwindigkeit appearance (or minimum) potential. -wertanzeiger minimum (voltage) indicator.

Minimalrelais no-load relay.

Minimum-auflösungsvermögen minimum separabile (of eye). **-bereinigung** zero clearing (df). **-peilung** minimum (or zero) signal direction-finding method. **-schärfen** zero cleaning, z. sharpening (df). **-verfahren** method based on minimum deviation (opt.) **-wertzeiger** minimum (voltage) indicator.

Minutenzeiger minutes hand (of timepiece).

mischbar miscible, mixable. **-, gegenseitig** consolute (miscible in all proportions).

Misch-barkeit, Nicht- non-miscibility, immiscibility. **-brett** mixing panel, monitoring p., supervising p.

mischen mix, mingle, blend, adulterate, alloy, combine, unite.

Mischer mixer, fader, blender, (ac., etc.), mixing booth, "aquarium" (m.p. studio slang) **-, fünfgliedriger** five-channel mixer. **Ton-** tone fader, t. mixer.

Mischerschaltung keying or mixer tube scheme (for adding sync and video signals).

Misch-gerät mixer bus. **-hexode, Fading-** a.v.c. mixer hexode. **-kristall** crystalline solid solution, mixed crystal. **-metall** mixed metal, alloy, "Misch-metall." **-molekül** mixed molecule (loosely associated molecular union). **-peilung** taking bearings on board, without directional receiver, from directive beacon. **-pult** mixing desk, monitoring d., supervising d. **-pult, fahrbares** dolly or cart-type mixer, teawagon console m. **-regler** mixing control. **-röhre** mixer tube. **-röhre, selbstschwingende** self-heterodyning mixer tube, autodyne. **-signal** signal spectrum (containing video and sync impulses, telev.) **-stufe** mixer stage, m.-first detector s. **-tafel** mixing panel, monitoring p., supervising p. **-tisch** mixing desk, monitoring d.

Mischungs-lücke miscibility gap. **-rechnung** alligation (math.)

Mischwalzwerk set of mixing rollers.

miss-farbig discolored, inharmonious in colors. **-lungener Teil** rejected take (m.p.)

Miss-ton distorted sound, off-pitch note. **-verhältnis** disproportion, asymmetry, disparity.

missweisend(er)Kurs uncorrected bearing, magnetic heading, m. course. **-Nord** uncorrected or magnetic north.

Missweisung, Orts- local declination.

Missweisungsgleichen isogonic lines, isogons.

Miszelle micelle.

Mit-arbeiter fellow worker, collaborator, co-worker, contributor, assistant. **-bewegung** relative motion (of atomic nucleus), comovement, associate motion. **-führungskoeffizient** (Fresnel) coefficient of drag.

mitgenommene, selbständige Kippmethode self-running controlled time-base method (telev.)

Mithör-einrichtung side-tone circuit. **-schalter, -schlüssel** listening key.

Mitkoppeln determining of course or fix, by dead reckoning (navigation).

Mitkopplung positive regeneration, p. feed-back; spurious or undesired f.

mitlaufende Spannung follower potential.

Mitlauter consonant.

Mitnahme synchronization, pull-in-step, locking-in; locking of circuit. **-bereich** range of forced oscillations, "entrainment" range, coherence r., pull-in-step r.

Mitnehmen (bei Entladung) drift (in discharge).
Mitnehmer part carrying, dragging or driving along another part· driver, carrier, gripper, catch, dog, nose, clutch, engaging piece, etc., entrainment means.
mit-reissen entrain, carry, drag or pull along. **-schwingen** co-vibrate, oscillate in resonance, sympathy or unison, be in sympathetic vibration. **-schwingend(er) Leiter** equi-frequent conductor, resonant or co-vibrant c. **-schwingend(e) Saite** sympathetic string or chord.
Mitschwingung co-vibration, sympathetic v.
Mittags-kreis,-linie meridian.
Mitte center (point), middle, midpoint (of tapping).
Mittel middle, medium, expedient, means, remedy, agency; mean (math), average.
Mittel, Füll- filler, filling, stuffing, loading material or compound. **Schalt-** circuit element, c. means.
Mittel-anzapfung center tap, mid (point) t. **-bildhelligkeit** average illumination, background i., mean brightness value of picture. **-blende** middle diaphragm. **-ecke** lateral summit (cryst.) **-feld** central field (of vision). **-frequenz(cf. Wellenabgrenzung)** medium frequency m. wave.
mittelgedämpfter Raum moderately live room.
Mittel-glied intermediate member (of a series). **-lage** tenor(of voice). **-lichter** half tones, average shading values. **-linie der Funkpeilstation** center line of approach sector. **-linienkondensator** logarithmic capacitor (in which relation capacity: angle obeys l. law). **-lot** median perpendicular.
mitteln average, ascertain mean value.
Mittelpunkt center, c. point, midpoint (of tapping). **Krümmungs-** center of curvature. **-tastschaltung** center-tap key modulator scheme.
Mittel-schenkel center leg, c. limb. **-senkrechte** median perpendicular. **-wert, quadratischer** virtual or root-mean-square (rms) value.
Mitten-abstand distance between successive rulings (in diffraction grating, indicated by grating constant). **-eindruck** binaural balance (in sound locating).
mittenrichtig centered, co-centric.
Mittenschärfe central definition, sharpness of center.
mittensymmetrisch centro-symmetrical.
mittig co-axial, concentric. **-, aussen-** eccentric, off-center.
Mittigkeit centrality, centricity.
mittler mean, average, medium, central, mid...**-(e) freie Weglänge** mean free length of path. **-(e) Helligkeit** average (background) illumination (telev.)
Mittungsstift centering pin.
mitziehen (s.a.Ziehvorgang und ziehen) draw, drag or pull along, pull in step, tug; warp.
Mn(Motorgeräusch im Norden) noise of motor north signal.
Modell, Atom- (R-B) atom model or conception. **Einteilchen-** single particle model. **Mutter-** master pattern.
Modellierton modeling clay.
Modellschreiner wood pattern maker.
modeln modulate by pulse-time modulation (t-m), amplitude m. (a-m), frequency m. (f-m) or phase m. (p-m); model, mold.
Modul, piezo-elektrisches piezo-electric modulus.
Modulation, additive upward modulation. **-, subktraktive** downward modulation.
Modulation auf Dunkel modulation to dark condition, negative m. **-auf Hell** modulation to light condition, positive m. **-der Stimme**

inflection of voice.

Modulation, Absorptions- absorption modulation, Heising m. **Dichte-** (charge) density modulation. **Elektronenstrom-** electron quantity modulation, m. of bunched e. stream. **Frequenz-** frequency modulation **Gegen-** opposite modulation. **Geschwindigkeits-** velocity modulation. **Gitter-** grid modulation, g. control (Brit.). **Intensitäts-** intensity, brightness or brilliance modulation. **Kreuz-** cross modulation. **Leitungsstrom-** conduction current modulation. **Nullphasen-** null phase modulation. **Parallelröhren-** Heising modulation, choke m., plate m., constant-current m. **Phasen-** phase modulation. **Puls-** pulse modulation (used in radar). **Quer-** cross modulation. **Reihenröhren-** constant-potential modulation. **Spannungs-** velocity modulation. **Zeit-** velocity (variable-speed) modulation; variable-speed scan.

Modulationsfrequenz modulation frequency, signal f.

modulationsgesteuerter Sender transmitter in which amplitude varies with AF amplitude, AF modulated transmitter.

Modulations-grad percentage modulation, m. percentage, degree of m., depth of m. **-gradkontrolle** modulation meter. **-röhre** mixer tube, modulator t. **-tiefe** cf. **M.-Grad.**

Modulator, besprochener voice-impressed or voice-actuated modulator.

Modulator, Licht,- light relay, any device adapted to reconvert current into light or brightness variations, e.g., television receiver lamps, Kerr cells, modulated neon lamps, c.-r. tubes etc. **Licht-,selbstleuchtender** light relay of the glow-tube, neon arc lamp, sodium lamp, etc. type. **Licht-, (Fremdlicht)** light relay of the Kerr cell, Karolus c., etc., types adapted to reconvert currents into brightness variations by the use of a constant independent ray source.

Modulierbarkeit modulability, modulation capability.

Modulus, Drillings-, Schiebungs- torsion modulus, shear m., rigidity m.

modulieren s. **modeln.**

Mohr, Platinum- platinum black. **Quecksilber-** black mercuric sulfide.

moiréartiges Geräusch moire pattern of noise.

Moiréeffekt moiré or watered-silk pattern (in photography).

moirieren water; cloud (fabrics).

Molekül, unstarres non-rigid molecule. **Ketten-** chain molecule. **Kreisel-** top molecule, spinning m. **Misch-** mixed molecule (in a loosely associated or l. bound molecular union).

Molekülaggregate molecular clusters.

Molekular-gewicht (Mol.-Gew.) molar weight, gram-molecular w. **-strahlen** molecular rays, m. beam.

Molekül-haufen molecular clusters. **-sieb** micro-filter. **-spektrum** molecular spectrum.

Molenverhältnis molar ratio, mole r.

Molisierung re-combination (of ions in gas).

molletieren mill, knurl (coins, etc.)

Mol-norm molal. **-verhältnis** molar ratio, mole r. **-volumen** molar volume. **-wärme** molar heat.

Moment, Brems- brake torque, retarding t. (in a meter). **Dreh-** torque, moment of rotation. **Kern-** nuclear (magnetic) moment. **Massen-** moment of inertia, rotational i. **Reibungs-** friction torque.

Moment-bild instantaneous photograph, snapshot. **-prozess**

m-process, evanescence p. (in decay of phosphorescence). -transport momentum transfer.
monaurales Hören monaural reception, m. listening.
Möndchen lune, meniscus.
mondförmig crescent-shaped, lunate, meniscal.
monoatomar monatomic.
Mono-chord monochord, sonometer (with one string) -chromasie monochromatism, monochromasia, (condition of electrons being of like wave-length and like speed).
monochromatisierend, selbst- auto-monochromatic (e.g., when an instrument becomes its own monochromator).
monomolekularer Film monolayer, monomolecular layer (adsorbed film).
Monopackverfahren mono-pack method.
Montage setting-up, mounting, erection, installation or assembly (work), fitting.
Montagekopie first answer print.
Morsepunkte, breite lengthened dots. -,spitze clipped dots.
Mosaik-block mosaic block (cr.) -platte target, mosaic plate (telev.). -röhre, doppelseitige image multiplier iconoscope. -schirm mosaic (of iconoscope). -schirm, Überfahren durch Strahl sweep out mosaic with beam.
Motor-geräusch im Norden(Mn) noise of motor North signal. -unterbrecher motor-drive make-and-break.
μ F microfarad (mfd, mf).
$\mu\mu$ F picofarad, micromicrof.
Muffe muff, sleeve, bush, socket joint, clamp, coupling box. Röhren- pipe socket.
Muffenverbindung socket joint.
Mühle, Licht- radiometer (of Crookes).
Mulde, Potential- potential well, p. hole.
Multipletten multiplets.
Mund mouth, opening, muzzle, vent, orifice.
Mundharmonika mouth harmonica, m. organ.
mündliche Verhandlung oral hearing.
Mund-öffnung, Trichter- mouth opening of horn, largest flare. -stück, Trichter- throat of horn. -stücköffnung chink between mouthpiece and reed (of clarinet).
Mündung mouth, aperture, orifice.
Mündungs-geschwindigkeit muzzle velocity. -wucht kinetic energy at the muzzle.
mürbe mellow, tender, soft, brittle, friable, short (of metal).
Musa-Empfang multiple-unit steerable antenna reception, musa r.
Muschel, Augen- eye cup. Hör-, Ohren- ear cap, earpiece. Okular- eyepiece cup.
Muschelglas globular lens.
muschelig shelly, conchoidal.
Musikwerk musical automaton, mechanical music instrument, "jute box"
Muster, Geräusch- noise pattern. Stör- spurious or false pattern (in iconoscope picture).
muster-gültig serving as a model, standard or ideal, classic. -haft standard, typical, exemplary,
Musterkopie master copy, m. print.
Mutter mother disk, master matrix (in phonograph record making).
Mutter, gesicherte lock nut, jam n. Daumen- thumb nut. Flügel- winged nut, butterfly n. Gegen- lock, clamp, check, jam or binding nut. Kordel- milled nut, knurled n. Schrift- type mold, matrix. Überwurf- castle nut.
Mutter-form master mold (cer.); parent (al) form. -frequenz master frequency. -gewinde female screw thread. -modell master pattern. -schlüssel (nut) wrench, spanner. -schraube bolt and nut. -sender key station, master s. (in chain broadcasting). -substanz parent substance. -uhr master clock. -verschluss nut lock.
Mutung, Funk- radio prospecting, r. metal locating.

N

Nabe nave, hub, boss.
Nabelpunkt umbilic.
nachahmen imitate, copy, mimic; counterfeit, adulterate.
Nach-anmeldung supplementary application.
nacharbeiten finish, dress, re-mill, re-trim, re-true (up or off).
Nach-aufnahme re-take. **-baratome, nächstnähere** next nearest neighbor atoms. **-belichtung** post-exposure. **-beschleunigung** after-acceleration, post-a., acceleration by second gun. anode. **-beschleunigungselektrode** second or additional gun anode or accelerator electrode, after-accelerator. **-bild** after-image; copy, imitation, replica.
nachbilden copy, reproduce, imitate; balance, simulate. **-,neu** rebalance.
Nachbildung, Grund- basic network. **Leitungs-** balancing network, artificial b. line **Neu-** rebalancing.
Nachbildwirkung persistence of vision (in form of after-image).
nacheilen lag, trail (behind).
Nacheilwinkel angle of lag.
nacheinander one after another, successively, sequentially, consecutively, seriatim. **-färben** dye again, counterstain.
Nach-fliessen after-flow (of metals), persistence of plastic f., elastic after-effect. **-folgekern** product nucleus. **-forschung** research, search, investigation. **-frist** time extension, respite. **-führung des Rahmens** re-setting or re-adjustment of frame or coil (d.f.)
nach-geben give way, yield, concede. **-geschaltete Stufe** higher stage, end s., power s. **-giebig** yielding, flexible, pliable.
Nach-giebigkeit compliance, yieldingness. **-glühen** after-glow, phosphorescence. **-glühcharakteristik** persistence characteristic. **-hall** echo, reverberation; resonance. **-halldauer, übertriebene** excessive reverberation period.
nachhallen dying down of sound (in an acoustically live room).
Nachhall-kammer, -raum reverberation room, enclosure or space. **-zeit** reverberation period.
nachhaltig lasting, enduring, persevering.
Nach-konzentration re-focusing, action of second lens. **-ladeerscheinungen** phenomena associated with absorption current and residual charge (of condenser).
nachlassen temper, anneal (glass, metals, etc.); slacken, relax, abate.
Nachlauf-einstellung automatic re-setting of direction-finder to zero. **-peiler** direction-finder in which readjustment to zero is effected automatically. **-profil** wake profile. **-stufe** (cathode) follower stage.
Nachleuchten after-glow, phosphorescence, persistence of light or of vision (on retina). **Verminderer des Nachleuchtens** killer, poison.
nachmessen check dimensions, check-up of measurements, test figures, etc.
Nach-prüfung re-examination, recheck, verification, review. **-reifung** maturing (of film).
Nachricht communication, intelligence, information, signals; news. **-übermittelung** signalling, intelligence transmission.
nachrichten re-align, re-straighten,

readjust, re-set.

Nachrichten-system, Mehrfach- multiplex signal system. -technik communication or signal art or technics, signal work.

Nach-schlagewerk reference work. -schlüssel master key, pass key. -schrift picture re-creation; postscript. -schub, Kohlen- carbon feed (arc lamp).

nachstellbar regulable, re-adjustable, resettable.

Nachstellung, Bild- framing of picture, racking of p. (m.p.), centering control (telev.) Bild-,falsche misframing, out-of-frame condition. Blenden- phasing of shutter (film work).

Nachstimmung, Rahmen- re-setting or re-check of frame or coil tuning (in automatic radio compass).

Nachstimmungskondensator aligning condenser, trimming capacitor; vernier c.

nächstnähere Nachbaratome next nearest neighbor atoms.

nachsuchen apply (for a patent), file (a petition).

Nachsynchronisieren post-scoring (a silent film), dubbing, scoring (first picture, then sound), re-recording.

Nachsynchronisierung (von Trickfilm mit synchronisch springendem Ball) bouncing ball synchronization.

nachtblindes Auge night-blind eye, nyctalopic e.

Nacht-blindheit hemeralopia, nyctalopia. -effekt night effect, polarization error, directional or quadrantal errors (in general).

nachteffektfrei night-effect-free, free from n.e. or polarization errors.

Nach-teil disadvantage, injury, prejudice, loss, detriment, shortcoming, demerit, inconvenience, difficulty. -transport take-up sprocket action.

Nachtstreckenbefeuerung route beacon for night flying service.

Nach-verstärker amplifier in stage above input stage, (mostly) power amplifier. -vertonung dubbing, re-recording of sound track.

nachweisen detect, discover, prove, demonstrate, identify; point out, refer to

Nachwickel-rolle take-up reel or spindle, hold-back sprocket. -trommel take-up sprocket.

Nachwickler (lower) take-up sprocket.

Nachwirkung after-effect, secondary e., accompanying e.; hysteresis, persistence, decay (period). -,elektrische electric after-effect (persistence of strain). -,magnetische magnetic fatigue, m. after-effect, residual loss (of metals). -,plastische plastic flow persistence, after-flow, elastic after-effect.

Nachwirkungs-strom absorption current, c. of residual charge (of a condenser). -verlust residual loss.

Nachwirkzeit decay period.

Nadel, Abfühl- selecting needle or pin, pecker. Abspiel- needle, stylus (riding in groove).

nadelförmig needle-shaped, acicular.

Nadel-funkenstrecke needle-point spark-gap, needle g. -geräuschfilter needle scratch filter (phon.) -reibung needle friction (causing stylus drag). -schalter vibrating reed rectifier.

Nadelton stylus or needle-recorded and reproduced sound. -ansatz sound on disk attachment or head. -verfahren sound on disk method. -verstärker phonograph amplifier. -zusatz sound on disk attachment.

Nähe nearness, vicinity, neighborhood, proximity, propinquity, closeness, adjacency, contiguity. Sonnen- perihelion.

Naheinstellung short-range focus.

naheliegen be or lie near, border (on), be adjacent or contiguous; be patent, obvious, evident or self-suggesting.
Näherung approach, approximation, proximation.
Näherungsformel approximate formula, a. equation.
nahestehend intimately related or connected, affiliated.
Nah-feld vicinity or short-range field (of antenna). **-peilung** close-range direction-finding. **-punkt** near point.
Nährboden nutritive substratum, nutrient medium.
Nahschwund radio fadeout or fading 60-100 miles around transmitter site. **-antenne** low-angle or short-range fading antenna.
Nah-signalmittel short-distance or s. range signal means. **-störungen** nearby interference (from sources close to receiver).
nahtlos seamless, jointless.
Nahtschweissung seam welding.
Nahverkehrsbezirk local control zone, approach c.z.
namhaft famous, well known, renowned, reputable, considerable, appreciable.
Napf bowl, basin, pan, cup.
Narbe scar, seam, grain (of leather and paper), pit, defect, flaw, pock. **Rost-** rust pit. **Schneid-** cutting scar.
Narbung pocking.
Nase catch, lug, nose, beak, cam, tappet; projection, spout, nozzle. **Klink-** detent, latch.
Nasenkeil jib, key.
Nassverfahren wet process.
Natriumleuchte sodium lamp.
naturähnlich realistic, true or similar to nature.
Natur-erscheinung natural phenomenon, n. process, n. action. **-farbfilm** natural color film. **-forscher** investigator of nature or in field of natural sciences, n. philosopher, naturalist.
naturharter Stahl self-hardening steel.
Naturlehre natural science, n. philosophy.
natürliche Zahl natural number.
Natürlichkeit der Wiedergabe faithfullness, fidelity or realism of reproduction.
Naturtreue fidelity, faithfulness, realistic properties.
Navigation, Flugzeug- avigation, aircraft navigation, aircraft piloting. **Koppel-** navigation by dead reckoning. **Richtungs-,Seiten-** directional navigation. **Strecken-** long-distance (air) navigation or avigation. **Vertikal-** vertical guidance, v. avigation.
Navigationsfeuer navigational or maritime radio beacon.
Nawimembrane cone diaphragm of curved profile, curvilinear cone, para-curve.
Nebel nebula (spectroscopy), mist, fog, veil (phot.). **Stör-** chief interference zone, maximum noise z. **-apparat** atomizer. **-kamera** cloud chamber. **-landung** blind landing, instrument l. **-spurmethode** Wilson cloud track or chamber method (streak of droplets produced by a particle).
Neben-achse lateral axis (of rhombic crystal); secondary a. **-apparat** accessory equipment, auxiliary apparatus. **-bestandteil** secondary ingredient. **-betrieb** by-operation, secondary process, side-line. **-beweis** secondary proof, additional evidence. **-bild** ghost image. **-bindung** secondary union, s. bond, linkage of the second order.
nebeneinander in parallel (electrical connection). **-stellen** juxtapose, place side by side or alongside.
Neben-entladung lateral discharge stray d., secondary d. **-erzeugnis** by-product. **-fläche** secondary face. **-geräusch** extraneous noise (m.p.); ambient n. (in sound locating). **-intervenient** co-intervener,

-kette side chain, subordinate c. -klage incidental action. -kläger co-plaintiff. -kopplungsschwingung spurious oscillation. -licht (auf Platte)light-shot (on film). -linie secondary line, branch line, satellite (spectr.) -maximum secondary radiation, lateral r., s. lobe or ear (in directional characteristic or space pattern). -patent sub-patent, collateral or subordinate patent. -peilstelle associate direction-finder ground station. -reaktion side reaction, secondary r. -rohr side or branch tube or pipe.

Nebenschluss, Ayrton'scher universal shunt, Ayrton s. Gitter- grid leak. -feder shunting spring, off-normal s. -kondensator by-pass condenser. -resonanz parallel resonance.

Neben-setzung juxtaposition. -sprechdämpfung cross-talk transmission equivalent, cross-talk attenuation. -stelle, Peil- associated direction-finder. -uhr secondary clock. -umstände collateral circumstances, c. factors. -weg by-path. -winkel adjacent angle, adjoining a. -zipfel minor lobe, m. ear (of space pattern).

Negativ, Meister-,Original- master negative, metal master (of phonograph record). Rahmen- rack negative.

Negativ-ebene negative plane, n. aperture (of film printer). -farbauszüge separation negatives. -gamma negative gamma (γ). -lack transparent negative varnish. -schwärzung negative density. -umfang intensity range or latitude of a negative. -widerstandsröhre dynatron type of tube with negative resistance (e.g.,kallirotron).

Nehmer, Lizenz- licensee, grantee.

Neigekopf tilting head (tripod).

Neigung inclination, tendency, slope,pitch,dip,trend. **Schwingungs-** tendency to oscillate or "spill-over" Strahl- inclination of ray.

Neigungs-differenz difference of vergence. -ebene inclined plane. -messer clinometer (to measure angles of slope), Abney level (to measure vertical angles). -messer (zum Messen von Neigungen im Meeresboden) clinometer for measuring sea bottom slopes. -messer, Kreisel- bank-and-turn indicator. -winkel inclination (angle), a. of slope, a.of dip. -winkelmesser clinometer.

Nenner, General-,Haupt- common denominator.

Nenn-reichweite rated range, nominal r. -wert nominal,rated,assessed, normal, expected or face value.

Neonflächenglimmlampe plate neon lamp (telev.)

Neonröhren-aussteuerungskontrolle, -lautstärkemesser neon tube volume indicator.

Nernstlampe, Mikro- Nernst microscope lamp.

Nerven-reiz nervous stimulus. -taubheit perception or nerve impairment of hearing, n. of perception deafness.

Netz (of Gitter) net, netting, gauze, network, reticle, graticule, reticulation, reticulum, grate, lattice, grid.

Netz, Anodenschutz- screen grid, plate shield. Anpassungs- matching network. Erddraht- ground mat. Grad- graticle, reticle, map grid. Hilfs- auxiliary (control) grid. Karten- graticle, map grid. Raumlade- space-charge grid. Schutz- shield or screen grid; protective network. Spannungs- space-charge grid. Strecken- air routes. Verstärker- amplifier channel.

Netz-anode plate fed from power unit or light socket. -anschlussgerät (strictly:) power pack, power unit; power converter set, electric s., mains-operated radio s. -antenne,Licht- light socket or lighting circuit antenna, lamp-socket a.

netzartig (s.a. Netz) netlike, reticular, reticulate, grate-, grid- or grating-like,
Netz-bedeckung coverage factor. **-bildung** reticulation, netting. **-brummen** mains hum, motor boating (of a-c), a-c pickup. **-drossel** hum eliminator coil, hum-bucking c. **-ebene** lattice plane, plane family of a crystal. **-ebenenabstand** grating constant (spectr.), inter-lattice plane distance, distance between successive planes indicative of lattice constant (cr.)
netzen wet, moisten, steep, soak.
netzförmig net-shaped, reticular, reticulate.
Netzgerät electric set, mains-operated set, power-pack or power-unit s., power-unit electric radio s.
Netzhaut retina. **-bild** retinal image. **-trägheit** retinal persistence. **-zentersehen** foveal vision.
Netz-mikrometer, Okular- micrometer with reticle eyepiece. **-stecker** power supply plug, light-socket p. **-struktur** cellular network, lattice, reticular structure, reticulation. **-ton** mains hum, line h. **-transformator** power transformer, mains t. **-unruhe** mains or supply-line ripple or fluctuation. **-verstärker, -vervielfacher** mesh multiplier, m. amplifier (with secondary emission, seriation or cascading of electron-permeable fine-mesh targets). **-werk** network, reticulum. **-werk, Trenn-** dividing network.
Neu-aktivierung der Kathode re-activation or rejuvenation of a cathode. **-gestaltung** re-organization, re-arrangement, modification. **-heit** novelty, originality.
neu-heitsschädlich anticipatory, rendering novelty negative.
-nachbilden re-balance.
neunwertig nonavalent.
Neusilber argentan, German silver
neutral einstellen adjust to neutral, set to n.
neutraler Schirm neutral screen.
Neutralisations-kondensator balancing or neutralizing condenser. **-schaltung** neutrodyne circuit organization.
Neutroneneinfang neutron capture.
nichtabgestimmt untuned, aperiodic.
Nicht-anerkennung non-acknowledgment. **-beobachtung** non-observance.
nicht-berechtigt disqualified, not entitled. **-bevorzugte Orientierung** random orientation, non-preferential o., non-privileged o. or direction. **-dissoziiert** undissociated, non-d.
Nicht-eisenmetalle non-ferrous metals, non-iron m. **-erfüllung** non-performance, non-fulfillment, default. **-erscheinen** non-appearance, non-attendance, absence, default, contempt.
nichtgerichtetes Mikrophon astatic microphone, non-directional m.
nichtig null, void, invalid, annuled.
Nichtigkeits-abteilung annulment department. **-beschwerde** plea of nullity. **-erklärung** declaration of nullity, annulment or cancellation. **-klage** action or suit for cancellation, annulment or invalidation.
nicht-ionisiert unionized, non-ionized. **-kommutative Algebra** non-commutative (vector)algebra. **-kondensierte Hochstromentladung** uncondensed heavy-current discharge. **-kristallisch** amorphous, non-crystalline.
Nichtleiter non-conductor.
nicht-leuchtende Flamme roaring flame. **-lineare Verzerrung** non-linear distortion. -

-linearität lack of linearity, non-l. **-metall** nonmetal.

nichtmischbar immiscible, nonmiscible; unsoluble, insoluble, insolvable.

Nichtmischbarkeit non-miscibility, immiscibility.

nicht-reproduzierbar non-reproducible, non-controllable. **-resonierend** aperiodic, non-resonant. **-rostend** rust-proof, stainless, rust-free. **-sphärisch** non-spherical, aspherical. **-ständig** non-permanent, temporary, transient, transitory. **-symmetrisch** asymmetric, unilateral. **-umwandelbar** inconvertible. **-vergasbar** non-gasifiable, non-volatile. **-wässerig** non-aqueous, anhydrous.

Nicht-zahlung default of payment, non-payment. **-zerfallbarkeit** non-disintegrability, non-fissionability.

nichtzerstörende Prüfung non-destructive test.

nieder-drücken press down, force d., depress or touch (a key or button). **-fallen** fall down, drop, precipitate, settle.

Niederfrequenz-drosselsatz low-pass filter. **-röhre** AF tube, low frequency t., "apple" (m.p. studio slang).

Nieder-führung downlead. **-gang** descent, decline, downward stroke or travel, lowering movement or stroke.

niederlegen lay down, put d., deposit.

Niederschlag precipitate, precipitation, (radio-active) deposit, sediment, settlement. **-arbeit** precipitation process, iron-reduction work. **-gefäss** precipitating vessel. **-mittel** precipitant.

Niederschlags-apparat precipitator (Cottrell). **-potential** deposition potential.

nierenförmig reniform, kidney-shaped (for instance, condenser plate).

Niete, Loch- hollow rivet, tubular r.

Niet-schaft rivet shank, r. stem. **-teilung** rivet spacing, r. pitch. **-verbindung** riveted joint.

Nipkowscheibe Nipkow scanning disk.

Nitrierstahl nitriding steel, nitralloy s.

Nitrierung nitrification, nitrogen, nitridation (steel).

Nitrobenzol nitrobenzene.

Niveau level.

Niveau, Energie- energy level, quantum state, term. **Grund-** background level (or average) illumination (telev.). **Stör-** noise level.

Niveau-fläche level surface; equipotential or isop. surface. **-rohre** level tube, leveling t. **-stufe** term, energy level (of electrons).

nivellieren equalize, smooth, bring down to an average level, grade, even.

Nivellierer, Spannungsspitzen- voltage peak limiter.

Nocken cam, lifter, tappet, disk or wheel with lobes. **-, zweistufiger** double-lift cam. **Leerlauf-** delayed pulse trip cam. **Schalt-** trip cam, trigger c. **Steuer-** sequence switch cam.

Noiseless, Gegentakt- split-wave noiseless film recording. **-blende** biasing or noise-reducing stop or shutter vane or gate. **-stromkreis** back-ground noise reducing circuit, noise gating c.

Nomogramm straight-line or alignment chart, self-computing c., monogram.

Nonius-einteilung, -skala vernier scale.

Nonne mold matrix.

Nord, missweisend uncorrected north, magnetic n. **-, rechtweisend** corrected north.

Norddrehfehler northerly turning error.

nördliche, Breite northern latitude.

Nordlicht aurora borealis. **-linie** green auroral line.

Norm standard, norm, model.
Norm, Vor- preliminary standard, tentative s.
Normal-bildkadre standard picture frame. **-essig** standard vinegar, proof v. **-flüssigkeit** normal or standard solution or liquid.
normalglühen normalize.
normalisieren standardize, normalize (brief heating slightly above A_3 point).
Normal-lehre normal gage, master g. **-potential** normal potential. **-schliff** standard (interchangeable) ground joint. **-schrift** standard track (sound recording).
normalsichtig normal sighted, emmetropic.
Normalweingeist proof spirit.
Normenaufstellung formulation or laying down of standard rules or norms, standardization or normalization.
Normierung standardization, normalization, gaging.
Normung des Frequenzganges standardization of frequency response or characteristic.
Not-amt temporary exchange, emergency e. **-anlage** provisional or emergency plant or equipment. **-behelf** makeshift, expedient, last resort, temporary or emergency means, improvisation.
Note, Triller- trill note, warble n.
Not-fall case of emergency, case of need. **-mittel** expedient, makeshift or emergency means. **-ruf, -signal, -zeichen** S O S call, (radio-telegraphic) distress call or signal, mayday (radio telephone distress call).
Novokonstant Cu-Mn alloy with Al Fe.
Nuance shade, tint.
Null zero, nought, cipher, null. **-frequenz** frequency at which phase shift is zero. **-leiter** neutral wire. **-linie** zero line, missing l., band center (spectr.) **-linienverlagerung** zero setting, electrical biasing (in film recording). **-methode** zero method, null m., balance m. **-phasenmodulation** zero phase or null phase modulation.
Nullpunkt zero point, null point, neutral p., origin (of a diagram). **-abweichung** zero error. **-anomalie** origin distortion. **-einstellung** zero adjustment. **-empfindlichkeit** zero-level sensitivity. **-fehler** origin distortion, ion cross effect. **-schärfung** zero point clearing, cleaning or sharpening (in df).
Nullstelle zero place, abscissa which gives zero value to Bessel function.
Nullung eines Störers connecting casing of interfering device with neutral (in d.c. mains).
nullwertig avalent, non-valent.
Nullzweiglinien zero branch lines.
numerische Objektivöffnung numerical aperture of objective.
Nummer, laufende serial number, consecutive n.
Nummernscheibe dial (switch), number plate, d. plate. **Ablaufen der -** return of the dial. **Aufziehen der -** winding up the dial.
Nummernwahl impulse action or stepping, selection.
Nurflügelflugzeug tailless airplane.
Nussbaum walnut.
Nute slot, notch, groove, recess, key-way, oilway. **-und Feder** slot and feather (union), slot and key, groove and tongue joint.
Nute, Schmier- oil groove or slot, worm.
nuten slot, groove, notch, flute, keyseat.
Nutenwellen slot ripples, tooth r.
Nutsche, Mikro- micro suction filter, m. funnel.
nutzbarer Linsendurchmesser useful or effective (clear) aperture of a lens.
Nutz-dampfung damping resulting in useful radiation or signal power, effective transmission equivalent. **-feldstärke** signal strength or

field intensity, "useful" field intensity. -kreis useful circuit, active c. -lautstärke signal level, sound s. volume. -leistung useful power, signal p., useful work or effect; efficiency. -niesser usufructuary, beneficiary. -signal intelligence signal, any s. with information.

-spannung useful potential, signal p. -strom useful current, signal c. -stromkreis signal circuit, utilization circuit. -widerstand (cf Wirkwiderstand) useful or signal resistance. -wirkung efficiency.

N.Z. (Neutral. Zahl) neutralizing number.

Ober-banden overtone absorption bands. **-begriff** introductory part of German patent claims setting forth prior art, broad class and nature of invention and disclosure. **-beleuchter** chief light electrician; scaffold or top light controller (studio).

Oberfläche, gasbehaftete gas-contaminated surface, s. with adsorbed gas. **-, spiegelnde** specular surface. **-, vertonte** sound-impressed surface, surface or area bearing sound track.

Oberflächen-ableitung surface leakage. **-abtaster** surface analyzer. **-beladung (mit Hg)** activation, sensitization or charging of a surface (with Hg). **-effekt** surface effect (of photo-cathode). **-einheit** unit of area.

oberflächengerichtet surface normal.

Oberflächen-härtung case-hardening (metals), surface cementing. **-integral** integral taken over a surface. **-normale** surface normal. **-phase** surface phase. **-schicht** surface layer, s. barrier. **-spannung** surface tension. **-spannungsmesser** od. **-spannungsprüfer, nach Abreissmethode** surface tensiometer of adhesion-balance type. **-taster** surface analyzer (to measure irregularities). **-widerstand** surface resistance (of insulation); skin-effect resistance (in a-c flow). **-wirkung** surface action, skin effect; surface work (cryst.)

oberflächlich superficial.

Oberharmonische harmonic, higher h. **-, dreifache** third harmonic, triple h. **-, gerade** even harmonic. **-, ungerade** odd harmonic.

Ober-leiter, künstlerischer art supervisor. **-licht** overhead scoop, scaffold light, skylight. **-strich** permanent maximum current or carrier, peak carrier amplitude, maximum or black-level picture signal (telev.) **-taste** sharp (mus.). **-ton** overtone (often a harmonic), upper partial (NOTE: harmonics and overtones are partials of a frequency above fundamental, a partial being a pure-note component of a sound or complex note).

Objekt, Übertragungs- object to be televised or scanned. **Untersuchungs-** test specimen, t. sample, t. piece. **-glas** slide, mount (micr.)

Objektiv (cf Linse, Lupe, Glas) objective, object glass, lens (receiving first light from object). **-, höchstlichtstarkes** ultra-rapid or ultra-high-speed objective or lens. **-, kurzbrennweitiges** short-focus lens. **-, lichtstarkes** high-speed objective. **-, zusammengesprengtes** broken-contact (uncemented) lens or objective. **Aufzeichnungs-** recording objective. **Fern-** telephoto lens. **Immersions-** immersion objective (electron microscope).

Objektiv-brett lens or objective panel or board. **-deckel** lens cap. **-fassung** objective or lens mount or barrel. **-gewindes, Ansatzflache des** screw shoulder or s. collar of objective mount. **-halter** lens holder. **-höhe** height of lens or objective. **-öffnung, numerische** numerical aperture of objective. **-offnungswinkel** angular aperture of objective. **-ring** lens adapter. **-satz** set of lenses, lens combination. **-spule** objective coil. **-wechsler** revolving nosepiece (micr.)

Objekt-kammer object chamber. -mikrometer stage micrometer. -schleuse object airlock. -sucher object finder (opt.) -tisch micrometer stage, object stage or stand, manipulator. -träger object slide, mount, stage or stand, o. support, o. carrier (plate), specimen holder (e.g., a celluloid film or skin). -trägerplättchen object support lamina or slide (film or skin). -umfang range of brightness (of subject).
Ofen, Etagen- storey furnace. -sohle furnace bottom, bed, sole or floor.
ofentrocken kiln- or oven-dried, d. in kiln or oven.
offen hollow (of sound). -(er) Schwinger open oscillator. -(e) Sprache open language, uncoded l. einseitig- unilaterally open (-ended).
offenbarer Unsinn manifest absurdity.
Offenbarung disclosure.
offenkundige oder öffentliche Vorbenutzung public use, prior use.
öffentlichen Hand, Betrieb der public utility.
öffnen, eine Röhre open, unlock, render conductive, allow discharge, cause breakdown of a tube, trigger a t. (say, a blocking oscillator). -, einen Stromkreis break or open a circuit or a line.
Öffnung aperture, orifice, opening, mouth, vent, leak, crack, slit, port; F number (opt.) -, numerische, des Objektivs numerical aperture of objective. Blenden- aperture of diaphragm or stop. Mundstück- chink between mouthpiece and reed (of clarinet). Schall- louvre (of bell tower). Spalt- chink (of clarinet). Verschluss- shutter aperture, s. opening.
Öffnungs-bild aperture image. -fehler apertural defect, error or effect. -verhältnis aperture ratio, relative a., field of view (picture projection). -weite flare, width of opening, (of horn, trumpet, etc.) -winkel, Halb- semi-apertural angle. -winkelmesser apertometer.
Ohm'scher Widerstand ohm resistance, steady-current r.
Ohr ear, lug, eye, catch, loop. -, vertäubtes deafened ear.
Öhr ear, handle, eye, catch, eyelet.
Ohrempfindungsgrenze loudness contour of ear, auditory sensation area.
Ohren-kappe ear muff. -kurve auditory sensation curve. -muschel ear, cap, e. piece.
Ohrhöhle ear cavity.
Ökonomiekoeffizient efficiency of economy.
ökonomisch economic(al).
Oktaeder octahedron.
oktaedrisch octahedral.
Oktave, eingestrichene one-stroked octave.
Oktavkoppel octave coupler (organ).
Okular, verzerrungsfreies orthoscopic eyepiece, distortion-free e. Augenlinsen- eyelens of eyepiece. Mess- micrometric eyepiece. Sucher- focusing lens.
Okular-aufsteckglas eyepiece correcting lens. -blende eyepiece diameter, e. stop. -kollektiv field lens of eyepiece. -muschel eyepiece cup. -netzmikrometer micrometer with reticle eyepiece.
ölbildend oil forming, olefiant. -(es) Gas olefiant gas (ethylene).
Öldämpfung oil damping (in vibrators); o. dashpot. -fänger oil catch, o. collector.
ölhaltig containing oil, oleiferous.
Ölharz oleoresin.
ölig oily, oleaginous.
ölkarburiert oil-carburetted.
Öl-nute oil groove, oil run, oilway. -pumpe oil pump, type of high-vacuum pump, e.g., "Hyvac". -schalter, Impuls- oil-blast circuit-breaker. -schiefer oil shale.

-schwungmasse rotary stabilizer (in film feed). -tröpfchenmethode oil drop method.

Opalglas opal glass, frosted g., light-diffusing g. -scheibe, Vollpot opal diffuser disk (phot.)

Operatorenrechnung operational calculus.

Optik optics (science), optic (optical means such as lenses, etc., placed in path of rays in physical apparatus, etc.)

Optik, Elektronen- electron optics, e. lenses (and their control of beams). Licht- light optics, physical o., geometric o. of light rays. Spalt- microscopic aperture, imaging optic means, slit "optic". Ton (film)- soundhead lens, s. optic (m.p.)

Optik-auswechslung optical system or assembly comprising interchangeable intermediate lens barrels. -tubus optical barrel.

optisch, spannungs- photo-elastic. -(e) Bank optical bench. -(e) Drehung optical rotatory power, o. rotation, rotatory polarization.- (e) Inaktivierung eines Quarzkristalles racemization.

ordentliche Komponente ordinary component (of rays, etc.)

ordnen order, arrange (methodically or systematically), regulate, settle, collect terms (math.)

Ordnung order, arrangement, organized disposition, class; power (math.) -, Kurve dritter cubic curve. -, n^{ter} of nth order.

ordnungsgemäss orderly, methodical.

Ordnungs-grad degree of order. -stellung, Ein, -stellung, Zwei- position of grating resulting in a one-order or a two-order spectrum.

ordnungswidrig irregular, disorderly, contrary to order, discipline or rule.

Ordnungszahl ordinal number; n. of a series, atomic n.

Orgel, Dompfaffen-, Vogel- bird organ. -zungenpfeife reed organ pipe.

orientiert, nicht bevorzugt randomly oriented, non-preferentially o.

Orientierung orientation, direction; trend, inclination. -, bevorzugte non-random orientation, privileged orientation.

orientierungsabhängig orientation-sensitive, o.- responsive, be a function of, or dependent upon, orientation, direction or angle.

Ort place, locus, region, locality, site, position.

Original-negativ master negative. -positiv master positive.

Ort, geometrischer vector point, geometrical locus.

Ort und Stelle, an in situ, on the site, on the spot.

Ort, Elektronen- position or locus of an electron. Flug- position of flying craft, fix. Koppel- fix, dead-reckoning position.

Orthochromatisierung orthochromatic processing.

Orthodrome straight (or rhombic) line portion of a great circle.

-ortig -place.

ortig, drei- three-place.

örtlicher Tagesgang local diurnal variation.

Orts-bestimmung position-finding. -diagramm locus diagram, circle d. -fehler geometric location error. -funktion position function. -gang local variation. -gebiet local area. -missweisung local declination.

ortsveränderlich non-stationary, movable, (trans)portable.

Ortung position-finding, orientation, taking of bearings, navigation, avigation. -, akustische sound location, acoustic orientation. Funk- radio position-finding, obtaining r. fix.

Ortungskompass navigation compass.

Öse lug, eye, ear, hook, loop, eyelet. Löt- soldering tag, tab.

Ösen-blatt tong, lip, flange. -bolzen eyebolt.

Oszillationsquantumszahl vibrational quantum number.
Oszillator, leitungsgesteuerter line-controlled oscillator. **Sperrkipp-** blocking oscillator, squegging o.
Oszillograph, Innenphotographie- internal photography oscillograph. **Saiten-** string oscillograph. **Schleifen-** loop, string or vibrator oscillograph.
Oszillographenschleife oscillograph loop, o. vibrator.
Oxydations-flamme oxidizing flame. **-mittel** oxidizing agent. **-stufe** stage of oxidation, degree of o.
Oxyd-beschlag coating of oxide. **-faden** oxide-coated cathode, oxide filament.
oxydhaltig oxidic, containing oxide.
Oxydhaut skin of oxide, film of o.
oxydierbar oxidizable.
Oxydkathodenröhre tube with oxide-coated filament, dull emitter t.
oxygenieren oxygenate, oxygenize.
Ozokerit ozocerite.
Ozonsauerstoff ozonized oxygen, activated oxygen in the form of ozone.

P

P Zweig P branch (spectr.)
Paarbildung pairing, formation of pairs, drawing together of lines (due to hum, in interlaced scanning).
Paare, Echo- paired echoes.
Paarerzeugung pairing.
paarig erscheinen (von Linien, im Bild) pairing of lines (telev.)
Paarigkeit pairing, formation of pairs (caused by hum, in interlaced scanning).
Pack pack, packet, bale, bundle.
Packung packing, serving, wrapping. Störschutz- interference eliminator kit.
Packungserscheinung packing phenomenon.
Paket, Federn- bank, assembly or set of springs.
Paneel panel.
Panoramabild panoramic picture or view, panning shot.
Panzer shield, ironcladding, armor (of a cable).
Panzerholz metal-plated plywood.
Panzerung shielding, screening, ironcladding.
Papageno-flöte, -pfeife Pandean pipe.
Papier, abziehbares transfer paper. -, selbsttonendes self-toning paper.
Papier, Koordinaten- ruled paper, coordinate paper, cross-section p.
Papier-bahn web of paper. -band tape of paper, web of p. -hohlraumkabel air-space paper-core cable, dry-core c. -pergament parchment paper. -strang web of paper. -streifen paper strip, paper web. -umhüllung wrapping of paper. -vorschub paper feed.
Papp-bogen sheet of pasteboard or cardboard. -deckel pasteboard, paperboard, carton.
Pappe pasteboard, paperboard, cardboard, carton, millboard.
Pappelholz poplar.
Pappenguss pressure-cast carton or cardboard (using paper pulp), papier mâché, millboard.
Pappmasse papier-mâché.
parabolisch, rotations- full parabolic.
Parabolspiegel parabolic reflector.
parachsial paraxial.
Paraffinbad paraffin bath.
Parallax, Sucher- seeker or finder parallax.
parallel, achs- axis-parallel, paraxial.
Parallel-endmass standard gage block, end-to-end s. bar. -impedanz leak impedance.
Parallelogramm, schiefwinkliges rhomboid, oblique parallelogram.
Parallel-resonanzkreis parallel-resonance circuit, rejector c. -röhrenmodulation Heising choke or plate modulation, constant-current m.
parallelschalten (connect in) parallel, shunt (across).
Parallel-schaltung parallel connection, p. circuit organization, multiple connection, paralleling. -verschiebung translatory motion, t. shift. -versuch parallel experiment, check-up e., duplicate determination. -weg bypass, shunt, shunt path.
Para-lysator anti-catalyst. -stellung para position.
Pardune back stay, guy wire, g. cable.
Partei, Gegen- opponent, adversary, adverse party, "the other side" (in a suit).
Partialdruck partial pressure.
Paschen-Back Verwandlung Paschen-Back (magneto-optic) effect.
passend fit, suitable, convenient, appropriate.
Pass-fläche seat, surface where two

parts fit together or engage (snugly). -sitz press-fit, snug f., seat, lodgment.

Passierrohr indicating plug, inside gage.

Passivierung passivation.

Passung fit.

Pastekathode pasted (oxide-coated) cathode.

Pastille pastil, button, bead, lozenge.

Patent,angemeldetes filed p., pending p., p. applied for. -, schwebendes pending patent. Abänderungs- re-issue patent. Apparat- patent covering and disclosing apparatus or machine, a.p. Beweise,die Gültigkeit eines...es Abbruch tun evidence impeaching, or prejudicial to, validity of a patent. Erfindungs- patent of invention. Haupt- parent patent, main or original patent. Klage- patent in issue, in suit or at bar. Stamm- parent patent. Stoff- patent covering a substance or material, product p. Verbesserungs- patent of improvement. Verfahrens- process patent, method p. Zusatz- patent of addition (to parent p.)

Patent-anmassung usurpation of patent rights. -anspruch claim of a patent. -beschreibung patent specification. -dauer life of a patent, continuance of a p., term of a p. -erlöschung expiry or expiration of a patent; annulment of a p. -erteilung allowance, grant or issue of letters patent. -gebühr patent fee, p. tax. -gesetz patent law, p. code, p. statutes. -gültigkeit anfechten challenge validity of a patent.

patentiert im In- und Ausland covered by domestic and foreign patents.

Patent-inhaber patentee, grantee, holder of letters patent. -klage patent suit, p. action. -kommissar stellvertretender deputy or acting patent commissioner. -streit patent action, p. litigation. -unteranspruch sub-claim of a patent. -verlängerung renewal of a patent. -vorwegnahme anticipation, anticipatory reference.

patentverletzendes Merkmal patent-infringing fact or feature.

Patentverletzungsklage action or suit for infringement of patent rights.

Patrize upper tup die, punch.

Patrone cartridge (for specimen in electron microscope); pattern, stencil, mandrel.

Patronenreihe string of cartridges.

Pauke kettle drum.

Paukenstimmschlüssel kettle drum wrench.

pauschen swell, refine (met.)

Pause copy, tracing; space, interval, intermission, pause. Dunkel- obscuring period, cut-off p., dark p. Gespräch- interval of no speech. Licht- photo(graphic) tracing, blueprint. Telegraphier- space (in keying or telegraphing work).

pausen trace, calk.

Pedal, Forte- loud pedal. Piano- soft pedal. Tonhaltungs- tone sustaining pedal.

Pegel level, gain (in db units); water gage. -,mittlerer mean power level (as indicated in volume indicator). Dialog- dialogue level. Leitungsbedarf- power reference level. Rausch- noise level (due to Schottky effect and thermal noise). Schwarz- black level. Stör-, Untergehen der Bildspannung im swamping of picture or video signal in noise level.

Pegel-ausgleich equalizing level, l. equalizer. -messer transmission measuring set. -schreiber level recorder. -strom pilot current, p. carrier (in suppressed carrier system).

Pegelung level or volume (gain) regulation.

Peil Q Gruppen Q code groups.

Peiler, direkt anzeigender direct-reading direction-finder. -,selbstanzeigender automatic direction-finder, direct-reading d.f.

Peiler, Bodenstation- ground station direction-finder. Nachlauf- direction-finder with automatic zero setting. Sicht- visual direction-finder, direct-reading d.f.

Peil-abschnitt, geeichter calibrated sector. -antennensystem directional antenna system. -antrieb frame drive, loop d. (handle, wheel and spindle). -aufsatz bearing plate (of d.f.). -empfänger direction-finding receiver. -fehler direction-finding error, bearing e. (due to night effects, etc.), distortion of bearing. -feld bearing field, -flugleiter ground direction-finding control operator. -funkempfänger radio compass, direction-finder. -hauptstelle main control direction-finder station (of an air communication safety district). -haus direction-finder station building. -kompass navigator's compass. -kurs, recht voraus head-on course (fore-aft axis or lubber line of craft directed toward objective). -kurve space pattern, directional diagram. -leitstelle main control direction-finder station (of an air communication safety district). -nebenstelle associated d.f. ground station, sub-control station. -punktmarke, schwarze black dot, b. mark (on scale). -rahmen coil antenna, loop a., frame a. -richtung direct course or line to transmitter or beacon. -ring direction-finder loop or frame. -scheibe, Funk- bearing plate (of radio direction-finder). -seite, eindeutige absolute direction. -seitenschalter direction-finding sense switch, sense-finding s. -spannung directional signal potential (picked up by directional antenna in df, as distinguished from auxiliary antenna potential). -stelle, Boden- ground df station, aeronautical g.s.-, Neben- associate ground station df. -strahl, raumgeradliniger orthodrome. -strahlwegablenkung distortion of bearing, deviation due to diurnal or seasonal factors, weather, terrain, local conditions, etc.; lateral deviation. -tabelle bearing chart.

peiltreue Bodenwelle ground wave resulting in undistorted bearing.

Peilung taking bearings, direction-finding. -, falsche false bearing, erroneous b. -, optische pilotage. -, wahre true bearing.

Peilung, Deck- alignment bearing. Eigen- board (aircraft) direction-finding, taking bearings with homing device on board. Einzel- bearing taken from one object. Falsch- erroneous bearing, false b. Flimmer- (cf Wechselkardioide) loop reversing switch method (d.f.) Fremd- ground direction-finding (directional data or radio bearing supplied on request to airplane from ground station). Grosskreis- great circle bearing. Hör- auditory direction-finding. Kreuz- cross bearing (taken from two or more ground df stations). Längs- taking bearings from df stations roughly in direction of travel or course. Maximal- maximum-signal strength (df) method. Minimum- minimum or zero signal strength (df) method. Misch- taking bearings on board (without directional receiver) from directive beam. Nah- close-range direction-finding. Standlinien- position-line bearing, great circle bearing. Standort- obtaining a fix, position or location finding (by two or more direction-finders), position-line bearing, great-circle b. Ziel-, missweisende Richtung Flugzeug-Boden magnetic reciprocal course (from airplane to ground station). Ziel-, rechtweisende true course to steer.

Ziel-,rechtweisende Richtung vom Flugzeug zur Bodenpeilstelle true reciprocal course (from airplane to ground station).

Peilungsbeiwert error compensation value.

Peil-verfahren,Impuls- impulse direction-finding method. **-verlagerung** bearing shift, b. error, b. displacement. **-vorsatz** d.f. means added to normal receiver equipment. **-zeichen** call sign, code signal (say, of a ground station). **-zeiger** pointer of df apparatus.

Pendel, Schwere- gravity pendulum.

Pendel-bewegung shutter movement (m.p.) **-feder** (im Zerhacker) vibrator blade, reed or spring (of chopper). **-fenster** internal pressure plate. **-frequenz** quench or bias frequency furnished from auxiliary oscillator circuit (in super-regeneration); electron-oscillation frequency. **-frequenzschaltung** (Armstrong) super-regeneration circuit organization. **-gewicht** pendulum weight, bob. **-gleichrichter** vibratory rectifier, tuned-reed r. **-hammer** (Charpy's) impact testing machine. **-kugellager** self-aligning ball bearing.

pendeln oscillate (of electrons), vibrate, swing (of phases), pendulate, undulate, pulsate, rock, hunt (around an average value).

Pendeln,Hinauf- (gradual) rise in resonance.

Pendel-rückkopplung super-regeneration. **-rückkopplungsspannung** positive quench or biasing voltage (impressed on grid in super-regeneration circuits). **-schlagwerk** (Charpy) Charpy impact testing machine. **-umformer** vibrating rectifier.

Pendelung,Elektronen- oscillating of electrons. **Last-** load fluctuation, hunting of load. **Luftstrom-** air-jet pendulation.

Pendel-vervielfacher reciprocating and accelerating secondary-electron multiplier. **-wechselrichter** vibratory inverter, inverted rectifier of vibrator type.

Peracidität superacidity, hyperacidity.

Perforationsgeräusch sprocket hole modulation or noise (m.p.)

Periode cycle, period. **Halb-** half cycle, h. period, semi-oscillation,alternation. **Schwebungs-** beat cycle. **Viertel-** quarter-period. **Zugwechsel-** field frequency in interlaced scanning (= 2 x frame f., telev.)

Periodenzeit time of vibration, t. of oscillation.

perlen form bubbles, drops or beads; sparkle, glisten like pearls.

Perlenwand directive or beaded screen.

Perl-glimmer margarite. **-mutterblech** crystallized tinplate, moiré métallique. **-mutterpapier** nacreous paper. **-rohr** bead tube (filled with glass beads to break up a fluid). **-schnur** maryante (cryst.); string of beads or pearls, line of droplets.

permanentdynamischer Lautsprecher dynamic or moving-coil loudspeaker using permanent magnet (to produce constant field), p-m dynamic l.

Permanentsatz permanence principle, sum rule.

Permeabilitätsmesser permeameter.

Perpendikel pendulum, perpendicular. **-ende oder -Gewicht** bob.

Personenabtaster mit wanderndem Licht spotlight scanner for persons.

Perspektive,Frosch- worm's eye view. **Füll-** plenary perspective. **Vogel-** bird's eye view.

Pesenantrieb belt drive.

Petent petitioner, applicant, suppliant.

Petrolkoks petroleum coke, oil c.

Pfahl stake, pile, stick, pole, post.

Pfeife, gedackte, gedeckte stopped pipe. **Doppelzungen-** double-tongued flute. **Galton-** Galton pipe. **Labial-,Lippen-** flue pipe, flute pipe. **Orgel-** reed organ pipe. **Quer-** fife. **Rohr-** reed pipe. **Signal-** siren, whistle. **Stimm-** tuning pipe, pitch p. **Vogel-** bird whistle. **Zink-** zinc pipe. **Zinn-** tin pipe. **Zungen-** reed pipe, r. flute. **Zwitscher-** bird warbler.

pfeifen whistle, sing, squeal, howl, self-oscillating (of a tube)

Pfeifpunkt-abstand,-sicherheit singing, whistling or stability margin (of radio tube).

Pfeil arrow, dart; strain.

Pfeiler pillar, post, column, upright, standard.

Pfeil-höhe sag or dip (of a line wire), height (of an arch, meniscus, etc.), camber (of airplane wing). **-rad** double helical or herring-bone gear wheel.

Pferdefleischholz Surinam wood.

Pflanzenwachs vegetable wax.

Pflichtenblatt (contractual) specification.

Pflichtwert specification value, contract v.

Pflock peg, pin, plug, stake.

Pfosten pillar, post, upright, column, standard.

Pfropfen cork,stopple, stopper; graft.

pH Wert pH value (hydrogen-ion concentration).

Phase phase; stage, epoch, era.

Phase,ausser de-phased, out-of-phase (condition). **gleichbelastete - (n)** balanced phases. **-,in** in-phase, co-phasal, in step. **-,verschobene** displaced phase, shifted p., dephased or out-of-phase condition.

Phase, Eintritts- entrance phase (of network or filter). **Oberflächen-** surface phase. **Verdunkelungs-** dark interval (m.p.)

Phasen-abgleichvorrichtung phase balance, p. changer, p. shifter. **-ablösung** phase switch (in polyphase operation of ignitron). **-aussortierung** phase focusing. **-drehung** phase rotation, p. shift. **-einstellung** phasing, phase adjustment, p. shift, p. regulation. **-entzerrung** phase correction, delay equalizer; delay or echo weighting term.

phasenfalsch misphased.

Phasen-fehler error of phase, p. distortion. **-fokus** phase focusing. **-geschwindigkeit** phase speed, p. velocity, wave v. **-gleichgewicht** phase (rule) equilibrium. **-gleichheit** phase coincidence, in-phase state, co-phasal s. **-hub** phase variation or fluctuation. **-kurve** phase curve. **-laufzeit** time of phase transmission. **-lehre** solution theory, doctrine of phases. **-mass** phase constant, wavelength c.

phasenmässig in proper phase relation; looked at from phase viewpoint.

Phasenmodulation phase modulation. **-nacheilung** phase lagging. **-raum** phase space, extension-in-phase. **-regel** phase rule. **-regelung** phasing, phase adjustment, p. shift. **-reinheit** freedom from phase shift or angular differences. **-resonanz** phase resonance, velocity r.

phasenrichtig in proper phase relation.

Phasen-rückdreher phase lagger, p. shifter. **-schwund** phase fading. **-spaltung** phase splitting. **-sprung** phase angle shift.

phasenstarr in locked phase relation, p. -locked.

Phasen-teiler phase splitter. **-transformator** phase shifting transformer. **-Übergang** phase transition. **-umkehrer** phase inverter, p. reverter. **-verdrehung** phase shifting, p. rotation.

phasenverkehrt wirken act in reverse polarity, act in opposite phases or in p. opposition.

Phasenverschiebung phase shift, p. displacement, p. advancing (if forward), dephased condition. **-von 90 Grad** quadrature relation of phases. **-von 180 Grad** phase opposition.

phasenverschoben dephased, out-of-phase, in p. quadrature (when 90 degrees), in p. opposition (when 180 degrees).

Phasen-verzerrung phase distortion. **-verzögerung** phase lag, p. delay. **-voreilung** phase lead, p. advance. **-vorschieber** phase advancer. **-welle** phase wave, de Broglie w., electron w. **-winkel** impedance angle, phase a., impedance factor, Q factor, quality factor.

phasig, falsch- misphased, dephased. **gleich-** co-phasal, equi-phased, in phase. **richtig-** true-phased, in correct phase (relationship).

Phonograph, Licht- sound track film player (without pictorial accompaniment).

Phonographen-drehplatte turntable. **-platte** phonograph disk, p. record. **-plattenteller** turntable.

Phoresezelle, Elektro- electrophoresis cell.

Phosphore phosphors, luminescent or fluorescent (screen or lamp) materials.

phosphoreszenzerzeugend phosphorogenic.

Phosphorogene phosphorogens, activating additions (promotive of phosphorescence).

photoelektrisch(e) Ausbeute, -(es) Emissionsvermögen photo-electric emissivity, p.-e. yield. **-(e) Schwellenfrequenz** photo-electric threshold frequency, critical f. **-(e) Wirkung** photo-electric effect, Hertz e. **-(e) Zelle** photo-electric cell, electric eye (either photo-conductive, p. -voltaic or p. -emissive, though the last-named should strictly be called a photo-tube, a light-sensitive tube or a photo-electric t.)

Photogramm photograph, photographic record.

Photographie, Kleinbild- microphotography, photomicrography.

photographisch festhalten record photographically (on a record sheet).

photographisch(e) Dichtemessung (photo)densitometry, photographic density measurement. **-(en) Schicht, Bremsvermögen der** stopping power of photographic emulsion. **-(em) Zeichendruck, Telegraph mit** photo-printing telegraph.

Photokathode photo-cathode, photo-electric c. **-, zusammenhängende** plain mirror or continuous (non-mosaic) photo-cathode (of Farnsworth dissector and superemitron). **-, Lichtstärkeverteilung auf** light intensity distribution on photo-cathode.

Photokopie photostatic copy, photographic c., photoprint, photostat.

Photometer, Bunsen-, Fettfleck- Bunsen photometer, grease-spot p., with translucent central spot in screen. **Flimmer-** flicker photometer. **Gleichheits-** contrast photometer, total-reflection p. **Kontrast-, mit Würfel** contrast photometer of cube-type with cubical cavity. **Kugel-** sphere photometer. **Ulbricht** integrating sphere. **Lummer-Brodhun-** contrast photometer, Lummer-Brodhun p. (based on equality of illumination). **Spektral-** spectrophotometer. **Topf-** photometric integrator. **Ulbricht-** sphere or globe photometer of Ulbricht. **Würfel-** cube photometer, integrating p. with cubical cavity.

Photometer-aufsatz photometer head. **-schirm** photometer screen.

Photo-metrierung photometric

evaluation or recording. **-physik** photo-physics. **-strom** photo-cell current. **-volteffekt** photo-voltaic effect (as in a cuprous-oxide cell).

Photozelle photo-electric cell (photo-conductive, photo-voltaic or photo-emissive), electric eye. **Mikro-** globule of mosaic screen.

Photozellen-kabel low-capacity shielded cable between photo-electric cell and amplifier of the video cable, co-axial or concentric cable type. **-verstärker** electron multiplier phototube; photo-cell amplifier. **-vorlicht** priming or bias illumination.

physikalisch-chemisch physical-chemical, physico-c.

π Glied (Π-Glied) pi (π) type section, filter mesh or network m.

Piccoloflöte piccolo flute.

Pickelbildung (corrosion) pitting (of metal).

piezo-elektrischer Modul piezo-electric modulus.

Piezo-kristall piezo-electric crystal. **-quarzzelle** supersonic light valve (predicated on Brillouin and Debye-Sears effect).

Pikkelflote piccolo flute.

Pilgerschrittverfahren Mannesmann, Perrins or reciprocating rolling process.

Pille, Getter- getter tab, g. pill, g. patch. **Gleichrichter-** oxide or metal rectifier of reduced size (say, 1 sq. cm), Westector r.

Pilz-isolator mushroom, petticoat or umbrella insulator. **-lautsprecher** omnidirectional (360 degree) exponential horn loudspeaker or sound radiator (mostly mounted on a pole top).

Pinne pin, peg, tack; peen (of a hammer).

Pinseldetektor cat whisker detector.

Pinzette pincette, tweezers, forceps, nippers, pincers.

Pistole, Spritz- spray gun, s. pistol.

Pitotröhre Pitot tube (to determine fluid or liquid velocity of flow).

Plan, Bedrahtungs- wiring diagram. **-biegeversuch** plane bending test.

Planck's Wirkungsquantum Planck's constant (quantum of action).

Plandrehen face planing.

Planetangetriebe planetary gear, epicyclic g. (train).

Plangitter plane grating (for light dispersion).

planieren plane, planish, smooth, level, even.

plan-konkav plano-concave. **-konvex** plano-convex. **-liegen in Fokusebene** flattened in focal plane. **-mässig** systematic, methodical, according to plan. **-parallele Platte** plane-parallel plate. **-schleifen** surface-grind. **-symmetrisch** planisymmetric.

Plasma, Rest- residual electrons, plasma (region without resultant charge).

Plastik plastic, stereoscopic or relief effect (pictures). **Bild-** plastic or relief effect, form of telev. picture distortion (consisting of multiple contours of diminishing intensity giving pseudo or erroneous three-dimensional appearance).

plastisch(e) Bildwirkung illusion of depth (of a picture or telev. image). **-(er) Film** stereoscopic film. **-(es) Fliessen** viscous flow. **-(e) Verformung** plastic deformation, p. working.

Platine plate, sheet bar, p. slab, p. bar.

Platin-mohr platinum black. **-papier** platinotype paper.

Plationröhre amplifier tube with filament disposed between control grid and plate (with low internal resistance, one model with photo-electric g.)

Plättchen lamella, small plate, platelet, scale.

Platte plate, sheet, slab, lamina, lamination, leaf (of a table), flagstone. **-, lichthoffreie** non-halation plate, p. free from halo.

Platten, Ablenk- deflection plates, deflector p. **Auffang-** target, collector or impactor plates, dynodes (in multiplier). **Doppel-** twin plates, twin crystal, composite c., biquartz. **Strichkreuz-** cross wire plates, reticles, graticules (telescope and other optical instruments).

Plattenfördermechanismus plate-feed mechanism (in electron microscope).

plattenförmig laminar, laminated, plate-like, p. shaped, lamellar, lamelliform.

Platten-geräusch (phonograph) record noise, surface n. **-halter** plate holder. **-kristall, Doppel-** divided plate crystal, biquartz. **-lampe** flat-plate (neon) lamp; tube with center plate instead of grid. **-schlag** disk wobble (of phonograph record). **-schnitt** plate contour, plate section (of condenser, etc.). **-spieler** disk-type phonograph or gramophone. **-staffeln, Glas-** (Michelson) echelon diffraction grating spectroscope. **-teller** turntable (of phonograph). **-wechsler** record changer.

Plattieren plating.

Plattine plate, mill bar, slab.

Platz, freier hole (in nuclear theory). **Gitter-** lattice place, l. point. **Lande-** landing field. airport.

Plätzchen little cake, lozenge, troche, tablet.

platzen burst, crack, break open, explode, implode (inward collapse or burst, as of a vacuum tube).

Platzwechsel transposition, crossing; exchange of places (of electrons, etc.)

plötzlich sudden, abrupt, precipitous. **-(es) Schwingen** spilling over (of an amplifier tube, attended with fringe howl).

Pochwerk steam stamp.

Pockholz lignum vitae, pockwood.

Pohl'sche Wippe Pohl commutator (double-pole double-throw switch).

Polarisation, Dreh- rotatory polarization, optical rotation.

Polarisations-apparat polarizing apparatus, polariscope. **-ebene** plane of polarization. **-ebene Drehung** der polarization error. **effekt** polarization error.

polarisationsfehlerfrei free from polarization error (d.f.)

Polarisationskapazität polarization capacitance.

polarisationsoptisch photo-elastic.

Polarisations-prisma polarizer. **-winkel** polarizing angle, Brewster a.

polarisiert, elliptisch elliptically polarized. **-, geradlinig** plane-polarized. **Kreisbahn-, zirkular -** circularly polarized.

Polarlicht aurora (borealis or australis).

Pol-bedeckung pole arc, circumferential length of pole face, p. pitch percentage. **bildung** pole formation, polarization. **-bogen** pole arc, p. pitch percentage. **-eck** summit (cryst.) **-effekt** pole effect.

Pole, ausgeprägte salient poles. **-, entgegengesetzte** opposite poles, unlike p. **-, gleichnamige** like poles, similar p. **-, ungleichnamige** opposite poles, unlike p. **Gegenkontakt-** co-operative contacts, opposite contacts.

polen, um- reverse polarity.

polieren polish, burnish, shine.

Polierschleifmaschine grinding and polishing machine.

Politur polish, gloss, shine, shining, lustre. **-messer** glossimeter, glossmeter.

Polklemme pole terminal. -papier pole paper, p.-finding p.

Polster,Luft- air cushion, a. pocket; dashpot.

Pol-suchglimmlampe polarity-finder glow-tube, p. indicator. -teilung pole pitch.

Polung,Gegen- opposite polarity, opposition of p., reversal of p.

Pol-wechsel reversal of polarity, change of p. -zwischenraum pole clearance, inter-polar space or gap.

Porenraum pore space, cell s., cell cavity.

Porzellan-erde porcelain clay, China c., kaolin. -masse porcelain body. -schiffchen porcelain boat. -tiegel porcelain crucible. -ton porcelain clay, kaolin.

Posaune trombone. -,durchschlagende trombone with free reed (organ). Alt- alto trombone. Diskant- soprano trombone. Zug- slide trombone.

posaunenartig verschiebbar shiftable trombone-fashion or like a telescope, telescoping.

Posaunenauszug telescoping means, extension m. (for tuning Lecher-wire system).

Positiv, Bromsilber- positive in silver bromide. Meister- mother record, master positive. Original- master positive.

positive elektrische Ladungseinheit unit positive charge.

Positron positron, positive electron, antielectron.

Postmikrophon solid-back transmitter, carbon-back microphone or telephone instrument.

Potential,Abstossungs- repulsion potential. Entstehungs- appearance potential, ionization p. Geschwindigkeits- velocity potential. Gleichgewichts- equilibrium potential (of mosaic). Grenzflächen- interface potential. Rückstoss- recoil potential. Strömungs- flow potential, streaming p.

Potential-berg potential barrier. -fläche, konstante equipotential surface, isopotential s. -gefälle, weltzeitlicher Anteil universal diurnal (variation) component of potential gradient. -hügel potential barrier. -kasten, -mulde potential hole, p. well. -schwelle potential barrier.

Potenz,in die n^{te} - erheben raise to the nth power.

Potenz-gefäll potential gradient. -schwelle(Kerndurchdringung) nuclear penetration factor.

potenzieren raise to higher power (by process of involution, math.)

Potenzreihe exponential series, power s. (math.)

Präge-bär ram (of a press). -film (cf Gravarfilm) engraved sound film (track cut in film).

prägen stamp, coin, imprint, impress, emboss.

prägepolieren roll polish, r. finish.

Prägepresse screw press.

Prägung,Rand- marginal embossing of film.

Präklusivfrist preclusion period.

Prall-elektrode electron mirror, dynode, reflecting electrode, impactor anode (of multiplier phototube or photo-electric electron-multiplier phototube), target electrode, rebound or reflecting impactor (of Weiss mesh multiplier).

prallen(cf prellen) rebound, bound, bounce, be reflected.

Prall-kraft elasticity, resiliency. -winkel angle of reflection, a. of ricocheting.

Präparier-arbeit dissecting work (micr.). -lupe dissecting lens.

prasseln crackle, rustle, rattle.

Pratze claw.

Praxis, in die - überfuhren reduce to practice or to a practical form, work a patent or the invention disclosed therein.

Präzisionswage precision balance, p. scales.

prellen (cf prallen) toss, make to rebound, thump or click (key), chatter, reflect.

prellfrei free from thumping or click (in key operation).

Prellkraft resiliency.

Press-bernstein ambroid. -bodenröhre loctal base tube. -druck ram pressure (in pressfits).

Presse, Fach- technical press, t. literature, professional men's press. Klebe- blooper, splicer (for film). Präge- screw press. Zieh- (cf Tiefzieh-) drop (forge) press (for drawing and shaping articles from sheet material).

pressen press, compress, stamp, mold (under pressure).

Presser compressor (in volume-range control).

Press-form pressure mold. -gasschalter (cross) airblast switch. -körper,-ling pressed object or article, blank after pressure application or molding. -luftlautsprecher compressed-air loudspeaker, pneumatic l. (without diaphragm). -luftschalter (cross) airblast switch (similar to gas b. s.) -masse fictile substance, moldable material, plastic m., synthetic, (thermo-) plastic. -matrize, negative negative stamper (phonograph record). -platten pressings (phonograph record disks). -prozess pressure-shaping, molding or fashioning method. -sitz press-fit, force or driving f. -spahn glazed cardboard, presspan insulation material, pressboard. -stempel press ram. -stock stamper, die (built up from mother record, in phonograph record manufacture). -stoff plastic, moldable or fictile material,

Primär-abstimmung pre-selector means. -empfang single-circuit reception.

Prinzip geringsten Zwanges principle of least strain or resistance.

Prinzip, Ausschliessungs- exclusion principle (formerly Pauli's equivalence p.) Auswahl- selection principle. Korrespondenz- Bohr's correspondence principle.

Prinzipal open diapason (organ stop).

Prinzipschaltung basic circuit diagram (showing underlying principle).

Prioritäts-beanspruchung priority claim. -belag priority proof. -jahr convention year.

Prisma prism, wedge (when weak). -, aneinandergesprengtes broken-contact p., two quartz prisms closely fitted together without cementing. -, doppelumkehrendes inversion prism, inverting p. (reverses in both axial directions of image). -, geradsichtiges direct-vision prism. -, stark fächerndes highly dispersive prism.

Prisma, Ablenkungs- deviating prism. Aufrichtungs- erecting prism. Mehrkant- polyhedral prism. Polarisations- polarizer. Stufen- echelon prism. Umkehr- (cf Bildumkehrung) inverting prism, reverting p. Vergleichs- comparison prism. Wollaston- Wollaston prism, double-image p., Zwillings- biprism, double-image p.

prisma-ähnlich prism-like, prismoidal. -förmig prism-shaped, prismatic.

Prismen-fläche prismatic surface, p. face. -sucher prismatic finder, p. seeker.

Privataufführung pre-view.

Probe (cf Prüfer, Prüfung, Messer, Messgerät, Messung) sample, specimen, test piece; trial, test. gereckte stretched specimen. -, vergütete quenched or tempered specimen or test piece. Beschuss- bombardment test, shooting t. Kapazitäts- capacity test. Kegeldruck- Brinell ball impression or indentation test, ball pressure t. Loch- drift test. Metall- assay, test for metal; m. testing. Risshärte- abrasion test, scratch t. Schlag- impact test, blow t., percussion t. Tiefzieh- cupping test, cuppability t. (piece or specimen). Ton- sound test; tonsil t. (in recording voice of a singer). Zerreiss- tensile test, breaking t., rending test (specimen).

Probe-abdruck proof, proof print, p. sheet. -aufnahme test picture, trial shot. -entnahme sample taking, sampling (at random). -machen testing, assaying. -nehmen sampling.

Probenglas specimen glass, s. tube.

Probe-objekt specimen. -saal rehearsal hall. -stück specimen, sample, test piece. -ziehen sampling.

Probiergefäss testing vessel, assaying v.

Profil, Feld(stärken)- field pattern (often mapped in an electrolytic tank). -Tragflügel- aerofoil section.

Profil-begrenzung contour line, boundary of a section or profile. -prüfer (electronic) profilometer. -stempel profile stamp (surface pressing).

Projektion, flächentreue equal-area projection (mapping) Dia- oder Diapositiv- lantern slide projection. Durch- back projection. Gross- large-picture projection work (with magnascope, etc.) Hintergrund- background projection. Kegel- conical projection (of a map). Merkatorkarten- Mercator projection, cylindrical p., orthomorphic p.

Projektions-anlage, Rück- back projection equipment; re-p. apparatus. -aussteuerungsanzeiger projected volume indicator. -bildschreibröhre projection-type television receiver. -fenster projection aperture, p. gate. -spule projection coil. -tubus projection tube (micr.)

Propeller, Verstell- feathered (variable-pitch) propeller or airscrew.

proportional, umgekehrt od. verkehrt inversely proportional.

Proportionalitätskonstante constant of proportionality.

Prozent, Gewichts- percent by weight. Raum- percent by volume.

Prozentgehalt percent content, percentage.

prozentuale Aussteuerung percentage modulation, m. percentage.

Prozess manufacturing process or method, action, lawsuit, cause, suit, trial, legal case, court c. (law). Anfechtungs- contested action.

Prozess-kosten law costs, c. of proceedings, c. of a legal action. -ladung writ of summons.

prüfen, nach- re-examine, re-check, retest.

Prüfer, Erichsen- Erichsen cupping machine. Glattheits- smoothness meter, glossimeter. Härte- hardness tester (Brinell, Rockwell sclerometer and scleroscope). Lichtsinn- visual photometer. Linsen- lensometer. Material-, elektronischer aniseikon. Oberflächenspannungs-(nach Abreissmethode) adhesion-balance type of surface tensiometer. Profil- (electronic) profilometer. Rauheits- smoothness meter (determines degree of polish), glossimeter, profilometer. Ritzhärte- sclerometer, scratch hardness tester. Röhren- tube checker, t. tester. Stromschluss- continuity tester. Tonhöhe- tonvariator. Universal- multiple-purpose tester, multimeter. Windungsschluss- (inter-)turn short-circuit tester.

Prüfbank (lens) test bench (opt.).

prüfen test, assay, probe, try, prove, examine, inspect, verify, scrutinize.

Prüf-feld test field, proving ground. -feld, Stoss- potential impact, flash or impulse testing field. -generator signal generator, test oscillator, all-wave o.

-gerät trouble shooting device, tester, set analyzer (radio). -ling test piece, t. specimen, t. model, t. sample. -pult test desk. -röhre für Fernseher phasmajector, monoscope, monotron. -schnarre test buzzer. -sender service test oscillator. -sonde, Klauen- claw-type test probe or prod, search electrode. -spitze penetrator (of hardness tester), indenter. -stabeinschnürung waist (forming in bar under tensile test). -stand test bed. -streifen buzz track (on a film). -verfahren prosecution (of a patent application), period during which Patent Office Examiner reviews novelty and merits of a disclosure; test method.

Prüfung (s.a. Probe, Prüfer, Messer, Messgerät, Messung). examination, testing, perusal, consideration, inspection, investigation, reviewing, surveying, verification, check, search, probe. -, nichtzerstörende, zerstörungsfreie non-destructive test. Abnahme- acceptance test. Dauerstandfestigkeits--(long-time) creep test. Eisenpulver- Magnaflux inspection (using iron dust). Impuls- s. Prüfung, Stoss-. Material- testing of materials. Rauhheits- testing with smoothness meter or glossimeter. Regen- rain test, wet t. (of an insulator). Stoss- impact, flash or impulse test (of insulation). Treib- cupping test. Verschleiss- abrasion test, t. to ascertain wear and tear. Zerreis-, Zieh- tensile test, breaking t., rending t.

Pseudo-dämpfung loss of selectivity by parallel internal resistance. -schwund pseudo-fading (due to swing or swaying of antenna).

Puffer grommet, cushion, pad, buffer, shock absorber. -betrieb floating battery operation. -lösung buffer solution. -stufe buffer stage, separator s.

Pulsation pulsation, beat, throb.

Pulser pulsator (fatigue tester)

Puls-verfahren, Ein- monopulse method. -verstärker, Bildsynchronisierungs- frame synchronizing (or low sync) impulse amplifier.

Pult, Misch- mixing, monitoring or supervising desk. Prüf- test desk. -befeuerung firing on stepped grate bars. -ofen black-flame hearth.

pulverartig pulverulent, pulverous, powdered, dust-like.

Pulverdiagramm powder pattern.

pulverig powdery, pulverulent, pulverous, dust-like.

Pulverkern, gepresster Eisen- compressed iron-dust core.

pulvern powder, pulverize.

Pumpe, Kreisel- centrifugal pump, turbine p. Lade- charging pump, compression p.

Pumpenlamelle vane (of Gaede vacuum pump).

Pumpstengel pumping lead (short piece of tube connecting vessel to be evacuated and vacuum pump), exhaust tube, e. vent.

Punkt, ausgezeichneter cardinal point (opt.) Berührungs- point of contact, tangency or osculation, touching p. Brenn- focal point (opt.) Einschnürungs- crossover (in electron-optics). Flucht-vanishing point. Haupt- principal point, unit point (opt.) Kardinal- cardinal point (opt.) Kehr- cusp. Knoten- nodal point (opt.) Morse- breiter, spitzer, lengthened dot, clipped dot. Nabel-umbilic. Wende- cusp.

Punktallinse toric lens.

punktförmig punctiform; crater-shaped. -(e) Induktivität lumped inductance, concentrated i. -(e) Quelle point (shaped) source or emitter (of light, electrons,

etc.), focus lamp.
Punkt(glimm)lampe point or crater neon lamp, tungsten arc lamp, Pointolite, focus l.
punktieren, strich- dot-dash.
Punkt-kathode point source cathode, crater c. -lagen point positions. -licht point (crater-shaped) light source. -losigkeit astigmatism. -marke spot reading (on dial of direction-finder). -raster point (dot or granule) raster, screen, plate or grating. -schreiber, Ein- single-point recorder. -schweissung spot welding. -singularität point singularity. -strich semicolon.
punktsymmetrisch having point or radial symmetry.
punktuelle Abbildung point-focal vision or imagery.
Punkt-verlagerung, Bild- spot shift (causing "plastic", telev.) -verteiler, Bild- picture scanner.
punktweise Auswertung point-by-point evaluation.
Pupille, erweiterte dilated pupil. Ausgangs-, Austritts- exit pupil. Eingangs-, Eintritts- entrance pupil.
Pupillen-grösse pupillary size, p. diameter. -spiel response of pupil -weitemesser korometron.
Pupinspule loading coil, Pupin c. (for reactance balance).
Puppe, Kohle- carbon rod.
Purpur, Seh- visual purple, rhodopsin.
Pustlampe blow-lamp.
putzen clean, cleanse, scour, polish, trim.
Pyramiden-stumpf truncated pyramid, frustum. -würfel tetrahexahedron.
Pyrometer, Glühfaden- optical pyrometer, disappearing-filament p., monochromatic p.

Q Faktor Q factor, magnification f.
Q Gruppen Q code groups.
Quadrate, Methode der kleinsten method of least squares.
quadratisch square, quadratic, tetragonal (cryst.), quadrangular. -(e) Abtastung square-law scan. -(er) Gleichrichter square-law rectifier or detector. -(e) Gleichung quadratic equation. -(er) Kondensator square-law condenser or capacitor. -(er) Mittelwert root-mean-square (rms) value, virtual v.
Quadrat-wellengenerator generator of square or rectangular waves. -wurzel ziehen extract square root.
quadrieren square, raise to 2nd power.
Quadrupolmoment quadrupole moment.
Quant, Licht- photon, light quantum.
quanteln quantize.
Quantelung, Raum-, Richt- directional quantization, space q.
Quanten-ausbeute quantum yield, q. efficiency. -bedingungen quantum conditions.
quantenhaft pertaining to quanta or the quantum theory.
Quanten-hypothese quantum hypothesis. -sprung quantum transition (abrupt transfer of a quantum system from one of its stationary states into another), quantum jump or leap. -theorie quantum theory.
Quantenzahl, innere inner quantum number. Oszillations- vibrational quantum number. Rotations- rotational quantum number.
Quantisierung quantizing.
quantitative Messung quantitative measurement, q. analysis.
Quantum, Planck's Wirkungs- Planck's constant, quantity of action.
Quantum-ausbeute quantum yield, q. efficiency. -berichtigung quantum correction. -gewichte quantum weights, statistical w. -mechanik quantum mechanics. -übergang quantum transition, q. jump, q. leap. -zustand quantum state, energy level.
Quarz, optisch inaktiver racemic quartz, racemate. Links- levogyrate quartz crystal, left-handed q.c. Rechts- dextrogyrate quartz crystal, right-handed q.c. Steuer- (frequency) stabilizing quartz.
quarzähnlich quartzose, quartzous. -artig quartzic, quartzous, quartzy.
Quarzfaden quartz thread, q. filament.
quarzgesteuerte Steuerstufe crystal-stabilized master or drive (oscillator) stage.
quarzig quartzy, quartzose.
Quarz-inaktivierung, optische racemization. -keil, Doppel- double quartz wedge. -kristall, fertiges piezoid, finished blank (including electrodes). -kristall, optisch inaktives racemate. -mehl quartz powder. -resonator piezo-electric resonator, quartz r. -sender (piezo-electric) crystal or quartz-stabilized transmitter. -zelle, Piezo- supersonic light valve (predicated on Brillouin and Sears-Debye effect).
quasi-elastisch quasi-elastic. -unendlich quasi-infinite.
Quecksilberdampflampe mit Heizkathode phanotron, tungar and similar types of tube (with argon filling). -mit Steuergitter mercury-vapor tube of the thyratron, ignitron, etc., type.
Quecksilber-faden mercury thread, m. column. -falle mercury trap (vacuum pump operation). -halogen mercury halide.
quecksilberhaltig mercurial, with Hg.

Quecksilber-hochdruck-Entladungsrohr high-pressure mercury discharge tube or vessel. -kathode mercury (pool) cathode. -legierung mercury alloy, amalgam. -lichtbogen--Kathodenfleck, fixierter anchored cathode spot of mercury arc. -masse mercury pool, m. sump (in bottom of a tube). -mohr black mercuric sulfide. -schalter mercury switch. -strahlunterbrecher mercury-jet. interrupter, m. break. -tropfelektrode dropping mercury electrode. -vorrat mercury pool (in discharge tube).

Quelle,einfarbige Leuchtstrahl- monochromatic illuminator source, monochromator. -,punktförmige point (-shaped) source or emitter. Elektronen-electron source, e. emitter. Licht- source of light, illuminant. Strom- source of current (supply).

Quellenbild,Licht- light source image.

Quellung swelling, welling, soaking, imbibition.

quer cross, across, transverse, transversal, traverse, diagonal, lateral, oblique, askance, athwart.

querab athwart. -Backbord abeam, on the port beam. -Steuerbord abreast, on the starboad beam. -Peilung taking bearings from athwart or abeam stations.

Quer-arm cross arm, shunt a. (of network). -aufnahme horizontal picture. -effekt transversal effect (of crystal).

Quer-entzerrer shunt-type attenuation equalizer or compensator. -feld transverse or cross (magnetic) field. - glied, in 1/2 -endender Kettenleiter mid-shunt terminated network. -kondensator bridging capacitor, by-pass c., shunt c. -kontraktionskoeffizient Poisson's coefficient, rho ratio (lateral contraction: longitudinal extension). -leitung leakage conductance, leakance; shunt line, cross l. -leitungsbrücke transconductance bridge. -magnetfeld transverse magnetic field. -modulation cross modulation. -pfeife fife. -profil cross-section. -riss transverse crack, cross c., fracture of mineral), cross-section. -schärfe definition in picture direction.

Querschnitt cross-section, transverse s., cross-cut; cross-sectional area. Streu- scattering cross-section. Wicklungs- cross-sectional area of winding.

Querschnitts-verminderung cross-section reduction (of area), draft, draught. -zusammenziehung contraction or reduction of area, formation of a waist (in test piece under tension or tensile stress).

Quer-schrift lateral-wave or Berliner type track (used in gramophone and modern phonograph records). -schwingung transverse vibration, shear v. (sometimes called flexural v.), contour vibration or oscillation (cryst.) -spulenbelastung leak load. -steuerröhre cross-control beam tube, drift t., tube operated on beam deflection. -verbindung,Funk- intercommunication channel. -verschiebung und Längsv. cross adjustment and axial adjustment (of electron microscope). -widerstand shunt resistance, cross r. (of network) -zusammenziehung lateral contraction. -zweige cross arms, shunt a. (of network).

Quetsche,Rollen- squeegee.

quetschen pinch, squeeze, crush, mash, squash.

Quetschfuss squash, press or foot (of a tube).

quetschfussfreie Röhre loctal base tube.

Quetsch-hahn pinchcock. -kondensator

compression-type condenser.
Q Faktor Q factor, magnification f.
Q Gruppen Q code groups.
quicken amalgamate.
quieken squeak, squeal.
Quinte violin E string.
quirlen twirl, stir with a whirling motion, turn.
Quotientenmesser quotient meter, ratiometer.

R

Rachenlehre gap gage.

Rad, Daumen- sprocket wheel. Farb- inking wheel, i. roller. Flügel- worm wheel, screw, s. propeller. Hemm- escapement wheel, brake w. Kamm- cog wheel. Kegel- bevel wheel, mitre w., cone w. La Cour'sches phonic wheel. Pfeil- double helical or herringbone gear wheel. Rayleigh-, Schallreaktions- Rayleigh disk (placed in a sound field). Schnecken- worm wheel, worm. Schrittschalt- step-by-step wheel, stepping w. Schwung- fly wheel, balance w. Sperr- ratchet wheel, cog w. Spiegel- mirror wheel, Weiller scanning w. or drum, drum scanner with peripheral mirrors. Steig- escapement wheel, ratchet w. Stern- star wheel. Stift- sprocket wheel. Stirn- spur wheel. Stufen- step wheel, cone pulley. Ton- tone wheel. Umlauf- planetary wheel. Vorschub- feed wheel. Zeilenspiegel- line scanning mirror wheel (of Scophony Co.)

Rad-bewegung rotary motion, wheel m., rotation. -buchse bushing.

Rädchen, Reiter- jockey wheel, j. roller.

Raddrehung rotation; torsion.

Rädern shredded wheat effect (m.p.)

radförmig rotate, spider-shaped, wheel-s., radial (of a structure).

radialsymmetrisch radially symmetric, in radial symmetry.

Radier-firnis etching varnish. -grund etching ground.

radioaktiv(e) Emanation radioactive emanation (known as radon, thoron and actinon). -(es) Folgeprodukt metabolon. -(es) Gleichgewicht radio-active equilibrium (being either transient or secular). -(er) Niederschlag radioactive deposit, film of r. matter, -(e) Produkte metabolons, successive disintegration products of radioactive materials.

Radioaktivität, künstliche induced or artificial radioactivity.

Radio-amateur radio fan, r. amateur, "ham". -meterflügel radiometer vane. -sonde(für Wetterforschung) radiometeorograph, radio sonde.

Radium-behälter radiode (glass or metal tube or capsule). -emanationseinheit curie unit quantity of radon.

Radius, Abrundungs- contour radius. Krümmungs- radius of curvature. -vektor radius vector.

radizieren extract the root, evolve.

Rad-kranz wheel rim, tire. -linie cycloid (in stroboscopic aberration), epicycloid. -phänomen wheel illusion. -reifen tire, rim. -satz axle plus two (shrunk-on) wheels. -stand wheel base. -taster wheel-operated means (causing signals, bell ringing, etc.)

Radtkeverfahren sound reproduction method of A.A. Radtke (invented 1917, USP issued 1938); cell impulse amplified, while maintaining and restoring grid potential by high impedance between cell cathode and negative end of source of potential.

Raff-aufnahme, Zeit- time-lapse-motion photography (low-speed shooting, high-speed projection.) -bild time-lapse picture, fast-motion p.

Raffer, Mikro- camera for stop-motion or time- lapse motion micrography. Zeit- time-lapse or stop-motion camera.

Raffinieranlage refinery, refining plant.

Rahmen stator, frame, former; loop

or coil (aerial), frame antenna. Im - der Erfindung inside the scope (or spirit) of the invention.

Rahmen,Druck- film trap (in projection). Eisen- iron-cored frame (d.f.) Enttrübungs- frame to make zero or minimum signal point sharper (free from night effects, etc.), zero clearing or sharpening frame. Entwickelungs- developing rack. Kopier- printing frame. Luft- air-cored frame (d.f.) Peil- coil antenna, loop a., frame a.

Rahmen-antenna coil antenna, loop a., frame a. -antenne,Kreuz- crossed coil antenna. -entwickelung rack development, tray d. -nachstimmung re-setting or re-check of frame or coil tuning (in automatic radio compass). -negative rack negatives. -sucher wire-frame view-finder, iconometer.

Rakete,Leucht-,Licht- flare.

Ramankomponente modified frequency of incident light, or Raman spectrum, in a scattering substance.

rammen ram, beat down, tamp, pound.

Rampenbeleuchtung footlights.

rampig shear-scarred (of glass), burred.

Rand edge, border, rim, margin, brim, hem, flange, boundary, periphery, brink. Bördel- flanged edge, beaded edge. Schirm- border, edge or margin of screen.

Rand-anmerkungen marginal notes, m. annotations, marginalia. -aussparung(für Tonstreifen) setting aside margin (of film strip) for sound track. -bedingungen boundary conditions, marginal, fringe, or edge c. -bemerkungen marginal notes, m. annotations, marginalia. -effekt edge effect, Eberhard e. (of film), fringe or marginal e.

rändeln edge, border, rim, mill, knurl, bead.

Rand-kante lateral edge. -kraft force (acting) at edge, marginal force. -prägung (des Films) marginal embossing of film (Pathé). -strahl marginal ray, peripheral r. -winkel angle of contact (between wall and liquid), wetting a. -zone,verstickte nitrided edge or marginal zone.

Rasierstein styptic stick.

raspeln rasp.

rasseln rattle, rustle, clatter.

Rast(e) (bei Fern- und Schaltersteuerung) notch, detention point, detent, arrester means (in telemetric, switch, etc., operation). Feder- spring catch, s. detent.

rasten,ein- lock in position (as by a detent).

Rasten-arretierung arrester means, locking m., detent. -schalter step switch, multiple-point s. -spreizen bracing members (of piano) -teilscheibe notched index wheel.

Raster scanning field, s. pattern or "raster," unmodulated light-spot scan or scanned area (of a television picture); mosaic (in photocathode tube and in iconoscope); grating or screen (of dot, ruled or line type, used in printing).

Raster,Druck- printer's screen or meshwork. Farb- color embossing. Fernseh- (cf Raster) television scan surface, scan area or "raster." Halb-,Ineinandergreifen der interlocking or meshing of fractional scans (in interlaced scanning). Linien- ruled plate, line screen or grating. Punkt- point, dot or granule type of grating, plate or screen.

Raster,Zwischenzeilen-,geradzahliger even-line interlaced scan. - -, ungeradzahliger odd-line interlaced scan.

Raster-aufnahme,Farb- color screen photograph. -blende scan hole, scan aperture, raster screen aperture. -element picture unit, p. element or elementary area. -farbverfahren,Linsen- lenticulated or lenticular screen color printing method. -feinheit resolution, raster detail or definition (of scanning pattern). -film,Linsen- lenticular or lenticulated screen film. -frequenz field frequency. -kathode photocathode, mosaic screen. -körner silver globules (of iconoscope mosaic), silver plugs (in barrier -grid m.). -mikroskop screen scan microscope, raster scan m., electron scan m. -platte scan pattern, s. field, s. area, photo-sensitized mosaic plate (of iconoscope); screen (dotted, granule, line or ruled). -punkt elementary scan area, e. picture point or area. -schirm mosaic screen.

Rasterung definition or resolution of a picture (say 300 lines); embossing or lenticulation (of a film surface). Fein- high-definition scan (of a picture).

Raster-wechselfrequenz field frequency (in interlaced scanning). -zahl number of picture points, p. units or elementary areas (determining definition or resolution of picture).

Rast-klinke detent, latch, pawl. -nase detent, latch. -polkegel herpolhode, space cone. -stern (im Wellenschalter) notched star (of wave-band switch).

Rastral music ruling pen.

Ratsche rattle. Fühl- micrometer friction thimble.

Rauch-fahne smoke streamer. -glas tinted glass, smoke g. -klappe vent, stack (of projection booth).

rauchschwach smokeless (of powder).

rauh rough, raw, roughened, granulated, knurled, burry.

Rauhen granulating, graining, roughing or roughening (superficially).

Rauhheit(Lautsprecher) raucousness, hoarseness (of loudsp.)

Rauhheitsprüfer smoothness meter (determines degree of polish, or lack of polish, on surfaces), glossimeter, profilometer.

Raum space, room, chamber, cavity, enclosure, volume; scope. -,luftleerer vacuum, vacuous space, exhaust s -,überakustischer exceedingly live room or space (with excessive reverberation period). Abhör- mixer room, monitoring booth, box. Anfach-, Arbeits- generator space, working or drift s. (in drift or beam tube). Aufnahme- studio, teletorium, telestudio (of audio and video broadcast station, etc.). Aufschluss- part of electrically prospected volume of soil or earth subject to current or displacement lines. Besprechungs- studio. Bild- image space. Ding- object space. Fall- cathode fall space (in which c. drop occurs). Geschwindigkeits- velocity space. Halb- half space. Hall- acoustically live room; reverberation chamber. Nachhall- reverberation room, r. space. Phasen- phase space, extension-in-space. Poren- pore space, cell s., cell cavity. Sammel- receiver, receptacle, collector space. Schlag- spark-gap chamber or space. Zuschauer- auditorium. Zwischen- intervening space, intermediate s., interstice, interspace, interstitial s., interval.

Raum-akustik stereo-acoustics, architectural a., space or room a. -beständigkeit volume stability. -bild stereoscopic picture, plastic p.; space diagram (d.f.). -bildung,Hohl- cavitation (in supersonic wave work, etc.). -bildvorführung stereoscopic picture projection. -chemie stereochemistry. -formel spatial formula. -gebilde space diagram; solid three-dimensional structure. -gefühl space or spatial feeling (in vision). -geometrie solid geometry.

raumgeradlinig orthodromic. -(e) Peilstrahlen orthodromes.

Raum-geräusch set noise (caused by insufficient sound insulation, in m.p. work), set noise (in radio receiver) comprising Johnson n., shot n., hum, etc. -gewicht weight by volume. -gitter space lattice, "raumgitter"- -gruppe space group, point g. -halleffekt liveness (ac.) -helligkeit volume brightness. -inhalt volume content, capacity. --integral space integral. -ladenetz space-charge grid. -ladungswelle space-charge wave. -ladungswolke concentration of space charges, accumulation of s.c.

räumlich spatial, concerning space, space-geometrical, three-dimensional. -(e)Behinderung steric hindrance (chem.) -(e) Rotationsgruppe space rotation group. -(e) Verteilung spatial distribution, geometric d. -(er) Winkel solid angle (in spheradian units)

Raum-mikrophon non-directional microphone, polydirectional m. -punkt space point. -quantelung space quantization, directional q. -resonator,Hohl- cavity resonator, chamber r., rhumbatron (as part of klystron). -sektor solid sector. -sinn space perception. -strahl, Welt- cosmic ray. -strahlenstoss burst or shower of cosmic rays. -strahlung cosmic ray radiation. -teil part by volume, volume. -toneffekt stereophonic sound effect, binaural e. -verhältnis volume relation, proportion by v. -verteilung spatial distribution, geometric d. -welle sky wave, space w., indirect w,; spherical w. -winkel solid angle (in spheradian units). -winkeleinheit steradian, spheradian, unit solid angle. -wirkung (cf Bildplastik) stereoscopic effect (of a picture); spatial e., auditory perspective, stereophonic e. (ac.)

raumzentriert body-centered, space-c. (cryst.)

rauschen rush, gurgle, murmur, rustle (in a tube), whistle, roar.

Rauschen (back) ground noise (in sound reproduction, radio, film, etc.). Röhren- tube noise, thermal agitation of tube, hiss. Schrammen- scratch noise. Staub- dust noise (m.p.).

Rausch-pegel noise level (Schottky effect, thermal noise). -spannung noise potential.

Raute rhombus, diamond.

Rauten-antenne rhombic antenna. -flach rhombohedron.

rautenförmig diamond-shaped, rhombic.

R.E. (Richtempfang) directional reception.

Reagenzglas test tube.

Reagenzien reagents, reactive means.

Reaktions-gleichung equation (of a reaction). -masse reaction mass, mass resulting from a reaction.

Rebromierung re-bromination.

Rechen-bild nomogram. -fehler error or mistake of calculation or computation. -knecht ready reckoner. -schieber slide rule. -tafel computation table.

rechnen calculate, compute, figure, reckon, count, estimate.

Rechner,Dreieck- triangulator (for trig. calculus). Verzugs-acoustic corrector (sound location).

rechnerisch arithmetic, mathematical, by calculation or computation, calculative.

Rechnung, Mischungs- alligation (math.) Störungs- perturbation calculation.

Rechnungs-beamter auditor, accountant. -legung making or submitting an accounting.

recht achteraus right astern. -voraus right ahead.

Recht,Eigentums- ownership right, proprietary r., property right, proprietorship.

recht,flucht- flush, aligned, in true alignment. winkel- right-angled, rectangular, orthogonal.

rechteckige Kurvenform der EMF square emf curve, flat-topped emf c. -Welle square wave.

rechten Winkel, im at right angles, orthogonally, normally, transversely.

Recht-fertigung justification, defense, vindication, exculpation. -mässigkeit lawfulness, legality, rightfulness, legitimacy.
Rechts-anspruch legal claim, title. -beugung miscarriage of justice. -drall right-handed lay or twist (of a cable).
rechtsdrehend dextrogyrate, dextrorotatory, of right-handed or clockwise rotation.
Rechts-drehung dextrorotation, right-handed polarization. -einwand demurrer, plea, objection. -fall (legal) cause or case, suit, litigation. -gewinde right-handed (screw-) thread. -grund legal ground, l. reason, argument. -händigkeit right-handedness.
rechtsichtig normal-sighted, emmetropic (of eye).
rechts-kräftig machen make legal or non-appealable, legalize, validate, ratify. -laufende Welle dextro-propagating wave.
Rechts-kurve right-hand curve, starboard c. -mittel aufheben avoid or remove a remedy of law. -nachfolger legal successor, assign, assignee. -punkt point in or of law. -quarz dextrogyrate quartz, right-handed q. or piezo-electric crystal. -schraube right-handed screw. -spruch judgment, decision, sentence, verdict. -streit(es), Gegenstand des matter in a controversy, case at bar, in issue or in suit. -system right-handed system. -zuständigkeit jurisdiction, competence.
recht voraus Peilkurs head-on course (with fore-aft axis of vessel or craft directed toward objective).
rechtweisender Kurs true (north) course. -Windkurs true heading.
rechtwinklig right-angled, rectangular, orthogonal. -, dreifach trirectangular.
rechtzeitig timely, opportune, punctual.
recken stretch, extend, elongate, pull, rack; subject to traction, strain (metal testing); shingle, tilt (metal).
redigieren edit (film).
Reduktions-getriebe reduction gear, step-down g. -transformator step-down transformer.
Reduktorlampe adapter-transformer lamp.
Referierdienst abstracting service, reporting s., press clipping s.
reflektierende Brennfläche catacaustic.
Reflexion, rückweisende retro-directive reflection. -, zerstreute diffuse reflection. Spiegel- specular reflection, mirror-like r. Wellen- wave reflection, w. echo (with one or more hops). Zickzack- staggered reflection.
Reflexions-brennfläche catacaustic. -ebene plane of reflection. -flecke flare ghosts, flares (phot.) -goniometer reflecting goniometer. -höhe level of reflection (of waves in ionosphere). -lichthof halation by reflection. -vermögen reflecting power, reflectivity, reflectance. -winkel angle of reflection. -zahl number of reflections (of waves from ionosphere involving single or multiple hop).
Refraktometer, Eintauch- immersion refractometer, dipping r. -grenzwinkel critical or limiting angle of a refractometer.
Regel, Dreifinger- right-hand rule, thumb r., Fleming's r. Faust- rule of thumb, rough and ready r. Phasen- phase rule.
Regel-glied regulating attenuator. -grad degree of dynamic-range or contrast regulation; maximum range of gain variation.
regellos geordnet with random orientation.
Regellosigkeitsproblem random problem, haphazard p.
Regelmässigkeit regularity.
regeln regulate, adjust, arrange, set, order, control, govern, regularize.
Regelschwankungen hunting (in making

adjustments or in regulation).

Regelspannung automatic volume control (avc) potential; dynamic-range or contrast regulation or control p., generally; control p. **Schwund-** automatic volume control (avc) potential.

Regelung, Nach- re-adjustment, re-setting. **Rückwärts-** indirect control, retroactive action, backward regulation.

Regelverstärker variable-gain amplifier, avc amplifier; noise or silencing a.; amplifier in which regulator or control potential operates.

Regen rain (in film).

regenbogenfarbig rainbow colored, iridescent

Regenbogenhaut iris.

Regenerat reclaimed product (rubber).

Regenversuch rain test, wet test (on insulators).

Regiepult mixing table, m. desk.

Regisseur director.

Register stop (organ). **-haltigkeit des Films** registration stability of film. **-übertragung** transfer of registry. **-zug** draw stop (of organ).

registrieren record (graphically), register, file.

registrierend, selbst- self-recording, automatically r. **-(er) Spannungsteiler** potentiograph. **-(er) Strommesser** recording ammeter.

Registrierung, Dauer- long-period recording, time record.

Regler regulator, governor, controller. **Druck-** barostat. **Dynamik-** volume- or dynamic-range control means, compandor, contrast regulator. **Film-** film traction regulator, pull r., stabilizer. **Fliehkraft-** centrifugal governor. **Haupt-** main regulator or governor (of a turbine). **Hörbarkeits-** audibility network; volume control means (avc). **Kälte-** cryostat. **Klangfarbe-** tone control, tonalizer. **L-** L-section attenuator. **Lautstärke-** audibility network; volume control means (avc). **Licht-** light relay, light modulator (Kerr cell, etc.); light control means. **Misch-** mixing control. **Saal-** theater fader. **Steck-** plug-in and socket (volume) control means. **T-** T section attenuator, double-L- s.a. **Temperatur-** temperature control, thermostat, thermo-regulator, thermautostat (electronic). **Tonbandbreite-** tone band-width control.

Regler-dose control aneroid (for engine). **-leitung** pilot wire. **-pult** control desk, monitoring d. (in sound recording). **-stellung** fader setting (in film recording).

Regulierband, Zeit- timing tape.

regulieren, über- overshoot, overrun, excessively regulate.

Regulierschraube adjusting screw, set s.

Regulierung, Bandbreite- band-width regulation, b. switching.

Reibahle reamer.

Reibe grater, rasp.

reiben rub, grind, triturate, rasp, grate, abrade.

Reiboxydation rubbing corrosion, frictional oxidation,

Reibung, innere internal friction, solid viscosity. **Schlupfungs-** sliding friction, slip.

Reibungs-arbeit frictional work, magnetic hysteresis. **-kegel** cone of friction. **-moment** friction torque. **-widerstand, Schall-** acoustic resistance.

reich, flächen- with many faces or sides, polyhedral. **linien-** rich in lines, with abundance of l.

Reichsgesetzblatt Imperial (German) Law Journal or Gazette.

Reichweite (eines Empfängers) (distance) range (of a receiver apparatus), distance-getting ability of a r.a. **Nenn-** rated range, nominal r.

Reichweitenverteilung range distribution.

reif ripe, mature,

Reif(en) ring, hoop, tire, collar.

Reifkeim digestion nucleus.
Reifung ripening, digestion of emulsion (of film). **Nach-** maturing (of film).
Reifungskörper ripening (accelerator) substance, sensitizer.
Reihe series, row, succession, sequence, train, rank, file, bank (of contacts, keys, etc.)
Reihe, an die - kommen take (one's) turn. **der - nach** sequentially, in sequence or succession. **gegensinnig in -** series-opposing. **gleichsinnig in -** series-aiding. **unendliche -** infinite series.
Reihe, Bild- sequence or series of pictures, frames or pictorial actions. **Fourier-, doppelte** double Fourier series. **Impuls-** series, succession or train of impulses. **Potenz-** exponential series, power s. (math.) **Tasten-** row or bank of keys.
Reihen-anordnung seriation or series connection (of resistances and condensers, etc.) **-bilder** sequence of pictorial actions or pictures, motion pictures. **-entwickelung** expansion in a (power) series, power seriation. **-fertigung** mass production, assembly-line p. **-folge** sequence, consecution, succession, series, order. **-folge energetischer Bevorzugung** sequence or order of energy preference. **-funkenstrecke** multiple spark gap. **-glied** series element (of a network). **-impedanzglied** series impedance element, s. impedor. **-röhrenmodulation** constant-voltage modulation. **-verlustwiderstand** equivalent series resistance (of a condenser).
rein(es)farbengleiches("einheitliches")Gelb psychologically unique yellow. **-(e) Skala** true scale. **-(e) Sprache** clear voice. **-(e) sinusförmige Bewegung** plain harmonic or sinuous motion. **-(er) Ton** pure note or tone, simple t.
Reinartzschaltung Reinartz circuit scheme (comprising both capacity and inductance or tickler-coil feedback), also; a special short-wave receiver circuit scheme.
reinigen purify, refine, clean, cleanse, wash, purge, scavenge, separate, rid.
Reiniger, Klang- acoustic clarifier (on ldspk. baffle). **Strom-** harmonic eliminator, filter, stopper or excluder, smoothing means.
Reinigungskreis smoothing circuit, filter c., low-pass f. (following a rectifier in a power unit, to remove ripples).
Rein-kathode bright emitter. **-kohle** solid carbon.
Reinton (cf **Heulton**) note with one mode of vibration, single-frequency tone. **-blende** biasing stop, noise-reducing s., shutter, vane or gate. **-schaltung** noiseless recording circuit organization. **-verfahren** noiseless recording (on film).
Reissblei graphite.
reissen tear, pull, drag; pluck (a string or chord).
Reiss-festigkeit resistance to tearing or breaking, tenacity, tensile strength. **-gebiet** discontinuity, cut-off or break-off region (of oscillations). **-kohle** crayon, charcoal crayon. **-länge** breaking length (cable and paper testing). **-verschluss** slide fastener, zipper.
Reiter slide contact, slider, cursor. **Gewichts-** rider. **-rädchen** jockey roller, j. wheel.
Reiz stimulus, excitation, stimulation. **Licht-** luminous stimulus, light s. **Nerven-** nervous stimulus. **Seh-** optical or visual stimulus or excitation. **-schwelle** threshold of sensation or stimulus, liminal value of stimulation.
Relais, elektro-optisches electro-optical cell, Faraday c., light relay depending on F. effect. **-mit mittlerer Ankerruhestellung**

neutral relay, r. with differentially moving armature. Bremzylinder- dashpot relay. Differenzial- differential relay, discriminating r. Distanz- distance relay. Elektronen- electron relay, thermionic r, Flacker- flashing relay. Fortschalt- stepping or impulsing switch or relay. Gegenstrom- reverse-current relay. Halte- holding relay, restraining r., guard r. Kupfermantel- copper-jacketed relay. Licht- light relay (converting current into light variations) (e.g., Kerr cell, Karolus c., modulated neon lamp, etc.). Licht-, elektrooptisches electro-optic cell, Faraday c., light relay depending on F. effect. Minimal- no-load relay, under-current r. Rückstrom- reverse-current relay. Schneideanker- knife-edge relay, Schnell- high-speed relay. Spannungs- under-voltage relay, low-v.r., no-v.r. Steuer- pilot relay. Stufen- relay with sequence action. Tast- keying relay, r. key. Tauchkern- plunger relay. Verriegel- locking relay. Verzögerungs- slow-acting relay, time-lag r.

Relais-aberregung de-energization of a relay. -anker relay armature. -vorwähler relay preselector.

Relaxations-schwingungen relaxation, sawtooth or ratchet oscillations. -zeit der Ionenwolke relaxation time of ionic atmosphere.

Relief (cf Raumwirkung und Plastik) plasticity (of pictures), plastic effect, three-dimensional appearance of telev. pictures. -schreiber embosser. -träger relief support (of embossed film).

Reportage news reporting or gathering work. -film newsreel, news film, topical f.

reproduzierbar, nicht- non-reproducible, non-controllable, non-repeatable.

Requisitenraum property room.

Reserve, Leistungs- margin of power, standby p. -lampe spare lamp.

Reservierungsmittel resist, reserve (in calico printing, metal etching, etc.)

Resonanz entgegenwirkend antiresonant.

Resonanz, in resonant, resonating or in resonance with, in unison with, in tune with, syntonized. In - bringen resonate, cause to be in resonce with. -, untersynchrone subsynchronous resonance, submultiple r.

Resonanz, Dreh- torsional oscillation resonance. Höhlungs- cavity resonance (of rhumbatron). Nebenschluss- parallel resonance. Schüttel- vibration resonance.

Resonanz-anzeiger resonance indicator (e.g., vibrating reed instrument), tuning i. -beschleuniger magnetic resonance accelerator (e.g., cyclotron comprising two dees or half hollow cylinders and spiral beam, and other forms of beam tubes using ions, positrons and electrons), induction electron accelerator, also called rheotron and i. accelerator. -boden sounding board; belly (of violin). -fluoreszenz resonance radiation (fluorescence). -frequenz natural frequency, resonance f. -frequenz, Unter- submultiple (resonance) frequency, subsynchronous f. -hohlraum resonance cavity or chamber (rhumbatron). -körper sound or sounding board (or other resonant structure). -kreis, Spannungs- series resonant circuit. -kreis, Strom- parallel resonant circuit, tank c.

Resonanzkurve, schiefe unsymmetric or skew resonance characteristic. -, zweihöckrige, zweispitzige, zweiwellige double-hump or double-

peak resonance curve.
Resonanzkurven-einsattelung,-senke crevass of resonance curve, dip.
Resonanz-röhre resonance or tuning indicator (mostly glow tube), flashograph. -schärfe selectivity, sharpness of resonance or tuning- -schwingung resonant vibration, sympathetic v., co-vibration. -spannung resonance potential, radiation p.,p. resulting in transition of electron from ground state to next orbit and emission of resonance line spectrum. -spitze resonance peak. -spitze, abgeflachte flat-topped resonance crest. -sprung transition of electron resulting in emission of radiation, -strahlung resonance radiation (of vapor lamp).
Resonator,Hohlraum- cavity resonator, chamber r., enclosure r. Quarz- piezo-electric resonator, tuning or resonance indicator.
Resotank micro-wave generator (electron-oscillation tube or magnetron t. confined in hollow space or sphere), cavity resonator magnetron.
Rest-abweichung residual aberration (of lens). -dämpfung net transmission equivalent, net attenuation, -flüssigkeit residual liquid. -härte residual hardness (of water). -keim residual nucleus. -plasma residual electrons, plasma (region practically without resultant charge). -strahl residual ray, r. radiation, "reststrahl".
Retortenhelm retort head, r. helm.
Reusenantenne cage antenna.
Reversionspendel reversible pendulum, Kater p.
Revolver revolving nosepiece (in microscope). Spulen- coil switching assembly (for wave-band change). -kopf lens turret (of camera).
Rezension review, criticism.
Reziprok,Widerstands- inverse resistance, resistance reciprocal (= conductance).
Rhodan thiocyanogen, sulfocyanogen.
Rhombendodekaeder rhombic dodecahedron.
Rhomboeder rhombohedron.
Rhombus rhomb(us), lozenge.
Rhythmus,im in tune, synchronism, rhythm or unison with; at the rate of.
Richt-antenne directional antenna, unilateral a. -bogen clinometer. -charakteristik directional characteristic, space pattern. -dorn mandrel. -effekt rectification effect; directional action. -empfang directional reception. -empfänger,Einpeilen auf Punksender tune directional receiver to radio station (to take bearings).
richten direct, turn, arrange, adjust, set, straighten (out), dress, align, lay (a gun), train or aim (at). -auf train on, aim at. sich - nach be guided by.
richtende Eigenschaft directional property (of a loud-speaker, screen, antenna, etc.)
Richter,Strom- rectifier of thyratron, ignitron or other mercury-arc type. Wechsel- inverter, inverse rectifier, d-c to a-c transverter.
Richt-fähigkeit directivity, directional property. -faktor rectification factor. -funkfeuer directive beacon station, d. transmitter s., radio range s., beam s. -gerät bombsight, sight.
richtig right, correct, just, true, ortho. -,farben- orthochromatic (phot.)
Richtigkeit correctness, accuracy, faithfulness, fidelity (of reproduction)
richtigphasig true-phased, in correct phase relationship.
Richt-kanonier gunlayer. -kennlinie rectification characteristic; directional diagram, space pattern. -konstante rectification constant. -kraft directive force, directing f.; versorial f. -lautsprecher directional loudspeaker. -linie guide line, guiding l., directive.

-mass standard gage. -mikrophon directional microphone (uni- d., bi-d., ultra-d., etc.) -moment torsional rigidity (couple re-quired to twist spring a certain angle). -platte orientation plate (micr.) -scheit rule, straight edge. -sendeanlage beam station, radio beacon s., radio range s. -spannung output voltage furnished by rectifier, d-c. potential of r. -strahler directional antenna, d. loudspeaker or projector of sound. -strom output current (d.c.) of a detector; rectified (d.c.) component of plate current.

Richtung,ausgepeilte bearing. Anflug- direction of approach, homing direction (of airplanes). Aus - falsche misalignment (of track or image). Gleit- slip direction (cryst.) Kristallachsen- crystallographic axis orientation. Vorzugs- privileged direction.

Richtungs-anzeige,eindeutige uni-directional direction-finding (with sense-finding). -anzeiger direction-finder, d. indicator; course indicating radio beacon. -bestimmung direction finding.

Richtungscharakteristik directive or directional characteristic or diagram, space pattern. -Doppel-Kreis- figure 8 characteristic, tangent-circle pattern.

Richtungseffekt head and tail effect (in developing machine); directional action.

richtungsempfindlich with directional response, dependent upon direction or orientation,non-astatic.

Richtungs-und Entfernungsbestimmung ranging (say, by echo method).

Richtungs-fehler deviation (of pencil or beam), directional distortion (telev.) -funkbake(zum Anfliegen) runway localizing beacon (terminal marker, used in approach procedure). -geber director, directive antenna. -hören aural determination of direction (of airplanes), airplane spotting or locating. -kurve directrix. -navigation directional navigation. -quantelung directional quantization. -sender beacon station, beam s., radio range s. -sinnbestimmung sense-finding, sensing(determines sense of direction or absolute d.)

richtungsunempfindlich non-directional, astatic.

Richtungs-vermögen directivity. -verzerrung direction distortion (in sound reproduction). -zerstreuung scatter of direction (beam).

Richt-vermögen directivity. -verstärker amplifying detector, plate-current d. -verstärker, Schirmgitter- screen-grid tube with plate-current rectification. -weiser radio beacon, r. range, beam station. -widerstand uni-directional resistance (to flow), valve effect, rectifier e. -wirkung directional effect, directivity. -zylinder directional cylinder.

Riefe groove, furrow, channel, flute, slot, milled knurl. Zieh- drawing groove, d. scratch.

riefeln mill. groove, knurl.

Riegel (sliding) bolt, lock, fastening.

Riemen strap, belt, band, thong. -,halbgeschränkter quarter-turn belt. Umhänge- shoulder strap.

Riemen-fett belt dressing. -scheibe belt pulley.

Rieseln rippling, gushing, trickling; swarming (of film), friable, non-caking condition (of materials).

Rieselturm scrubber, wash tower, spray t. or column,

Riffel-faltenlautsprecher loudspeaker with folded and corrugated horn. -lautsprecher Riffel loudspeaker (with diaphragm corrugations and substantially horizontal radiation.) -membrane pleated

diaphragm, d. with circular corrugations.
riffeln furrow, channel, groove, ripple (flax).
Rille furrow, groove, track (of phonograph disk).
Rille,Führungs-,Schall- groove or track with undulations (of phonograph record or disk).
Rindenschicht cortical layer.
Ring,geteilter split ring, segmented r. Absaug- (cf. Saugspannung) annular or ring-shaped collector electrode. Aufschraub- lens flange. Befestigungs- ring fastener, fastening r. Debye-Scherrer- D.-S. ring or circle, powder pattern, Hull r. Dichtungs- gasket ring, gland. Lauf- race, raceway, ball-race (in bearing). Peil- direction-finder loop or frame. Schleif- slip ring, collector r. Teil- index ring. Verschluss- locking collar (of magnifier).
Ring-bildung ring formation. -bolzen eyebolt. -elektroden, abbildende focusing rings. -entladung ring discharge. -figur zone plate (of Fresnel), Huygens z.
ringförmig ring-shaped, annular, cyclic, toroidal, toric.
Ring-kopf annular recording head. -leitung,Antennen- multiple-receiver connection to antenna. -lochscheibe film scan disk (with apertures arranged circularly). -übertrager repeating coil.
ringungesättigt cyclically unsaturated, containing an unsaturated ring.
Ringwulst tore, torus, toroid, "doughnut".
Rinne channel, groove, gutter, trough.
rinnen run, flow, trickle, leak.
Rippe rib, fin, vane, ripple, flute, corrugation, undulation, section (of a radiator). Kühl- cooling fin, c. flange, radiator fin.
Rippentrichter ribbed funnel.
Riss fissure, crack, flaw, rent, tear, gap, opening, elevation (say, front or side).
Riss,Haar- micro-flaw, hairline crack,craze, capillary hair fissure. Schatten- silhouette. Seiten- lateral elevation, side e. Spalt- cleavage crack, c. fissure.
Riss-bildung cracking, fissuration. -festigkeit crack strength, c. resistance. -härteprobe abrasion test, scratch t.
Rissigkeit,Schweiss- chinking or fissuration due to welding.
Ritze rift, slit, fissure, crack, chink, cleft, scratch. Stimm- glottis, glottic catch or cleft.
Ritzhärte scratch hardness, sclerometric h. -prüfer sclerometer, scratch hardness tester.
Ritzspalt scratched slit (for instance, in silver coat, in sound recorders).
Roentgen... s. Röntgen...
roh raw, coarse, rough, unrefined, unwrought, unworked, unfinished, in blank form; broad (radio tuning).
Roh-block (raw) ingot. -bramme slab, ingot. -film blank film, raw stock. -formel empirical formula. -gang abnormally fast drive or run, irregular working (metallurgy). -ling blank, unworked, unfinished or unfashioned piece or part.
Rohmannshaut Rohmann's skin.
Rohr (s.a.Röhre) tube, valve (radio); pipe, duct, conduit. -,hartgeschlossenes rigidly terminated tube. Abzugs- vent stack (of an arc-lamp) Befehls- pilot or master thyratron (firing first in a trigger circuit). Beobachtungs- observation tube (of electron microscope). Dunst- ventilating pipe. Folge- piloted thyratron (fires second in a trigger circuit).

Hebe- syphon (tube). **Hosen-** syphon or Y-pipe. **Hüll-** encasing tube, jacket pipe. **Neben-** side or branch tube or pipe. **Perl-** bead tube (filled with glass b. to break up a fluid). **Skalen-** scale tube (of spectroscope). **Stern-** astronomical telescope. **Übersetzungs-(in Schwundregelung)** converter tube (in avc circuits, between regulator and regulated tubes). **Zähl-** counter tube (Geiger - Müller).

Rohr-bildung tubing or piping action. **-blatt** reed (of clarinet).

Röhrchen tubelet, capillary tube. **Schreib-** pen, syphon (of a recorder).

Röhre (cf Rohr) tube, valve, pipe, duct, conduit. **-,abgeschmolzene** sealed off tube. **-,fremderregte, fremdgesteuerte** master-excited tube (with independent drive circuit). **-,fussIose** loctal tube. **-,gassgefüllte** gas-content tube, gaseous t.; gassy t. **-,harte** hard tube, highly evacuated t. **-,indirekt geheizte** heater-type tube, indirectly heated t. **-negativen Widerstandes** dynatron oscillator. **-,quetschfussfreie** loctal base tube. **-,selbsterregte** self-excited or self-oscillatory tube (in a direct drive circuit). **-,weiche** soft valve, ionic v., gas-filled tube, gassy t. **-,zerlegbare** demountable tube.

Röhre,Abstimm- tuning indicator glow-tube (e.g., flashograph), c.-r. indicator t. or magic eye. **Achtelektroden-** octode. **Amplituden(glimm)-** glow-tube amplitude indicator, resonance i., tuning i. of neon type. **Ballast-** absorber valve, a. tube (takes energy during spacing periods). **Beeinflussungs-**modulator tube. **Bildabtaster, mit mechanischer Blende** Farnsworth dissector or electron camera tube, Dieckmann & Hell magnetic scanner, Campbell-Swinton c.-r. scanning or pickup tube. **Bildschreibe-** televisor tube, viewing t., picture-reproducing tube using cathode rays. **Bildspeicher-** signal- or charge-storage tube, normal iconoscope. **Bildspeicher- und Bildwandler-** supericonoscope. **Bremsfeld-(cf Bremsfeldschaltung)** retardation valve, Barkhausen-Kurz retarding-field or positive-grid t. **Doppelgitter-** bigrid tube, b. valve, double-grid t. **Doppelsteuerend-** power pentode wherein cathode grid is connected with control g. rather then with cathode. **Doppelzweipol-** duodiode. **Doppelzweipol-Dreipol-** duodiode-triode tube. **Doppelzweipol-Vierpol-** duodiode-tetrode tube. **Dreielektroden-** triode. **Dreifach-** three-unit tube, three-purpose t. **Eichel-** acorn tube. **Einschluss-** sealed tube. **Elektronen-** thermionic tube, electronic t., electron tube, e. valve. **End-** power tube, end t. **Entladungs-** discharge tube, vacuum t. **Exponential-** exponential tube, variable-μ t. **Fünfpol-** pentode tube (either of power type or of screen-grid type). **Gasentladungs-** gas discharge tube or valve. **Geräuschunterdrückungs-** squelch tube, noise-reduction t. **Glättungs-** glow-tube stabilizer. **Habann-** type of split-anode magnetron, dynatron oscillator (strictly, dynatron uses secondary electrons). **Haupt-** main oscillator (in independent drive system). **Ionensteuerungs-** thyratron. **Isolier-** buffer tube, isolating t. **Kathodenstrahl-** thermionic tube (obs.); beam t.; Lenard cathode-ray t.; Braun and other c.-r. tubes working with beam and screen such as oscilloscope, iconoscope, emitron, kinescope, Farnsworth dissector and oscillight, etc. **Kippentladungs-** time-base discharge tube. **Klein-** miniature tube. **Knopf-** shoebutton tube, acorn t. **Kopplungs-** coupling tube (RF input t. associated with antenna). **Kraftschwing-** power oscillator. **Lautzeit-**drift tube, beam t., tube involving transit-time or phase effects. **Lenard-** Lenard c.-r. tube (used in radiology and experiments on phosphors). **Magnetfeld-** magnetron type of tube working with constant magnetic field (strictly, a m. operates with variable magnetic field), permatron. **Mass-** measuring tube, gage t., graduated t., burette. **Mehrfach-** multi-valve, dual- or multi-purpose tube, multi-unit t. **Mehrgitter-** multigrid tube (say, tetrode, pentode, etc.). **Misch-** mixer tube. **Misch-, selbstschwingende** self-heterodyning mixer tube. **Modulations-** mixer tube, modulator t. **Mosaik-, doppelseitige** image-multiplier iconoscope. **Negativwiderstands-** kallirotron or dynatron types of negative-resistance tube. **Niveau-** level or leveling tube. **Oxydkathoden-** tube with oxide-coated filament, dull emitter t. **Plation-** amplifier tube with filament placed between control grid and plate; (one model has photo-electric g.) **Pressboden-** loctal base tube. **Prüf-, für Fernseher** phasmajector, monoscope, monotron. **Quecksilberhochdruck-** high-pressure mercury (-vapor) discharge tube. **Quersteuer-** cross-control beam tube, drift t., t. predicated for its operation on transverse action on beam. **Resonanz-** resonance indicator, tuning i. (mostly glow-tube), flashograph. **Röntgen-, mit Heizkathode** hot cathode X-ray tube, Coolidge t. **Schirmgitter-, Schutzgitter-** shield grid tube, shielded-plate t., screen-g. t. **Schwing-, fremderregte** master-excited oscillator tube (in an independent drive circuit). **Schwing-, selbsterregte** self-excited or self-oscillatory tube (connected in direct drive circuit). **Sechspol-** hexode. **Siebenelektroden-** heptode. **Sondenbildfang-** Farnsworth dissector tube. **Speicher-** (signal) storage or storing tube (telev.) **Speicher-und Bildwander-** super-iconoscope. **Stab-** arcotron. **Steuer-** master, drive or pilot oscillator, exciter tube. **Trenn-** isolating tube, buffer t. **Trennstrom-** spacing valve. **Trift-** drift tube, beam t. (e.g. Klystron). **Übersetzungs- Umkehr-** converter tube (used in avc schemes between regulator and regulated tubes). **Vakuum-** vacuum tube, discharge t. **Vervielfältigungs-** multiplier, multipactor. **Wasserkühl-** water-cooled tube. **Wiedergabe-, Braun'sche** Braun cathode-ray or electronic picture reproducing tube. **Zeichenstrom-** marking valve. **Zweielektroden-** diode. **Zweifachzweipol-** duodiode tube. **Zweifeld-** drift tube working with two fields (Heil types of tubes). **Zwischerstrom-** spacer valve.

Röhren-empfang mit gleichzeitiger HF und AF Verstärkung dual or reflex (valve) reception. **-fassung, federnde** cushioned socket or anti-microphonic tube or valve-holder. **-fuss** pinch, press, squash, stem or foot of a tube or valve. **-fuss, umgebördelter** re-entrant squash or press. **-kapazität** interelectrode capacitance. **-kappe** top cap (to bring in control grid lead). **-kolben** bulb of a tube, vessel or container of a tube or valve. **-konduktanz** transconductance. **-löscher** tube quench. **-lot** pipe solder. **-mehraufwand** tube complement. **-prüfer** tube tester, t. checker. **-rauschen** tube noise, thermal agitation of a tube, t. hiss, valve rustle. **-sockel** tube base. **-spannungsmesser** vacuum tube voltmeter. **-sperrung** cut-off of tube. **-stativ**

telescope tripod. **-stempel** press or squash of a valve or tube. **-summer** vacuum tube AF oscillator. **-träger** tube or pipe support. **-übersteuerung** tube overloading (causing blasting). **-voltmesser** vacuum tube voltmeter. **-voltmesser, selbstgleichrichtender** self-rectifying tube voltmeter. **-wippe** kipp relay, trigger circuit. **-wulst** tubular tore, "doughnut". **-zerplatzen, nach innen** implosion of a vacuum tube. **-zwischenstecker** tube adapter, valve a.

Rohr-holz cane. **-krümmer** pipe bend, p. elbow, p. knee. **-leitung, Hohl-** co-axial or concentric line, pipe l.; wave guide (dielectric or conducting). **-pfeife** reed pipe. **-schlange** coil (of pipe), worm, spiral tube. **-schlüssel** pipe wrench, **-sitz** cane seat. **-streifen** skelp or strip (for forming tubes). **-stutzen** short piece of pipe (serving as an opening, outlet, socket, or connection), nipple, short cylindrical piece, bush or sleeve. **-verlegung** pipe laying. **-verstopfung** pipe clogging, stop-up, choke. **-welle** dielectric wave.

Rohstück blank.

Rollbahn cycloidal path (of electrons); roller type conveyor; runway or taxiway (of airfield). **-kreis** cycloidal path (motion of electrons).

Rolle roll, roller, pulley, reel, spool, coil, sprocket; records, file-wrapper, file, docket.

Rolle, Abzieh- pull-down spool or sprocket. **Auflauf-** take-up reel. **Auftrags-** application roller, inker, inking r. **Belichtungs-** printing drum (of film printing machine); sound recording d., scanning point, translation p. (in s. film reproduction). **Druck-** impression roller, presser or pad r. (in film feed). **Farb-** inking roller, i. wheel. **Film-** reel, take-up r. or spool. **Filmtransport-** film feed sprocket. **Führungs-** sprocket (wheel). **Gummiandruck-** rubber-tired presser or pad roller. **Haupt-** (part played by) leading star. **Kordel-** milled or knurled roller. **Leit-** idler, guide roller. **Nachwickel-** take-up sprocket. **Schnur-** grooved roller, pulley. **Schwungmassen-** rotary stabilizer, roller with fly-wheel action, impedance wheel. **Seil-** pulley. **Spann-** tension roller (m.p.). **Sprossen-** sprocket drum or wheel. **Titel-** part played by leading star. **Transport-** sprocket wheel (film feed). **Umlenk-** flanged idling roller, deviating r. **Vorrats-** magazine, m. roll. **Vorwickel-** upper feed sprocket, pull-down s. (m.p.). **Zacken-** sprocket wheel.

rollen curl (of film).

Rollen-darsteller, Haupt- principal, star. **-fenster** roller film gate. **-käfig** roller cage. **-laufbahn mit Gummiüberzug** rubber-tread of a pad roller. **-nummer** docket number, register n. **-quetsche** squeegee. **-scheibe** chain sheave. **-zug** block and tackle.

Roll-glasschneider wheel glass cutter, g. cutting w. **-kassette** film roll darkslide. **-kreis** cycloidal path (of electrons). **-schweller** crescendo pedal (organ). **-treppe** escalator. **-walze** roller squeeze.

römische Kerze Roman candle, magnesium torch, smoke pot.

Röntgen-aufnahme röntgenography, X-ray picture, skiagraph. **-beugungsaufnahme** X-ray diffraction exposure or picture.

-**bremsstrahl** Röntgen rays caused by collision and checking. -**dosismeter** X-ray dosimeter. -**einheit** roentgen (international X-ray quantity unit), Xu, X unit, Siegbahn u. (wavelength). -**messer** roentgenmeter, iontoquantimeter. -**photographie** röntgenograph, X-ray photograph, skiagraph. -**röhre mit Heizkathode** hot-cathode X-ray tube, Coolidge t.-**röhre, fensterlose** windowless X-ray tube. -**rohre,Hartwerden** hardening of X-ray tube (due to clean-up and loss of vacuum). -**strahlenhärtemesser** penetrometer, qualimeter. -**strahlenmesser** röntgenmeter, iontoquantimeter, dose meter, dosimeter, intensimeter,

roentgenstrahlen-undurchlässig radiopaque.

Rose,Kompass- card of compass, scale of c.

Rosettenbahn rosette-shaped path (of electrons).

Rost rust; grate, gridiron. **Etagen-** step grate.

Rostanfressung corrosion, honey combing, pitting, tubercular or channel-shaped corrosion.

rösten calcine, roast, sinter.

Rost-festigkeit corrosion resistance, non-corrodibility. -**grübchen**, -**narbe** pit.

rostsicher non-corrodible, corrosion-resistant, rust-proof, non-rusting.

rot,hell- bright red, of bright-red incandescence. **kirsch-** cherry red, of cherry-red incandescence.

Rotation,behinderte hindered or inhibited rotation (of molecules).

Rotations-bewegung rotational motion, rotary m. -**gruppe, räumliche** space rotation group. -**korper** body or solid of rotation or revolution.

rotationsparabolisch full-parabolic.

Rotations-quantenzahl rotational quantum number. -**schwingungsband** rotation vibration band. -**schwingungsspektrum** rotation vibration spectrum.

rotationssymmetrischer Körper body presenting rotation symmetry.

Rot-auszug red record (color film). -**bruch** red-shortness (of metal). -**bruchversuch** hot breaking test. -**buchenholz** red beech.

Roteltonung russet toning, brown-reddish t.

rotempfindlich red sensitive.

Rotglühhitze red heat (of incandescence).

rotierend(er) Dämpfer rotating stabilizer (film feed). -**(e) Umlaufverschlussblende** rotary flicker shutter, r. light-cutoff.

Rotorfunkenstrecke spark-gap rotor.

Rotterdammgerät radar.

Rotverschiebung shift of spectral lines toward red or longer waves.

Rouleauverschluss roller blind shutter.

Ruck jerk, jolt, sudden shift or motion, tug.

Rückbewegung back stroke, return s., retrogression, flyback (of beam, telev.)

rückbilden form again, re-form.

Rück-bildung back (reversible) reaction, regression; re-formation. -**dehnung** damping capacity (of metals). -**dreher, Phasen-** phase lagger, p. shifter. -**druck** reaction pressure, back p.

Rückenwind tail wind.

Rück-flussdämpfung structural return loss. -**flusskühler** reflex condenser, return c., return-flow cooler. -**führfeder** retractile spring, restoring s. -**führschnecke** return screw conveyor, return worm. -**führung**

return lead, r. path, r. circuit; flyback, retrace (telev.) **-gang** return, retrogression, decline, backstroke; flyback (of pencil).

rückgekoppeltes Audion regenerative grid-current detector, ultraudion.

Rück-gewinnung recovery, recuperation (of current or energy), reclamation, regeneration. **-griff(cf Durchgriff)** inverse grid transparency. **-heizung** extra or additional heating (of cathode, from other electrodes). **-hörbezugsdämpfung** side tone reference equivalent. **-hören** side tone. **-hub** back-stroke, return s. **-kehrpunkt** cusp. **-kippgeschwindigkit** flyback speed, time-base unlock s.

rückkoppeln regenerate, feed back (a tube), put in reactive coupling relation (either positive or negative).

Rückkopplung regeneration, reactive coupling, feedback c., reaction. **-mit Hilfsfrequenz** super-regeneration (using quench potential).

Rückkopplung, Ionen- ion feedback. **Mehrfach-** multiple regeneration. **Pendel-** super-regeneration (with quench potential). **Reinartz-** feedback (of Reinartz) using both capacity and inductance. **Schall-** acoustic feedback, a. regeneration.

Rückkopplungsaudion regenerative grid-current detector, ultraudion.

rückkopplungsfrei non-regenerative, free from feedback action.

Rückkopplungs-kreis regenerative circuit, reactive c., feedback c. **-spannung** positive bias potential, p. quench potential (impressed upon grid, in super-regenerative circuit organization). **-sperre** reaction suppressor, "vodas," feedback s. **-spule** tickler coil, feedback c. **-unterdrückung** decoupling, balancing out, tuning out, reduction or suppression of feedback.

Rücklauf return of pencil, flyback or retrace of cathode spot or beam. **-der Kippspannung** flyback of sweep potential, return of time-base to zero.

rücklaufend recurrent, retrograde, in return or opposite direction. **-(er)Blitz** return lightning stroke.

Rücklauf-verdunkelung flyback or return trace elimination, blackout of flyback, gating (by blanking pulse). **-zeit** retrace period, flyback p.

Rück-leiter return wire, r. lead, r. conductor. **-leitung, Erd-** ground return (circuit). **-meldefeld** revertive signal panel. **-melder** revertive signal or position repeating means, check-back position indicator (in telemetric or remote-control work). **-nahmestreit** action for withdrawal.

Rückprall rebound, bump, reaction, repercussion, strike-back, back-stroke, recoil, return, back-fire. **-elektrode** (cf **Prallelektrode**) rebound electrode, target e., impactor e. (in Weiss mesh multiplier).

Rück-projektionsanlage reprojection equipment; back projection e. **-reaktion** back reaction. **-schlag** see **Rückprall**. **-schlag, Bogen-** arc-back. **-schlagventil** check valve.

rückschreitende Welle regressive wave, reflected w.

Rück-seite back side, posterior face, rear surface, reverse. **-sicht, mit - auf** with due regard for, because of, in the light of. **-spiegelung** specular reflection, regular r. -spielen playback, dubbing. -sprung (cf Rückprall) rebound, ricochet.

rückspulen re-wind.

Rückstand, in - kommen fall behind, be in arrears. Gas- residual gas, remnant of g. (or vapor). Kondensator- residual charge of condenser. Trocken- dry residue.

rückständig in arrears, backward, not progressive, retrogressive, reactionary.

Ruckstellkraft restoring force, resiliency, retractility, elastic f., restitutive f. or pressure.

Rückstoss repulsion, rebound, backward push, recoil, back stroke, reaction, shock. -atome recoil atoms. -bahn recoil track. -elektron recoil electron, Compton e. -potential recoil potential.

Rück-strahl echo, reflected wave, reflection. -strahlbrennfläche catacaustic. -strahldiagramm back reflection photogram. -strahler reflector, re-radiator. -strahlung, spiegelnde specular reflection, regular r. -strahlungsfaktor reflectance, reflection factor. -strahlungsmesser reflectometer. -strahlverfahren (Ultramikr.) reflection method (ultra or electron microscope using a thin foil). -strom inverse current, backward c. (in rectifier). -stromrelais reverse-current relay.

rücktreibende Kraft repelling force, reflecting f. (acting on electron pencil).

Rückverwandlung reconversion.

rückwärtige Stromstossgabe revertive pulsing.

Rückwärts-leitung high-resistance direction. -regelung indirect control, retroactive control action, backward regulation.

Rückweg return path.

ruckweise by jerks, jars or jolts, intermittently, discontinuously, spasmodically.

rückweisende Reflexion retrodirective reflection.

rückwerfende Brennflache catacaustic.

Rück-werfer rejector, reflector, throwback means. -wirkung, Anoden- anode or plate feedback or reaction. -wirkung, Blind- reactance (either in the form of inductance or capacitance). -wirkung, Schall- acoustic impedance (comprising resistance and reactance, the latter being either in the form of inertance or compliance, corresponding to inductance and capacitance in electricity). -wirkungsleitwert reactive admittance, susceptance (either capacitive or inductive). -zugfeder retractile spring. -zugtaste back-spacing key. -zündung backlighting, arc-back, backfiring; backlash (imperfect rectification, in valves).

Ruf ring, ringing, call (signal). Not- SOS or distress call, mayday (radio telephone d. call).

Rufer developer.

Ruf-horn signal horn. -zeichen, Funkstellen- station call, code sign, code signal, "signature".

Ruhe rest, repose, quiescence, silence, Q. state, unmodulated, unexcited or neutral condition, stationary equilibrium.

Ruhe....(in Zusammensetzungen) no-signal..., when in unexcited or unelectrified state, Q or quiescent condition (of a valve).

Ruhe-arbeitspunkt Q point, quiescen condition (of valve). -belichtung unmodulated lighting, no-modulation l., average or steady illumination, no-sound l. (in variable-density sound recording). **-elektrizitätslehre** electrostatics. **-kontakt** spacing contact, rest or inoperating c. (closure occurs upon de-energization of relay), contact of back part of key lever. **-lage** equilibrium position, balanced p. **-lage, Strahl-** spot zero. **-licht** unmodulated lighting, no-modulation l., average illumination, steady i., no-sound i. (in variable-density sound recording). **-masse (der Teilchen)** rest mass (of particles).

ruhendes Bild still picture, "still", unanimated p., ordinary photograph.

Ruhe-punkt point of rest, fulcrum, quiescent or Q p. (on tube characteristic). **-schwärzung** unmodulated density, no-sound d., (in variable-density sound recording). **-spannung** bias potential, steady p., no-signal p., repose p., Q point p. **-stellung, Anker-, Relais mit** neutral relay. **-stellung, Tonstreifen-** zero modulation of track. **-streifen** unmodulated track. **-strom** no-signal current, quiescent c., anode-feed current (d.c. component of anode c.) **-strombetrieb** closed-circuit operation. **-trägerfrequenz** center frequency, resting f. (in f - m). **-transparenz** unmodulated transmission (of film). **-wert (cf Ruhezustand)** steady, static, stationary, no-signal, neutral or quiescent value, Q point (e.g., when a tube furnishes no oscillations or no potential difference exists between grid and filament). **-winkel** angle of repose, a. of friction (rarely used). **-zustand** state of rest, s. of quiescence, Q, neutral or repose condition.

Ruhrapparat stirring apparatus, agitating a., stirrer, agitator.

rühren stir, agitate, puddle, pole, rabble.

Rühr-scheit paddle, rake, stirrer. **-stab** stirring rod, s. pole, paddle, rabble.

Rumpf body (of engine, etc.); trunk, hull (of a ship); fusel lage (of airplane); core (of ions), kernel (stable inner electron group); negative unvaried portion of luminous glow (in g. tube). **Ionen-** ion core.

Rumpfelektron inner electron (forming part of core, kernel or rumpf e. group).

Rundblick-aufnahme panoramic view, p. photograph, panning shot, pan, pam (m.p. slang). **-fernrohr** panorama sight, panoramic telescope. **-kamera** panoramic camera.

runderhaben convex.

Rundfunk, Leitungs- AF wire broadcast, AF rediffusion.

Rundfunk-tonfilm sound telecast, s. telecine (of film), broadcast of television program with sound accompaniment (or of video and aural signals simultaneously). **-welle (cf Wellenabgrenzung)** broadcast wave, medium-frequency w., b. band (frequencies between 550 and 1600 kc in USA).

rundhohl concave.

Rund-kerb round notch, Charpy n. (with root radius). **-laufen der Ionen** circulation of ions. **-strahlantenne** polydirectional antenna, omni-d.a.

runzelig wrinkled, puckered, shriveled; mottled, reticulated (of film gelatin).

Runzeln mottling, wrinkling.

Rüssel nose, snout, nozzle.

Rute, Wünschel- dousing or divining rod.

Rutengänger, Wünschel- douser.

Rutherford-Bohr Atommodell R.-B. atom model or conception.
Rutil rutile, titanium dioxide.
Rutsche chute, shoot, slide.
rutschen glide, slide, slip, chute.
Rutsch-erscheinung slip phenomenon (cryst.). **-kupplung** slip clutch.
Rüttelformmaschine jolt molding machine, jar ramming m.
rütteln shake, jolt (by jerky motion).
Rütteltisch jarring, jolting or bumping table.

S

S. A. (Selbstanschluss-System) automatic telephone system.
Saal,Probe- rehearsal hall.
Saalregler theater fader.
Sache subject(-matter), cause, case, theme, topic, matter, affair, docket, in re.
sachgemass appropriate, in an expert way.
Sach-kenntnis,-kunde expert knowledge, expertness, competence, knowledge in an art.
sachkundig experienced, expert, skilled, trained in an art or profession.
Sachlage state of affairs, factual situation.
sachlich objective, material, neutral, real, in a matter of fact way.
Sach-verhalt factual situation, facts of a case. **-verständiger** expert, authority, specialist (in technological art or science, etc.) **-verzeichnis** subject index. **-wörterbuch** encyclopedia.
Sack bag, sack, pocket, pouch, sac. **Wind-** wind cone, w. hose (airport).
säge-artig saw-like, serrated. **-förmige Spannung,doppel-** double sawtooth potential.
Sägemehl saw dust.
Sägezahn-generator sawtooth generator (telev.) **-kurve,Steilabfall der** abrupt drop of sawtooth wave. **-spannung** sawtooth voltage, ratchet v.
saigern liquate.
Saite,ausgespannte stretched chord, s. string. **-,geschlagene** percussed chord or string. **-,gezupfte** plucked string. **-,mitschwingende** sympathetic string, co-vibrant s. **Begleit-** accompaniment string(zither).
Saiten-galvanometer string galvanometer, vibration g. **-summer** chord buzzer. **-zug** string tension.
salbig salvy, unctuous.
Salz,Fixier- fixing salt.
salzgetränkte Kohle mineralized carbon, impregnated c.
Salzkristall,Seignette- Rochelle salt crystal.
sämisches Leder Chamois leather.
Sammel-anode collector anode, gathering a., output a., ultimate a. **-anschluss** p.b.x. line, private branch exchange. **-anschlussteilnehmer** subscriber with several lines. **-elektrode** output electrode, collector e., gathering e., catcher (klystron). **-gefäss** receiver, reservoir, collecting vessel, receptacle. **-glas** converging lens; preparation tube, specimen t. **-linse** condensor lens, converging l., convergent l., positive l. **-linse,Haupt-** second lens (in electron gun). **-linse,Vor-** first focusing lens, cathode l. (like Wehnelt cylinder, apertured disk, etc., in cathode-ray tube).
sammeln concentrate, focus, collect, gather, accumulate, assemble, catch.
Sammel-raum receiver, receptacle, collector space, catcher space (in Klystron). **-rohr** collector pipe, mains. **-schiene** busbar, omnibus bar. **-spule** focusing coil. **-system** collective or condensor system or element (projector). **-zylinder** focusing cylinder.
Sammler accumulator, storage battery. collector. **Wind-** air reservoir, compressed-air tank.
Samtdichtung velvet trap.
Sandform sand mold (foundry).
sandführend sand-bearing, sandy, areniferous, arenaceous.
Sandguss sand casting.
sandig sandy, areniferous, arenaceous.
sanft soft, gentle, mild, smooth.
Saphir sapphire. **-spat** cyanite.
satinieren satin, glaze, burnish.

Satiniermaschine,heisse hot burnishing press.
Sattdampf saturated steam, saturated vapor.
Satte bowl, dish.
Sattel crevass, dip (of resonance curve).
sättigen,über- supersaturate.
Sättigungs-aktivität saturation activity. **-apparat** saturator.
sättigungsfähig saturable, capable of saturation.
Sättigungs-kapazität saturation capacity. **-strom** saturation current. **-wert** valence.
Satz deposit, sediment, settlings, set (of things); sentence, principle, law, rule, theorem, axiom. **-vom ausgezeichneten Lichtweg** law of extreme path.
Satz,Federn- spring bank, s. assembly. **Flächen-** theorem of conservation of areas. **Glas-** glass batch, g. composition, g. charge. **Gleichverteilungs-** principle of equipartition of energy (Maxwell-Boltzmann). **Haupt-** fundamental principle, f. law, axiom (math.) **Impulserhaltungs-** theorem of conservation of momentum. **Knall-** detonating composition. **Kontakt-** bank of contacts. **Lehr-** abstract theorem or doctrine. **Leit-** guide rule, guiding principle, basic p. **Linsen-** lens combination, system of lenses. **Lücken-** vacancy principle. **Objektiv-** set of lenses, system of l., lens combination. **Permanent-** permanence principle, sum rule. **Rad-** axle and two wheels. **Summen-** sum rule, permanence principle. **Verschiebungs-** displacement law. **Verteilungs-** principle of distribution, d. law. **Vorder-** antecedent, premise. **Wechsel-** exchange principle.
sauber clean, neat.
säuern acidify, sour.
Sauerstoff-ion anion, negative ion. **-messer** eudiometer. **-pol** oxygen pole, anode. **-verbindung** oxygen compound.
Saugapparat suction apparatus, aspirator.
saugen suck(up), absorb, imbibe, aspirate.
Sauger suction apparatus, aspirator, exhauster. **Wellen-** wave trap, smoothing choke coil, series reactor.
Saug-fähigkeit absorptive capacity, absorptivity, imbibition power. **-feld** positive field, suction f., f. which draws away. **-festigkeit** resistance to suction, s. strength. **-filter** suction filter. **-flasche** filter flask. **-geschwindigkeit** rate of evacuation (of vacuum pump). **-gitter** space-charge grid. **-hahn** suction cock. **-heber** syphon. **-höhe** suction head, s. height, absorptive height. **-kreis** acceptor circuit; impedance or absorption wave trap. **-leitung** suction pipe, s. piping. **-luft** vacuum. **-messer** vacuometer. **-näpfchen** suction cup. **-spannung** anode, driving or positive potential (of photo-cell), saturation p. **-spule** smoothing coil. **-stutzen** pump intake, p. nozzle. **-transformator** booster transformer, sucking t. **-ventil** suction valve.
Säule column, pillar, prism (cryst.); pile (electr.). **Glimm-** light column, glow c. (of tuning indicator). **Licht-** light column. **Quecksilber-** mercury column. **Volta'sche-** voltaic pile.
Säulenachse prismatic axis.
säulenartig columnar, prismatic (cryst.)
Säulen-lampe pillar (lamp), batten. **-tischfernsprecher** desk (stand) telephone set.
Saum seam, hem, edge, fringe. **-,farbiger** color fringe.
säumen hem, edge, border.
Säureballon acid carboy.
säure-beständig stable to acid action, fast to acid. **-bildend**

acid forming.- -haltig containing acid, acidiferous.
Säureschwemmverfahren acid flotation method.
säurewiderstehend resistant to acid, fast to a.
Säurezahl acid number.
sausen rush, whiz, whistle, hum.
Schaar,Kurven- family, group or system of curves.
schaben scrape, shave, grate, rub, abrade.
Schablone pattern, template, stencil, mold, form, model, former.
Schablonenstreifen controlling strip (automatic film printing).
Schabsel scrapings, parings, shavings.
Schacht,Licht- light tunnel (of sensitometer).
Schadenersatz festsetzen assess damages or indemnity. -klage action for damages.
schädlich spurious, stray (of radiations), noxious, injurious, dangerous, harmful, detrimental. -machen vitiate, contaminate.
Schadloshaltung indemnification, indemnity, compensation.
Schaft shaft, shank, stock, stem, handle, stalk. Niet- rivet shank, r. stem.
Schake link (of a chain).
Schäkel shackle.
Schale dish, basin, cup, pan, bowl, scale, tray; husk, shell, rind, skin. Abtropf- drainer, drip, jar, dish. Anoden- anode segment (of magnetron). Dampf- evaporating dish, e. basin. Elektronen-, beinahe geschlossene nearly closed electron shell. Kugel- partial sphere, cup, spherical shell, cup-shaped or hemispherical part, calotte. Unter- sub-shell (of electrons). Zweier- duplet rings.
Schaleabschirmungskoeffizient shell absorption coefficient (shielding function).
Schalen-bau der Elektronenhülle structure of electron shell.
-entwicklung tray development, dish d. -guss chill casting. -gussform chill. -kupplung sleeve coupling. -lack shellac. -windstärkemesser cup-type anemometer,
Schall(s.a. Akustik,Geräusche, Klang & Ton) sound, acoustic action, sound a., ring, peal, noise (=disturbing s.)
Schall,Körper- material conduction of sound (say, through solids). Luft- air conduction of sound. Tritt- impact sound (transmission).
Schall-abstrahlung sound radiation or projection. -abwehr noise abatement, n. suppression, sound attenuation, s. absorption. -aufnahme,akustische direct mechanical recording (as distinguished from electrical r.) -aufnehmen sound recording; s. pickup, s. collection, s. receiving (by a microphone). -aufnehmen,Glimmröhren- glow-tube or g.-lamp sound recording (e.g., by Aeolight). -aufzeichner sound recorder. -ausbreitung, vertikale vertical spread or diffusion of sound. -ausschlag displacement of a particle (ac). -bandaufnahme magnetic steel tape recording. -becher bell(wind instrument). -bestrahlung,hohe high radiation distribution efficiency. -brett baffle(board). -bündel pencil of sound. -dämmung sound reduction,s. attenuation, s. deadening, s. absorption, silencing (of noise). -dämmzahl sound transmission characteristic.
schalldämpfend sound absorbing, s. attenuating, damping, deadening (for sound).
Schall-dämpfer silencer, sound absorbing means, muffler. -dämpfungseinheit unit of sound absorption (e.g., Sabin unit). -dampfungsfaktor sound absorptivity, s. absorption coefficient. -dämpfungsmittel sound absorbing,

s. attenuating, s. damping, s. deadening means or material, sound absorbent or absorber, s. insulation. **-deckel** sounding top (of mus. instrument), sounding board, sound b., abat-voix. **-druck, effektiver** effective sound pressure, rms (root-mean-square) acoustic p. **-durchlässigkeit** sound transmission, s. transmittance. **-einfall,streifender** glancing incidence of sound wave. **-empfang** sound collection, s. reception (of a microphone). **-empfänger** sound pickup, sound collection device (e.g., microphone), s. receiver. **-empfindungsschwelle** threshold of audibility, t. of acoustic perception or percipience (represented by audiogram). **-ereignis** sound action, acoustic a.

schallen sound, resound, ring.

Schallfärbung timbre, tone color, tone quality.

schallharte Trichterwand rigid or non-absorbing horn wall.

Schall-härte acoustic stiffness, sound hardness (denoting ratio pressure amplitude to velocity amplitude as a function of frequency). **-impedanz** acoustic impedance. **-ingenieur** sound technician, acoustician. **-leistung** response (of an acoustic system). **-mikrophon** acoustic microphone. **-öffnung** louvre (of bell tower). **-ortung** sound location. **-platte für Rundfunk** 16" transcription record for broadcast studios. **-platte,Bild-** sound and picture on disk. **-platte,Mess-** frequency record, phonograph disk bearing sliding notes or constant notes; parlophone record (produces warble note). **-plattengeräusch** record noise, surface n. **-plattensteg** wall, land or barrier between record grooves. **-quelle** sound generator, sound source, source of acoustic energy. **-reaktionsrad** Rayleigh disk, etc. (placed in a sound field). **-reibungswiderstand** acoustic resistance (dissipative component of acoustic impedance). **-reiz,überschwelliger** super-threshold (sound) intensity level. **-rille** groove or track (on a disk). **-rückkopplung** acoustic feedback, a. regeneration. **-rückwirkung** acoustic impedance (comprising resistance and reactance, the latter either in the form of inertance or of compliance, corresponding to electrical inductance and capacitance.). **-sammler** sound or microphone concentrator. **-scheinwerfer** sound reflector, acoustic radiator. **-schirm** baffle. **-schlucker** sound absorbent, s. deadening material. **-schluckung** sound absorptivity, acoustic a. **-schnelle** volume current (ac.). **-schnelleempfänger** velocity microphone. **-schreiber** sound recording stylus, s. recorder. **-schutz** sound proofing, s. insulation. **-schwellenstärke** threshold sound intensity (of audibility and of feeling).

schallsicher sound-proof.

Schall-stärkemesser phonometer, sound level meter. **-strahlbündel** pencil of sound. **-tilgung** sound absorption, s. suppression, s. deadening, s. damping, s. attenuation. **-tilgungsmittel** gobo, nigger, baffle (in form of a wall, etc., in m.p. work).

schalltot non-resonant, non-oscillable, acoustically inert or inactive.

Schall-treibwerk loudspeaker operating mechanism, motor element. **-übertragungsanlage** (public) address system; electroacoustic transducer (rare). **-umfang** sound range, s. contrast. **-umwandlungseinrichtung** electro-acoustic transducer. **-verstärkungsmittel** sound re-inforcing system (in halls, auditoria, etc.) **-verzugs,Ausschalten des** acoustic correction (sound locator).

-wahrnehmung des menschlichen Ohres human tone or sound perception, auditory percipience. -wand(cf Tonführung) baffle (partition or board), loudspeaker screen (to prevent acoustic feedbacks), sound panel (for either s. absorption or s. reflection). -weglange sound path length.

schallweich sound absorbent.

Schallwelle,Über- supersonic wave, ultrasonic w.

Schallwellen-aufzeichner phonodeik (records on film). -ausgleich neutralization of sound waves (prevented by baffles). -widerstand acoustic impedance; complex quantity comprising acoustic resistance corresponding to internal friction responsible for energy dissipation, and acoustic reactance either in the form of inertance (=electric inductance) or of compliance (=electric capacitance).

Schall-widerstand ratio pressure amplitude: velocity a. -wiedergabe, Radtkeverfahren s. Radtke verfahren. -wiedergabequalität realistic or faithful sound reproduction (with "atmosphere" or "room tone"). -wirkungswiderstand acoustic resistance.

Schalt-band controlling strip (of printing machine). -bild circuit diagram, wiring d., hook-up. -buchse jack. -element circuit element, c. means.

schalten,gegeneinander connect in opposition, c. differentially. -, in Brücke bridge (across). -, vielfach multiple. -, vorwärts step forward, notch f.

Schalter,halbversenkter semi-sunk or semi-recessed switch. -mit 5 Ausgängen switch with five points, five-point s. -, sechsteiliger, mit zwei Stellungen two-position six-point switch. Bogenlösch- Deion circuit breaker. Dämmerungs- twilight switch. Dreieck-Stern- delta-star switch. Dreiwege-three-way switch. Druckgas- cross gas-blast switch. Druckluft- autopneumatic circuit breaker, air blast switch. Expansions- hydroblast or expansion circuit breaker or switch. Einaus- off-on switch. Hebel-, zweipoliger double-lever switch. Hörsprech- talk-listen switch (of intercommunication system). Impulsol- oil blast circuit breaker. Kipphebel- toggle switch. Messer- knife (blade) switch. Mithör-listening key. Nadel- vibrating reed rectifier. Peilseiten- direction-finder sense-switch, sensing or sense-finding s. (operates df and sense-finding loops). Pressgas- gas-blast switch. Pressluft- airblast switch. Quecksilber-mercury switch. Rasten- step switch, multi-point s. Schnapp- snap switch, tumbler s. Sprech- speaking key. Stern-Dreieck star-delta switch. Steuer- sequence switch, master s., control s. Stöpsel- plug switch. Stromstoss- impulsing switch, stepping s. Stufen- multi-point switch, multicontact s., step s. Thermo- thermal lag switch. Trenn- isolator, isolating switch, disconnecting link. Überwachungs- monitoring key. Verstimmungs- wave-change switch, w. changer (in a transmitter). Verzögerungs- time-lag switch. Wahl- selector switch. Wasser- hydroblast switch. Wechsel- double-throw switch. Wellen(bereich)- wave-band switch. Wende- reversing switch, r. key. Zeitschnell- high-speed switch, quick-break s.

Schalter-arbeit switch work. -kamm sequence switch cam. -leistung cf Abschaltleistung.

Schalt-kapazität circuit capacitance (radio). -leistung,Ab- rupturing capacity, circuit-breaking c. (of a power switch). -mittel circuit element, c. means; switching device. **-nocken** trip cam, trigger c. **-periode** shutter period, shift p., moving p., feed stroke (m.p.) **-rad,Schritt-** stepping wheel. **-rolle** sprocket wheel (film feed). **-schema** diagram of connections, circuit d., hookup. **-spannung** bias reducing potential, unblocking potential (silent tuning). **-strom** current on contact. **-trommel** intermittent (motion) sprocket (film feed).

Schaltung (s.a. Stromkreis und Kreis) connection, circuit arrangement, c. organization, c. scheme, hookup, switching.

Schaltung,bildweise intermittent film feed or movement. **-,stete** continuous film feed or movement.

Schaltung,Abwäge- comparator circuit organization. **Bild-** feed or movement of frame or picture (m.p.), frame ratcheting. **Blink-** blinking arrangement, ratchet, relaxation or sawtooth generator scheme with RC time-base and neon lamp. **Bremsfeld-** retarding-field circuit organization (with positive grid and negative plate), Barkhausen-Kurz circuit, oscillating-electron or electron-oscillation scheme with reflecting electrode. **Brücken- H. and T. see Entzerrungskette.** **CW-** capacity-resistance circuit scheme, c.-r. network. **Doppelröhren-** tandem-tube circuit arrangement, push-pull c.a. **Doppelweg-** full-wave rectifier circuit organization. **Dreipunkt-** Hartley (oscillator) circuit scheme, potentiometer circuit scheme. **Entlade-** circuit causing fast drop of sawtooth voltage. **Film-** pull-down film movement, film feed. **Filmfort-** film feed, f. travel, f. movement. **Glimmlampenkipp-** neon-lamp time-base circuit scheme. **Graetz-** four electrolytic rectifier bridge-arrangement. **Greifer-** claw film feed. **Gruppen-** series - parallel connection. **Hintereinander-** series connection or arrangement, seriation, cascading a., tandem a., concatenation (of motors). **Kaskaden-** cascading connection, concatenation, tandem arrangement (of motors). **Ketten-** relay chain circuit. **Lade-** circuit means causing slow rise of sawtooth voltage. **Mischer-** keying and mixer tube scheme (for adding synchronizing and video signals). **Mittelpunkttast-** center-tap key modulator arrangement. **Neutralisations-** neutrodyne circuit organization. **Parallel-** parallel connection, p. circuit organization, multiple connection, paralleling. **Pendel-** (Armstrong) super-regeneration circuit. **Reinartz-** regeneration circuit (comprising capacity and tickler coil). **Reinton-** noiseless recording circuit organization. **Schlager-** dog or plunger movement or mechanism. **Schleifen-** circuit looped back and forth. **Spar-** circuit organization designed to economize (plate) current (e.g., Nestel's). **Zeilen-** vertical stepping-down movement, picture traversing.

schaltungstechnischer Aufbau constructional details of circuit organization or wiring.

Schalt-vorgang switching operation, s. process. **-wähler, Schritt-** step-by-step selector.

schaltweise Bewegung intermittent movement or feed, discontinuous m.

Schalt-welle wiper shaft. **-werk** film feed, intermittent or shuttle mechanism (m.p.); control switch m. or gear, trip or stepping m. **-werk,Fort-,-werk,Schritt-** step-by-step mechanism, stepping or switching m. **-zeichen** symbols of radio circuit elements, legend (list of symbols) letters or numerals).

Schamottestein firebrick.

scharf strident, shrill, piercing (of sound), peaked (of a curve). **-begrenzt** highly selective (of resonance).

Schärfe resolving power, resolution, definition; sharpness (of selection, distinction, discrimination or tuning), degree or measure of distinctness (e.g., of optical actions), acuity.
Schärfe, gestochene microscopic sharpness. Abstimm- sharpness of tuning or of resonance, selectivity. Bild- picture definition, p. resolution. Fleck- spot focus, sharpness of f. Gehör- acuity of hearing. Längs- definition in line direction. Mitten- central definition, sharpness of center. Quer- definition in frame direction. Resonanz- selectivity, sharpness of resonance or of tuning. Seh- visual acuity, sharpness of vision, visual focus. Tiefen- depth of focus, d. of field. Trenn- selectivity, filter discrimination.
Scharfeinstellung sharp focusing (opt.), automatic tuning indication.
Schärfen, Minimum- zero cleaning or clearing, z. sharpening (d.f.)
Scharfenfeld (in Bezug auf Öffnung, Entfernung und Brennweite) focal field (in respect of aperture, distance, and focal length).
Schärfenfeld-grenzen limits of sharpness or of definition; camera lines (delineating good focus area). -tiefe depth of field.
Schärfen-fläche surface of sharp or distinct vision. -messung, Gehor- audiometry. -tiefe depth of vision, definition in d., depth of focus. -winkel angular resolving power. -zeichnung sharpness of delineation.
Schärferegelung, Trenn- automatic bandwidth selection.
scharfgängig property of a screw cut with triangular thread.
Schärfung, Nullpunkt- zero or minimum point cleaning or clearing, sharpening of minimum (d.f.).
scharf-winklig acute-angled.
-zeichnende Linse achromatic lens.
Scharnier hinge, joint, articulation. Dreh- swivel pivot.
scharren scrape, scratch.
schartig notchy, nicked, jagged, serrate.
Schatten shades, shadows(of picture); umbras. Funk- radio shadow, r. pocket, r. dead spot. Halb- penumbra, half-shade or shadow, partial shadow. Kern- umbra.
Schatten-anzeiger shadow tuning indicator, shadowgraph. -apparat, Halb- half-shade apparatus, half-shadow a. -bild silhouette. -grenze, Halb- penumbra boundary. -kompensation, Halb- penumbral compensation. -mikroskop shadow microscope. -riss silhouette. -veränderung change in shade or shading values. -verhältnis (zwischen Drähten und freier Gitterfläche) cover ratio (say, 1/3), transparency. -wurf sound shadows. -zeiger shadowgraph (tuning indicator).
Schattierung shading, hatching, shade, tint.
Schattierungsverfahren variable-density recording method.
Schaubild diagram, graph; hookup; perspective view.
Schauer, Höhenstrahlen- bursts or shower of cosmic rays (traced by hodoscope).
Schauerstrahlen shower radiations.
Schaufel blade, bucket (of turbine), shovel, scoop, beater, vane. Lauf- runner blade, r. bucket (turbine.)
Schaufelabstand circular pitch (of turbine).
Schau-glas display glass; specimen g., sample g. -hörerschaft aural and visual audience, persons listening to and seeing simultaneously audio and video actions.
schaukeln oscillate, rock, reciprocate, move to and fro, swing.
Schau-linie graph, curve. -linse viewing lens. -loch peephole,

inspection h., bezel, observation port (for projectionist, to view the screen).
Schaum, chromatischer chromatic soft focus, c. bleeding. **Farb-** fringing (of film). **Glas-** glass gall, sandiver.
Schaum-abheber froth skimmer, drosser. **-gold** Dutch metal. **-gummi** sponge rubber. **-löffel** skimming spoon, skimmer. **-ton** Fuller's earth.
Schau-versuch lecture experiment, demonstration e. **-zeichen** indicator signal, visual s., optical s., telltale means.
scheckig dappled, spotted, mottled.
Scheibchen, Beugungs- diffraction disk. **Zerstreuungs-** circle of confusion, blur circle.
Scheibe disk, slice, pane, dial; pulley, wheel. **-, stroboskopische** stroboscope, stroboscopic pattern wheel. **Ablese-** dial (of direction-finder). **Abtast-** Nipkow scanning disk, exploring d. **Ausgleichschwung-** rotary stabilizer, roller with fly-wheel action. **Blenden-** auxiliary rotary shutter disk, stop d. with spiral slot cyclically co-operating with holes of quadruple scanning disk. **Daumen-** cam (with lobe). **Dreh-** turntable (part of m.p. studio equipment). **Exzenter-** cam. **Fall-** drop shutter, d. annunciator. **Finger-** dial switch, finger disk, f. wheel. **Funkbeschickungs-** compensator cam, zero-clearing c. **Kettenrollen-** chain sheave. **Klinken-** ratchet disk, notched d. **Kombinator-** combiner disk, c. wheel. **Korrektur-** compensating cam (of d.f.) **Kurbel-** crank disk. **Leer-** idler wheel. **Linsen-** lens disk (scanner). **Loch-** Nipkow scanner disk. **Matt-** ground glass, frosted g. (plate or pane). **Mehrfachspiralloch-** multi-spiral disk (for scanning). **Nipkow-** N. scanning disk. **Nummern-** dial switch, number plate, dial p. **Peil-** bearing plate. **Riemen-** belt pulley. **Rollen-** chain sheave. **Sektoren-** spinning disk (in flicker test). **Sperr-** ratchet wheel. **Spiralloch-** Nipkow scanning disk, spiral d. **Stufen-** cone pulley, tapered p., stepped p. **Teil-** index disk, i. plate. **Übersetzer-** combiner disk, c. wheel. **Unterlag-** washer; supporting disk. **Vielkant-** polyhydral mirror (of scanner). **Wähler-** dial type selector switch. **Wechselblenden-** auxiliary rotary disk with spiral slot. **Zwischenlag-** washer.
Scheibenanode, Loch- apertured disk anode, ring a.
scheibenformig disk-shaped, discoid.
Scheiben-isolator disk insulator, dish i. **-linse, Loch-** aperture disk lens. **-strahl** disk ray.
scheidbar separable, analyzable.
Scheide-anstalt refinery. **-linie** boundary line, separating l.
scheiden separate, isolate, analyze, decompose; pick, sort, sever, divide.
Scheide-verfahren refining method, separation m. **-wand** partition, separator (wall), diaphragm, septum, baffle (ldspk.). **-wasser** nitric acid, aqua regia.
Schein lustre, light, shine, glow.
scheinbares logarithm. Dämpfungsdekrement equivalent logarithmic decrement.
Schein-dämpfung apparent attenuation. **-echo** pseudo echo.
Scheinergrad Scheiner degree, d. of S. scale.
Schein-leistungszähler apparent power meter (in volt-ampere units). **-leitwert** admittance. **-verbrauchszähler** apparent power meter, trivector. **-werfer, Linsen-** spotlight. **-werfer, Facettenspiegel-** segmented or facetted searchlight reflector.
Scheinwiderstand apparent resistance, impedance, impedor (part having i.).

Blindkomponente d. - (es) reactive component of impedance (either inductive or capacitive). Wirkkomponente d. - (es) active, watt or dissipative component of impedance.
Scheit, Richt- rule, straight edge. Rühr- paddle, rake, stirrer.
Scheitel vertex, summit, crown, top, apex. Brucken- bridge gap. Glas- lens vertex.
Scheitel-brechwert, Grund- primary vertex refraction. -linie vertical line, zenithal l. -punkt vertex, zenith, apex.
scheitelrecht vertical, perpendicular, plumb.
Scheitel-spannungs-Voltmesser peak voltmeter, crest v. -tiefe depth of vertex, vertical d. -wert peak value, crest v., maximal v., amplitude. -winkel vertex angle, opposite a., vertical a.
Schelle clip, clamp, shackle. Erd- ground clamp.
Schellenbaum bell tree, crescent.
Schellhammer riveting snap, s. hammer.
Schema schema, scheme, diagram, blank, form, model, pattern. Schalt- diagram of connections, circuit d., hookup. Term- term diagram. Wirk- actual or practical work diagram.
Schenkel leg, limb, foot, shank, side, side piece, branch. Achs- (axle) journal. Aussen- outer leg, o. limb. Magnet- limb or leg of a magnet. Mittel- center leg.
Schenkel-drossel, Drei- three-legged reactor or choke. -rohr bent tube, b. pipe, V-tube, elbow tube, e. pipe.
schenklig legged or limbed.
Scherenspreizer lazy tongs.
Scher-festigkeit shearing strength. -schwingungen shear vibrations.
Scherung, Bild- shearing of image. Kurven- correction (of induction curve, etc.)
Scherungs-modulus shear modulus, torsion m., rigidity m. -winkel angle of shear.
Scheue, Licht- photophobia.
Schicht layer, stratum, bed, coat, film, skin, lamina. -auf Schicht emulsions placed together so as to face each other.
Schicht, atmosphärische atmospheric layer, sphere (troposphere, tropopause; stratosphere or isothermal region; ionosphere or Heaviside layer or Kennedy-Heaviside layer, Appleton l.). -, gerasterte mosaic (photosensitized) coat or screen. Bremsvermögen der photographischen — stopping power of photographic emulsion. Feinkorn- fine-grain emulsion. Filmbild- film emulsion, f. coat. Grenz- interface, boundary layer. Hohl- air space. K.-H.- Kennedy-Heaviside layer. Oberflächen- surface layer, s. barrier. Sperr- barrier layer, stopping l., blocking l. (of cell).
schichten arrange in layers, stratify, pile (up), stack.
Schichtengitter, Schichtgitter stratified lattice, layer l.
Schicht-Kathode cathode bearing, layer or coat (of oxide, light-sensitive substance, etc.) -körper stratified structure, s. body. -länge path length (of bands). -seite emulsion or sensitized side or face (of a film). -textur stratified texture, sheet intergrowth.
Schichtung stratification, arranging in layers, piling or stacking (up).
Schiebekontakt adjustable slide, slide contact, cursor.
schieben push, shift, thrust, shove, crab (of airplane).
Schieber slide, slider, slide bar; slide valve, damper (regulating draft).
Schieber, Belichtungs- exposure lid, e. shutter. Kontakt- adjustable slider, slide contact, cursor.

Rechen- slide rule.
Schieber-kassette roller blind darkslide. -korper slide valve or gate.
Schiebe-rohr sliding means to tune an oscillator (by C or L variation).
Schieber-rahmen lantern slide carrier. -verschluss, Fall- drop shutter. -widerstand slide rheostat.
Schiebkraft thrusting power, thrust, shear.
Schiebungs-fläche slip plane (cryst.) -modulus torsion modulus, shear m., rigidity m.
Schieds-richter arbitrator, referee, umpire, arbiter. -spruch arbitration, arbitrament, award.
schief oblique, sloped, slanty, askew, skew, crooked, warped, unsymmetric, skew-symmetric (of curves).
schief-achsig oblique axial. -symmetrische negative Paare skew-symmetric "negative" pairs.
Schiefe dissymmetric condition; skew.
Schiefeinfall grazing incidence, oblique i.
Schiefer slate, shale, schist. -bruch slaty, scaly or laminar fracture.
schief-liegend inclined, sloping, obliquely positioned. -symmetrisch skew-symmetric (of curves). -winklig oblique-angled. -winkliges Parallelogramm rhomboid, oblique (angled) parallelogram.
Schielen (beim Goniometer) squinting (minimum signal points are displaced an angle other than 180 degrees).
Schielwinkel angle of strabism, a. of squint (eye).
Schiene rail, strip, bar; rim, tire (of a wheel); splint, splice. Andruck- pressure pad or guide. Druck- pressure shoe (film feed). Kontakt- contact bar. Sammel- bus bar, collector b. Strom- third rail.
Schienenstoss rail joint.
schiessen shoot, dart, emit, dash, bombard, set off.
Schiffchen, Porzellan- porcelain boat. Verbrennungs- combustion boat.
Schild label, sign, plate, shell, shield.
Schiller lustre, shine, iridescence, play of colors, opalescence.
schillerfarbig iridescent with metallic color or lustre display.
Schirm screen, shield (of dissector tube), shade, umbrella, visor (of a cap), can (of a screened or shielded coil, etc.)
Schirm, kapazitiver electrostatic screen or shield. -magnetischer magnetic shield, m. screen, can (of a coil). -, neutraler neutral screen. -, ununterbrochener, zusammenhängender continuous photosensitized screen or photocathode (in Farnsworth dissector and superemitron).
Schirm, Bild- projection screen. Blenden- gobo (of sound absorbing material). Durchsichts- (der Speicherröhre) transparent screen (of storage type television tube). Leucht- fluorescent screen, phosphor s., luminescent s., luminous s., target. Mosaik-, Überfahren d. sweeping out of mosaic (with beam). Raster- mosaic screen. Schall- baffle. Streifen- rod screen. Ton- sound screen. Verstärkungs- intensifying screen (X-ray work).
Schirm-aussteuerung full utilization of screen (up to border, in cathode-ray tube, etc.). -breite screen width.
schirmende Spule, selbst- astatic coil, self-shielded c.
Schirm-gitter-Richtverstärker screen-grid tube with plate-current rectification. -gitterröhre screen-grid tube, shield-grid t., shielded-plate t.

-luminophore phosphors, fluorescent substances (for cathode-ray screens, etc.). **-membrane** screened diaphram (of microphone). **-rand** margin of screen, s. border or edge. **-substanzen, Leucht-** phosphors, fluorescent substances (for cathode-ray screens, etc.) **-träger** carrier, foundation or support for screen phosphor material (in CR tube).

Schlacke slag, cinder, scoria, dross, clinker.

schlacken slag, form slag, scorify.

Schlacken-bildung slag or ash formation, scorification. **-wolle** rock wool, slag w., mineral w.

schlackig slaggy, drossy, clinkery, scoriaceous.

schlaff slack, loose, flabby, flaccid, soft, exhausted, tired.

schlaffes Trumm slack (or driven) side or end of a belt.

Schlag stroke, blow, percussion, impact, knock, shock, kick, beat, pounding; wobble (of phonograph disk); click, thump or chatter (of a key or contact); lay (of a cable). **Kalt-** cold heading die. **Platten-** disk wobble (of phonograph). **Tast-** key click or thump, chatter.

Schlag-beanspruchung shock, blow or impact load or stress. **-biegeprobe** shock, blow or impact bending test.

Schlägel mallet, maul, beater, pestle, hammer, drumstick.

Schlägerschaltung dog or plunger movement.

Schlag-feder impact spring. **-figur** percussion figure. **-fläche** impact surface, striking s. **-gold** leaf gold. **-instrument** percussion instrument (mus.). **-länge** lay, length of twist (of cable). **-lot** hard solder, copper-zinc brazing mixture. **-lot, Messing-** brass solder. **-probe** impact, blow or percussion test. **-probe, Dauer-** continuous impact or shock test (piece). **-raum** spark-gap chamber or space. **-ton** strike note (of bell, followed by hum n.) **-weite** spark, flashover or striking distance (of a gap.) **-werk** breaking machine, impact tester (of Charpy); signal device, striking mechanism. **-werk, Dauer-** continuous impact testing machine. **-zeug** percussion instrument (mus.)

Schlamm mud, sludge, pulp, slime, slush, sediment.

Schlämmkreide precipitated chalk, prepared c.; whiting.

Schlange, Kühl- cooling coil, spiral condenser pipe or tube. **Rohr-** spiral tube, coil.

Schlangen-holz snake wood, letter w. **-kühler** spiral condenser, coil c. **-kurve** serpentine.

Schlauch hose, tube, tubing, pipe (of flexible material). **Sprech-** speaking tube.

schlechte Abstimmung mistuning, off-tuned position.

schleichen sneak, creep.

Schleier veil, haze, slight cloudiness, mist, fog, turbidity (in liquids); reflected, double or ghost image causing veiling (in images or pictures).

Schleier, dichroitischer dichroic fog, silver f., red f. **-, zweifarbiger** dichromatic fog. **Entfernungs-** distance fog. **Grund-** inherent fog, ground f. (of emulsion, resulting in background noise). **Kalk-** chalk fog. **Luft-** air fogging (chemical action of atmosphere).

schleierig foggy, fogged, hazy, misty.

Schleier-korn fog grain (of film). **-schwärzung** fog density. **-wert** fog value.

Schleife loop, noose, knot, slide, chute. **Bahn-** orbital loop. **Erd-** earth circuit, ground c. **Film-** film loop. **Leitungs-** loop, metallic circuit. **Oszillographen-** oscillograph loop, o. vibrator.

schleifen grind, sharpen, polish, cut, slip, slide; loop (a line).

Schleifen,Kristall- lapping, precision grinding of quartz crystal plates. **Plan-** surface grinding.

Schleifen-bildner loop setter. **-fänger** feed and take-up sprocket mechanism. **-flug ("U")** turn, loop. **-messung,Erddraht-** loop test. **-oszillograph** loop, string or vibrator oscillograph. **-schaltung** circuit looped back and forth.

Schleifer sliding wiper, copper wire interrupter (sliding over nickel disk to break up c.w. and render them audible).

Scheif-feder wiper, slide spring. **-kunst,optische** optical grinding artifice or art. **-maschine** grinding machine, grinder; abrasion or hardness tester. **-mittel** abradant, abrasive. **-pulver** grinding powder, polishing p. **-rille** groove, furrow. **-spalt** slit stop (opt.) **-wirkung** abrasive action, attrition.

Schleisse splint, splinter.

schlenkern swing, sling, fling.

Schlepp-antenne trailing (wire) antenna (of airplane). **-kontakt** trailing contact, make-before-break c., shorting c.

Schleuder centrifuge, centrifugal machine; sling. **Elektronen-** electron gun.

Schleuder-fuss gun press. **-guss** centrifugal casting.

Schleuse,Licht- light valve. **Objekt-** object airlock chamber, air-lock (of electron microscope). **Vakuum-** airlock (of electron microscope).

schlicht smooth, even, plane.

schlichten smooth, plane, dress, adjust, arbitrate or settle (a dispute).

Schlichtgleitsitz plain sliding fit.

Schlieren streaks, schlieren, striae (of glass), cords (when heavy). **-methode** striae method. **-zone** dead zone, skip z., shadow z., shadow.

schliessend,gut- tight-fitting, tightly or hermetically sealed.

Schliessung und Unterbrechung make and break (of a circuit).

Schliff grinding, sharpening, polish, cut (of a gem); ground section, g. specimen, g. slide, micro-section surface (as of metals, for etching); ground-in joint.

Schliff,Atz- ground section (for etching metal). **Fett-** greased joint. **Kern-** male part of ground-in joint. **Kugel-** ground spherical ball-and-socket joint. **Normal-** standard (interchangeable) ground joint. **Metall-** polished section, micro-s.; metal filings.

Schliff-kolben,-stopfen ground(-in) stopper. **-verbindung,Glas-** ground glass joint.

Schlinge loop, noose, tie. **Bahn-** orbital loop (of electrons).

Schlingerstativ cardan-suspended tripod.

Schlitten slide, sliding carriage, slide rail, chariot (in Hughes apparatus). **Führungs-** slide carriage. **Kreuz-** compound slide.

Schlitz slit, slot, fissure, cleft, incision, split. **Aufzeichnungs-** recording slit, r. aperture (of contact or of optical type). **Ausgangs-** exit slit (in spectrograph). **Ton-** reproducer aperture, r. slit.

Schlitz-blende slit stop. **-brenner** batswing burner. **-effekt** apertural effect, slit e. **-kopiermaschine** slit-type printer, contact p. (for film). **-lampe** slit lamp (of microscope). **-magnetron,Zwei-** two-split magnetron, two-segment m. **-verschluss** focal plane shutter, slit-type rotary disk shutter.

Schluckbeiwert absorptivity.

schluckend,licht- light absorbent, optically a. or absorptive.

Schlucker,Schall- sound absorbent

material, s. deadening m., s. insulation. **Wellen-** wave trap (of absorptive or impedance type).
Schluckerzahl,Schall- sound absorption coefficient, acoustic or sound absorptivity (e.g., in sabin units).
Schluck-grad sound absorptivity, s. coefficient. **-stoff** absorptive material, absorbent, imbibent.
Schluckung,Schall- sound absorptivity, acoustic a.,s. absorption.
Schlupf slip, slippage.
schlupffreier Antrieb (s.a. **zwangsläufig**) non-slip drive, geared d. (free from backlash).
Schlupfkupplung slip clutch, s. coupling.
Schlupfrigkeit lubricity, slipperiness, oiliness.
Schlüpfung slip, slippage.
Schlüpfungs-messer slip meter. **-reibung** sliding friction, slip.
Schluss short-circuit. **-, magnetischer** magnetic shunt. **Eisen-** path closed or passing through iron (for magnetic flux), magnetic or iron path, magnetic shunt, keeper (of a permanent magnet). **Elektroden-** electrode short-circuit. **Erd-** earth connection; (arcing) ground. **Wasser-** water seal, trap.
Schlussblende final diaphragm (opt.)
Schlüssel key, switch (for make and break); code. **Mithör-** listening key. **Mutter-** nut wrench. **Nach-** master key, pass k. **Sprech-** speaking key. **Strom-** key switch, make and break. **Uberwachungs-** monitoring key. **Verwürfelungs-** jumble code.
Schlüssel-bolzen clutch key. **-gruppen** (Q) code groups.
Schlüsselung coding, cryptographing.
Schluss-fehler,Erd- ground fault. **-gebühr** final fee. **-kondensator** terminating condenser or capacitor, short-circuiting c. **-löschvorrichtung,Erd-** ground fault neutralizer. **-prüfer,Strom-** continuity tester. **-prüfer,Windungs-** (inter-) turn short-circuit tester, continuity t.
Schmal-film substandard film, narrow f. **-kante** narrow edge, n. side.
Schmelz enamel, glaze; fusion, melting bath, melt.
schmelzbar fusible, meltable, non-refractory. **-, schwer-** difficultly meltable.
Schmelze melting, fusion; smelting, smeltery, fused mass, melt; melting heat, blow run; fluid solution (metallography); batch (of glass).
Schmelz-fluss fused mass, melt. **-führung** smelting practice or schedule, conduct of heat, run. **-glasur** enamel. **-kirsche** igniting pellet. **-mittel** flux. **-schweissung,Gas-** gas torch autogenous welding.
Schmelzung heat (run). **Durch-** burnout (causing open circuit), blowout (of fuse). **Uber-** superfusion, enameling.
Schmelzungslinie fusion curve.
Schmerz-grenze, -schwelle threshold of feeling (of ear).
schmiedbar malleable, forgeable, capable of being wrought.
Schmiede,Gesenk- drop forge.
schmiegsam flexible, pliant.
Schmiegungsebene osculating plane.
Schmier-loch oil hole, oil run. **-nute** oil groove, oilway.
Schminke make-up.
Schmoren scorching, freezing or melting together (of contacts); mike stew, frying (in radio).
Schmutzgeräusch dirt noise.
schmutzig dirty, soiled, contaminated, unrefined, polluted, filthy, smutty.
Schnabel bill, beak, nozzle, nose.
Schnapp-schalter tumbler switch, quick-action s., snap s. **-schuss** snapshot. **-stativ** self-locking tripod.
Schnarre,Prüf- test buzzer.
Schnarrwecker buzzer (alarm).
Schnauze snout, mouth, nose, nozzle,

spout. **Ausguss-** pouring lip, p. nozzle, p. mouth.
Schnecke worm, endless screw, volute, spire. **Eintrag-** feed worm. **Ruckführ-** return screw conveyor, r. worm.
Schneckenfeder coil spring, coiled s.
schneckenförmig spiral, helical, worm-shaped.
Schnecken-gangfassung helical lens mount. **-gewinde** helix. **-rad** worm wheel.
Schneidbrenner cutting torch.
Schneide knife edge, edge (of a blade). **-haltigkeit** ability to preserve keenness, cutting power or edge.
Schneiden, Film- editing of film. **Gewinde-** thread cutting, tapping. **-ankerrelais** knife-edge relay. **-aufhängung** knife-edge suspension. **-lagerung** knife-edge suspension or supporting.
Schneider, Rollglas- wheel-type glass cutter, g. cutting w.
Schneide-stichel cutter, cutting stylus, c. tool, engraver, cutting head (phon. record). **-ton** edge tone, flue stop (of f. pipe).
Schneid-narbe cutting scar, burry condition. **-winkel** cutting angle (known as dig-in or drag a., when other than 90 degrees) (in record making).
Schneise radio beacon course, equisignal track, sector or corridor. **Anflug-, Einflug-** approach track, a. sector, a. corridor.
Schnell-anlauf high-speed start, rapid starting, high acceleration. **-binder** quick-setting cement.
Schnelle quickness, velocity, speed, fastness, celerity, swiftness. **Schall-** volume current (ac.). **-empfänger, Schall-** velocity microphone.
schnellflüssig easily fusible, readily meltable.
Schnell-kopie quick print, hurry p. **-kraft** elasticity. **-methode** rapid or high-speed method, quick procedure. **-photographie** instantaneous photograph, snapshot. **-relais** high-speed relay. **-schalter, Zeit-** high-speed circuit breaker. **-verfahren (Film)** high-speed film processing.
schnelltrocknend quick-drying, siccative.
Schnitt cut, cutting, incision, section, slice; cutting and editing (m.p.)
Schnitt, Frequenz- crossover frequency (in woofer-tweeter combination at which equal power is delivered to both ldspks.) **Haupt-, erster (I.)** primary plane, tangential p., meridional p. (opt.) **Haupt-, zweiter (II.)** secondary plane, sagittal p. (opt.) **Kegel-** conic section. **Kristall -** see **Kristallschnitt.** **Strahlen-** jet contraction, contractio venae.
Schnittbild-entfernungsmesser split-field range-finder. **-versetzung, Ausschalten der** halving adjustment (of coincidence range-finder).
Schnitt-brenner slit burner, batswing b. **-fanger** section lifter (micr.) **-fläche** sectional plane, section, surface of a cut or section.
schnittig, wind- streamlined.
Schnitt-kopie editorially cut print, edited p. **-punkt** crossover (beam), point of intersection. **-schraube** grub or headless screw. **-weite** distance between back lens and image, intercept length. **-weite, konjugierte** conjugate intercept, c. distance. **-zeichnung** cross-sectional drawing, illustration, view, picture or figure (Fig.)
schnitzeln, schnitzen cut, carve, chip, whittle.
Schnitzer knife, whittle, paring tool.
Schnorchel (cf **Luftmast**) air intake, breathing mast, air funnel, a. mast (on modern undersea craft, for submerged Diesel operation).
Schnur string, cord, chord, line,

twine, tape, band. **-rolle** grooved roller, pulley.
Schnürspur squeeze track, matted t. (sound film). **-blende** mask to make squeeze track.
Schnürung, Ein- squeezing (film), compression, bindup, constriction.
Schoopieren spraying with Schoop gun.
Schoop'sche Spritzmethode Schoop spray (gun) method, aerograph m.
schöpfen draw (liquid, etc.), scoop, dip; create.
Schornsteinwirkung stack effect, chimney e,
Schott bulkhead, partition, separation.
Schottky Effekt Schottky effect, shot e., autoelectronic e.
Schottky'sche Gerade Schottky line.
schraffieren shade, hatch, line.
schräg oblique, sloping, slanting, inclined, beveled, chamfered, skew, askance, canted.
Schräge slope, slant, bevel, obliquity, diagonal, cant.
Schräglage, Kraftlinien- slope of flux lines.
schrägstellbar inclinable, tiltable.
Schräg-stellen tilting (of electrode in a magnetron). **-strahlen** skew rays.
Schramme optical scratch, shadow s. (film), slash, scar.
Schrammenrauschen scratch noise.
Schranke, Licht- light barrier. **Potential-** potential barrier.
schränken cross, put across; set (a saw).
Schraubdeckel screw cover, s. cap, s. lid.
Schraube, Achter- octet ring. **Einstell-** set screw, adjusting s., leveling s. **Feinstell-** micrometer screw, vernier. **Flügel-** thumb screw, wing nut. **Klemm-** clamp screw, binding post, terminal. **Knebelgriff-** tommy screw. **Lappen-** thumb screw. **Links-** left-handed screw. **Luft-, versetzbare** aircrew or propeller of variable pitch (feathered). **Maden-** headless screw, grub s. **Mutter-** bolt and nut. **Rechts-** right-handed screw (thread). **Regulier-** adjusting screw, set s. **Schnitt-** headless screw, grub s. **Senk-** countersunk screw. **Spann-** turnbuckle screw, (magnetic) chuck (in machine tool). **Spiegel-** mirror screw (scanner). **Stell-** set screw, adjusting s. **Stift-** tap bolt.
Schrauben-bohrer screw tap, auger, twist drill. **-feder** coil spring, helical s. **-kopf, versenkter** countersunk screwhead. **-lehre** micrometric calipers. **-linie** helix, helical line. **-nut** helical groove, h. slot, worm. **-quetschhahn** screw pinchcock. **-zahngetriebe** helical gear, spiral g., worm.
Schraub-lehre micrometer, micrometric gage. **-lehre, Uhr-** dial micrometer. **-sockel** screw base (of lamp). **-stöpsel** screw plug.
Schrecksekunde time elapsing between shock on molecular system and reaction.
Schreibempfang visual reception, recorder r.
schreiben write, record (graphically), type, delineate, trace, reproduce (an image or picture).
schreibende Stimmgabel tuning-fork chronoscope.
Schreiber, Bild- picture reproducer, p. delineator, p. recreator, p. receiver, video-signal receiver, televisor, viewing tube. **Einpunkt-** single-point recorder. **Farb-** inker, printer. **Fern-** teletype, teletypewriter, telegraph printer; t. operator. **Heber-** syphon recorder. **Hell-** Hell type recorder. **Kurven-** oscillograph, oscilloscope (when no graphic record is made), curve tracer. **Mehrpunkt-** multipoint recorder. **Pegel-** level recorder. **Relief-** embosser. **Schall-** sound recording stylus. **Tintenstrahl-** ink vapor jet recorder (in picture telegraphy).
Schreib-feder stylus, recording pen,

style. -fehler clerical error, typographic e. -fläche record sheet or surface. -instrument graphic instrument, recording i. -kopf recorder head. -röhrchen pen, syphon (of a recorder). -röhre,Bild- picture receiver tube, viewing t., electronic reproducing device. -stift cutter, stylus, cutting tool, engraving t. -thermometer thermograph, recording thermometer.

Schreien,Über- overmodulation.

Schreiner,Modell- wood pattern maker.

Schrift,Amplituden- variable-area or variable-width recording or track. Beschwerde- plaint, appeal papers. Dichte- variable-density recording or track. Gegentakt- push-pull track or record. Nach- picture recreation, postscript. Normal- standard track. Quer-,Seiten- lateral track, Berliner t., horizontal cut (gramophone). Sprossen- variable-density record or track. Tiefen- hill-and-dale (Edison) track or recording, vertical sound groove. Tonfolge- sound track or record (on film, disk, etc.) Verteidigungs- defense, plea. Vieldoppelzacken- multiple double-edged variable-width track. Vielzacken- multilateral sound track. Zacken- variable-area track, v.-width t. Zweizacken- bilateral track.

Schrift-boden recording base. -material literature references (e.g., of prior art), bibliography, documentation, file wrapper (of patent case), records. -metall type metal. -mutter type mold, matrix. -satz written pleadings, memorial, brief. -stück bill, brief, papers, document. -tum literature (sources), bibliography, documentation, source material, references (from magazines, patents, etc.) -tumverzeichnis bibliographic list.

Schritt,Erfindungs- object of invention, step of i. Hebe-,Höhen- vertical step (in automatic telephony). Wählerhebe- vertical stepping of selector.

Schrittschaltwähler step-by-step selector.

schrittverkürzte Wicklung short-pitch winding.

Schrittweite interval (between spectral lines).

Schrodinger-de Broglie Materiewelle de Broglie elementary particle (electron, proton), electron wave, phase w.

schroten granulate, bruise, rough-grind, crush coarsely.

Schrott-effekt (small) shot effect, Schottky e., auto-electronic e. -geräusch shot noise (due to Schottky effect).

schrumpfen shrink, contract, shrivel.

Schrumpf-mass amount of shrinkage or contraction. -spannung contraction or shrinkage strain or tension.

Schrumpfungsausgleich shrinkage compensation.

Schub shove, push, thrust, throw, shear. -kurbel crank. -lehre sliding gage or calipers. -spannung shear stress. -stange thrust rod, push(er) r. -vektor thrust vector. -weg finite distance.

Schuh,Gleit- pressure pad, p. guide. Kabel- cable eye, c. socket, connector lug, sweated thimble.

Schuhmacherwachs cobbler's wax.

Schuldigerklärung verdict of guilty, conviction.

Schuppe scale, flake.

schuppig,fein- fine-scaled.

Schurre chute.

Schüttel-herd bumping table. -mikrophon granule microphone.

schütteln shake, agitate, churn, toss.

Schuttelresonanz vibration resonance.

Schüttwinkel angle of repose, (sometimes) angle of friction.

Schutz screen, shield, protection, guard, preserving or conserving

means, safeguard.
Schütz relay, contactor.
Schutz, Berührungs- protection against electric shock hazard. **Schall-** sound proofing, s. insulation. **Übersteuerungs-** overload limiter.
Schutz-einrichtung, Fehl- fault-clearing device. **-funkenstrecke** spill gap. **-gerät, Knall-** crack silencer, pistol shot s., noise killer. **-gitter** shield (in shield-grid thyratron). **-gitterrohre** screen-grid tube, shield-g.t.; shielded plate t. **-hülle** protective cover, p. coat, p. sheath. **-kappe, Licht-** light hood. **-korb** arcing ring, guard r. (of insulator). **-marke** trade-mark. **-masse** resist (used in printing, etc.) **-mittel, Lichthof-** anti-halation means or substance. **-netz, Anoden-** screen grid, plate shield. **-rechte** patent rights, trade-mark r., etc. (granted by or registered in Patent Office). **-wulst, Sprüh-** anti-corona collar or tore, guard-ring.
schwach, rauch- smokeless (of powder).
schwach ausgesteuerter Tonstreifen low-modulation track.
schwächen weaken, enfeeble, impair, attenuate, dilute, dim, absorb (light), mute (sound).
Schwächer, Fünfstufen- five-step weakener (phot.) **Stufen-** step-type weakener (spectr., phot.)
schwachsichtig weak-sighted, amblyopic.
Schwächung, Licht- light absorption, l. dimming, l. drop, l. loss, cutting down or reduction of l. intensity, l. diffusion.
Schwächungs-glied attenuator. **-periode** fading period. **-widerstand** gain controller, g. regulator, potentiometer.
Schwanenhals goose neck.
schwanken fluctuate, oscillate, vary, shake, rock, sway or swing (of an antenna).
Schwankung, Frequenz- frequency flutter, f. drift, lilt (when slow); swing or deviation of f. (above and below assigned resting or center f., in f-m). **Lautstärke-** fading, fade out, volume or intensity fluctuation. **Regel-** hunting (in making adjustments or in regulating). **Ton-** pulsation of reproduced sound intensity, flutter (known as wowows when up to 6 cycles per second, as flutter when 6-30, as gargles when 30-200, and as whiskers when over 200 cps.) **Volumen-** volume flutter.
Schwankungs-frequenz (cf Tonschwankung) frequency of flutter. **-messung, Ton-** flutter measuring instrument. **-schwächer, Frequenz-** discriminator (converts frequency drift into d.c. potential). **-theorem** fluctuation theorem.
Schwanz, Blind- arrangement making resistance alternately zero and infinite, in dipole feeders; stub. **Film-** trailer or run-out (piece of blank film attached to film end). **Wellen-** wave tail (in facsimile).
Schwanzbanden tail bands (in afterglow).
schwanzloses Flugzeug tailless airplane.
Schwanzstück, Kurven- toe region of film characteristic (or of gamma curve of emulsion).
Schwärme, Molekül- swarms of molecules.
schwarz(e) Temperatur black-body temperatur, cavity t.
Schwarz-druck black printing, p. in b. **-pegel** black level. **-sender** unlicensed transmitter, illicit t., "underground" t. station. **-strahl** black (-body) radiation, cavity r., Planckian r. **-temperatur** black-body temperature, cavity t.
Schwärzung blackening, density. **Kopier-** printing density. **Ruhe-** unmodulated density, no-sound d. (in variable-density recording). **Schleier-** fog density.
Schwärzungs-abstufung density graduation. **-bereich** density range (γ) (say, from .3 to 3.0). **-dichte** density (of film). **-einheit** unit

of (film) density. **-hof** halo. **-kurve** characteristic film curve, H & D (Hurter & Driffield) curve. **-messer** densitometer. **-messer, Kreiskeil-** circular wedge densitometer. **-umfang** density scale, range or latitude of d. variations. **-verfahren** variable-density method.

Schwarzweissbildsendung black and white facsimile picture or phototelegraphic transmission.

Schwarzwert black level value (telev.)

Schwebefähigkeit buoyancy, floating power, suspension p.

schwebende Stimme voix céleste.

Schwebe-stoff,-teilchen suspended particles, s. substance, s. matter.

Schwebung beat, interference action (opt., ac.) **Helligkeits-** flutter of brightness.

Schwebungs-amplitude surging amplitude, beat a. **-empfang** beat reception, heterodyne r. **-frequenz,Null-** zero beat frequency. **-null** zero beat. **-periode** beat cycle. **-summer** audio oscillator, beat-frequency o. **-ton** beat note.

schweifen curve, chamfer, tail.

Schweige-trichter cone of silence (marker). **-zone** skip distance, shadow region.

Schweiss,Glas- glass gall.

Schweiss-backe welding die. **-blase** arc crater. **-druck** welding upset. **-düse** welding tip. **-eisen** weld iron.

schweissen,stumpf- jump weld, butt-w. **-,übereinander-, überlappt-** lap-weld, scarf-w.

Schweiss-fuge welding seam, w. line, w. joint, shut. **-hitze** welding heat. **-löcher** pores (of wood). **-naht** shut, weld, weld seam. **-raupe,breite** spread bead. **-rissigkeit** "chinking" or fissuration due to welding. **-stelle** welded or welding area, shut. **-tropfen** welding dribble, w. beads. **-umformer** welding converter. **-umspanner** welding transformer.

Schweissung,Kehlnaht- fillet welding. **Naht-** seam welding. **Punkt-** spot welding.

Schwelle,Empfindlichkeits-(Schall) threshold of audibility or acoustic perception. **Potential-** potential barrier, p. threshold. **Potenz- (Kerndurchdringung)** nuclear penetration function. **Schmerz-** threshold of feeling. **Sichtbarkeits-** threshold of vision or visibility.

Schwellen-frequenz threshold frequency, critical f., photoelectric threshold. **-reiz** threshold or liminal (stimulation) value. **-stärke,Schall-** threshold sound intensity (of audibility or of feeling). **-wert,Farb-** liminal value or threshold of intensity of a color. **-wert der Platte** exposure factor of plate (phot.)

Schweller,Roll- crescendo pedal (of organ).

schwellig liminal, concerning threshold. **unter-** subliminal.

Schwell-kasten swell box. **-pedal** swell pedal (organ). **-werk** swell organ.

Schwelung low-temperature carbonization, slow burning, smoldering. **Dampf-** destructive distillation with steam.

Schwemm-stein porous brick (from clay and gravel). **-verfahren** flotation method.

schwenken swing, turn around, pivot, fulcrum, swivel, rotate, revolve, slew, tilt (telev. camera).

Schwenkungswinkel angle of traverse (gunnery).

Schwerbetrieb heavy-duty operation or service.

schwerdurchlässig difficultly permeable, of low permeability or transmittance, highly opaque.

Schwere heaviness, weight, gravity. **-feld** gravitational field. **-messer**

barometer, gradiometer (measures gradient of earth gravity field). -pendel gravity pendulum.
Schwererde heavy earth, baryta.
schwer(es)Elektron heavy electron, barytron, dynatron, mesotron, penetron, x particle. -(er) Wasserstoff deuterium (with mass of 2), tritium (with mass of 3).
schwerflüchtig difficultly volatile, d. volatilizable.
Schwermetallaktivator heavy-metal activator or phosphorogen (in phosphors).
Schwerpunkt mass center, c. of gravity. Leucht- luminous "center of gravity," photometric c., p. centroid.
Schwerpunktsystem center of gravity system, c. of mass system.
schwerschmelzbar difficultly fusible, refractory.
Schwimmen lateral flicker (in overspeedy panorama work).
Schwimmer float, pontoon, swimmer.
Schwimm-kraft buoyancy. -methode flotation method. -regel, Ampère's Ampere's right-hand rule. -spannung blurring potential, p. causing sway of picture; floating, spurious or stray p.
schwinden shrink, contract, fade.
Schwing-amplitude der Membran amplitude of diaphragm excursion. -anker oscillating armature. -audion oscillating detector, autodyne, endodyne. -audionwellenmesser autodyne wave-meter or frequency m.
Schwinge link.
schwingen vibrate, oscillate, swing, rock, pivot, reciprocate (to and fro), surge (back and forth).
Schwingen, plötzliches spilling over (of an amplifier), sudden undesired oscillating (attended with fringe howl), trigging of a tube.
schwingen, durch-(cf Ballaströhre) swing through (operating condition in which wave is not interrupted or cut off). mit- co-vibrate, oscillate in resonance, sympathy or unison, experience sympathetic vibration.
schwingender Draht swinging of filament (in magnetron).
Schwingentladung oscillatory discharge, oscillating d.
Schwinger oscillator, vibrator. -, offener open oscillator. Frei- free radiator (loudspeaker). Kipp- time-base oscillator. Sperr- blocking oscillator.
schwingfähig oscillatory, vibratory, oscillable, vibrant, vibratile.
Schwing-feder vibrator(y) spring, reed (of a chopper). -kennlinie oscillating characteristic, c. of an oscillatory system, resonance characteristic. -kontaktgleichrichter vibrating-reed rectifier. -kristall crystal oscillator. -löcher dead spots (trouble in direct-coupled antenna). -relais trigger relay.
Schwingröhre, fremderregte master-excited oscillator valve (connected in an independent drive circuit). -, selbsterregte self-excited or self-oscillatory valve (connected in direct drive circuit). Kraft- power oscillator.
Schwing-spiegel oscillatory mirror, vibrating m. -spiegelabtaster vibrating mirror scanner (of Mihaly's Telehor). -spule moving coil, signal-current c., voice c., speech c. (of loudsp.)
Schwingung vibration, oscillation, undulation, swing, wave, high-frequency a.c., "oscar"(m.p. studio slang). -durch Nebenkopplung stray or spurious oscillation. -, erzwungene forced oscillation, constrained o. -, freie free oscillation. -, gedampfte damped oscillation, decadent o. -, ungedämpfte undamped oscillation, persistent o., sustained o. -, zusammengesetzte complex (harmonic) wave.
Schwingung, Biegungs- flexural

vibration, lateral v. **Bild-** video or picture impulse or signal. **Dicken-** thickness vibration (cryst.) **Drehfeld-** rotating-field oscillation. **Drehungs-, Drillungs-** torsional vibration. **Gitter-** lattice vibration (cryst.) **Grund-** fundamental wave, first-harmonic w. **Intermittenz-** relaxation, sawtooth or ratchet wave or oscillation, time-base o. **Ionen-** ionic oscillations, fringing effect, waving in the breeze (across image). **Kipp-** see **Intermittenzschwingung**, **Längs-** longitudinal vibration or oscillation. **Laufzeit-** electron oscillation (of magnetron). **Magnetron-** magnetron oscillation (of three kinds: electronic, negative-resistance and rotating-field oscillations). **Mit-** resonant vibration, sympathetic v., co-vibration (acting in unison). **Nebenkopplungs-** spurious oscillation. **Quer-** transverse vibration, shear v. **Resonanz-** resonant vibration, sympathetic v., co-vibration (acting in unison). **Relaxations-** sawtooth, relaxation or ratchet oscillation. **Scher-** shear vibration. **Ubertragungs-** signal oscillation, s. wave. **Umfangs-** circumferential oscillation or vibration. **Unter-** subharmonic oscillation, subfrequency o. **Unterbrechungs,- Unterdrückungs-** quench oscillation (in super-regeneration). **Wärme-** heat vibration. **Zwangs-** forced oscillation, f. vibration.

Schwingungs-abreissen quenching, suppression or discontinuance of oscillations. **-abschnitt** "bit" of vibration (in sound recording investigation). **-amplitude, Membran-** excursion amplitude of diaphragm. **-anfachung** wave generation, wave excitation. **-aufzeichner** vibrograph (to record mechanical vibrations). **-ausschlag (einer Saite)** deflection, excursion or swing (of a chord). **-band, Rotations-** rotation vibration band. **-bauch** loop or anti-node of an oscillation or vibration, internode. **-bogen** amplitude of oscillation. **-dämpfung** vibration or shock absorption, damping (out) of oscillations or vibrations. **-dauer, Kipp-** relaxation period. **-einsatz, harter** hard start of oscillations. **-einsatz, weicher** gentle or smooth start of oscillations.

schwingungserzeugende Kraft vibromotive force.

Schwingungserzeuger, selbstsperrender self-blocking oscillator. **-, Vielfach-** multivibrator.

Schwingungserzeugung, selbständige self-oscillation, spontaneous or undesired oscillating, spilling over (attended with fringe howl and whistling in amplifier).

schwingungsfähig oscillable, vibratile, oscillatory, vibratory.

Schwingungs-festigkiet vibration strength. **-form** mode of vibration, form of v.

schwingungsfrei, eigen- aperiodic, deadbeat.

Schwingungs-knoten oscillation or vibration node or nodal point. **-kreis** oscillatory circuit, oscillation c., tuned c., tank c. **-neigung** tendency towards, or on the verge of, oscillating or spilling over (with fringe howl and whistle). **-schleife** loop, antinode or internode of stationary wave or oscillation. **-spektrum** vibrational spectrum. **-transformator (Tesla)** Tesla coil, T. transformer, oscillation transformer. **-versuch, Dreh-** oscillation torsion test. **-weite** amplitude of oscillation or vibration. **-zahl, Umlauf-** rotation frequency. **-zeit** time of vibration.

Schwund, Nah- fading or radio fade-out 60-100 miles around transmitter. **Phasen-** phase fading. **Pseudo-** fading due to swinging or swaying of antenna.

Schwund-ausgleich volume control, fading compensation. **-automatik** automatic volume control means (avc). **-regelspannung** (automatic) volume control potential, avc p. **-verminderungsantenne** anti-fading antenna.

Schwung vibration, oscillation, swing. **-bahn** sound take-off drum (m.p.) **-gewicht** pendulum. **-hebel** balance beam, rocker, rocking lever. **-kraft** centrifugal force, vibratory power, vibrating p. **-maschine** centrifugal whirler. **-masse, Öl-** rotary stabilizer (in film feed). **-massenrolle** rotary stabilizer, roller with fly-wheel effect (film feed), impedance wheel. **-moment** moment of rotation, moment of inertia; fly-wheel action. **-rad** fly-wheel, balance w. **-radkreis** parallel resonance circuit, fly-wheel c. **-scheibe, Ausgleich-** rotary stabilizer, roller with fly-wheel action (film feed), impedance wheel.

S.E. (**Siemens-Einheit**) Siemens (resistance) unit.

S.E. Faktor secondary-emission factor.

sechs-eckig hexagonal. **-fach** sixfold, sextuple.

Sechs-flach, **-flächner** hexagon, hexahedron. **-polröhre** hexode.

sechsseitig six-sided, hexagonal.

Sechstelkreis sextant.

sechs-wertig hexavalent, sexivalent. **-winklig** six-angled, hexangular, hexagonal.

seegrün marine green.

Seelicht marine phosphorescence.

Segerkegel Seger cone (for temperature determination), fusion c.

Segment, Wähler- bank of stationary contacts.

segmentförmig segmental.

Sehen, einäugiges monocular vision, non-stereoscopic v. (eye, opt.). **-, zweiaugiges** binocular vision, stereoscopic v. **Farben-** color perception, c. vision, **Netzhautzenter-** foveal vision.

Seher, Gegen- two-way television or video apparatus (as in video telephone.

Seh-feldblende field diaphragm, f. stop. **-funktion** visual function. **-kegel** angle of vision, cone of v. or view. **-lehre** optics. **-linie** line of sight, l. of vision, l. of collimation.

Sehne chord (geometry); fiber (metal), fillet.

Seh-nerv optic nerve. **-nervenapparat** organ of vision. **-purpur** visual purple, rhodopsin. **-reiz** optical stimulus, visual s. or excitation. **-schärfe** visual acuity, sharpness of vision, resolving power (of eye). **-schärfegrenzwinkel** angular resolving power, critical angle of visual acuity. **-schärfeneinteilung** focusing scale. **-schwäche, vorgetäuschte** simulated amblyopia. **-tiefe** depth of focus, depth of field. **-vermögen** resolving power (of eye), visual faculty, power of sight, p. of vision. **-weite** distance of vision, visual range; distance between scale and mirror (times 2 in telescopic measurements). **-winkel** visual angle.

Seiden-faden silk fiber, s. thread **-papier** tissue paper.

Seifen-bildung formation of soap, saponification. **-lamelle** plane soap film or membrane. **-leim** soap glue, s. paste, size (paper).

seigern liquate.

seignettesalzkristall Rochelle salt crystal.

Seiher strainer, filter. **Einsteck-** plug-in inlet strainer.

Seil rope, cable, chord, line. **Abspann-** guy rope, span r.

Seilrolle pulley, sheave.

Seite, Emulsions- face bearing emulsion, coated side (of film).

Seiten-abrutschanzeiger side slip indicator. **-abweichung** lateral deviation. **-ansatz** side tube, lateral arm, l. appendage. **-ansicht** lateral elevation, l. view, side e. **-arm** side tube, lateral arm, l. appendage (of a vessel); side branch. **-aufriss** lateral elevation, side e. **-bandubertragung, Ein-** single-side-band transmission. **--Zwei-** double-side-band transmission. **-bandunterdrückung** side-band suppression. **-bestimmung** sense-finding or sensing (determination of absolute direction or sense of direction); determination of side towards which airplane is off course. **-eck** lateral summit. **-elektroden** end electrodes, wing e. (of a magnetron). **-fehler** lateral deviation. **-fläche** lateral face, flat side, facet. **-frequenz** side frequency (one of sum or difference frequencies). **-führung (von Film)** lateral guidance, side guiding (of film strip in gate). **-gleitflug** side-slip (of airplane). **-kante** lateral edge. **-navigation, -ortung** directional navigation, d. avigation. **-riss** lateral elevation, side e. **-schalter, Peilungs-** sense-finding switch, d.f. sense s. **-schrift** lateral track, horizontal t. or recording, Berliner track (of gramophone). **-strahlungen** secondary radiations, lateral r., stray or spurious lobes (of space pattern). **-veränderungen** turns and banks (avigation). **-verhaltnis, Bild-** picture aspect ratio. **-wind** wind on the beam, side or cross w., w. blowing across course. **-windabtrift, Kurs mit Vorhaltewinkel bei** course heading into wind, crabbing (due to drift).

...seitig -sided, lateral,... angular.

seitlich lateral. **-(es) Abrutschen** side-slip (of airplane). **-(e) Führung** lateral guidance. **-(em) Minimum, Antenne mit** directive antenna. **-(e) sphärische Abweichung** transverse spherical aberration.

Sektor, Anflug-, Einflug- sector of approach, a. sector, a. track, corridor of a. **Raum-** solid sector.

Sektoren-membrane sectorial cone (loudspeaker). **-scheibe** spinning disk (in flicker test). **Verschluss** sector shutter.

Sektormembran (non-rigid) sector diaphragm or cone.

Sekundär-emissionsfaktor (SE) secondary-emission factor. **-emissionsvervielfacher** electron multiplier tube, e.m. phototube (with plurality of secondary-emission reflecting electrodes, known as dynodes in the e.m. photot.) **-struktur** secondary structure gridwork (cryst).

Sekunde, Sonnenzeit- solar second.

Sekundemeterkerze meter-candle-second.

Sekundenuhr time piece with seconds hand.

selbstabgleichend self-balancing.

selbständige Entladung spontaneous discharge, self-sustained d., unassisted. d. **-Funktion** self-consistent function. **-mitgenommene Kippschwingmethode** self-running controlled time-base method.

selbstangetrieben self-propelled.

Selbstanschluss-System automatic telephone system.

selbstanzeigender Peiler automatic or direct-reading direction-finder.

Selbstauslösung automatic release, a. tripping action.

selbstdichtend self-sealing, s. tightening, s. packing.

Selbst-einstellung self-adjustment, s. setting. **-energie** self-energy (of electron). **-entzundung** spontaneous combustion, self-ignition. **-erhitzung** self-heating, spontaneous h.

selbsterregender Sender self-oscillatory transmitter.

selbsterregt(e) Schwingröhre self-excited oscillator tube. **-(er) Trägerwelle, Superhetempfänger mit** homodyne receiver apparatus.

Selbsterregung self-oscillation, self-sustained o.; spontaneous (undesired) o., spilling over (with attendant fringe howl) of regenerative receiver.
Selbst-farbe self-color, solid c. **-gang** automatic feed, a. operation, a. action.
selbstfarbig entirely of the same tint, self-colored.
selbst-gehendes Erz self-fluxing ore. **-gleichrichtendes Röhrenvoltmeter.** self-rectifying tube voltmeter.
Selbst-kapazität self-capacitance, distributed c. of a circuit (due to proximity effects). **-konzentration** self-focusing, gas. f. **-laut** vowel. **-lautdeutlichkeit** vowel articulation.
selbstleuchtender Lichtmodulator light modulator of the glow-tube, neon lamp, sodium l. type, light relay with inherent light.
Selbst-leuchter self-luminous substance, fluorescent s., phosphor. **-leuchtverfahren** self-luminous or s. -emissive method (using electrons arising on object, in electron microscope). **-lötung** autogenic soldering.
selbst-rechnender Komparator computing recording comparator. **-registrierend** self-recording, automatic r. **-schirmende Spule** self-shielded coil, astatic c. **-schwingende Mischröhre** self-heterodyning mixer tube, autodyne.
Selbstschwingungsvorrichtung thermionic trigger device.
selbst-sperrender Schwingungserzeuger self-blocking oscillator. **-ständige Entladung** spontaneous discharge, self-sustained d., unassisted d. **-ständige Funktion** self-consistent function. **-ständige mitgenommene Kippschwingmethode** self-running controlled time-base method.
Selbststrahlungsverfahren (Elektronenmikroskop) self-emissive electron microscope or self-illuminating method (based on electrons arising on surface of object).
Selbsttönen ringing or squealing (of a tube).
selbstonendes Papier self-toning paper.
Selbst-überlagerung autodyne, self-heterodyne. **-umkehrung** self-reversal (of spectral lines). **-unterbrecher** trembler, self-interrupter, buzzer. **-zünder** pyrophorus lighter, self-igniting l., automatic l.
Selektion, Amplituden- amplitude selection, a. filtering.
Selektivkreis selective circuit, selector c.
Selen-chlorid selenium chloride. **-cyanid** selenocyanate. **-halogen** selenium halide.
selenhaltig containing selenium, seleniferous.
Selen-kupfer copper selenide. **-metall** metallic selenide. **-zelle** selenium (photo-conductive type of) cell, photronic c.
Seltenheit rarity, scarcity, infrequency.
Sende-anlage, Richt- beam, beacon or radio range station. **-bild** outgoing subject copy. **-kanal** transmission channel, signaling c.
Sender, gepeilter tuned-in beacon (whence bearings are taken). **-, modulationsgesteuerter** transmitter with carrier being varied with AF amplitude; AF modulated transmitter. **-, selbsterregender** self-oscillatory transmitter. **-, tonender** musical-spark transmitter, singing-s.t.
Sender und Empfanger, Kombinierter transceiver (radio), microtelephone, handset.
Sender, Ball- re-broadcast station, repeating s., station belonging to a network. **Bild-** video transmitter, facsimile t. **Diapositiv-** film transmitter. **Einflug-** boundary marker beacon. **Einstrahl-** beam transmitter, unidirectional t. **Fernseh-** television transmitter (comprises strictly both the aural and video signal transmitting means).

Gemeinschafts- chain broadcast station, network s. **Haupt-** main oscillator (in independent drive system). **Hilfs-** auxiliary transmitter, local oscillator (in heterodyne apparatus). **Kontroll-** monitor transmitter, check-up t.; service test oscillator. **Leitfunk-,Leitstrahl-** radio beacon, beam station, r. range. **Markierungs-** marker beacon, marker. **Maschinen-** alternator transmitter. **Mutter-** key station, master s. (in chain broadcasting). **Prüf-** service test oscillator. **Quarz-** (piezo-electric) crystal or quartz stabilized transmitter. **Richtungs-** beam station, beacon s. **Schwarz-** unlicensed transmitter, illicit t., "underground" t. **Sperrkipp-** blocking oscillator. **Steuer-** master oscillator, drive o. (working on main transmitter in independent drive). **Stimmgabel-** tuning-fork oscillator, vibrating-reed transmitter. **Stör-** interfering station, jamming s.; shadowing signal, spurious irregular shading signal (telev.). **Telegraphie-,tonmodulierter** AF or tone-modulated telegraphic signal transmitter. **Telephonie-, sprachgeschalteter** voice-modulated telephony (signal) transmitter, v.-m. radiophone. **Tochter-** repeater or affiliated station (related to key station, in chain broadcasting and in b. networks). **Ton-** modulated c.w. transmitter, audio or aural transmitter (of sound part of a telev. program). **Tonfunken-** singing or musical (quenched) spark transmitter (telegraph). **Zeichen-** signal transmitter, s. station. **Zwischen-** re-transmitter, repeater station.

Sender-abstand interference guard band, tolerance frequency. **-durchmodulierung,hohe** operation of transmitter station at a high modulation percentage.

sendereigner Ton note or pitch peculiar to a beacon station, code note; tuning n.

Senderohre,frequenzgesteuerte frequency-stabilized transmitter tube.

Sendung transmission or emission (of radio waves or signals), program.

sengen singe, scorch, parch, flame.

Senke crevass, dip (of resonance curve), trough, valley, depression, negative source.

senken depress, lower, sag, slope, droop (of curve), de-accentuate (a band).

Senkkörper sinker, bob.

Senkrechte vertical, normal, perpendicular, plumb (line). **-fällen** let fall a perpendicular. **-errichten** erect a perpendicular. **Mittel-** median perpendicular.

senkrechte Beleuchtung scaffold lighting, overhead l., top l.

Senkspindel specific gravity spindle, hydrometer.

Senk-rechtstellung vertical position, perpendicular p., plumb line p. **-schraube** countersunk screw.

Senkungswinkel depression angle.

Senkung sag, droop, depression, dip.

Sensitometer sensitometer (for comparing and grading film), densitometer.

Separatabdruck reprint, separate impression, special i.

Sequenzillusion sequence illusion.

Serien-apparat series camera. **-entwicklung** series expansion (math.)

serienmässige Herstellung quantity production, large-scale manufacture.

Serienresonanzkreis series-resonance circuit, acceptor c.

sicher,feuchtigkeits- damp-proof, moisture-p. **hochspannungs-** safe against high-voltage breakdown, possessing high puncture strength, proof against high-potential shock.

Sicherheit,Pfeifpunkt- singing margin, whistling m., stability m. (of radio tube). **Stand-** stability or steadiness (to shock or vibration).

Sicherheits-koeffizient safety factor.

-lampe safety lamp, s. light.

sicherstellende Vorkehrungen preventive measures, precautions, prevautionary steps.

Sicherung, Grob- spark gap or similar arrester. **Lage-** locking in position. **Luftleerspannungs-** vacuum arrester.

Sicht visibility, sight.

Sichtbarkeit, überschwellige superthreshold visibility.

Sichtbarkeits-messer visibility meter (e.g., Lukiesh-Taylor). **-schwelle** threshold of visibility, t. of vision.

sichtbarmachen render visible or visually perceptible, visualize, indicate visually or optically.

Sicht-messer visibility meter. **-peiler** visual or direct-reading direction-finder. **-weite** direct-sight range or distance, line of sight, optical distance. **-winkel** angle of sight. **-wirkung** sifter effect, sorting, classification. **-zielflug** visual homing flight, flight depending on visual indication or reading.

Sicke reinforcing seam, crease or pleat, r. fin (bent, pressed or otherwise fashioned in metal sheeting).

Sieb filter, network, sieve, screen, bolter, strainer. **Amplituden-** amplitude filter, a. discrimination selector.

siebeneckig heptagonal.

Siebenelektrodenröhre heptode.

siebenwertig heptavalent.

Siebgebilde selective system, filter means, network.

Siebkette ladder-type filter, band-pass f. or transmission network. **-, mehrgliedrige** multi-mesh filter, multi-section f. **-von grosser Lochweite** broad-band filter.

Sieblochbreite, Frequenz- transmission band, transmitted b., b.-width of filter, spacing between cut-off points.

Siebschale dish with perforated bottom.

Siebsteilheit sharpness of selective network.

Siebung screening, shielding, sifting, straining, filtering, sorting, bolting, classifying.

Siebwiderstand contact resistance (of barrier cell).

Siede-erleichterer boil adjuvant. **-kolben** boiling flask. **-kühlung, Kathode mit** vapor-cooled cathode. **-messer, Wasser-** hypsometer.

Sieke reinforcing seam, pleat, crease or fin (bent, pressed or otherwise fashioned in metal sheet).

Siemenseinheit (S.E.) Siemens unit (of resistance).

Signal, tönendes tonal signal, musical spark s.

Signal, Abschattierungsstör- spurious signal, shading s. (in iconoscope operation, known also as "shading," "black spot," "tilt and bend," etc.) **Bild-** video signal, picture s. **Drehlicht-** revolving beacon. **Entwarnungs-** all-clear signal. **Fernseh-** video signal, television s. **Freigabe-** clearance signal. **Gleichlauf-** synchronizing signal, sync impulse; tripping s. **Haupt-** inner marker signal, main entrance s. (airport). **Kenn-** tuning note (characteristic audio signal or note for tuning). **Misch-** signal spectrum (comprising video and sync impulses). **Not-** distress signal, SOS s., mayday (radio telephone). **Nutz-** intelligence signal, s. conveying information. **Stör-** shadowing signal, shadow s., spurious s. (iconoscope operation); jamming signal, interference s. **Synchronisier-** synchronizing signal, sync impulse, tripping s. **Vor-** caution signal, warning s., first or outer marker s. (airport). **Warn-** alarm signal, caution s., alert, raid alarm.

Signal-horn bugle. **-lampe** signal lamp, telltale l. **-lampentafel** signal lamp or annunciator board or panel. **-mittel** signaling means, communication m. **-pfeife** siren, whistle. **-platte** signal plate (of iconoscope). -überdeckung blanketing or swamping of signals. -verlauf signal spectrum, s. pattern. -wand wall of signals, curtain of s.

Silben-abschneidung clipping, obliteration or mutilation of syllables (in signal art). -frequenzgerät syllable vodas (of Bell Co.) -umkehrung inversion of syllables (in secret, garbled or scrambled telephony). -verständlichkeit syllable articulation, syllable intelligibility.

Silber-belegung silver coating, s. plating, s. film, silvering. -blech silver foil, (thin) silver plate. -glätte litharge. -glimmer common mica, muscovite. -halogen silver halide. -lot silver solder. -papier silver paper, silvered p.; tinfoil, tinfoiled p.

Sims molding, shelf, cornice.

Simultanleitung earthed phantom.

singende Bogenlampe singing arc lamp, Duddell arc.

Singulette singlets.

sinken sink, fall, decline, droop (of a curve).

Sinkstoff deposited substance, settled s., settlings, sediment.

Sinn, im bejahenden in the affirmative (sense). Ladungs- nature of charge, sign.

sinnähnlich substantially synonymous, in an analogous way or sense.

Sinn-bestimmung, Richtungs- sense-finding or sensing (determination of sense of direction or absolute direction). -bild symbol, emblem.

sinn-gemässe Anwendung analogous, rational, logical, corresponding or equivalent application or use. -getreu faithful (e.g., in translation).

Sinnverständlichkeit intelligibility.

sinn-verwandt synonymous. -voll intelligent, clever, ingenious, very suitable, logical. -widrig absurd, senseless, contrary to sense or reason, illogical.

Sinterkohle sintering coal, non-caking c.

sintern sinter, form s., concrete, frit, consolidate, vitrify (cer.)

Sinter-quarz siliceous sinter. -schlacke clinker.

Sinterung sintering, fritting, fusing, vitrification, baking, caking, agglomeration, consolidation.

Sinus, hyperbolischer hyperbolic sine (sinh).

sinusförmige Bewegung, rein- simple or plain harmonic motion, sinusoidal m.

Sinus-linie sine curve. -schwingung sinusoidal wave, sine w. -winkel sine of angle.

Sirene, Kondensator- condenser chopper or siren (produces pulsating current). Licht- light chopper, c. wheel. Zahn- tooth-wheel synchronizer, tone-w.

Sirutor Westector, rectifier of the copper or oxide type.

Sitz seat, fit, lodgment. Press- press-fit, force fit, driving f. Schlichtgleit- plain sliding fit. Treib- driving fit.

Skala, reine true scale (mus.) Entfernungs- distance scale. Frequenz- frequency spectrum, f. scale. Linear- slide-rule dial (with straight scale).

Skalen-intervall scale division, s. interval. -rohr scale tube (of a spectroscope).

Slip side slip (of airplane).

smaragdgrün Guignet's green.

Smaragdspat green feldspar, amazonite.

SMK (Sekunden/Meter/Kerze) meter-candle-second.

Snellius'sches Brechungsgesetz Snell's law of refraction.

Sockel base or cap (of lamp or tube); pedestal. Röhren- tube base, valve b.

socken crystallize out, contract (of metal).

Soffittenlampe overhead scoop, scaffold light or lamp.

Sog,Hohl- cavitation.
soggen crystallize out, contract (of metal).
Sohle sole, bottom, bed, floor, level, base.
solarisationsfreie Platte antihalation dry plate (phot.)
Sole brine, salt water.*
Solinglas crown glass.
Soll-kurs charted course, prescribed c. **-wert** nominal, rated, prescribed, assighed, normal, face or theoretical value.
Sonde probe, search electrode, test prod, sound, sonde (radio). **Abtast-** scanning aperture, hole of Farnsworth dissector tube. **Hitzdraht-** hot wire anemometer. **Klauenprüf-** claw-type test prod or probe. **Prüf-** test prod, probe, search electrode. **Radio-,für Wetterforschung** radio meteorograph, radio sounding, sonde or sondage devices (for weather recording).
Sondenbildfangröhre Farnsworth type of pickup tube, F. dissector t. (with a scanning aperture).
Sonder- special, separate, exclusive. **-abdruck** reprint, separate p.
sondern separate, sever, segregate, sort (out).
Sonne (Klieg) sun, sun arc, Klieg lights. **Gegen-** parhelion, mock sun.
Sonnen-ausbruch solar eruption. **-bahn** ecliptic (apparent path of sun). **-bestrahlung** insolation, irradiation by solar rays, exposure to sunlight. **-blende** sun shield, black screen. **-ferne** aphelion. **-fleckperiode** sunspot period, s. cycle. **-nahe** perihelion. **-schutz** sunshade. **-strahlungsmesser** solarimeter, pyroheliometer, actinometer (for solar flux density). **-tätigkeit** solar activity. **-wärmemesser** solarimeter, pyroheliometer. **-zeitsekunde** solar second.
sorbierender Stoff sorbent, sorptive.
sorbierter Stoff sorbed substance, sorbate.
sortieren sort, assort, size.
Sortierung,Phasenaus- phase focusing (in beam or drift tube such as Klystron).
sossig spotty, bleary (of a picture).
Sourdine mute.
Spähne,Eisen- iron turnings, borings, filings or chips.
Spalt crack, gap, split, rent, fissure, slit, interstice, aperture, lacuna.
Spalt,Aufzeichnungs- recording slit, r. aperture. **Stufen-** step slit (phot.) **Ton-** sound recording aperture or slit.
Spalt-ausleuchtung slit illumination. **-bild,optisches** optical slit image. **-bildgerät** photographic soundhead. **-blende** slit, apertured stop. **-breite** width of slit (of spectroscope, etc.), apertural width. **-brüchigkeit** cleavage brittleness. **-ebene** cleavage plane, c. face. **-effekt** aperture effect, apertural e. **-fläche** cleavage face, c. plane. **-flächenzeichnung** cleavage plane or crystal marking.
spaltfreie Beleuchtung apertureless or slitless illumination.
Spalt-körper cleavage substance, c. product. **-offnung** chink (in clarinet). **-optik** microscopic aperture-imaging optical means or "optic". **-prozess** fission process. **-rohr** collimator. **-riss** cleavage crack, c. fissure. **-streuung** slot leakage, s. stray. **-ton** slit tone, jet t.
Spaltung,Phasen- phase split. **Strahl-** beam splitting, split beam.
Spaltungs-kristall cleavage crystal. **-produkt** fission product, cleavage p. **-untergrund** fission background.
Span chip, splinter; shaving, boring, turning (mostly used in plural form).
spanabhebende Bearbeitbarkeit free-cutting machinability. **-Formgebung** shaping or fashioning by machine tool, with incidental removal of chips.

Spange clasp, buckle, clip, staybolt.

spannbar tensile, extensible, ductile, capable of being tensed or tensioned.

spannen stretch, strain, set, wind, tense, tension, tighten, span, clutch, grip.

Spann-kraft tensional or elastic force, tension, load, extensibility, expandibility, elasticity. **-rolle** tension roller, spring-loaded idler (m.p.). **-schraube** turnbuckle, turn screw, chuck (magnetic).

Spannung (cf Potential) potential, voltage, tension, pressure (in electricity); strained state or condition; strain (of molecules).

Spannung, abgeflachte flat-topped potential (wave). **anlaufen gegen eine-** buck against a potential. **an - liegender Draht** negative wire. **elastische-** elastic strain. **magnetische-** magnetic potential (line integral of field intensity). **mitlaufende-** follower potential. **sägeformige-** sawtooth potential, ratchet p. **-verriegelnde-** cutoff biasing potential.

Spannung, Ablenk-, deflection potential, sweep p. **Abreiss-** stopping potential, cut-off p. **Anregungs-** exciting potential, stimulating p., pre-ionizing p., p. causing electron to be raised to a higher level out of its atomic bond. **Ausgleich-** transient potential; compensating p. **Beschleunigungs-** gun potential, accelerator p. **Bild-, Untergehen der** swamping of video or picture signal potential (in noise level). **Bildwellen-** video (signal) wave potential. **Bogenzünd-** ignition potential, arc-striking p. **Brems-** negative anode potential (in electron-oscillation tube). **Brenn-** constant glow potential (is slightly lower than striking p.), normal running p. (of photo-glow tube, etc.) **Bruch-** ultimate stress, breaking s. **Brumm-** hum potential, ripple p. **D-** Dee voltage (of cyclotron). **Druckeigen-** residual compressive stress. **Durchschlag-** breakdown, rupture, disruptive, puncture or flash-over potential (el. engg.). **Eigen-** residual stress, inherent or internal tension or strain. **Einsatz-(cf Zündspg.)** sparking-over or flash-over potential; breakdown p. (in grid glow tube), starting p. **Empfangs-** signal potential, incoming p. **Feder-** spring tension, elasticity. **Film-** film tension, f. pull. **Gipfel-** peak potential, crest p. **Glas-** strain or internal tension of glass. **Gleich-, wellige** ripple potential, pulsating p. **Glimm-** glow potential, breakdown p. (indicates incipient glow discharge in grid-glow tube); anode p. (at which glow discharge begins in photo-cell), stopping potential (lies slightly below glow p. in photo-glow or grid-glow tube, and is the value to which p. must be dropped to stop glow discharge once it has started). **Grenzflächen-** interfacial tension. **Haft-** adhesive stress. **Ionisierungs-** ionizing potential, ionization p., p. sufficient to detach electron from its atomic bond. **Kipp-, Rücklauf der** flyback of sweep potential. **Kreis-** "circle" voltage. **Lösch-** stopping potential (at which glow discharge is stopped; s. p. is lower than glow p. in photo-cell and glow tube), extinction p., critical p. (at which t. goes out); cut-off p. (of thyratron). **Nutz-** useful potential, signal p. **Oberflächen-** surface tension. **Peil-** directional signal potential (picked up by directional antenna, to be distinguished from auxiliary a.p.) **Regel-** automatic volume-control (avc) potential; dynamic-range or contrast regulating p., generally. control p. or regulator p. **Resonanz-** resonance potential, radiation p., p. resulting in transition of electron from ground state to next orbit, with incidental emission of resonance spectrum. **Richt-** output voltage of rectifier, d.c. v. of rectifier. **Ruhe-** bias or steady potential, no-signal p., repose or quiescent p., Q point p. **Sägezahn-** sawtooth potential, ratchet p. **Saug-** anode potential, driving p., positive (d.c.) p. (of photo-cell), saturation p. (if all electrons are drawn to anode). **Schalt-** variable bias potential (in silent tuning and automatic volume control). **Schrumpf-** contraction or shrinkage strain or tension. **Schub-** shear stress. **Schwimm-** floating, spurious or stray potential, blurring p., p. causing sway of picture, in telev. **Schwundregel-** automatic volume control (avc) potential. **Sperr-** cut-off potential, quench p., stopping p., biasing p. **Stoss-** impulse potential, shock p. **Streck-** yield stress. **Teil-** partial potential, componental p. **Tonwellen-** audio(signal) wave potential. **Trübungs-** potential causing quadrantal error (QE). **Über-** overvoltage, excess-voltage, overvolting. **Umlauf-** magnetomotive force (mmf) (being the line integral of magnetizing force). **Verkettungs-** interlinked voltage. **Versorgungs-** supply (source) potential, voltage s. **Verzögerungs-** delay voltage, delayed-action v. (in avc). **Vor-** biasing potential (impressed on a grid, magnet, etc.); prestress (of metals); inherent bias tension (of a spring) to restore, say, to normally tensed or tensioned condition, anode, drive or positive p. (of photo-cell); normal or initial tension (in materials). **Vor-, verriegelnde** cut-off biasing poential. **Vorglimm-** sustaining potential (in pre-glow range, slightly below critical discharge p.) **Wiederkehr-** recovery voltage. **Zeichen-** signal potential. **Zünd-** firing potential, striking p. (of ignitron, thyratron, etc.); p. at which in glow-tube discharge changes from assisted to self-sustained form. **Zusatz-** booster potential, additional p.

spannungführender Draht live wire, charged w., hot w.

Spannungs-änderung change of strain (in metal). **-aussteuerung, Anoden-** plate voltage excursion. **-Dehnungsdiagramm** stress-strain diagram, stress-deformation d. **-differenz, Zündlösch-** striking-extinction potential difference. **-eisen** iron in voltage circuit (of a meter). **-entlastung** stress relief, anti-fatigue or anti-vibration means (in transmission lines). **-erhöher** voltage step-up means (e.g., transformer), booster. **-festigkeit** puncture strength, dielectric s.

spannungsfrei Glühen strain relief anneal or thermal treatment. **-(es) Glas** strain-free glass.

Spannungs-glättung smoothing device (filtering out ripples), potential stabilizer. **-knoten** voltage node, potential n. **-kreis** potential circuit (of a meter). **-kurve, Herz-** electro-cardiogram.

spannungslos free from tension or strain, untensed; dead (absence of electrical potential).

Spannungs-messer, Oberflächen- surface tensiometer. **-messer, Röhren-** vacuum tube voltmeter. **-modulation** velocity modulation. **-netz** space-charge grid.

spannungsoptisch photo-elastic.

Spannungs-prüfer, Oberflächen-, nach Abreissmethode adhesion-balance type of surface tensiometer. **-punkt** terminal, tap. **-reihe** contact potential series, electromotive s., electro-chemical s. **-reihe, Thermo-** thermo-electric series. **-relais** undervoltage relay, low-v.r., no-v.r. **-resonanzkreis** series-resonance circuit, voltage r.c. **-rücklauf,Kipp-** flyback of sweep potential, return of time-base p. to zero. **-schutz, Berührungs-** protection against electric shock hazard. **-sicherheit** breakdown strength, puncture proofness. **-sicherung,Luftleer-** vacuum arrester. **-spitzenanzeiger** crest-voltmeter, peak-voltage indicator. **-spitzennivellierer** voltage peak limiter. **-sprung** voltage step, v. leap, abrupt change of v., discontinuity. **-spule** voltage coil (in two-wattmeter method). **-stoss** impulse voltage, potential i. or pulse. **-strom** current flowing in potential circuit (of a meter). **-teiler** voltage or potential divider, potentiometer. **-teiler,Glimm-** glow-tube potentiometer, g.-t. voltage regulator. **-tensor** stress tensor. **-verlust** drop of potential. **-wechsel** stress alternation, s. variation, s. cycle. **-zeiger** voltmeter; voltage vector.

Spannvorrichtung jig, chuck, holding, gripping or clamping device.

Sparren rafter, spar (of airplane).

Sparschaltung (Nestel) circuit organization designed to economize plate current.

sparstoffarme Legierung alloy low in scarce or critical materials.

Spartransformator auto transformer.

Speckstein soapstone, steatite, talc.

Speichelkammer moisture chamber (of soldering iron).

Speicher accumulator, means to store up electrons (or energy in general). **Stromstoss-** impulse storer, digit storing register. **Wärme-** heat accumulator, regenerator.

Speicher-platte storage pickup. **-röhre** signal or charge storing or storage tube (telev.) **-röhre, Bild-** signal or charge storage tube, normal iconoscope. **-wandlerröhre,Bild-** super iconoscope. **-wirkung** signal storage, charge accumulation.

Speise-apparat feed apparatus. **-hahn** feed cock. **-leitung** (aerial) feeder lead, downlead, energy-feeder cable or wire, transmission line. **-rohr** feed pipe, supply pipe or tube.

Speisung supply, s. source, feeding, excitation, energization.

spektral zerlegtes Licht spectroscopically dispersed or separated light.

spektralanalytisch spectroscopic, spectrometric.

Spektral-apparat spectroscopic apparatus, spectroscope. **-bereich** spectral range, s. region. **-darstellung** representation by spectra. **-linie** spectral line, spectrum l. **-linienfaltung** "faltung" of spectral lines. **-photometer** spectrophotometer. **-tafel** spectral chart.

Spektrograph,Geschwindigkeits- velocity spectrograph, v. analyzer. **Massen-** mass spectrograph, m. spectrometer. **Vakuumgitter-** vacuum grating spectrograph. **Zweikristall-** double-crystal spectrograph.

Spektroskop,Luftplatten- Fabry-Pérot interferometer, étalon (when not variable).

Spektrum,Elektronensprung- electron transition (or jump) spectrum. **Feinstruktur-** fine structure spectrum, micro-s. s. **Flammen-** flame spectrum. **Funk-** see Wellenabgrenzung. **Grosstruktur-** macro-structure spectrum. **Magnetorotations-** magnetic rotation spectrum. **Streifen-** band spectrum. **Übersichts-** general spectrum. **Wellen-(cf Wellenabgrenzung)** spectrum of electromagnetic waves; radio wave s.

Sperr-bereich suppression band or range, attenuation b. exclusion b. **-blende** blocking light stop, limiting s. **-dämpfung** stop band attenuation.

Sperre (cf Sperrer) wave trap, stopper, rejector; lock, seal, baffle. **Band-**

band rejection filter, elimination f., exclusion f. **Rückkopplungs-** reaction suppressor, feedback s., "Vodas". **Spiegelfrequenz-** image-frequency stopper. **Stör-** noise gate.

sperren shut, shut off, block, stop, bar, space (apart), guard (a position), stopper, exclude, preclude, lock, cut-off or choke (a tube).

sperrend, selbst- self-locking, automatically l.

Sperrer (cf. **Sperre**) trap, suppressor, excluder; lock, block. **Echo-** echo killer, e. suppressor.

Sperr-feder click spring. **-filter** rejection filter, r. circuit, stopper c., suppressor f. **-flüssigkeit** sealing liquid. **-gleichrichter** barrier-film rectifier, blocking r., electronic r., metal r., cuprox r. **-glied** suppressor. **-greifer** pilot pin. **-hahn** stopcock. **-haken** pawl, latch, catch, trip, dog, click, detent, ratchet. **-hebel** pawl, latch, arresting lever, lock l. **-holz** plywood. **-kegel** pawl, detent, click. **-kette** low-pass filter. **-kipposzillator, -kippsender** blocking oscillator. **-klinke** pawl, latch, catch, trip, dog, click, detent, ratchet. **-kondensator** stopping condenser, insulating c., blocking c. **-kreis** rejector circuit, parallel-resonance c., wave trap c. **-kreis, HF-** low-pass filter or selective circuit. **-kreis, NF-** high-pass filter or selective circuit. **-kreiskopplung** parallel resonance coupling. **-membrane** stop diaphragm. **-rad** ratchet wheel, cog w. **-relais** locking relay, guard r. **-richtung** high-resistance direction, backward d. (of a rectifier). **-scheibe** ratchet wheel. **-schicht** barrier layer, blocking l., stopping l. **-schichtzelle** barrier-type cell, photo-voltaic c., photronic c., rectifier c. (operating with a blocking layer). **-schwinger** blocking oscillator. **-spannung** biasing potential (impressed upon grid), cut-off p., quench p. (in super-regeneration). **-strom** backward current, inverse c. (of rectifier). **-topf** wave trap.

Sperrung, Röhren- cut-off of a tube, quenching of discharge. **Strahl-** beam cut-off, b. eclipsing, b. gating (telev. tube, by blanking pulse).

Sperr-ventil stop valve, check v. **-wirkung** (cf **sperren**) brake action, retarding a., delay a., lagging, valve a. **-zeit, Teil-** partial restoring time (in echo suppression). **-zelle** barrier-type cell (working with blocking layer).

Spezialfilm hoher Steilheit high-contrast emulsion film.

spezifisch(e) Ladung specific charge, charge-mass ratio. **-(er) Widerstand** resistivity.

spezifizieren specify, itemize, particularize.

sphärisch überkorrigiert spherically overcorrected. **-unterkorrigiert** spherically undercorrected.

sphäroidaler Zustand spheroidal state.

Sphärokristall spherical crystal, sphero-crystal.

Spiegel mirror, reflector, speculum, polished or reflecting surface, any deposited layer or coat (inside a tube, etc.), silvering; surface or level (of a liquid, etc.)

Spiegel, Augen- ophthalmoscope, skiascope. **Fang-** collecting mirror. **Geräusch-** noise level. **Getter-** getter patch, g. film. **Hohl-** concave mirror, c. reflector. **Kipp-** oscillating mirror. **Kugel-** spherical mirror, s. reflector. **Lamellen-** strip-like mirror (in screw scanner). **Parabol-** parabolic reflector. **Schwing-** vibratory mirror, oscillating m. **Stör-** noise level.

Spiegel-belag mirror coating, reflecting film, silvering (of a mirror). **-bild** speculum image, mirror i.;

flare ghosts or spots (camera), double or reflected (ghost) image (in telev.).

spiegel-bildlich homologous, mirror-picture condition, specular. **-bildlich(es) Verhältnis** inverse relationship. **-blank** highly polished.

Spiegel-bogenlampe reflector arc-lamp. **-feld** revertive-signal panel, check-back position indicator (giving readings of remote instruments, in telemetering work). **-frequenz** image-frequency (in radio). **-frequenzsperre** image-frequency stopper or rejector. **-glas** plate glass. **-gleichheit** mirror symmetry.

spiegelig specular, mirror-like.

Spiegel-kranz row, array, rim or set of mirrors (in scanner), drum scanner. **-kreuzlibelle** reflector-type cross-level (panoramic telescope). **-metall** speculum metal.

spiegeln reflect, shine, glitter, be reflected, glare (from glass surfaces).

spiegelnde Oberfläche specular surface. **-Reflektion** specular reflection, mirror-like r.

Spiegelneigungsmesser Abney level (of reflecting type, to measure vertical angles).

Spiegel-rad mirror drum (e.g. of Weiller). **-rad,Zeilen-** line scanning mirror wheel (of Scophony). **-reflexion** specular reflection, regular r. **-schraube** mirror screw (scanner). **-symmetrie** mirror symmetry, specular s.

Spiegelung reflection, specular effect, picture mirroring; replica (of curves), homologous condition or relation. **Dreh-** combined rotation and reflection. **Rück-** specular reflection, regular r. **Viel-** multiple of image.

Spiegel-welle reflected wave. **-wirkung** cut-off, reflecting or mirror effect (of electron lens when electrons are unable to cross barrier).

Spiegler reflector.

Spiel play, playing; working, operation, action (of mechanical parts), cycle of action; backlash, shake, play or lost motion (of wheels, gears, etc.); deflection, throw or excursion (of meter or instrument needle); set (of things).

Spiel,Flanken- backlash (of screw-thread). **Glocken-** carillon (set of bells).

Spiel-art touch (mus. instrument); variety. **-aufbau,Ionisierungs-** Townsend structure or buildup.

Spieler,Platten- disk-type phonograph or gramophone. **Rück-** playback, dubbing device.

Spiel-haus,Licht- motion-picture theater or playhouse, movie. **-leiter** director of play. **-raum** play, latitude, margin, range, scope, freedom of action, elbow room, clearance, backlash. **-raum,Belichtungs-** latitude of exposure. **-saiten** fingerboard strings (zither). **-zeugkino** home movie.

Spiesskant diamond.

Spin,Einheits- unit spin. **Kern-** nucleus spin.

Spinbahn-aufspaltung spin orbit splitting. **-kupplung** spin orbit coupling.

Spindel spindle, pivot, axle, arbor, shaft, axis of revolution.

spindelförmig spindle-shaped, fusiform.

Spindelführung screw control (of recording cutter).

Spindeln hydrometer test.

Spindeltrieb worm drive, w. gear.

Spin-drehimpuls spin angular momentum.

Spinelle spinel (cryst.)

Spin-glieder spin terms. **-impulskopplung** spin orbit coupling. **-momentdichte** spin momentum density. **-multipletten** spin multiplets.

Spinne,Lautsprecher- spider of loudspeaker (inside or outside, to keep voice or signal coil centered).

Spin-vektor spinor (two-dimensional vector). **-verdopplung** spin doubling. **-verteilung** spin distribution (math.)

Spiralantenne (flat) spiral aerial or antenna, extensible s. antenna, s. loop.

Spirale spiral, helix, coil, twisted structure.

Spirallochscheibe Nipkow scanning disk with helical array of apertures, spiral d. scanner. **Mehrfach-** multi-spiral scanning disk. **Vierfach-** quadruple scanning disk.

Spiral-steigung pitch or lead of a spire or spiral. **-tauschung** spiral illusion.

Spitze point, peak, crest, tip, top, apex, vertex, summit. **Prüf-** penetrator (of hardness tester). **Resonanz-** resonance peak or crest. **Stecker-** tip of telephone plug (fitted into tip jack or pup j.)

Spitzen-anzeiger,Spannungs- crest or peak voltage indicator, c. voltmeter. **-aufhängung** point suspension, pivot s. **-elektrode** needle electrode, point e. **-entladung** point discharge, needle (-gap)d.

spitzengelagert journaled on points.

Spitzen-glas reticulated glass. **-lagerung** jeweled bearing, pivot jewel, point suspension, p. support. **-leistung** record (achievement, performance, or accomplishment), peak output. **-nivellierer, Spannungs-** voltage peak limiter. **-zähler** needle counter. **-zirkel** compasses, dividers.

spitzer Morsepunkt clipped dot.

spitzig pointed, acute, sharp, tapering.

Spitzlichter bright lights, highlights (in a picture); tangential lighting.

spitzwinklig acute-angled.

Splint sap, sapwood, pin, peg, key, splint, cotter, split pin. **-holz** sapwood, sap.

Splitter,Kristall- crystal chip, c. grain.

Sprach-aufnahme speech, voice, vocal or dialog recording. **-beeinflussung** voice control, modulation by v. **-deutlichkeit** articulation, intelligibility (of vowels and consonants, etc.)

Sprache speech, voice, articulate sound, language; code.

sprachgeschalteter Telephoniesender voice-modulated telephony (signal) transmitter, v.-m. radiophone.

Sprachlaute speech sounds, articulate s.

sprachmoduliert speech-modulated, voice-m.

Sprach-untersuchung speech study, s. analysis. **-verständlichkeit** articulation, intelligibility of speech or articulate sound (ascertained, e.g., by logatoms). **-verstärkungsanlage** (public) address system. **-wiedergabe** voice reproduction; demodulation of speech.

Spratzen flicker, sputter or spitting (of a cathode).

Sprechaufnahme speech, voice, vocal or dialog recording.

sprechende Bogenlampe speaking arc, Simon arc, Duddell a.

Sprecher,synthetischer voder. **Fernseh-** television telephone, video t.

Sprechfaktor,Fern- telephone influence factor. **-frequenz** voice frequency, speech f. **-hörer** microtelephone, telephone handset, transceiver. **-kapsel** condenser microphone. **-kopf** soundbox, recording soundhead. **-kopf,Abnehmer-** sound pickup, reproducing head. **-leistung** response of an electroacoustic system. **-schalter** speaking key. **-schlauch** speaking tube. **-schlüssel** speaking key. **-strom** voice, speech, sound, telephone or signal current. **-stromkreis** circuit carrying voice or AF signal currents (in loudspeaker). **-transformator** speech transformer, AF t. **-verkehr, Fernseh-** video telephone traffic, television telephone communication.

Spreize spreader, spacer, strut, outrigger.

Spreizenkamera, Spreizkamera strut camera, extension c.
sprengen explode, blow up, blast, burst; sprinkle (with water).
sprenkeln sprinkle, speckle, mottle, spot.
Spriegel hoop.
springen burst, break, crack, fracture; spring, jump, leap.
Spring-federwage spring balance. **-kraft** springiness, elastic force, elasticity, power of recoil or rebound, resiliency. **-punkt** critical point (where a break or discontinuity occurs).
Spritze, Elektronen- electron gun. **Licht-** recorder lamp with constricted neon arc (e.g., Ewest "light gun," pointolite, etc.)
spritzen squirt, spurt, spout, sputter, spray, exude, inject (syringe).
Spritz-entladung initial discharge caused by rapid surge of ions, needle-gap d. **-guss** (pressure) die-cast. **-pistole** spray gun, s. pistol (of Schoop), aerograph.
spröde brittle, short. **-machen** embrittle.
Sprödigkeit, Kalt- cold brittleness.
Spross, Sprosse shoot, sprout, germ; rung (of a ladder).
Sprossen-rolle sprocket drum. **-schnurspur, Gleichtakt-** variable-density single squeeze track. **-spur, Gleichtakt-** single variable-density track. **-trommel** sprocket drum or wheel.
Sprossung germination.
sprühen spray, spit, scintillate, scatter, produce corona (on wires).
Sprüh-festigkeit spray proofness. **-ionen** spray ions. **-schutzwulst** anti-corona collar, guard ring. **-verlust** corona form of discharge, brush d. loss.
Sprung (cf **Sprungentfernung**) crack, chink, fissure, flaw, fault; jump, bounce, leap; break, discontinuity (of a curve), transition (in spectra and quanta); hop (of waves). **Absorptions-** absorption discontinuity. **Barkhausen-** Barkhausen effect (abrupt changes of magnetization). **Bild-** picture repetition frequency, p. cycle; break or shift of vision. **Phasen-** phase angle shift. **Quantum-** quantum transition, q. leap, q. jump. **Spannungs-** voltage step, v. leap, abrupt change or discontinuity of v.
Sprung-entfernung skip zone, s. distance (between beginning and end of wavehop). **-faktor, Zeilen-** interlace factor or ratio.
sprunghaft irregular, non-sequential, sudden, by leaps and bounds.
Sprung-höhe pedestal level or height, amplitude of return or flyback (telev.). **-methode, Zeilen-** interlaced, interleaved or intermeshed scanning method (with odd or even line interlace). **-punkt** (cf. **Springpunkt**) transition point. **-spektrum, Elektronen-** electron transition or jump spectrum. **-stelle** point of sudden irregularity, unsteadiness, change, or discontinuity. **-wahrscheinlichkeit** transition probability, leap p.
sprungweise (cf **ruckweise**) by steps or stages, by leaps and bounds, intermittently, discontinuously, non-sequentially.
Sprung-winkel, Aus- angle of reflection. **-zeit** transit time, transtion t. (of electrons).
Spule, angezapfte tapped coil. **-, angezapfte, in der Mitte** center-tapped coil, mid-tap c. **-, flächenhafte** laminar or areal coil, non-filamentary c. **-, gepanzerte, wenig streuende** iron-core deflection yoke unit (with low spot and pattern distortion). **-, gleichstromvormagnetisierte** d.c. controlled, saturable coil (of loudspeaker). **-, kleine** bobbin. **-, körperlose** structure with entire mass of vibratory system contained in oil. **-, selbstschirmende** self-shielded coil, astatic c.
Spule, Ablenk- deflection coil, deflector c., d. yoke. **Abwickel-** feed spool, f. reel. **Achter-**

figure-8 coil. Auflauf- take-up reel. Aufsteck- plug-in coil. Auslöse- trip coil. Beruhigungs- filter choke, smoothing coil. Betätigungs- working coil, tripping coil, differential c. Binokular- binocular coil (form of closed-field c.) Dämpfungs- damping coil, amortisseur c. Dreh- moving coil, rotating c., rotor c. Dreh-, mit einem Lager unipivotal moving coil. Eisenblätterkern- laminated iron-core coil. Eisenstaubkern- iron-dust-core coil, Ferrocart core c. Feld- field coil, magnetizing c. Felderreger- field exciting coil (in electro-dynamic loudspeaker). Filmvorrats- film magazine roll, magazine. Honigwaben- honeycomb coil. Induktanz-, eisengeschlossene closed-core coil. Induktions-, variable variometer. Käfig- canned coil, shielded c. Kondensor- condensor coil (electron microscope). Kopplungs- coupler coil. Kopplungs-, veränderliche vario-coupler. Korbboden- spider-web coil, basket-type c. Kreis- tuning coil. Litzen- litzwire coil. Luft- air-cored coil. Luftdrahtverlängerungs- aerial loading coil, lengthening c. Magnet- solenoid, field coil. Massekern- compressed iron-core coil, dust-c.c., Ferrocart c.c. Mehrlagen- multilayer coil, banked c., pile c., honey-comb c. with banked winding. Objektiv- objective coil. Projektions- projection coil. Pupin- loading coil, Pupin c. (for reactance balance). Quer- leak coil. Ring- toroidal coil. Rückkopplungs- tickler coil, retroactive c. Sammel- focusing coil. Saug- smoothing coil. Schwing- moving, speech, voice or signal-current coil. Spannungs- voltage coil (in wattmeter). Steck- plug-in coil, detachable c.; demountable reel (film). Striktions- focusing coil. Strom- current coil (in two-wattmeter method). Such- search coil, probe c., rotating c., exploring c. (of d.f.), flip c. (in magnetic field tests). Tauch- moving coil (of loudspeaker), telescoping c. Teil- fractional coil, component or subdivision of a coil. Variometer-, drehbare rotor of variometer. Variometer-, feste stator of variometer. Verlängerungs- serial loading coil, lengthening c. Vorrats- magazine, m. roll (m.p.) Waben- honeycomb coil, lattice-wound c., duolateral c. Zeilen- line scan coil, l. sweep c.

spulen wind, reel, coil up.

Spulen-abzweig coil tap. -achse bob of spool, core of s. (m.p.) -anzapfung coil tap. -arm magazine or spool arm or support. -durchflutung magnetic potential of coil. -fahrstuhl band-width regulator with reciprocating auxiliary coupler coil parallel to oscillating c. -halter, Film- film roll holder. -kern bob of spool, core of s. (m.p.) -kette low-pass filter. -körper coil form, c. former, spool. -mark molded coil core. -revolver coil switch mechanism (for wave-band change). -topf can, canned coil. -träger magazine or spool arm or support. -windungsschlussprüfer inter-turn short-circuit or continuity tester.

Spund stopper, plug, bung; tongue (of dovetailed board).

Spur track, trace, trail, gutter, channel. -, vorionisierte pre-ionized track. Brumm- buzz track. Ton-, eingeprägte track embossed on film strip. Ton-, eingeschnittene track engraved in film strip.

spurenweise in traces.

Spur-faktor tracking factor (phonograph recording). -lager thrust bearing. -methode, Nebel- Wilson cloud track method (using a streak of droplets and an ionizing particle). -verzerrung tracking error, t. distortion (phonogr.) -zapfenlager (end) thrust bearing.

-zeit, Bild- Watkins factor, development f.
Stab rod, bar, stick. -antenne rod antenna, whip a.
Stäbchen rod (of nerve endings of eye), rodlet.
Stabfeder, Dreh- torsion bar spring.
stabilisieren stabilize, steady, make constant, uniform or normal, regularize.
Stab-kopf rod eye. -mikrophon carbon stick microphone. -röhre arcotron.
Stachel sting, thorn, spine, prickle, prong.
Staffeln, Glasplatten- (Michelson) echelon spectroscope.
Stahl, naturharter self-hardening steel. Edel- refined or superior alloy steel.
stahlband-armiert, -bewehrt steel-tape armored, ribbon wrapped.
Stahl-bandaufnahme magnetic steel tape recording. -drahtaufnahme sound on wire recording system. -kiesstrahlen, -kugelblasen steel shot blasting. -stechen steel engraving.
Stamm-baum flow sheet (of working process). -farbe primary color. -holz stem, heart or trunk wood. -körper parent substance, p. body. -patent parent patent.
Stampfe stamp, stamper, pestle, punch, rammer, tamper.
stampfen ram, tamp, pave, pound.
Stampfwerk stamp mill.
Stand, heutiger, der Technik present state of the art (or technology).
Stand, Achs- wheel base. Führer- control cabin, pilot's cockpit, p. compartment (airpl.), tank commander's position. Prüf- test bed.
Stand-anzeige dead beat indication or reading (of an instrument). -anzeiger bearing indicator means, d.f. dial.
Standarte lens carrier, base board.
Stand-bild still film picture, inanimate p. -entwicklung slow development, stand d.
Ständer stand, standard, pillar, post, pole, pedestal, upright, column, housing (of rolls), stator, field system (of dynamo). Abtropf- drainer. Lupen- lens stand.
Standfestigkeit stability, rigidity, steadiness, solidity, resistance to deformation, creeping strength (of materials). Dauer- long-time creep strength.
Standfestigkeitsprüfung (long-time) creep test.
standhaft steady, constant, firm, stable.
Stand-linie base line, position line. -linienpeilung position-line bearing, great circle b. -linienverfahren base-line method of measurement.
Standort station, stand, position (d.f.), fix. -bestimmung obtaining a fix, position or location finding (by two or more d.f. stations), fix, position-line bearing, great circle b. -daten positional information. -peilung see Standortbestimmung.
Standsicherheit stability or steadiness (to shock and vibration).
Stange stick, rod, bar, pole, post. Abspann- stay, terminal or strain pole or mast. Kreuzungs- transposition pole. Schub- thrust rod, push rod, pusher r. Zahn- rack rod.
Stanniol tinfoil.
stanzen punch, perforate, stamp or blank out (from a solid sheet).
Stanz-matrize cutting dieplate, punching die. -presse stamping press, punch. -verfahren stamping method, punching m.
Stapel staple, pile, heap, stack.
stark belastet heavily loaded, highly l., carrying a heavy load.
Stärkegrad intensity, degree of strength or of concentration.
Starkstromstörung interference from power systems; power (supply) failure.

starr stiff, rigid, non-yielding,
Starrheit rigidity, stiffness.
Startbahn runway.
starten,durch- "open up" and repeat landing procedure (in ZZ landing method).
Startweite (der Elektronen) distance from origin (of electrons), distance from object to lens, object distance.
stationär(e)Messung deadbeat measurement. **-(er)Zustand** steady, stationary or stable state.
Stations-melder station indicator, tuning i. **-wähler** station selector.
Stativ stand, support, tripod. **Kurs-** class microscope. **Röhren-** telescopic tripod. **Schlinger-** cardan-suspended tripod. **Verfolgungs-** running tripod, follow-shooting t.
Stativlupe stand magnifier.
staubdicht dustproof, dust tight.
Staub-figuren powder pattern, dust figures (of Lichtenberg, Bitter, Debye, Scherrer, Hull, etc.). **-kern** dust core, molded c. **-rauschen** dust noise (of film). **-welle** acoustic dust pattern (showing nodes and antinodes). **-zähler** dust counter.
Stauchdruck compression.
stauchen compress (by blow), knock, beat, upset, clinch (a rivet).
Stauchmatrize,Kalt- cold upsetting die.
stauen stow, dam up, choke, baffle.
Stau-gerät baffle type fluid flow gage. **-gitter** suppressor grid, baffle grid. **-körper** baffle, barrier.
Stauung,Elektronen- electron accumulation or cloud (in virtual cathode).
Stauwirkung baffle or accumulator effect, damming e.; resistance.
stechen stick, prick, pierce, sting, tap, engrave.
Stechkolben pipette.
Stecker,berührungssicherer shockproof plug. **-,dreiteiliger** three-point plug, three-way p. **-,unverwechselbarer** non-interchangeable plug.
Stecker,Blind- dummy plug. **Dreifach-** three-pin plug, triplug. **Klinken-** jack, plug switch. **Netz-** power supply plug, light socket p. **Zwischen-** socket adapter, a. plug.
Stecker-buchse,-hülse plug socket, connector s. **-säule** multiple socket and plug device, multi-way s. and p.d. **-spitze** tip (of telephone plug).
Steck-regler plug-in and socket (volume) control. **-spule** demountable magazine or reel (m.p.); plug-in coil, detachable c. **-umschalter** plug-in switch.
Steg path, small bridge (of string instrument), cross piece, strap, bar; material or wire portions of a grid (between meshes); any narrow strip of solid material between holes or recesses, as in m.p. film between perforations, or between tracks or grooves in phonograph disks, called barrier or land. **Druck-** pressure bar, harmonic b. **Gitter-** grid stay; solid portions of g. mesh, g. strips used for wires.
Steg-abstand distance between grid wires (or g. supports or stays). **-gitter** grid in which all wires are parallel to axis and surround cathode cage-fashion. **-zahl** number of grid wires, stays or supports.
Stehbild still picture, non-animated p. **-verfahren** lantern slide projection method.
Stehen des Bildes steadiness of image. **--,mangelhaftes** jumping or unsteadiness of image.
stehende Figur stationary pattern, s. figure. **-Welle** standing wave, stationary w.
Steh-kolben flat-bottomed flask. **-lager** pillow block, pedestal bearing.
steif stiff, rigid, firm, non-yielding.
Steife prop, strut, brace; rigidity, stiffness.

steigen rise, ascend, mount, increase.
Steigerer, Wucht- dynamic expander.
Steigerung increase, increment, raise, boost. Dynamik- dynamic-range expansion. Kontrast- Callier effect, print contrast (phot.)
Steig-höhe elevation (in capillary tube), height of rise or of ascent, pitch (of a screw). -leistungsmesser rate of climb meter ("variometer" measuring both rise and descent). -rad ratchet wheel, escapement w.
Steigung, Gewinde- pitch (of screw thread). Spiral- pitch of a spire, lead of a s.
Steigungsmesser, Luftschrauben- propeller pitch indicator.
Steigwinkel rise angle.
steil, zu too high a gamma (in film printing).
Steilabfall der Sägezahnkurve abrupt drop of sawtooth wave.
Steilheit gradient, mutual conductance (mutual characteristic), slope (of tubes); contrast (film).
Steilheit, Emulsions- steepness of gradation. Entwicklungs- slope (of D and H curve). Film-, hohe high contrast (of film). Filter- sharpness of selective network. Flanken- width of transition interval between transmission and attenuation bands; steepness of sides or slopes of a curve or of upstroke and downstroke of an impulse. Gradationskurven- gamma value or slope of response line. Sieb- sharpness of selective network. Transponierungs- slope or mutual conductance (of a mixer tube). Umwandlungs- mutual conductance, slope (of mixer tube, etc.)
Steilheitsmesser derivator, tangent meter (for curves).
Stein-bildung, Kessel- boiler scale formation, incrustation. -gut white ware (with white absorbent body and soft glaze). -kohlenschlacke coal cinders. -lager jeweled bearing. -öl petroleum. -porzellan hard porcelain.
Stelle spot, site, locale, location, place. an - und Stelle in situ.
Stelle, abgeschirmte dead spot, radio shadow, r. pocket. Dezimal- decimal place. Locker- loose place, l. spot (cryst.) Null- zero place. Peil- control station (in group of d. f. stations).
stellen place, put, set, regulate, adjust.
Stellenzahl position number, atomic n.; index (math.), number of digits.
Stellhahn regulating cock.
stellig, drei- three-place, t.-figure (value). fünf-(e) Genauigkeit five-figure accuracy. sechs- with six decimal places.
Stell-kraft, Rück- restoring, retractile, elastic or restitution force or pressure. -schraube set screw, adjusting s.
Stellung, Arbeits- working position, operative p.
stellvertretender Patentkommissar deputy or acting patent commissioner.
Stempel stamp, stamper, die, punch, pestle, piston, matrix. Lampen- press or squash of a lamp. Press- press ram.
Stempelrohr plunger tube, piston t. (used in acoustic impedance tests).
Stengel, Pump- pumping lead, pump tube (short piece of tubing connecting with vacuum pump), exhaust vent.
Stereogrammetrie stereophotogrammetry.
Stern-Dreieckschalter star-delta switch.
sternförmig star-shaped, stellate.
Stern-gliedkette T (-mesh) network. -rad star wheel. -rohr astronomical telescope. -tag sidereal day.
stet, stetig continuous, constant, stable, steady.
stetiger Bildwechsel continuous, steady, non-intermittent feed or motion (of picture strip).
steuerbar modulable, capable of being modulated or controlled; steerable (of antenna charactistic and polarization); manageable, manoeuvrable, controllable.

Steuerblende modulating electrode, modulation shield, m. grid (in cathode-ray tube).

Steuerbord achteraus starboard aft. -voraus starboard bow. -kurve starboard (right-hand) curve (airplane approach).

Steuer-daumen sequence switch cam. -drossel magnetic modulator, m. choke, Heising modulator. -frequenz pilot frequency, synchronizing f. -generator drive oscillator, master o. -gitter control grid, shield (of a c.-r. tube). -kurs steered course, compass c. -mann,Ton- operator controlling sound volume (in theaters).

steuern steer, regulate, control, modulate, guide, direct. herunter- regulate down, decrease carrier amplitude.

Steuer-nocken sequence switch cam. -quarz (frequency) stabilizing quartz crystal. -relais pilot relay. -röhre master oscillator tube, drive t., drive o. -röhre, Quer- cross-control tube, beam tube (depending for its operation on b. deflection). -schalter master switch, control s., sequence s. -sender master oscillator tube, drive t., d.o. (working upon main transmitter, in independent drive). -stelle sound gate (film), impulse receiving place or point (in remote-control, telemetering, etc.) -stelle,Ton- tone-control point, exciter p., sound gate. -strich dot or pointer to observe or read bearings. -stufe,fremdgesteuerte, quarzgesteuerte crystal-stabilized oscillator stage, master or driver stage stabilized by quartz crystal.

Steuerung auf Dunkel modulation to dark condition, negative m. -auf Hell modulation to light condition, positive m. Ausblend- obturator modulcation, cut-off m. Dichtigkeits- charge-density modulation. Dunkelwert- adjustment to value of darkness. Fern- remote-control action, telemetric c. (with selsyn motor and synchro system). Frequenz- von Senderschwingungen frequency stabilization of transmitter oscillations. Geschwindigkeits- velocity modulation. Gitter- grid sweep, g. swing, g. excursion, g. excitation; g. control (in ignitron, etc.) Helligkeits-,Intensitäts- brilliancy modulation, intensity m., brightness m. or control. Kristall- crystal stabilization, quartz control. Ladungs(dichte)- charge density modulation. Leitungsstrom- conduction-current modulation. Licht- light scanning, l. modulation, l. control. Linien- velocity modulation. Zeilen- velocity modulation, variable-speed m.

Steuerungs-bereich drive range, swing of amplifier grid, operation inside straight part of characteristic. -einrichtung,Licht- light valve, light relay.

Steuer-verstärker amplifier between power and input stage (operating on pure voltage amplification). -welle control wave (in chain broadcasting), pilot w. -zylinder focusing grid, Wehnelt g. or cylinder (in cathode-ray tubes).

Stich prick, puncture, stab, sting, stitch; engraving, tap, tapping, pass (rolling mill). Form- shaping pass (rolling).

Stichel,Schneide- cutter. cutting stylus, c. tool, c. head, engraver (phonograph recording), scriber.

Stich-flamme fine pointed flame, dart of f. -leitung tie line, tap l. -loch tap-hole. -pfropf tap-hole plug. -probe sample taken by tapping, random sampling, assay of tapped metal. -wortverzeichnis (key-word) index.

Stickstoffhalogen nitrogen halide.

Stiel handle, shaft, stem, stalk.

Stift pin, peg, tack,stud, pencil. Gewinde- headless screw, grub s. Justier- pilot pin, registration p. Mittlungs- centering pin. Schreib- stylus, style, cutting or engraving tool. Tast- tracer point, feeler.

Stift-lagerung pin suspension, p. supporting. -rad sprocket wheel. -schlüssel pin wrench. -schraube tap bolt.

Stilb stilb (unit of luminosity in Hefner candles per square cm.

stille Entladung corona, silent discharge, effluve. -Zone skip zone, dead spot.

Still-abstimmung silent tuning, quiet t. -einstellung silent tuning, q.a.v.c. tuning. -leben still life.

still-schweigend tacit, implicit, implied, silent, quiet, calm. -stehen stand still, stop, arrest, be stationary.

Stimme, gemischte mutation stop (organ). -, schwebende voix céleste. Grund- foundation stop (organ). -, Lautstärkenbereich der menschlichen human voice intensity range. -, Modulation der inflection of voice.

Stimmflöte tuning pipe, pitch pipe.

Stimmgabel, schreibende tuning fork chronoscope. -sender tuning-fork buzzer oscillator, vibrating-reed transmitter. -steuerung tuning fork control (in phonic wheel). -unterbrecher tuning-fork interrupter. -zinken prong or tine of a tuning fork.

stimm-hafte Konsonanten sonant, voiced or vocal consonants. -lose Konsonanten surd non-vocal or breathed consonants.

Stimm-pfeife tuning pipe, pitch p. -platte reed board, r. plate (of accordion). -ritze glottis, glottic catch or cleft. -schlüssel tuning key, wrench (of kettle drum) -stock wrest plank (piano). -ton tuning pitch. -wirbel wrest pin.

Stimmung, hohe high pitch. -, Flote mit flute with tuning slide. Tremolo- tremolo.

Stimmungsbild key picture, sentiment p. -, dunkles low-key picture. -, überhelles high-key picture.

Stirn, Wellen- wave face, w. front, w. head.

Stirn-ansicht end view. -bügel headrest (of magnifier). -fläche end face; face, front. -holz cross-cut wood, end-grain w. -rad spur wheel. -streuung overhang leakage flux. -verbindung face connector. -wand front wall, plate or panel; end w. -welle onde de choc, impact w., bow w. -widerstand (front) end resistance, leading-end r. -zapfen trunion, end journal.

Stocheisen poker (bar or iron), fire p., stirrer, stoker, rake.

Stock stick, staff, stock.

stocken stop, slacken, hold up, arrest, mold, decay.

Stockflöte stick flute.

Stockung stoppage, stagnation, obstruction, standstill, jam, "bottle neck".

Stoff substance, matter, material, stuff, fabric, cloth. Trocken- siccative, drying substance.

Stoff-aufwand expenditure or requirement of material. -bahn web of fabric or cloth (textile). -bedeckung, Wandbekleidung draping, gobo, tormentor (when in the form of a portable wall with sound-absorbent material, in m.p. studio). -mangel shortage of material, underfill (in rolling). -patent patent covering substance or material, product p., substance p. -teilchen particle of matter, corpuscle. -verbindung composition of matter.

Stöpfel, Stopfen stopper, plug, cork.

Stopfen, Griffel- stopper with thumb piece. Schliff- ground-in stopper (of glass).

Stopf-ton (der gedackten Trompete) stopped tone (of s. trumpet). -werg oakum.

Stoppbad short-stop treatment.

Stöpsel, geschlitzter split plug. Blind- dummy plug. Glas- glass stopper.

Stöpsel-glas stoppered glass. -hahn stopper cock, c. stopper. -körper plug body.

stöpseln plug in, insert a p. falsch- misplug.

Stöpsel-rheostat resistance box (plug operated). -schalter plug switch.
Stör-amplitude noise level. -anfälligkeit susceptibility to trouble, "X", interference or noise, trouble incidence. -befreiung elimination of interference, jamming, fading, strays, static or atmospherics. -befreiung, Einrichtung zur static eliminator, X stopper. -breite detuning "width" (detuning required to reduce resonance amplitude to 1 percent).
Storchschnabel pantograph.
stören interfere, jam, disturb, trouble, perturb, derange.
Störer, durch -' verdeckte Zeichen swamped signals.
Störernullung connection of case of interfering device with neutral wire (in d.c. mains).
Stör-faktor, Fernsprech- telephone interference factor. -feld stray field, interference f. -festigkeit immunity from noise, disturbance or distortion. -gebiet mush area (in chain broadcasting), disturbed area. -geräuschatmen fluctuation of noise, waxing and waning of n. -gries low noise in picture background due to thermal effects, Brownian movement, mechanical properties of tube, and low-frequency fluctuations of local emission density, i.e., shot effect. -lautstärke noise level. -muster spurious pattern, false p. (telev.) -nebel chief interference zone, maximum noise z. -niveau noise level. -pegel, 'ntergehen der Bildspannungen im swamping of picture signals in noise level. -schutz interference eliminator, trouble elimination, anti-parasitic means. -schutzpackung noise suppression, shielding harness (for airplane engines), interference eliminator kit. -sender shadowing signal (telev.), interfering or jamming station. -signal, Abschattierungs- spurious or shadowing signal (in iconoscope operation; known also as "shading," "black spot," "tilt and bend" etc.). -signal-Kompensation compensation of spurious signal in iconoscope operation by shading circuit. -sperre noise gate. -spiegel noise level. -stelle center of disturbance (cryst.), point of disturbance or perturbation, discontinuity. -suchgerät interference locator, signal tracer (used stage by stage in receiver).
Störung disturbance, perturbation, noise, distortion, interference, statics, atmospherics, jamming (in signal transmission), disarrangement, derangement, trouble. -, atmospärische atmospherics, static, stray, "X's", QRN. -, ausserirdische extra-terrestrial or interstellar noises or disturbances. -, luftelektrische statics, atmospherics, "X". Eigen- internal trouble. Gitter- lattice distortion, lattice dislocation. Laufzeit- delay distortion. Lichthof- halo disturbance. Nah- nearby interference, local i. (from sources close to receiver site). Starkstrom- interference from power systems, power failure.
Störungs-rechnung perturbation calculation. -suche trouble "shooting", interference location. -theorie perturbance theory. -welle parasitic or interfering wave, jamming w., garbling w. (in secret telephony). -wellenweg path of interference wave.
Stoss impact, shock, collision, pulse, count or burst (e.g., of cosmic rays), impulse, push, stroke, jolt, recoil, thrust; joint, junction (of rails). Durchbruch- burst pulse. Höhenstrahl- cosmic ray burst or shower. Rück- back stroke, recoil. Schienen- rail joint. Spannungs- potential impulse, p. pulse, impulse voltage. Treppen-

lap joint.

Stoss-dämpfung damping or attenuation of pulsations or vibrations, shock absorption. -dauer duration of collision or shock. -durchschlagfestigkeit impulse-voltage breakdown strength.

Stösse, Dreier- triple collision, three-fold c. Dunkel- background counts (in Geiger-Müller tube).

Stössel, Stösser pestle, stamper, rammer, tamper, stem (of valve), impulsing, pushing or impacting rod (in loudspeaker), driver rod.

stossen push, thrust, hit, knock, impact, ram, pound.

Stoss-erregung shock, impact or collision excitation. -feder, Strom- impulse spring (in dial). -festigkeit shock, impact or impulse (breakdown) strength. -fuge joint, junction. -galvanometer ballistic galvanometer. -generator impulse or test wave generator. -kraft percussive power, motive p. -kreis impulsing circuit. -linie boundary line, joining l. -messer, Kraft- ballistic pendulum. -prüffeld (potential) impact or impulse testing field. -sender impulse exciter, i. transmitter. -spannung impact potential, shock p. -stelle joint area or point, junction p. contact point. -ton beat note, throb. -verbreiterung impact broadening. -wage ballistic pendulum. -welle percussion impact or impulse wave, w. impacting upon a joint. -winkel butting angle, a. of joint. -zahl number of collisions, impacts, or bursts.

Strafe, bei - der Nichtigkeit on pain of nullity.

Straf-gesetz penal code. -recht criminal law. -summe penalty, fine.

Strahl (cf Strahlung) ray, radiation, beam (in d.f., telev., etc.), pencil, jet (of fluid), flash (of lightning); straight line, radius (geometry).

Strahl, achsenparalleler paraxial ray. -, ausfallender emergent ray. -, ausserachsialer extra-axial ray, abaxial r. -, ausserordentlicher extraordinary ray. -, durchfallender transcident ray, transmitted r. -, einfallender incident ray. -, ordentlicher ordinary ray. -, schräger skew ray, oblique r. -, streifender glancing ray. -, ultraroter infrared ray, ultrared r. -, ultravioletter ultraviolet ray, vita-r. (between 2900 and 3200 A.)

Strahl, Ab- reflected ray. Anoden- anode ray, canal r., positive r. Brems- (Röntgen) X-rays due to collision or checking. Dunkel- obscure radiation, heat r. (from a light source), invisible (actinic) r. beyond the violet. Elektronen-, weisser heterogeneous beam of electrons. Faden- pencil ray (electrons), beam or pencil of non-uniform cross-section, in gas-focused c.-r. tube, due to positive ions and secondary electrons, thread beam. Fahr- radius vector (math.). Gegen- reflected ray, reflection. Gesichts- visual ray Grenz- border-line ray, infra-Röntgen r. H- H-ray (consisting of positive hydrogen ions and protons). Kanal- canal ray. Knoten- cf Strahl, Faden-. Leit- guide ray, g. beam, radio beacon, radius vector (math.) Licht-, maximale Aussteuerung des - (es) clash point of light valve (in sound recording). Luft- air jet. Molekular- molecular ray, m. beam. Peil-, raumgeradliniger orthodrome. Rand- marginal (zone) ray, peripheral r., extra-axial r. (opt.) Rest- residual ray, r. radiation, reststrahl. Röntgenbrems- X-rays due to collision or checking. Schräg- skew ray. Schwarz- black (body) radiation, cavity r., Planckian r. Streu- scattered radiation, s. signal, stray radiation, stray light.

Weltraum- cosmic ray. Zentral- central ray, axial r. (opt.)

Strahl-abblender means to mask, stop down, occult, eclipse, gate or diaphragm rays, beam douser, blanking or blanketing means, dimmer. -ablenkung ray deflection, pencil d., beam d. or deviation; ray refraction.

Strahlabtaster (mit unsichtbaren ultraroten Strahlen) noctovision scanner (using micro- or infrared rays). Elektronen- electron scanning pencil, beam, brush or lever. Kathoden- electron beam scanner. Licht- spotlight scanner, brush or pencil.

Strahl-abtastung scanning or sweeping with a light or electron pencil. -apparat jet apparatus, injector (of steam).

strahlartige Entladung needle-point corona or streamer discharge, leader-stroke d.

Strahl-ausfall emergence of beam. -brechung refraction or splitting of rays. -bündel beam, pencil, bundle, bunch or brush of rays. -bundel,Kathoden- cathode-ray beam or pencil, (loosely) gun.

strahlen emit rays, radiate, be radiant or radiating.

Strahlen-brechungsmesser refractometer. -bündel(s.a.Strahlb.) bundle of rays or radiations, cone of r. -bündel,Schall- pencil of sound. -einfall, streifender glancing incidence of rays. -gang path of rays, trajectory of r., ray tracing, geometric configuration of rays. -härtemesser penetrometer, qualimeter (for X-rays). -kegel cone of rays. -messer radiometer (heat rays), actinometer (actinic r.), solarimeter (sun r.). -optik geometric optics. -reichweite range of rays or corpuscular emission. -schnitt jet contraction, contractio venae. -stoss, Hohen-,Raum- burst or shower of cosmic rays, cosmic ray track (traced by hodoscope). -teiler beam splitter.

Strahlentladung,Faden- thread ray discharge.

Strahlentstehungsort source or origin of electron pencil or gun.

Strahlenzieher protractor.

Strahler radiator (of radio waves, loudspeaker, etc.); heater (of a cathode). Massen- mass radiator. Mehrfach- bank or pillar of lamps, battens. Richt- directive antenna, directional a., directional loudspeaker or projector, radiator. Rück- reflector, re-radiator. Temperatur- incandescent light source, luminous radiator.

Strahl-erweiterung spread of beam (telev. tube). -erzeugerfuss gun press. -erzeugungssystem gun (comprising electron source, first and second lenses); ray radiator system. -fläche radiant area, radiator surface, radiation-emissive s. -hebelarm pencil leverage, deflectibility of electron beam.

strahlig,lang- elongated, needle-like (cryst.)

Strahl-intensitatskontrolle modulation of beam intensity (by grid, shield or Wehnelt cylinder, telev.) -methode,Atom- atom or molecule ray method. -neigung inclination of ray. -quelle,Leucht-,einfarbige monochromatic illuminator, m. source, monochromator. -röhre, Elektronen-(cf. Kathodenstrahlrohre) beam tube, cathode-ray tuning indicator t. or magic eye. -ruhelage spot zero. -schauer,Hohen- cosmic ray shower or burst. -schreiber, Tinten- ink vapor jet recorder (facsimile telegraphy). -sender, Leit- beacon, radio range or beam station. -spaltung split-beam, b. splitting. -sperrung beam cut-off or eclipse, gating of the b. -stösse,Höhen- cosmic ray bursts.

Strahlung,aktinisch undurchlässige adiactinic radiation. —,falsche stray radiation. Grau- grey radiation. Hohlraum- cavity radiation, black-body r. Raum- cosmic ray radiation. **Resonanz-** resonance radiation (of vapor lamp). **Rück-,spiegelnde** specular reflection, regular r. **Schwarzkörper-** black-body radiation, cavity r. **Seiten-** secondary radiation, lateral r., spurious lobe (of space pattern). **Temperatur-** thermactinic radiation. **Über-** irradiation. **Ultra-** cosmic rays. **Vernichtungs-** annihilation radiation.

Strahlungs-belastung radiation load. **-dämpfung** radiation damping, r. resistance. **-diagramm** radiation diagram or characteristic, space pattern, radiation p. **-druck** radiation pressure. **-empfänger** radiation-sensitive or responsive pickup, r. receiver, record sheet or surface (on which radiations impinge for measuring color temperature, etc., consisting of photographic plate, selenium cell, bolometer, etc.) **-fläche** (electron) emitting surface, emitting area, radiator s. **-formel, Planck'sche** Planck's equation (of spectral energy distribution). **-gegendruck** radiation reaction. **-gleichgewicht** radiation equilibrium. **-messer** actinometer (for actinic rays and solar flux density), radiometer (thermal rays), solarimeter (sunrays). **-messer, Sonnen-** pyrcheliometer, pyrheliometer, solarimeter, actinometer (for solar flux density). **-potential** radiation potential, resonance p. (to excite atom or molecule). **-temperatur** radiation temperature, effective t. **-verfahren,Selbst-** self-emissive or s.-illuminating method (used in electron microscope) based on electrons arising on object surface. **-verluste** corona losses, radiational l. **-widerstand** effective resistance (loudsp., antenna, etc.), radiation r., characteristic impedance. **-werfer** beacon or beam station.

Strahl-unterbrecher,Quecksilber- mercury (jet) interrupter.

-vergenz vergency of rays (either convergence or divergence). **-verlauf** path of rays, course of r., trajectory of r., ray tracing. **-wegablenkung,Peil-** distortion of bearing, lateral deviation (due to diurnal or seasonal factors, terrain, local conditions, etc.)

Strang rope, chord, cord; web (of paper, cloth, etc.), strand; line (of drive shafts).

Strasse,Feinblech- sheet rolling mill. **Wirbel-** vortex avenue.

Strebe,Gitter- grid stay, g. support.

strebenlos unsupported, unstayed, unpropped.

streckbar extensible, extensile, ductile, malleable.

Strecke distance, space, stretch, extent; path (e.g., grid-filament).

Strecke(z.B.a-b) interval a-b, element of length a-b.

Strecke abschneiden lay off a distance.

Strecke,Gas- gaseous path or gap (in a tube). **Gitter-Kathoden-** grid-filament path or circuit.

strecken stretch, extend, spread, draw, elongate, pull.

Strecken-befeuerung,Nacht- route beacon for night flying service. **-dienst,Lang-** long-range or long-distance avigation (navigation) service. **-element** linear element (math.) **-feuer,Haupt-** main or trunk route beacon, airway b. **-funkfeuer** range beacon. **-funkfeuer, Haupt-** main route radio beacon. **-karte** route airway map. **-leitstrahl** long-distance or route avigation (navigation) beam. **-messung** end-to-end measurement, straight-away m. **-navigation** long-distance avigation, route a. **-netz** air routes. **-spektrum** continuous spectrum. **-teilchen** linear element (math.)

Streckfestigkeit resistance to stretching, r. to elongation. **-figur** flow lines, lines of stress, surface bands. **-grenze** yield point, limit of elongation. **-spannung** yield stress. **-stahl** rolled steel. **-werk** rolls, rolling mill.

streichen stroke, rub; paint, coat, varnish; erase, cancel, bow or stroke (a string).

Streich-instrument bowed or string instrument. **-schale** oilstone, whetstone.

Streifen strip, stripe, ribbon, tape, fringe (in light interference), band, streak, stria, vein.

streifen graze (as in a shock or collision), glance, skirt, brush against; stripe, streak, striate.

Streifen, Abdeck- squeeze unit, mask, mat(m.p.); barrier (between frames). **Absorptions-** absorption bands, a. spectrum. **Bitter'sche** Bitter (magnetic powder) pattern or bands. **Brumm-** buzz track; hum (broad transverse bands appearing on c.-r. tube screen). **End-** trailer (blank end of a film). **Interferenz-** interference fringe (opt.). **Kontroll-** monitor slip. **Längs-** longitudinal stripes, striae or streaks. **Leucht- (in Entladung)** luminous streamers (in discharge). **Löt-** tag strip, terminal s., connection s. **Mach'sche** Mach bands. **Rohr-** skelp, strip (for forming pipes). **Ruhe-** unmodulated track. **Ton-** sound track.

streifender Einfall grazing, oblique or glancing incidence (of rays or sound waves).

Streifen-filter, dreifacher tricolor banded filter. **-gefüge** laminated structure, banded s. **-keil** photometric wedge. **-lichtverfahren** variable-density recording method. **-schirm** rod screen. **-spektrum** band spectrum. **-verschmälerung, Ton-** squeezing of sound track, matting.

Streifungen striations, striae (metal coatings, etc.)

Streifungen, Translations- translation gliding striae.

Streit, Patent- patent litigation, p. action, p. suit. **Rücknahme-** action for withdrawal.

Streitanmerkungen annotations of litigation.

Streiter contestant to a suit, litigant, disputant.

Streit-fall,-frage,-punkt,-sache issue, question at issue, contested case, dispute, controversy, matter at bar, m. in issue, m. in suit.

strengflüssig viscous, semi-fluid, difficultly fusible; refractory,

streuen scatter, strew, fluctuate, disperse, "spread" or scatter (of values).

Streu-faktor scattering factor (of x-rays), s value. **-feld** stray field, leakage f., extraneous f. **-impedanz** leakage impedance. **-induktivität** stray inductance. **-kapazität** stray capacitance, spurious c. **-kopplung** stray coupling, capricious c., spurious c. **-licht** stray light. **-querschnitt** scattering cross-section. **-richtung** scattering direction. **-strahlen** scattered signals, s. radiations, stray radiations, stray light.

Streuung scattering effect, diffusion e., dispersion, stray, leakage, spread (d.f.) **Flanken-** side leakage, s. stray. **Spalt-** slot leakage, s. stray. **Stirn-** overhang leakage flux.

Streu-winkel angle of scattering. **-zelle** scattering cell.

Strich stroke, streak, stria, line, dash, mark, grain. **Ab-** downstroke or trailing edge of a synchronizing impulse (telev.). **Auf-** upstroke or leading edge of a synchronizing impulse (telev.) **Dauer-** long dash, permanent note (N-A interlock in d.f.). **Gitter-** grating line. **Kompass-** point of compass. **Ober-** maximum or peak carrier current, m. picture signal. **Steuer-** dot or pointer to observe and read bearings. **Teil-** graduation line, division l., gage mark (of a scale).

Strichdiapositiv line transparency, l. diapositive.

Stricheinstellung, Bild- framing (by observing frame lines). **Bild-, falsche** misframing, out-of-frame condition.

stricheln shade, hatch, streak.

Strich-fokus line focus (of x-rays). **-formulierung** chemical formula using dashes or lines (to represent bonds). **-geräusch** frame-line noise (form of motor boating). **-gitter** ruled grating, simple line g. **-kreuzplatte** cross wire plate (of telescope), reticle, graticule.

strichliert broken (of a line), dash-lined.

Strich-mass line standard. **-methode** stroke method, line m. (paper test). **-punkt** semicolon.

strichpunktiert broken (dot-dash) (line).

Striktionsspule focussing coil.

stroboskopische Scheibe stroboscopic disk, s. pattern wheel.

Strohgeige Stroh violin (with diaphragm or contact microphone attachment).

Strom, abgehender outgoing current. **-, ankommender** incoming current, input c. **-, eingeschwungener** steady-state current. **-, phasenverschobener** out-of-phase current, phase-displaced c., dephased c., quadrature c. (when shift is 90°). **-, vagabundierender** leakage current, stray c. **-, verzweigter** branched current, branch c. **-, welliger** ripple current, undulated c., pulsating c.

Strom, Ableitungs- leakage current, stray c. **Arbeits-** watt current, working c. **Ausbreitungs-** dispersion current. **Ausgleich-** balancing current (of a bridge), compensating c. **Ausschwing-** decay current, decaying c. **Bild-** video current, picture c. **Blind-** reactive current, reactance c., wattless c. **Brückenausgleich-** bridge balancing current. **Dauer-** persistence current. **Dunkel-** dark current (flowing in photocell, in absence of light). **Einschwing-** building-up current. **Erst-** primary current. **Gespenster-** ghost current. **Halte-** holding current, retaining c. **Kreis-** circular current, c. flowing in a circuit. **Kriech-** creep current, sneak c., surface-leakage c. **Lade-** charging current (anomalous and normal, of condenser); charging c. (of battery). **Leck-** leakage current, leakance c. **Leit-** conduction current; pilot c. (in wired radio). **Leitungswechsel-** conduction alternating current. **Licht-** light flux, l. current (in lumens); lighting c. **Nachwirkungs-** absorption current, residual-charge c. (of condenser). **Nutz-** useful current, signal c. **Pegel-** pilot current. **Photo-** photo-cell current. **Quer-** leak current. **Richt-** output current (d.c.) of detector, rectified d.c. component of plate c. **Rück-** reverse current, backward c. (in rectifier). **Ruhe-** anode-feed current (d.c. component of plate or anode c.); no-signal c., quiescent c. **Sättigungs-** saturation current. **Schalt-** current on contact. **Spannungs-** current flowing in potential circuit (of a meter). **Sperr-** backward current, reverse c. (of rectifier). **Ton-** voice signal current, sound c. **Träger-, plötzlich veranderter** wobbled carrier. **Verdrängungs-** capacitance current, displacement current, dielectric c. **Verlust-** watt current, active c., energy c. **Verschiebungs-** capacitance current, displacement c. dielectric c. **Vor-, dunkler** dark pre-sparking current. **Vormagnetisierungs-** biasing current (for magnetic polarization). **Wand-** wall current (of amplifier). **Wechsel-** alternating current (a.c.) (mostly single-phase current is meant specifically), indirect c., any periodic c. **Wellen-** pulsating current, ripple c. **Zeichen-** marking current (telegr.) **Zwischen-** spacing current (telegr.)

Strom-abnehmer tap, collector of current. **-arbeit** Joule heat. **-aufnahme, Blind-** drawing or taking wattless, idle, or reactive current. **-aufnahme, Gitter-** drawing of current by grid. **-aussteuerung, Anoden-** plate current excursion. **-bahn, flächenartige** laminar or areal path of current, sheet c. **-belag** current coverage, c. distribution (in terms of amps. per centimeter of periphery). **-dämpfung, Echo-** echo current attenuation, active c. loss. **-dichte** current density (in amps. per unit of cross-section); luminous flux (in lumens).

stromdurchflossen alive (of a wire), traversed by current.

Strom-einsatz, Gitter- point where grid draws current, incipient g. c. flow. **-faden** streamer (in discharge), current path; stream line. **-flächenformel** current sheet formula.

stromführend current-carrying, c.-conducting or conductive, live, charged, "hot".

Strom-gattung kind of current, system of c. **-gebiet, Anlauf-** range or region of incipient current flow (e.g., at grid). **-gitter** atomic lattice.

Stromkreis (s.a. **Kreis**), **am Ende kurzgeschlossener** short-circuited circuit or line, end-shorted c. **-, am Ende offener** open-ended circuit. **-, unbelasteter** unloaded circuit. **-, verzweigter** branched circuit, divided c. **Abstimm-** tuned circuit, tuning c., tank. **Bezugs-** reference circuit. **Geräuschunterdrückungs-** noise reduction circuit, squelching c., silencing c. **Grundgeräusch-** (back) ground noise reduction circuit. **Impedanz-** impedor, circuit with impedance. **Nutzsignal** circuit, utilization, working or useful c., active c. **Reinigungs-** smoothing circuit, filter c., low-pass filter (following rectifier in power unit). **Sprech-** circuit carrying voice or signal current (in loudspeaker). **Stillabstimmungs-** silent tuning circuit, q.a.v.c. circuit.

Strom-leistung electric power (wattage). **-linienaufzeichner** device (often electrolytic tank) to map or plot potential, fields, lines, tubes of force, etc., on a model.

stromlos currentless, de-energized, dead.

Strommesser, registrierender recording ammeter. **Hitzdraht-** thermal ammeter, hot-wire a.

Strom-modulation, Elektronen- electron quantity modulation. **-reiniger** harmonic eliminator, h. excluder (filter), smoothing means, stopper. **-resonanzkreis** parallel resonance circuit, tank c. **-richter, gittergesteuerter** rectifier of thyratron type, mercury r. or ignitron with control grid. **-schiene** third rail, busbar. **-schlüssel** key switch, make and break. **-schlussprüfer** continuity tester. **-spule** current coil (of a wattmeter).

Stromstoss current pulsation, pulse or impulse. **Anlass-** starting rush or impulse of current. **-feder** impulse spring (of dial switch) **-gabe** impulsing, pulsing. **-gabe, rückwärtige** revertive pulsing. **-schalter** impulsing switch, stepping relay. **-übertrager** impulse repeater. **-verhältnis** impulse ratio.

Strom-tor thyratron. **-tor-Ignitron Kombination** mutator. **-übernahme** transfer of current, tapping of c.

Strömung current, stream, flow, flux, circulation, fluid movement; tendency, trend. **-, wirbellose** flow free from vortices, eddies or turbulence, irrotational flow.

Stromungs-potential flow potential, streaming p. **-widerstand** flow resistance (of a test piece); aerodynamic r., drag.

Strom-verdrängung skin effect, current displacement, Heaviside e. **-verzweigung** network of conductors, dividing or branching of currents. **-wage** ampere balance. **-wandler, Durchführungs-** bushing-type current transformer. **-wechsler** commutator, reverser. **-weiche** current divider (network or filter). **-wender** commutator, reverser. **-wert** current value, amperage. **-zeiger** ammeter; current vector. **-zuführung** current supply lead.

Strudel eddy, vortex, whirl (pool), maelstrom.

strudellos irrotational, non-eddying.

Struktur, feinschuppige magnetische fine-scaled magnetic structure. **-, hyperfeine-, überfeine** hyperfine structure. **-, zellenartige** honeycombed structure, pitted s., cellular s. **Fein-** fine structure, micro-s. (cryst.) **Sekundär-** secondary structure gridwork (cryst.). **Über-** superlattice; superstructure. **Überfein-** hyperfine structure. **Verzweigungs-** lineage structure.

Struktur-analyse, Grob- gross or macro structure analysis. **-chemie** structural chemistry.

strukturlos structureless, amorphous.

Stück, Flächen- area element, areal e., unit area. **Leiter-** conductor element, portion, part or (short) length of conductor.

Stückfärbung tissue staining, staining in toto (microscopy).

stückweise piecemeal, piece by piece.

Stufe step, stage, degree, grade, gradation; digit (automatic telephony), point (in multipoint switch). **-, nachgeschaltete gesteuerte** main oscillator stage (in independent drive system).

Stufe, Beschleunigungs- target (of multiplier), acceleration stage. **Einer-** units digit. **Farb-, Farben-** color gradation, tint, shade. **Graukeil-** intermediate shading value. **Helligkeits-** (intermediate) brightness stage, i. grey value (between black and white), shading or tonal v. **Lautstärken-** loudness unit, sensation u. **Licht-** printer step (copying machine). **Misch-** mixer stage, m. first detector s. **Nachlauf-** (cathode) follower stage. **Niveau-** term, energy level (of electron). **Oxydations-** stage or degree of oxidation. **Puffer-** separator stage, buffer s. **Überlagerungs-** mixer stage, first detector (in superheterodyne receiver). **Zehner-** tens digit.

stufenartig graduated, step-like, graded, gradual, echeloned, stepped.

Stufenfolge succession or sequence of steps or stages, graduation.

stufenförmige Echos echelon formation of multiple echoes.

Stufen-gitter echelon grating. **-keil** step tablet, stepped photometric absorption wedge (in x-ray, etc., work). **-linse** echelon lens, Fresnel l. **-messer, Farb-** tintometer. **-methode** gradient method (of cooling, etc.). **-platte** sensitometer tablet. **-prisma** echelon prism. **-rad** step wheel, cone pulley. **-relais** relay with sequence action. **-schalter** multi-point switch, m.-contact s. **-scheibe** cone pulley, taper p., stepped p. **-schwächer** step-type weakener (spectroscopy, photography). **-spalt** step slit (phot.).

stufenweise in stages or steps, stepwise, by degrees, successively, sequentially, gradually.

Stulpe cuff.

stülpen turn inside out, t. upside down, put over or upon in inverted position.

Stummfilm silent film.

Stumpf stump, butt-end. **Kegel-** truncated cone, frustum-cone,

frustum. **Pyramiden-** truncated pyramid frustum. **Wellen-** butt end of a shaft.
stumpfe Verbindung butt joint.
stumpfeckig blunt cornered.
Stumpfkegel truncated cone, frustum.
stumpfschweissen butt-weld, jump-w.
Stumpfungfläche, Ab- truncating face (cryst.)
stumpfwinklig obtuse-angled.
stürmisch stormy, turbulent, violent, vivid, agitated.
Sturz plunge, fall, sudden drop or decline, chute.
Stürzung, Bild- toppling of picture or image (pictures turned an angle in relation to one another).
Stütze support, stay, prop, bearing. **Trag-** supporting stay or prop.
stützen support, prop, stay, rest or bear on.
Stutzen, Ausblick- objective lens socket (opt.) **Rohr-** short piece of pipe or tubing serving as an opening, socket, or connection; nipple, short cylinder, bush or sleeve. **Saug-** pump intake, p. nozzle.
Stütz-isolator pin insulator, rod i. **-pfeiler** supporting pillar, s. column, bedding pile. **-punkt** point of support, bearing surface, fulcrum; airbase, base airport.
Substanz, Mutter- parent substance.
subtraktive Modulation downward modulation, positive m.
Suche, Störungs- trouble "shooting", t. locating, interference location.
Such-elektrode probe electrode, exploring e., collector e. (in vacuum t. work). **-empfang** straight-ahead reception, single-circuit r., direct r., non-heterodyne r.
Sucher searcher, locator, detector, finder, probe, view-finder (opt.) **Durchsichts-** direct-view finder, eye-level f. **Fehler-** aniseikon (for detecting cracks electronically), electromagnetic crack detector (by flash magnetization), materiologic devices (with X-rays). **Objekt-** object-finder (opt.) **Prismen-** prismatic seeker or finder. **Rahmen-** wire-frame view-finder, iconometer. **Ziel(auf)-** radar (radio detection and ranging apparatus), sonar (underwater sound navigation, detection and ranging).
Sucher-bild seeker picture, monitoring p. **-korn** backsight of view-finder. **-okular** focusing lens. **-parallax** seeker parallax, finder p.
Such-gerät, Stör- interference locator, signal tracer (traces progress of signal by stages). **-licht** searchlight. **-spule** search coil, rotating c., exploring c. (of d.f.); flip c. (used in magnetic field tests). **-tonanalysator** heterodyne analyzer. **-tonverfahren** beat-note method (using heterodyning search frequency).
südliche Breite south latitude.
Südlicht aurora australis.
Summand term of a sum (math.).
Summar (Leitz) Summar objective.
Summe, Licht- light sum, total l. emission. **Linien-** line integral.
summen hum, buzz; sum up, add (up).
Summen-frequenz summation frequency, sum f. **-geber, Fernmess-** telemetric integrator. **-gleichung** summation equation. **-satz** sum rule, permanence principle. **-ton** summation tone. **-wirkung** combined or compound action or effect.
Summer buzzer, small AF oscillator. **Dynatron-** dynatron oscillator. **Glimm-** glow-tube type of buzzer oscillator. **Magnet-** magnetic interrupter or buzzer-type of AF oscillator. **Röhren-** vacuum tube AF oscillator. **Schwebungs-** audio-frequency (AF) oscillator, heterodyne buzzer o., beat.f.o. **Stimmgabel-** tuning-fork AF buzzer oscillator. **Überlagerungs-** heterodyne AF warbler oscillator.
Summer-erregung buzzer excitation. **-wellenmesser** buzzer wave meter.
Summierungsgerät integrator.
Summ-spur, -streifen buzz track (on film).

Sumpf swamp, marsh, bog; pit, sump, pool (of mercury in a vessel), pouring basin.
Super-leitfähigkeit supra-conductivity, super-c. -het(erodyne)-empfang mit selbsterregter Trägerwelle homodyne reception.
Supra-leitelektronen super-conduction electrons. -leiter super conductor.
Surrerscheinung motor-boating (in radio receivers).
Süsswasser fresh water.
Symmetrie symmetry, balanced condition. -ebene plane of symmetry, crystallographic plane.
symmetrieren, symmetrisieren balance, symmetricize, symmetrize, align, make symmetric, neutralize.
Symmetriezentrum inversion center.
symmetrisch, achsen- axially symmetric, in axial symmetry. elektrisch- homopolar. höhen- symmetric about vertical axis (cryst.) kugel- spherically symmetric, spherosymmetrical. mitten- centro-symmetrical. plan- planisymmetric. punkt- having point or radial symmetry. radial- radially symmetric, in radial symmetry. rotations- in rotation or revolution symmetry. schief- skew-symmetric. spiegel- mirror symmetric, with specular symmetry. zentral- centro-symmetric.
Synchronisator, Zahnrad-, magn. cogged or tooth-wheel synchronizer. Zeiger- dial scoring machine (m.p.)
Synchronisieratelier scoring stage.
Synchronisieren s. a. Gleichlauf.
Synchronisieren, lokal geregeltes local synchronization (by tuning fork, etc., in telev.)
-, selbständiges self-running synchronization (telev.)
-, unselbständiges distance- or signal-controlled synchronization (telev.) Nach- post-scoring (of silent film), dubbing, re-recording (first picture, then sound). Nach-(von Trickfilm, mit springendem Ball) bouncing-ball synchronization (m.p.). Vor- pre-scoring (first sound, then picture). Zwangs- locked synchronization.
Synchronisierimpuls, Bild- frame synchronizing impulse or signal, low sync i. Zeilen- line synchronizing impulse or signal, high sync i.
-Abtrennung oder -Aussiebung synchronizing separation (separating control or sync impulses from video signals), amplitude separator, clipper.
Synchronisier-klappe clappers to mark sound and picture track for synchronization. -lücke, Zeilen- line synchronizing period. -zeitmarke marking of time, synchronizing m., time scale. -zwang locked synchronism, l. synchronization.
Synchronisierverfahren, Lücken- gap, underlap, interstice or interval synchronization method (sync signal introduced between end of line traversal and beginning of next line).
synthetischer Sprecher voder.
Systematik systematology.
Szenen, Atelier- regular stage scenery or sets. Effekt- scenics, light effects, etc.

T

T Regler T type section attenuator.

tabellarisch tabular, in tabulated form.

Tabelle, Belichtungs- exposure time table. **Peil-** bearing chart. **Umwandlungs-** conversion table.

tabellisieren tabulate, tabularize,

Tabletten, Farbe- varitone tablets.

Tafel table, tablet, plate, slab, sheet (of metal), cake (of chocolate), pane (of glass); slate, blackboard, panel.

Tafel, Einstell- focusing board, test chart. **Farben-** color chart, c. scale. **Fluchtlinien-** alignment chart. **Misch-** mixing panel, monitoring p. **Rechen-** computation table. **Signallampen-** signal panel, annuciator board or p. **Spektral-** spectrum chart. **Umrechnungs-** conversion table, reduction t., conversion scale (say, meters to kilocycles). **Zahlen-** numerical table, n. tabulation, table of figures.

Tag, Stern- sidereal day.

Taglindheit hemeralopia.

Tages-gang, örtlicher local diurnal variation. **-lichtfaktor** day-light factor (occasionally called window efficiency ratio, phot.), **-lichtkassette** daylight loading magazine (m.p.)

täglich daily, diurnal.

Takt time, cadence, measure (mus.), tempo, rate, rhythm, stroke (of engine).

Takt-funkensystem timed-spark system. **-geber** impulsing means, impulser, time-valve, time tapper, t. beater, metronome. **-geber, Kristall-** crystal monitor, c. stabilizer. **-impuls** synchronizing impulse, timing i.

taktmässige Ablenking timed deflection, t. sweep.

Tal, Wellen- trough, valley or dip of a wave.

Talg tallow, (solid) fat, grease.

Talwegkurve thalweg.

Tambourinbecken tambourine jingles.

Tangens, hyperbolischer hyperbolic tangent, tanh (math.)

Tangente, Wende- tangent to reversing point.

Tannenbaumantenne christmas tree antenna.

Tannenbaumkristall dendrite (cryst.)

Tanz, Antreten zum formation of electron groups, bunching (in beam tube), phase focusing. **Elektronen-** oscillatory (erratic) movements of electrons (caused by electric or magnetic actions).

Tanzen, Bild- unsteadiness or jumping of picture.

Tarnung camouflage.

Tasche pocket, pouch, bursa, trap. **Ausfrier-** liquid-air trap (vacuum pump).

Tastarm wiper (arm).

Tastatur, Universal- universal keyboard.

Tast-blende, Ab- scanning aperture, s. hole (Farnsworth tube). **-bolzen** gage finger, tactile part. **-drossel** magnetic modulator.

Taste, Abstands- blank key, spacer k. **Lösch-** cancel key, clearing k., release k. **Ober-** sharp (mus.), black key. **Rückzug-** back-spacing key. **Unter-** long key, white k.

Tasten keying, k. modulation, feeling, exploring, scanning, probing.

Tasten, Hinunter- carrier caused to drop to low value (zero) by synchronizing signal.

Tasten-feder, Brumm- spring for bass keys (accordion). **-feld** keyboard. **-reihe** bank of keys, row of k.

Taster calipers, feeler, keying device (radio), tactile part or device; scanner, explorer (telev. picture).

Taster, Aussen- outside calipers. **Fadendicke-** calipers with jaws measuring thread gage capacitively. **Gewinde-** thread calipers. **Innen-**

inside calipers. Rad- wheel operated means causing signals, bell ringing, etc. Tiefen- depth gage.

Taster-skala feeler scale. -zirkel gaging calipers.

Tastgerät stabilizer; keyer and limiter leveling amplitude variations in incoming video signals (telev.) Vibrations- vibro-tactile device.

Tast-geräusch key chirp (occurring after key has been opened), key thump, k. click (caused on closing key). -klick key thump, k. click. -prellen see -geräusch. -relais keying relay, r. key. -schaltung, Mittelpunkt- center-tap key modulator circuit. -schlag key click, k. thump, chatter. -stift tracer point, feeler.

Tastung (cf Abtastung) control, keying, modulation by key action; impulsing, pulsing. Einkanal- single-channel pulsing (telev. synchronization). Hoch- working impulse transmitted on voltage substantially above normal. Spiral- spiral scanning.

Tatbestand factual findings.

Tatigkeit, Sonnen- solar activity.

Tatsachen, Erfahrungs- empirical experience, practical facts or data.

Tatze paw, claw, cam.

Taubheit, Leitungs-(des äusseren Ohres) conduction deafness or impairment of hearing (outer ear). Nerven- (d. inneren Ohres) nerve deafness, perception d., nervous perceptive impairment of hearing (inner ear).

Tauch-bahn (der Elektronen) dip orbit (of electrons). -batterie plunge battery.

tauchen immerse, dip, plunge, steep.

Taucher-aufnahme submarine shooting. -kolben plunger.

Tauch-kernrelais plunger relay. -lack dipping varnish. -spule moving coil, signal coil (of loudspeaker); telescoping c. -spulenmikrophon moving-coil microphone. -zylinder plunge cylinder, plunger.

tauglich fit, proper, appropriate, useful, serviceable, usable, good, suited.

Taumelbewegung tumbler movement, t. action.

Taupunkt dew point, thaw p.

Tauschung, Augen- optical illusion. Spiral- spiral illusion.

Technik engineering art, technology, technics, technique. Funk- radio technology, r. engineering art. Nachrichten- signal art, communication art. Stand der- state of the art.

Techniker, Aufnahme- camera technician, c. man.

technisch commercial (grade, etc.), technical, technological, practical, industrial. glas- glass technological, vitrological.

technische Frequenz commercial frequency. -Wechselströme commercial alternating currents, industrial a.c. -Werkstoffe industrial, commercial or technical materials.

teigig doughy, pasty, mellow, plastic, kneadable.

Teil...sub-..., fractional, divisional, partial, componental.

Teil-abtastung fractional scan, partial s., coarse s. -bild compound image (of color record).

Teilchen particle, small part, corpuscle, element (math.) -, harte primaries. Grund- smallest particles, fundamental p., atoms. Schwebe- suspended particles. Stoff- particle of matter, corpuscle. Strecken- linear element (math.)

Teilchen-dichte particle density. -feld particle field. -geschwindigkeit particle velocity.

Teildruck partial pressure.

Teiler divider (of potential, etc.), divisor (math.), submultiple. Frequenz- frequency divider, demultipler or submultipler of f. Phasen- phase splitter. Spannungs- voltage divider, potentiometer. Strahlen- beam

splitter (m.p.)
Teilerdkapazität direct earth capacitance.
teilerfremd aliquant.
teilfremd prime to each other.
Teil-frequenz component frequency (say, of a spectrum). **-kanal** sub-channel. **-kopf** index head. **-kraft** component force, componental f. **-kreis** pitch circle, p. line, divided or graduated circle, dial, graduated circular plate. **-maschine** dividing engine, d. machine, ruling e. **-nehmer, Sammelanschluss-** subscriber with several lines. **-raster** fractional scan, f. sweep (in interlaced scanning). **-ring** index ring. **-scheibe** index circle, i. plate. **-scheibe, Rasten-** notched index wheel. **-spannung** partial voltage, component v., submultiple v. **-sperrzeit** partial restoring time (in echo suppression). **-spule** fractional coil, component or subdivision of a c. **-strich** graduation line, division l., d. mark (on a scale). **-ton** partial, p. note (either a higher harmonic or a sub-h.) **-trommel** divided drum, graduated d.
Teilung division, graduation, partition, pitch, scale, submultiplication. **Gleich-** equipartition. **Kern-** nuclear division. **Niet-** rivet spacing. **Vier-** quadripartition. **Zeilen-** line pitch (telev.) **Zwei-** bipartition, bisection.
Teilungs-ebene plane of division. **-fehler** faulty pitch (telev.), line p. defect (e.g., underlap of lines). **-gesetz** law of partition. **-koeffizient** distribution coefficient.
Teilvierpol section of recurrent structure.
Teleaufnahme telephotographic work.
Telegraphie, Bild- picture telegraphy, radio or wire picture transmission, phototelegraphy, facsimile, wire-photo. **Drahtwellen-** wired wave telegraphy, wire radio t., "wired wireless". **Unterlagerungs-** sub-audio frequency telegraphy system, composite operation (telegraphic and telephonic signals passed over one wire, often attended by thump noise).
Telegraphier-pause space (in telegraphic work). **-sender, tonmodulierter** AF or tone-modulated telegraphic transmitter.
Telegraphon télegraphone (Poulsen, Blattnerphone, etc.)
Telephon, Fernseh- television telephone, video t. **Funk-** radiophone. **Zungen-** reed or Brown-type telephone receiver.
Telephonie, Geheim- scrambled, garbled or secret telephony (based upon band transposition or inversion, use of scrambling circuits, speech inverters, etc.) **Mehrfach-** multiple telephony.
Telephonie-drossel magnetic modulator. **-sender, sprachgeschalteter** voice-modulated telephony (signal) transmitter or radiophonic transmitter.
Teller, Auflage-, Platten- turntable (of phonograph).
Teller-fuss plate- or dish-shaped foot, press or base. **-gerät, Zwei-** double turntable type of player. **-isolator** dish or disk insulator for antenna supporting.
Temperatur, charakteristische (Debye's) characteristic temperature. **-, schwarze** black-body temperature, cavity t. **Anlass-** tempering temperature, annealing t. **Farb-** color temperature.
temperaturabhängig temperature responsive, t. dependent, be a function of temperature.
Temperatur-beständigkeit temperature stability, t. constancy, condition of being unaffected by temperature (changes). **-feld** temperature gradient, t. "field" pattern. **-gang** variation of temperature, function or effect of t. **-leitfähigkeit** temperature diffusivity, thermal conductivity.

-**regler** temperature control, thermostat, thermo regulator, (electronic) thermautostat. -**strahler** incandescent light source, i. luminous radiator. -**strahlung** thermactinic radiation.
Tempoknopf speed regulator (in shape of a knob).
Tensor, Spannungs- stress tensor.
Tensoralgebra tensor algebra.
Terme, anharmonische anharmonic terms or constants.
Termin zur Verhandlung einer Sache anberaumen set a date for a hearing or a case, appoint a h.
Term-schema term diagram, t. scheme. -**verschiebung** term shift.
Tertiärempfang three-circuit reception.
Terzflöte third flute.
Tetraeder tetrahedron.
Teufe depth.
Textur, Schicht- sheet intergrowth, stratified texture.
tg α tan α.
T-Glied T-type section or filter mesh, T-type network m.
Thallium-sulfidzelle, -zelle thalofide (photo-conductive) cell.
Theaterkopie release print.
Thema theme, subject, subject-matter, topic.
thermische Fehlordnungserscheinung formation of holes (cryst.)
thermoelektrischer Effekt thermo-electric effect, Seebeck e. -**wärmemesser** thermel.
Thermo-element, Vakuum- vacuum thermocouple. -**galvanometer** Duddell's thermo-galvanometer. -**kraft** thermo-electric force. -**kreuzbrücke** thermo-couple or junction, thermal cross.
Thermometer, thermo-elektrisches thermel. **Registrier-, Schreib-** thermograph.
Thermoschalter thermal-lag switch.
Thorerde thoria (thorium oxide).
thorieren thoriate.
thoriertenKathode, Wiederbelebung der re-activation or rejuvenation of thoriated cathode, flashing.
thorisch toric.
Thoriumemanation thoron.
thoriumhaltiger Wolframfaden thoriated tungsten filament.
Tief-atzen intaglio etching (whole metal coated with resist, pattern cut out with scriber). -**decker** low-wing airplane. -**druck** intaglio printing process.
Tiefe depth of focus (of eye). **Eindring-** depth of penetration. **Modulations -** s. Modulationsgrad. **Schärfen-** depth of vision, d. of focus, definition in d. **Scharfenfeld-** depth of field. **Seh-** depth of focus, d. of field.
Tiefen bass-notes, low-frequency n., low-pitch n.; dark picture portions.
Tiefen-bestimmung, Fehler- stereoscopic radiograph for locating defects, depth determination of flaws. -**hervorhebung** bassy condition (low AF overemphasized). -**kompensation** reed armature arrangement to accentuate or emphasize low frequencies (in loudspeaker). -**konus** low-frequency cone loudspeaker, woofer. -**messer** fathometer, ocean sounding meter (e.g., supersonic), sonic depth finder. -**schärfe** depth of focus, d. of field. -**schrift** hill and dale track, Edison t. or sound groove, vertical recording (of phonograph disks). -**taster** depth gage. -**unschärfe** insufficient, or lack of depth of, focus. -**unterscheidungsvermögen** power of differentiating depths. -**winkel** angle of elevation below horizontal, a. of depression. -**wirkung** plasticity of image, plastic or stereoscopic effect (of picture).
tiefer Ton low-pitch sound, bass s. or note.
Tiefpassfilter low-pass filter.
tief-rund concave. -**schmelzend** low-melting. -**schwarz** jet black.
Tieftemperaturvergasung low-

temperature distillation or carbonization, partial c.

Tieftonkonus low-frequency cone loudspeaker, woofer.

Tiefung cupping (or drawing in drop press to make concave or dished articles).

Tiefungswert cupping value, ductility v.

tiefziehen deep draw, cup, dish.

Tiefzieh-fähigkeit capability of being cupped or drawn into article with deep concavity in drop press, etc., cuppability. **-probe** cuppability test (specimen), Erichsen test. **-verfahren** technological method of drawing sheet articles in drop press, cupping method.

Tiegel,Porzellan- porcelain crucible, p. pot.

Tiegel-brennofen crucible oven. **-flusstahl** crucible cast steel. **-form** crucible mold.

Tierleim animal glue, a. size.

Tikker ticker, tikker; radio chopper, c.w. train interrupter.

tilgen eradicate, destroy, cancel, erase, blot out, delete, efface, obliterate.

Tilgung (und Ausleuchtung) extinction or quenching (of fluorescence), evanescence.

Tilgung,Schall- sound absorption, suppression, damping, deadening or attenuation, abatement of noise.

Ton, dumpfer all-bottom sound. **-,gehaltener** sustained sound. **-,gleitender** glissando, sliding note. **-,hoher** h.-frequency note or sound. **-,hohler** boomy sound, dull s. **-,krächzender, kreischender** all-top sound or voice. **-,reiner** pure note, p. tone, simple t. **-,sendereigner** note or pitch peculiar to a beacon station, code note. **-,tiefer** low tone, low-pitch tone, bass note. **-,unreiner** impure tone, ragged note or t.

Ton,Anblas- Aeolian tone. **Ausfluss-** jet tone, slit t. **Dauer-** permanent note, tone, sound or signal. **Differenz-** difference tone. **Dinas-** Dinas clay. **Farb-** color tone, tint. **Feuer-** fireclay. **Fremd-** alien frequency or tone (in sound reproduction). **Ganz-** whole tone. **Halb-** semi-tone, sharp (mus.) **Heul-** sound with large number of modes of vibration. **Hieb-** Aeolian tone or note. **Interferenz-** beat note, interference n. **Kammer-** "A" of tuning fork. **Kapsel-** sagger clay, capsule c. **Keller-** tunnel effect, boom e. **Kombinations-** combination tone (either difference or summation t.) **Licht-** photographic sound recording. **Magnet-** magnetic sound recording. **Mis-** distorted note, off-pitch sound. **Nadel-** stylus- or needle-recorded and reproduced sound. **Netzbrumm-** motor boating (of a.c.), a.c. pickup, mains hum. **Ober-** overtone (often a harmonic), upper partial. **Rein-** note with one mode of vibration, single-frequency tone. **Schlag-** strike note (in a bell, followed by hum n.) **Schneide-** edge tone, flue stop tone (of flue or organ pipe). **Schwebungs-** beat tone, interference t., b. note. **Spalt-** slit tone, jet t. **Stopf-(einer gedackten Trompete)** stopped tone. **Stoss-** beat note, throb. **Summen-** summation tone. **Teil-** partial note, partial (being either a subharmonic or a higher harmonic or overtone).. **Triller-** trill, quaver, warble. **Übergangs-** transient. **Wechsel-** warble note, w. sound (air alarm). **Zwischen-** medium or intermediate tone or shade (of color).

Ton-abnahmestelle sound aperture, s. gate. **-abnehmer** pickup (of phonograph), soundhead (m.p.). **-abnehmertrommel** sound take-off drum, scanning d. **-abstufung** grading of tones, graduation of tonal intensities. **-abtastspalte** sound scanning slit. **-abtaststelle** sound gate, s. pickup point, s. scanning slit. **-analysator,Such-**

heterodyne analyzer of sound. -angeber,chromatischer chromatic pitch pipe. -ansatz,Licht- photographic sound film head. -ansatz,Nadel- sound-on-disk attachment or head. -arm tone arm. -art tonality, mode (major or minor), key. -atelier sound stage, studio, teletorium. -aufnahme-Kofferapparat box equipment, trunk unit. -aufnahme,Licht- sound film recording. -aufnehmer sound recorder. -aufzeichner, magnetischer magnetic sound recorder, magnetophone, Poulsen telegraphone, Blattnerphone, etc. -bandbreiteregler (tone) band width control. -belichtungsstelle sound gate, point where sound track is scanned, scanning light. -beseitigungsdrossel hum-bucking coil, hum eliminator choke. -bild tonal pattern, tone spectrum. -bildwand transoral screen (m.p.). -blende tone-control means, tonalizer; fader (continuously variable or otherwise). -blende,Rein- biasing or noise-reduction stop, gate, vane or shutter. -darbietung sound entertainment, acoustic e. -effekt,Raum- stereophonic sound effect, binaural e. -empfang modulated reception, tonal r. -empfänger,Bild- audio-video receiver combination. -empfindung acoustical perception, sound sensation, auditory percipience.

tönend(es) Signal tone s., tonal s., musical s. gleich- unisonant, consonant. selbst- ringing or squealing (of a tube, in radio apparatus).

Tonerde alumina, argillaceous earth.

tonerdehaltig containing alumina, aluminiferous.

Ton-erdesalz, aluminum salt. -farbe timbre, tone quality. -färbemittel tone shading means, tonalizer. -fenster sound gate, s. slit.

Tonfilm,Rundfunk- broadcast of video and aural signals, sound telecine or s. telecast, television program with s. accompaniment. Trick- sound cartoon. -optik soundhead lens or optic (m.p.), sound "optic". -rundfunk audio and video film broadcast, sound telecine. -wagen sound truck, location t., s. van. -zusatzgerät sound film head or attachment.

Ton-fixierbad toning or fixing bath. -folgeschrift sound track, s. record. -formstück,mehrzügiges multiple tile or clay conduit. -frequenz audio frequency, audible f., voice f., a-f. -frequenzen, höhere treble, high a-f or audio frequencies (handled by tweeter loudspeaker). -frequenzen,niedere bass frequencies, low a-f (handled by woofer loudspeaker). -führung labyrinth (of loudspeaker). -funkensender singing or musical (quenched) spark transmitter. -funkenstrecke quenched spark gap.

tongebend sound generative, acoustic, sonant.

Ton-gehör,absolutes absolute (genuine) pitch. -generator a-f generator, tonal or musical g. -gerät,Licht- sound film head or attachment. -haltungspedal tone sustaining pedal. -helligkeit pitch (of sound sensation). -höhe pitch. -höhevergleicher tonvariator, (sometimes) pitch pipe. -kamera sound recording camera. -kameramann recordist. -klebestelle blooping patch or splice (of sound film). -kodaskop sound kodascope. -kontrolle,Kopfhörer- monitoring with headset. -kopf sound film head or attachment. -kopie sound film used for reproduction. -lampe exciter lamp(mostly incandescent light-source) operating photocell. -laufwerk sound film feed mechanism. -leiter musical scale (natural, diatonic or just s., and tempered s.) -lichtlampe exciter lamp (mostly incandescent) operating photo-cell. -masse paste (cer.) -meister monitoring operator (working at control or mixer desk), recordist, sound engineer, director of sound. **-messer** tonvariator. **-mischer** tone fader, t. mixer.

Tonmodulierter Telegraphiersender tone- or AF-modulated telegraphic transmitter.

Tonmuffel clay retort.

Tonnenfehler barrel distortion, positive d.

Ton-optik soundhead lens, s. "optic" (m.p.) **-papier** tinted paper. **-probe** sound test; tonsil t. (for recording voice of a singer). **-punkt, Licht-** elementary shading value or density v. **-rad** tone-wheel.

Tonrichtiger Filter orthochromatic, true-color or correct-tone filter (opt.)

Ton-röhre singing valve. **-rolle** recording roll (of sound film). **-schiefer** clay slate or schist, argillaceous slate, argillite. **-schirm** sound screen. **-schlitz** reproducer slit, r. aperture, sound s. **-schwankung** pulsation of reproduced sound intensity (known as wowows up to 6 cycles per second, as flutter between 6 and 30 cps, gargles 30-200 and whiskers over 200 cps), generally called flutter regardless of cps (sound film). **-schwankungsmesser** flutter meter, f. measuring instrument (sound film). **-sender** modulated c.w. transmitter, tonic t.; audio, aural or sound t. (accompanying video program). **-spalt** sound slit, s. aperture. **-spektrum** sound spectrum. **-spur, eingeprägte** track embossed on film strip. **-spur, eingeschnittene** track engraved in film strip. **-spurbreite, Zacken-** width of variable-area sound track. **-steuermann** operator controlling sound volume in a theater. **-steuerstelle** control or exciter point (sound film).

Tonstreifen sound track, s. recording. **-, doppelspuriger** double-edged variable-width sound track. **-, einspuriger** single-edged variable-width sound track. **-, schwach ausgesteuerter** low-modulation track.

Tonstreifen, Licht-, veränderlicher Breite squeeze track (bearing photographic sound actions). **-verschmälerung** sound track squeezing, matting.

Ton-strom sound current, signal c., voice-signal c. **-träger** sound-track support. **-transportrolle** sound-film feed roller or sprocket wheel. **-trickfilm** sound animated cartoon. **-trommel** scanning drum, sound take-off d. **-überblendung** sound fading, s. fade. **-überlagerung** AF modulation. **-umfang** tonal range, bandwidth, compass. **-umschaltungseinrichtung** fading or shading device.

Tönung density value, shading v. **Beiz-** mordant toning. **Wärme-** heat tone (thermochemistry), h. quantity, h. effect.

tönungsrichtiges Bild picture of proper shading and contrast.

Tönungswert tonal value, gradation (of image).

Tonveredler high-frequency cutoff, h.f. filter, h.f. de-accentuator, sound clarifier, tone control comprising blocking condenser shunted to loudspeaker, tonalizer.

Tonverfahren, Klar- noiseless film method. **Magnet-** magnetic sound recording, telegraphone r. method, Blattnerphone m., etc. **Nadel-** sound on disk method. **Rein-** noiseless film method.

Ton-volumen, mittlerer power level (read in volume indicator), average l. **-wagen** sound truck, location t. or unit, lorry set, s. van. **-wahrnehmung** acoustical perception, sound or auditory sensation. **-welle** recording drum shaft. **-wellenspannung** audio wave potential (of sound accompaniment, in telev.) **-wender** tone converter. **-wiedergabe** sound reproduction. **-wiedergabeoptik** soundhead lens or "optic". **-zeitdehnung** acoustic slow motion (Silka method). **-zelle** sound cell (used in talking film). **-zentrum** tonal or acoustic center (of "gravity") of frequencies. **-zerlegung,-zersetzung** analysis of sound. **-zusatz,Licht-** photographic sound on film head or attachment. **-zusatz,Nadel-** sound on disk attachment or head. **-zuschlag** aluminous flux.

Topf can, shield; pot. **Sperr-** wave trap.

Topf-kreis tank circuit. **-magnet** shielded magnet, ironclad m. **-photometer** photometric integrator.

Tor unit of millimeter mercury column.

Tor,Strom- thyratron.

Torfkohle peat charcoal.

torkeln wobble, tumble, stagger.

Torsions-kopf torsion head. **-modulus** rigidity modulus, torsion m., shear m. **-wage** torsion balance.

tot(es)Feld dead zone (of a ribbon microphone), plane of ribbon. **-(er)Gang** lost motion, backlash. **-(es)Material** inert material, stable m.; neutral m., inactive m. **-(e)Wiedergabe** dead reproduction. **-(e)Windung** dummy turn, d. spire, idle turn, dead end. **-(e) Zone** skip zone, dead z.

tot,schall- non-resonant, acoustically inactive or inert, non-oscillable.

Totale medium-long shot, full shot, full-length figure (m.p.) **Halb-** (close)medium shot.

totbrennen overburn.

Töter,Krach- automatic silent tuning means, noise suppressor; n. gate (film recording).

Touren-Belastungskennlinie speed-load characteristic.

Trabanten satellites (spectr.).

träge inert, inactive, neutral, sluggish, slow-responding, lagging.

Träger carrier, bearer, supporter, foundation, bed, supporting means, base (of something), vehicle, backing; beam, girder.

Träger,positiver positive carrier, ion. **Aufzeichnungs-** recording or sound track or support (sound on film, on disk, etc.). **Blenden-** diaphragm carrier. **Film-** film base, f. support, f. foundation. **Hilfs-** auxiliary carrier, sub-carrier. **Ladungs-** charge carrier. **Lichtton-** photographic sound track support. **Mikroskop-** microscope stage. **Objekt-** object support, o. carrier (plate), specimen holder (e.g., celluloid skin or film), slide, mount, stage, stand. **Schirm-** carrier, foundation or support of screen (coated with phosphor substance, in cathode-ray tubes). **Wicklungs-** former, skeleton form, winding frame, coil form. **Zusatz-** missing or suppressed carrier re-introduced in receiver (in side-band signal transmission). **Zwischen-** sub-carrier.

Träger-folien supporting foils. **-frequenzfernmessung** carrier telemetering (e.g., metameter of GE Co.). **-plättchen,Objekt-** object support, o. lamina, o. slide. **-seite** film base side or face. **-strom,zeitlich veränderlicher** wobbled carrier current, warbled c.c. **-unterdrückung** carrier suppression (in quiescent c. telephony). **-wegtastung** carrier gating, c. suppression, c. interruption (in synchronizing work). **-zusatz** re-insertion of carrier.

Trag-fähigkeit buoyancy, lift, bearing, carrying or supporting strength, carrying capacity; yield, productiveness. **-flächenholm** wing spar. **-flügelprofil** aerofoil section.

Trägheit inertia, sluggishness, inertness, slowness, persistence (of eye or retina), time-lag. **Netzhaut-** retinal persistence.

Trägheitsmoment moment of inertia.

Trag-körper(des Magnetsystem) chassis (of magnet system). **-stütze** supporting stay, s. prop.

tränken steep, soak, saturate, impregnate.

Transformator,Abwärts- step-down transformer. **Anpassungs-** matching transformer. **Anzapf-** tapped transformer (especially center-tapped t.), split t. **Aufwärts-** step-up transformer. **Ausgleich-** hybrid- transformer, three-winding t., balanced t., differential t. **Dreiwicklungs-** hybrid transformer, three-winding t. **Käfig-** shielded transformer. **Netz-** mains transformer, power t. **Phasen-** phase shifting transformer. **Reduktions-** step-down transformer. **Saug-** booster transformer, suction t., sucking t. **Spar-** auto transformer. **Treiberröhre-** driver tube transformer. **Zwischen-** inter-stage transformer.

Transformator-gehäuse transformer shell, t. container, t. tank. **-kurve** amplification curve of a transformer.

Translations-fläche translation plane, slip p. (cryst.) **-linie** slip band (cryst.) **-streifung** translation gliding striae, crystal ridges.

Transparenz transparency, transmittancy, transmissivity (for light). **-schwankungen** modulus of transmission (film).

Transponierungs-empfang heterodyne reception. **-steilheit** slope or mutual conductance (of mixer tube).

Transport,Film- film feed, f. movement, film travel. **Moment-** momentum transfer.

Transporteur vernier, protractor.

Transport-kette conveyor chain, transfer c. **-mechanismus** feed mechanism, conveyor m. **-querschnitt** transport cross-section. **-rolle, Film-** picture film feed sprocket. **-rolle,Ton-** sound film feed sprocket. **-theorie,Impuls-** momentum transport theory (of turbulent flow). **-theorie, Wirbel-** vorticity transport theory. **-trommel** intermittent sprocket.

Transversalmethode variable-area method (of sound recording).

Trapezfehler keystone, trapezium or trapezoidal distortion (due to dissymmetry to earth of deflection plates, to electron beam angle, etc.)

Trefferzahl number of impacts or shocks.

Treffweite (=Brennweite in Optik) focal length (of electrons = f. distance in opt,)

T Regler T-type section attenuator.

Treiber driver, punch; propelling means. **-röhre-Transformator** driver tube transformer. **-system** driver system.

Treib-mittel driver, driving, motor or motive fluid or substance, fuel, propellant. **-probe** cupping test. **-sitz** driving fit. **-werk,Schall-** loudspeaker operating or driving mechanism or means, motor m.

Tremolostimmung, Tremulant tremolo, tremulant effect.

trennen separate, sever, sunder, divide, resolve, decompose.

Trenn-faktor separating factor. **-festigkeit** separating strength, static crack s. **-filter** dividing filter, separating f. **-frequenz** cut-off frequency. **-netzwerk** dividing network. **-röhre** isolating tube, buffer t. **-schalter** isolating switch, disconnecting link, isolator. **-schärfe** selectivity, filter discriminator. **-schärferegelung** automatic bandwidth selection. **-stromröhre** spacing valve (in telegraphy).

Trennung selection, filtering, separation (by filtering network, etc.); sorting (of ions, in mass spectrometer).

Trennungs-fläche parting plane, cleavage p., surface of separation. **-linie** dividing line, boundary l., partition l., demarcation l., l. of separation. **-winkel** angle of separation (of solid moving through fluid.)

Treppe, Roll- escalator.

treppenförmig (s.a. **Stufen**) in the form of stairs, stepped, echeloned, terraced.

Treppen-rost step grate. **-stoss** lap joint.

treppenweise echeloned, scalariform, stepwise, by stages, by degrees, gradually.

Tretkurbel foot pedal, treadle.

Treue, Farb- color fidelity (of color film), c. match, orthochromaticity.

treue Wiedergabe faithful reproduction, fidelity of r., orthophonic r. (of sound).

Trichter horn (of loudspeaker, with throat at narrow end and mouth at end with largest flare), funnel, hopper, cone, gate (in foundry); electromagnetic horn (of wave guide).

Trichter, Exponential-, abgestufter multi-flare exponential horn. **-, aufgewundener** twisted or curled exponential (loudspeaker) horn. **-, gefalteter** folded exponential horn.

Trichter, Riffel- corrugated horn. **Rippen-** ribbed funnel. **Schweige-** cone of silence marker (for airplane landing).

Trichter-einlage filter cone. **-hals** throat of (loudpeaker) horn.

trichterloser Lautsprecher hornless loudspeaker, direct radiator l.

Trichter-mundoffnung mouth (opening) of horn, largest flare. **-mundstück** throat of horn, narrow inlet of h. **-mündung** mouth opening of horn, largest flare. **-rohr** funnel tube, tube funnel. **-vorhof** air chamber of loudspeaker (between cone and horn). **-wand, schallharte** rigid or non-absorbing horn wall.

Trick trick, artifice, slight of hand, stratagem; trick effect or "stuff" (m.p.). **-bilder** trick shots, t. pictures. **-film, Ton-** sound animated cartoon. **-film, Zeichen-** animated cartoon. **-filmzeichner** animator. **-kamera** trick camera. **-tonfilm** sound cartoon.

Trieb, Spindel- worm drive. **Zahnrad-** rack and pinion drive, toothwheel drive mechanism.

Trieb-element motor element (of loudspeaker) **-gehäuse** gear housing, g. casing, g. box. **-knopf** milled head, pinion head (in microscope). **-kraft** moving force, motive power, propelling f. **-mittel** driving or motor means; fuel, propellant. **-schraube** coarse adjustment (in microscope). **-werk** machine, engine, drive mechanism, motor element (loudspeaker, etc.); gearing, transmission gear.

Triftröhre (cf **Laufzeitröhre**) drift tube, beam t. (e.g., Klystron).

Trillerklappe shake key (wind instrument).

trillern trill, quaver, warble.

Trillernote (cf **Wechselnote, Heulton**) trill note, warble note.

Triograph electrocardiograph with CR tube.

Tripelpunkt triple point (in metallography).

Tritt kommen, in (cf **Phasenaussortierung**) come in step or phase, bunching, p. focussing (of electrons).

Trittschall impact sound (transmission). **-dampfung** foot fall sound attenuation.

Trocken-apparat drying apparatus, drier, desiccator, dehydrator. **-entgasung** dry distillation. **-gerüst** drying frame, d. rack, **-gleichrichter** copper rectifier, cuprous-oxide r., selenium-oxide r., metal r., dry r. **-kompass** dry compass. **-kondensator** dry electrolytic condenser. **-linse** dry objective, non-immersion o. **-mittel** drying agent, drier, siccative, desiccative. **-rückstand** dry residue. **-stoff** siccative, drier, drying substance.

Trog, elektrolytischer electrolytic tank (for field pattern mapping).

Trogbatterie trough battery.

Trommel drum, cylinder, tympanum. **-, grosse** bass drum. **-, kleine** snare drum.

Trommel, Belichtungs- exposure drum, recording d. (on which film is exposed to modulated light beam). **Linsen-** lens drum (scanner). **Loch-** scanning drum (of Jenkins). **Mess-** graduated drum. **Nachwickel-** lower take-up reel. **Schalt-** intermittent sprocket (m.p.). **Sprossen-** sprocket drum. **Teil-** divided drum, graduated d. **Ton-** scanning drum. **Transport-** intermittent sprocket. **Vorwickel-** upper pull-down sprocket. **Wirbel-** high side drum. **Zacken-** sprocket wheel.

Trommel-blende scanner drum, d. with scanning apertures. **-fell** drum-head, tympanic membrane, tympanum (of ear). **-flote** fife. **-schlägel, -stock** drum stick. **-welle, Antriebs-** drive sprocket shaft.

Trompete trumpet.

tropfbar capable of forming drops or of dropping or dripping, liquid.

Tröpfchen globule, spherule, droplet, small bead. **-methode, Öl-** oil drop method.

Tropf-düse dripping nozzle. **-elektrode, Quecksilber-** dropping mercury electrode.

tropfen drop, drip, trickle.

Tropfen, Schweiss- welding dribble, w. bead.

Tropfenmesser stalagmometer, stactometer.

Tropf-gewichtmethode drop weight method. **-hahn** dropping cock. **-trichter** dropping funnel.

Tropfungszeit dropping time.

Trübe dross, pulp, slush, slurry, slime; turbidity, cloudiness.

trübe turbid, cloudy, dull, thick, veiled, foggy, frosted, dim, opaque, muddy.

Trübglas frosted glass, opal g., glass diffusing light.

Trubung (cf **Enttrübung**) tarnishing (on metal surface), clouding (of glass), opacity, night and other effects tending to impair sharpness of zero in d.f., wavering of signal beam, blurring (of minimum). **Funk-** blurring of minimum signal or of zero point in radio d.f. or radio bearings, uncleared zero.

Trübungs-grad degree of turbidity. **-koeffizient** coefficient of turbidity, Ångstrom c., degree of t. **-messer** (cf **Belichtungsmesser**) turbidimeter, opacimeter, nephelometer (type of photometer). **-spannung** potential causing QE (quadrantal errors).

Trug-bild illusion, phantom, fata morgana. **-schluss** erroneous conclusion, fallacy, fallacious argument.

Trumm film strip. **-, schlaffes** slack end (or driven portion) of a belt, rope, etc. **-, straffes** driving or taut end (of a belt).

Trümmer fission products, f. fragments, remains.

Tubus, Ansatz- extension tube. **Auszieh-** draw tube (of microscope). **Einstell-** micrometer adjusting means, adjuster tube (in electron microscope). **Optik-** optical barrel. **Projektions-** projection tube (micr.)

Tubus-schlitten tube slide. **-trager** tube support.

Tunneleffekt tunneling effect.

Tupfelanalyse spot analysis, drop a.

tüpfeln dot, spot, speckle, stipple, mottle.

Turbine, Anzapf- bleeder turbine. **Elektronen-** cyclotron and inverted (or Hollmann) forms thereof, magnetic resonance accelerator, induction electron accelerator (also called rheotron).

Turbinen-dichtung turbine (labyrinth)

seal. -dichtung, Kämmen der - rubbing of turbine seals.
-unterbrecher turbine interrupter, t. break.

Turm, wickel- mandrel.
Turmalinkristall tourmaline crystal.
Türöffner, lichtelektrischer photo-electric door opener.

überakustischer Raum(mit übertriebener Nachhalldauer) abnormally live space (having undue reverberation period).

Überanpassung, Widerstands- overmatching of impedance (load resistance greater than internal resistance).

überanstrengen over-exert, overstrain.

Über-anzeige overswing (of volume indicator). **-belichtung** overexposure. **-besserung** over-correction, o.-compensation.

überblasen over-blow (of a flute).

Überbleibsel residue, residuum, remainder, remnant.

überblenden dissolve, fade-over, mix, lap.

Über-blender fader, means to cause change-over from one projection to the next (m.p.); sound fader, means for fading in and out, blending and mixing (sometimes with central tap); fading mixer (telev.) **-blendung, Ton-** sound fading, s. fade. **-blendungsblende** dissolving shutter, lap-dissolving s. (for gradual transition from one scene into another).

überbrücken shunt, bridge, by-pass.

Überbrückungskondensator by-pass condenser, bridging capacitor.

über-dauern outlive, outlast, survive. **-decken** swamp, blanket, mask, cover up, eclipse, conceal.

Überdeckung, Zeilen- line overlap.

Über-drehen super-speed operation (in film work). **-drehzahl** excessive speed. **-druck, Atmospharen- (Atü)** atmospheres excess pressure.

Übereinandergreifen overlap (in diffraction pattern, scanned lines in telev., etc.); engage over one another.

Übereinanderlagerung superposition, superimposition.

Übereinanderliegend superjacent, superposed.

Über-einanderschweissung lap weld. **-einkommen** agreement, arrangement, convention, contract, compact.

über-einstimmen agree, coincide, harmonize, acquiesce in, correspond, register (one part with another), be in unison or synchronism (of phases). **-endlich** transfinite. **-entwickelt** overdeveloped, overdone, "cooked" (m.p. slang). **-exponiert** overexposed.

Überfahren des Mosaikschirmes durch Strahl sweep out mosaic with beam.

überfällig overdue.

Über-fallrohr overflow pipe. **-fang** covering. **-fanggläser** flashed opal glasses. **-fangschicht** glass liner, g. lining.

Über-feine Struktur hyperfine structure. **-flüssig** superfluous, overabundant, in excess, excessive; overflowing. **-fuhren** convert, transport, transfer, convey, reduce (to practice or to a practical form); convict.

Überführung, Weg- overpass, overhead crossing.

Überführungs-geräte(Kabelkasten) terminal equipment (cable box). **-kasten** test box; cross-connecting terminals. **-zahl** transport number (of ions), transference n.

Übergang transition, passage, crossing, (gradual) change (from one place or condition to another), blending, shading (of colors). **-von...zu** transition or change from....to; replacement of....by, substitution of....for.

Übergang, erlaubter permitted transition (in accordance with selection principle). **-, erzwungener** forced transition, non-spontaneous t. **-, unerlaubter, -, verbotener** forbidden transition. **-, verflauter**

softly shaded transition (lacking definition) (telev.). Bewegungs- motional transient. Phasen- phase transition. Quantum- quantum transition, q. leap.

Übergangs-einrichtung change-over (from one projector to another). **-glieder** transition types, t. members. **-leitwert** transconductance, transfer characteristic. **-punkt** transition point, place of t. or transfer. **-schicht** transition layer (opt.) **-stelle** transition point, place of t. or transfer. **-töne** transients. **-verlust** contact loss, transition l., junction l. **-wahrscheinlichkeit(der Elektronen von einem Niveau zum andern)** transition probability (of electrons to leap from one level to another). **-widerstand** contact resistance. **-zahlen** transference numbers (of solution). **-zustand** intermediate stage, intermediary s., transition s., transiency.

Übergewicht overweight, extra w., unbalance, desequilibrium, preponderance, predominance.

Über-glasen glaze, overglaze. **-greifen** overlap, engage over, encroach, transgress, cross a boundary. **-helles Stimmungsbild** high-key picture. **-hitzen** superheat (beyond saturation), overheat.

Überhöhung bank or cant (in a curve), hump, super-elevation (of rails), lobe (as on a cam).

Überholen overreach, outstrip, outrun, overtake, catch up with.

Über-holungsgebiet der Elektronen catchup, overtake or bunching range of electrons (in phase focusing, in beam-tube drift-space). **-hörfrequenz** supersonic, ultrasonic, ultra audible, superaudible or ultra-audion frequency.

Über-individuell super-individual. **-kalten** supercool, undercool.

Über-korrektur overcorrection, overcompensation. **-kreuzungspunkt, Elektronen-** electron crossover (point where principal rays from first lens cross axis = exit pupil in optics; in electron optics, crossover lies between cathode and first anode). **-krümmung** excessive curvature. **-kühlung** supercooling, overcooling, underc. **-lader** supercharger (of engine). **-lagerer** heterodyne, local oscillator (based on beat principle) e.g., of endodyne, autodyne or self-heterodyne type. **-lagerer, Selbst-** autodyne, self-heterodyne.

Überlagerung heterodyning, production of difference frequency or beat; superposition, impression of one wave or f. upon another. Ton- AF modulation, tonal m.

Überlagerungs-empfang heterodyne reception, beat r. **-kreis** circuit in which heterodyne or beat action is produced, heterodyne c. **-stufe** mixer stage, first detector (in superheterodyne receiver). **-summer** heterodyne warbler oscillator.

überlappt schweissen lap-weld, scarf-weld.

Überlappung over-lap (of lines, in telev. image).

überlastet overloaded; overrun (by working a lamp in overvolted condition).

Überlastung overload, overcharge, overtax; overrunning(of a lamp by operating it at voltage far above rated), overvolting.

Über-legung consideration, reflection, deliberation. **-leitfähigkeit** superconductivity, suprac. **-lichtung** over-exposure. **-mass** allowance (in press fits). **-mikrometer** ultra-micrometer. **-mikroskop, Durchstrahlungs-** transmission type ultra-microscope, super-m., electron m. **-mikroskop, Selbststrahlungs-** self-emissive electron microscope, self-illuminating m. (based on electrons arising on object surface). **-nahme, Strom-** transfer of current, tapping of c., c. tap.

-regulierung overshooting, excessive regulation, e. compensation, over c.
über-rot infrared, ultra-red. -sättigen supersaturate.
Überschall-lichtzelle supersonic light relay or valve. -wellen supersonic, ultra-sonic or u.-audio waves. -wellenanhäufung (zone of) accumulation of ultrasonic waves. -zelle supersonic light relay or valve.
Uberschiebung transvection. -schlag nose over (of an airplane). -schlag,Funken- sparkover, flashover, breakdown (say, of a spark-gap). -schmelzung superfusion, enameling. -schneidung overlapping, intersection (of lines), intercept; overcutting (when adjacent record grooves touch). -schreien overmodulation. -schreitung exceeding (a certain limit, mark, level or value); transgression. -schrift heading, caption, title.
überschwellige Schallreize super-threshold (sound) intensity levels. - Sichtbarkeit super-threshold visibility.
Übersee-Empfanger transocean receiver, transatlantic r.
Übersehen oversee, overlook, survey, sweep (with eye).
Ubersetzung,Frequenz- frequency transformation.
Übersetzungs-rohr converter tube (used in avc schemes between regulating and regulated tubes). -verhältnis transformation ratio, gear r.
Übersicht survey, review, summary, extract, synopsis, abstract, digest.
über-sichtig long-sighted, hypermetropic. -sichtlich easily followed up, easily understandable, visible, clear and simple, not complicated or intricate.
Ubersichtsspektrum general spectrum.
überspannt(er)Betrieb overvoltage operation, overvolted o., over-running. -(e)Entladung overvolted discharge.
Überspannung overvoltage, excess v., overrunning, overvolting.
Über-spannungsfunkenstrecke surge arrester, s. gap, s. absorber. -spielen re-recording. -sprecherscheinung cross-talk, babble (when from several channels), "monkey chatter"; distortion, lack of sensitivity.
über-stehend projecting, protruding, salient, standing over, surmounting; surnatant, supernatant (of liquids). -steuern overshoot, overload, overmodulate. -steuert over-"shooted" (m.p.)
Übersteuerung overloading (of tubes, causing blasting); over-modulation (in film recording), sound overshooting.
Übersteuerungs-abschneider overload chopper. -messer overload indicator. -punkt overload point (of a recorder). -schutz overload limiter (for film).
Über-strahlen der Ramanlinien swamping or outshining of Raman spectral lines. -strahlung irradiation. -streichen des Leuchtschirmes sweep out of television screen. -streichung eines Frequenzbereichs durch einen Kondensator sweeping of a frequency range by a condenser, coverage of f. band, scanning a range. -struktur superlattice; superstructure.
übertragen transfer, transmit, transport, convey; translate, transcribe; assign, vest in, delegate, confer, grant, cede.
Übertragender assignor, grantor, transferor.
Ubertrager,Brücken- differential transformer. -Stromstoss- impulse repeater.
Übertrager-kondensator grid condenser (in audion or grid-current detector). -kreis intermediate circuit, transformer c.
Übertragung,mehrwegige multipath transmission, multichannel t. Bild- fade-over (m.p.). Einseitenband- single-side-

band or vestigial s.-b. transmission. **Leitungs-** wired radio, w. wireless, carrier télegraphy or telephony, wire broadcast.

Übertragungs-band signal band. **-einrichtung,Schall-** (public) address system, electroacoustic transducer. **-erklärung** assignment (declaration), deed of transfer. **-faktor** transfer factor, image t. constant. **-frequenzlinie** transmission-frequency characteristic. **-gegenstand** teleview object, televised o., o. to be televised or transmitted by video signals. **-güte** transmission performance (in rating). **-kanal** transmission channel. **-kurve** transmission-frequency characteristic. **-mass** image transfer, attenuation constant. **-mittel** transmission medium, t. means. **-schwingungen** signal oscillations, s. waves. **-urkunde** deed of assignment, assignment. **-weg** transmission channel, t. path, t. medium.

Über-tretung contravention, offense, violation, transgression. **-verdichtung** supercharging. **-vergrösserung** extra or supplementary magnification, over-m.

übervIolett ultraviolet, of vita-ray nature (when of wave-length 2900-3200A).

Über-wachsungen overgrowths. **-wachung** tell-tale device, monitoring, supervision, watching. **-wachungsschalter od.-schlüssel** monitoring key.

überwiegend preponderant, predominant, prevalent.

Überwucht unbalance, imbalance.

überziehen cover, coat, line, plate, overlay, put on or over.

Überzug flashing, coat, cover, skin, plating, lining, envelop; incrustation, crust. **Gummi-,auf Rollenlaufbahn** rubber-tread of roller.

üblich conventional, customary, usual, common, ordinary.

u.f.,uff. and the following, et seq(q).

Uhr,Film- film footage counter. **Haupt-,Mutter-** master clock. **Neben-** secondary clock.

Uhrschraubenlehre dial micrometer.

U K W (Ultrakurzwellen) (cf Wellenabgrenzung) ultra-short waves, micro w., midget w., uhf w., quasi-optical w.

Ulbricht'sche Kugel Ulbricht sphere type photometer.

ultrakurze Wellen (cf Wellenabgrenzung) ultra-short waves, uhf w., quasi-optical w., micro w., midget w.

ultrarot infrared, ultrared.

Ultrarot-abtastung noctovisor scan. **-durchlässigkeit** diathermancy, ultrared or infrared transmittancy.

ultrarotundurchlässig athermanous, opaque to infrared.

Ultra-schallzelle supersonic light valve. **-schwärzung** blacker-than-black condition, infra-b.c.

Ultrastrahlung,kosmische cosmic rays or radiations.

ultraviolette Strahlen ultraviolet rays, vita r. (between 2900 and 3200 Å).

Ultra-wasser optically empty water. **-zentrifuge** ultra-centrifuge, high-speed c.

umändern alter, change, modify, amend (an application).

Umänderungsprozess,Kern- nuclear disintegration.

um-biegen bend over, round or back, double-back, crimp.

-bilden re-model, re-form.

-bördeln bead over, flange (all around). **-drehen** turn around, twirl, rotate, revolve.

Umdruck reprint, circular. **-papier** transfer paper.

um-fällen dissolve and re-precipitate. **-falzen** crimp over, bead.

Umfang circumference, periphery, circuit, circle, contour; scope, extent, compass, range, dimension, latitude, size, volume, confines. **Aufzeichnungs-** recording range. **Helligkeits-** brilliance, brightness or contrast range, key (of a picture), brightness relationship.

Lautstärke- volume range. **Licht-** range of light oscillations. **Negativ-** intensity range of negative. **Objekt-** range of brightness (in subject). **Schall-** sound intensity range, s. contrast. **Schwärzungs-** range or latitude of densities, density scale. **Ton-** tonal range, compass.

Umfang-kopieen,Gross- "hi-range" or wide-range prints. **-Kopien,Klein-** "lo-range" prints. **-kraft** tangential force, circumferential f., peripheral f.

Umfangs-geschwindigkeit circumferential or peripheral speed. **-geschwindigkeitsmesser** tachometer (e.g., stroboscopic, with strobotron). **-schwingungen** circumferential oscillations or vibrations.

umfassen embrace, comprise, span, encompass.

umfassend comprehensive, extensive, embracing.

Umfläche circumferential, contour or ambient surface.

umfliessen flow around or in contact with, circumcirculate, skirt in flowing.

Umformbarkeit re-formability, deformability.

Umformer converter, inverter (from d.c. to a.c.), thyratron; transducer. **Mess-** measuring transducer. **Pendel-** vibrating rectifier. **Schweiss-** welding converter.

Um-gangsklappe by-pass (clack) valve. **-gebung** surroundings, environment, ambient, ambiency.

umgehen by-pass, circumvent, obviate, dodge, avoid, evade, go around, circumnavigate.

Umgehungsschaltung by-pass circuit, b.-p. connection.

umgekehrt proportional inversely proportional.

umgekehrter oder umgestulpter Fuss re-entrant squash or press (of a tube).

umgiessen re-cast; cast around (something).

Umgrenzungslinie contour, boundary, peripheral or circumferential line.

Umgrösserung enlargement or reduction, change in size or dimension.

Umgrösserungsverhältnis ratio of enlargement, r. of magnification.

Umguss transfer (by pouring), decantation; recasting, recast.

Umhangeriemen shoulder strap.

Umhüllung covering, wrapping, casing, envelope, jacket, sheathing, shrouding.

Umhüllung,Papier- paper wrapping.

Umhüllungslinie envelope, contour line.

Umkehr,Selbst- self-reversal (of spectral lines)

Umkehr-absorptionskanten reverse absorption edges or limits. **-bad** reversing bath.

umkehrbar reversible, revertible, capable of being turned over or out.

Umkehrentwicklung reversal processing, r. developing.

Umkehrer,Phasen- phase inverter, p. reverter.

Umkehr-erscheinung(bei Beleuchtung von Schichten) photographic reversal or solarization (in layer exposure). **-film** film resulting from developing by reversal. **-prisma(cf. Umkehrungspr.)** inverting prism. **-punkt** reversal point. **-röhre** converter tube.

Umkehrung,Band- speech band inversion or transposition (in garbled or scrambled secret telephony). **Bild-** image inversion, i. reversion (when two axes, i.e., right and left, top and bottom simultaneously interchanged), solarization (reversal of gradation sequence, due to overexposure) **Silben-** syllable inversion (in garbled telephony).

Umkehrungsprisma,Doppel- inversion prism (inverts in both axes of image).

umkippen unlock time base.

Umklappen reversal (say, of motion).

um-kleiden line, sheath, cloak, surround. -kloppeln braid, put braiding around.
Umkopie optical reduction print, o. p. of altered dimensions.
Umkopieren optical printing.
Umkreis periphery, circumference; perimeter; surroundings.
umkreisen circle, encircle, revolve, spin, rotate, gyrate (around).
Umkristallisieren re-crystallizing (resulting in offsetting in tungsten wires).
Umladung change in charge, c. in sign (of molecules and atoms).
um-lagern re-arrange, re-group. -laufen revolve, rotate, circulate, gyrate (spirally). -laufend circulating, circular, gyrating, revolving.
Umlauf-frequenz rotational frequency (of electrons). -getriebe planetary gear, sun and planet g. -integral contour integral. -räder planetary wheels. -schwingungszahl rotation frequency. -spannung magnetomotive force, mmf (line integral of magnetizing force). -verschluss rotary shutter. -verschlussblende, rotierende rotary (flicker) shutter, light cut-off.
umlegbarer Klappsucher reversible or folding finder.
umlegen reverse, tilt, tip, throw (a switch), lay over or around, wrap.
Umlenk-prisma deviating prism. -rolle flanged idler roller, deviating r.
um-lernen learn anew, unlearn and re-learn. -manteln jacket, sheath, case, encase.
Um-modulierung re-modulation. -polung reversal of polarity.
umpressen press (-fit) around, put or apply (around something) by pressure.
Umrandungskurve envelope.
Umrechnung re-calculation, re-computation, conversion.
Umrechnungs-faktor conversion factor, reduction f., change ratio. -tafel conversion scale (say, meters to kilocycles).
Um-richter (cf Wechselrichter und Gleichr.) transverter (to change a-c to d-c or vice versa). -risszeichnung linear drawing, sketch, outline illustration, -roller re-winder. -satz, Leistungs- energy exchange, transduction.
umschaltbar adapted to circuit changes or changes in connection, switchable, reversible (by switch or similar action).
Umschalter(cf Schalter) reverser, commutator switch, switch, reversing gear; switchboard or panel. Gabel- hook switch. Steck- plug-in switch.
Umschaltung, Lauf- reversal of motion.
Umschaltungseinrichtung, Ton- fading or shading device, fader.
Umschlag cover, covering, wrapper, envelope; sudden change; hem, collar. Anker- armature travel, a. excursion, a. transit.
Umschlagspunkt transition point.
umschlingen wind around, wrap a., clasp, loop, cling to, embrace.
Umschlingungswinkel looping angle.
umschmelzen re-cast, re-melt, refound; melt around (something).
Umschnurung, Faden- serving of thread (on a cable).
Umschreiben re-writing, transcription, circumscription, description (math.); re-recording, electric transfer of sound track onto film or disk.
umschreibender Kreis circumscribing circle.
Umsetzung transposition, double decomposition, conversion, change, transformation, transduction.
Umsetzungsgerät transducer, transductor.
Umspanner, Schweiss- welding transformer.
um-spielen re-record, play back. -spinnen cover, braid or spin around.
Um-springen jumping, abrupt

reversing, arising of sudden irregularity or unsteadiness. **-spulen** re-winding (of film).
umspülen flow around or in contact with, skirt, circumcirculate.
Umstände,mildernde extenuating circumstances.
umstellen re-adjust, re-adapt, reverse, transpose, invert.
Umstellungspunkt transposition point.
umströmen flow around, f. in contact with.
umstülpen overturn, invert, put upside down or inside out.
Umwandler mutator, converter, transformer, transducer, sink. **Energie-** transducer, sink.
Umwandlung transmutation, conversion, transformation, change. **Kern-** nuclear transmutation.
Umwandlungs-einrichtung,Schall- (electro-) acoustic transducer. **-punkt** transformation point, transition p. **-steilheit** mutual conductance or slope (of tube). **-tabelle** conversion table. **-vorrichtung** transducer,sink. **-wärme** heat of transformation.
Um-wegleitung by-pass (lead). **-welteinfluss** environmental influence.
um-wenden turn, turn over, about, sideways or upside-down, invert, revert. **-wickeln** wrap round, cover, wind on or over, re-wind.
Umwindung convolution.
unabgestimmt untuned, aperiodic, unresonated, non-resonant.
unabhängig,frequenz- free from frequency effect, independent of f., not a function of f.
Unabhängigkeit,Frequenz- frequency independence, condition of a quantity of being independent of, or unaffected by, frequency.
Unabhangigkeitssatz independence theorem.
un-auffällig inconspicuous, concealed, attracting no attention. **-aufgeloste Linienbilder** unresolved line patterns. **-ausdehnbar** inexpansible, non-ductile, inextensible. **-bearbeitet** raw, crude, unwrought, unfinished, in blank form or b. condition.
Unbefugter intruder, unauthorized person.
un-belastet unloaded. **-bemannter Ballon** unmanned balloon. **-benannt** unnamed, anonymous; indefinite (math.) **-beschadet** without prejudice to. **-beschichteter Blankfilm** uncoated plain film base. **-besetztes Band** empty band. **-bestimmt** undetermined, indeterminate, indefinite.
Unbestimmtheitsrelation indeterminacy relation, uncertainty r.
un-bewaffnetes Auge unaided eye, naked e. **-bezogene Farben** unrelated colors. **-bunte Farben** hueless colors, achromatic c., c. devoid of hue. **-dehnbar** inextensible; non-ductile. **-dicht** untight, pervious, leaky. **-durchdringlich** impermeable, impervious, impenetrable. **-durchlassig** impermeable, impervious, opaque.
undurchlässig für aktinische Strahlen adiactinic. **-fur Ultrarot.** athermanous (to infra-red rays). **-fur Roentgenstrahlen** radiopaque. **licht-** light-tight, opaque or impervious to light, non-diaphanous. **warme-** athermanous or impervious to heat, heat insulating.
un-durchsichtig opaque, impervious, non-transparent, non-diaphanous. **-echt** pseudo..., non-genuine, imitated. **-echte Längenmessung** indirect measurement of length. **-edel** base, not noble or rare. **-elastisch gestreut** unelastically scattered. **-empfindlich** insensitive, non-reactive, non-responsive, immune from, neutral, inert, unsusceptible. **-empfindlich,richtungs-** non-directional, astatic.
Unendlicheinstellung infinity adjustment (opt.)
un-endliche Reihe infinite series.

-entflammbar flame-proof, non-inflammable, non-ignitable, non-combustible. **-entflammbarer Film** safety film. **-entgeltlich** gratuitous, gratis, free (of charge). **-entwickelt** implicit (math.) **-erlaubter Übergang** forbidden transition. **-erwünscht** undesirable, objectionable.

Unfähigkeit disability, inability, incapacity, incapability.

unfrei constrained, engaged, tied, bound.

Ungänze, kleine minor discontinuity or defect.

un-gebundene Ladung free charge. **-gedämpfte Anzeige** ballistic or non-aperiodic reading or indication (of instrument). **-gedämpfte Schwingung** undamped oscillation, sustained o., continuous wave (c.w.), persistent w. **-gefasstes Glas** rimless lens. **-geordnete Geschwindigkeit** velocity of agitation. **-geordneter Zustand** disordered state, d. arrangement. **-gepufferte Lösung** unbuffered solution. **-gerade** uneven, out of true, not straight, non-linear, crooked, odd (harmonic, multiple). **-geradwertig** of odd valence. **-gerichtet(e) Antenne** non-directional antenna. **-(es) Mikrophon** astatic microphone, non-directional m. **-geschaltet** unwired. **-gesehnte Wicklung** full-pitch winding. **-geteilter Ring** unsplit ring, solid r.

Ungewissheitsprinzip uncertainty principle, indetermination p., indeterminacy p.

ungleich unequal, unlike, uneven, dissimilar, odd, not uniform, different, unmatched.

Ungleichgewicht desequilibrium, unbalance, imbalance.

un-gleichnamig unlike, opposite (of poles). **-gleichseitig** scalene (triangle).

Un-gleichung inequality, non-equality. **-gültigkeit** invalidity, nullity, state of being void, annuled or not in force.

unhaltbar untenable. **-hämmerbar** non-malleable.

Universal-gelenk universal joint, Cardan j., ball and socket j. **-glimmlampengerät** universal neon tube tester. **-nebenschluss** Ayrton shunt, universal s. **-prüfer** multiple-purpose tester, multi-meter. **-tastatur** universal keyboard.

unkennbar undiscernible, unrecognizable.

Unkosten, General- general expense, overhead e.

un-lauterer Wettbewerb unfair competition. **-lösbar** undissolvable, indissoluble; undetachable, locked. **-messbar** immeasurable, incommensurable (math.) **-mischbar** immiscible. **-mittelbare Betrachtung** direct viewing. **-okular** monocular. **-paarwertig** of odd valence. **-periodisch** aperiodical, non-periodic, deadbeat, non-recurrent.

Unruhe fluctuation, irregularity, instability (of physical quantities); disquiet, unsteadiness; balance (of time piece). **Anoden-** fluctuation of anode feed current, rippled condition in plate c. **Heiz-** fluctuation in heating or filament current, f. c. ripples. **Netz-** mains or supply line fluctuations (of potential or load); mains ripples.

un-rund non-circular, cornered, out of round. **-scharf** unfocused, poorly defined, out of focus, blurred, fuzzy or ghosty (of image or picture).

Unschärfe insufficient focus, lack of f., out-of-focus condition, blur or fuzziness, lack of definition (of picture); flat tuning condition. **Tiefen-** insufficient depth of focus, lack of d. of f.

unselbständige Entladung assisted discharge, non-self-sustaining d., non-spontaneous d. **-fremdgesteuerte Kippschwingungen** distant-controlled or signal-c., non-self-running time-base (telev.)

Unselektivität lack of selectivity.

unsichtbaren Strahlen, Abtastung mit noctovisor scan with invisible (or infra-red) rays.

Unsinn, offenbarer manifest absurdity or nonsense.

un-starre Moleküle non-rigid molecules. **-stät, -stetig** unsteady, unstable, variable.

Unstetigkeit im Filmzug flutter in film pull.

Unstetigkeitsstelle point of unsteadiness, instability or discontinuity.

Unstimmigkeit discrepancy, disparity, lack of harmony.

unstreckbar non-extensible, non-ductile, non-malleable.

Unsymmetrie asymmetry, dissymmetry; (condition of) unbalance, unilateralness.

Unter... sub-..., fractional, partial.

unterabteilen subdivide, make sub-compartments or partitions.

Unter-anpassung under-matching (of impedance). **-anspruch** sub-claim. **-anzeige** underswing (of volume indicator). **-art** subvariety, subspecies. **-belastung** underload. **-belichtungsbereich** toe range (of under-exposure).

unterbrechen interrupt, break, stop, arrest, make discontinuous.

Unterbrecher, Motor- motor-driven make-and-break. **Quecksilberstrahl-** mercury (jet) interrupter. **Selbst-** trembler, self-interrupter, buzzer. **Stimmgabel-** tuning-fork interrupter. **Turbinen-** turbine interrupter or break.

Unterbrechung und Schliessung break and make.

Unterbrechungs-bad short-stop bath, interruption b., short-s. trough. **-funken** break spark, circuit-opening s., wipe s. **-schwingungen** quench oscillations (in super-regeneration). **-vermögen** circuit-breaking, rupturing or opening capacity or power (of a switch).

unterdrehen low-speed shooting.

Unterdruck reduced pressure (below atmosphere), partial vacuum.

Unterdrückung, Fluoreszenz- quenching or suppression of fluorescence (by killer, poison, etc.) **Seitenband-** side-band suppression. **Tatsachen-** concealment or suppression of facts. **Träger-** suppression of carrier (in quiescent telephony).

Unterdrückungs-schwingungen quench oscillations (in super-regeneration). **-stromkreis, Geräusch-** noise reduction, quelching, suppression or silencing circuit.

Unter-fläche under surface, under face, under side, base. **-form** sub-variety, subspecies. **-formanten** sub-formants. **-führung, Weg-** underpass.

Untergehen swamped state, swamping (of a station). **-der Bildspannung im Störspiegel** swamping or blanketing of picture signal by noise level.

untergeordnet subordinate, subsidiary, secondary (in rank, quality or order), immaterial.

Untergericht inferior court, lower c. (of law).

untergeschoben spurious, supposititious.

Unter-glasur underglaze. **-grund** background, subsoil, subterranean soil, underground. **-grund, Spaltungs-** fission background. **-gruppe** subsidiary group, subordinate g. **-haltungswert** entertainment value. **-hörfrequenz** sub-audio frequency, infrasonic f. **-hülle** sub-shell.

unterirdisch subterraneous, subterranean, underground.

Unter-korrektur undercorrection, undercompensation. **-kühlung** supercooling, undercooling, overcooling.

Unterlage support, base, substratum, backing, foundation, lining, bed, carrier. **Filz-** felt base, f. support, f. under-layer, f. foundation.

Vergleichs- basis of comparison.
unterlagert subjacent, infraposed, underlying.
Unter-lagerungstelegraphie subaudio-frequency telegraphy system, composite operation (telegraph and telephone signals passed over one line wire; often attended with thump noise). -lagscheibe washer, supporter disk.
unter-lassen omit, abstain or refrain from, fail, default, neglect. -legen lay under, underlay, put under, infrapose.
Unter-legescheibe washer, supporting disk. -licht horizontal light.
unter-liegend subjacent, underlying, placed below or underneath, infraposed, by subject to. -ordnen subordinate, make secondary or of minor rank.
Unter-ordnung subordination, submission. -patent sub-patent. -resonanzfrequenz submultiple resonance (of a subsynchronous frequency). -sagung inhibition, prohibition, forbidding, interdict, interdiction, restraint, injunction. -schale sub-shell (of electrons). -schallwelle infrasonic wave, sub-audio w.
unter-schatzen underestimate, undervalue, underrate. -scheidbar distinguishable, discernible, discriminable, differentiable.
Unterscheidungs-markierung distinctive marking, telltale m. -vermögen, Farb- color discrimination faculty of power. -vermögen, Tiefen- power of differentiating depths.
Unter-schiebungsklage suit for passing off goods as though of another person. -schied, Gang- phase difference, path d. (of rays, waves, etc.)
unter-schneiden undercut (groove too shallow or lateral movement of stylus too small). -schreiten fall below or short of (a certain value). -schwellig subliminal, below a threshold.
Unterschwingung subharmonic oscillation, subfrequency o.
unter-setzender Verstärker scaling down vacuum tube. -setzt geared down. -spannen subtend, sub-volt. -spannter Betrieb undervolted operation.
Unterspannung under-running voltage, undervolting.
unterstellen presuppose, assume, allege, impute, charge.
Unter-steuerung under-modulation, insufficient m. -stützung eines Einspruches argued support of an opposition.
untersuchen analyze, examine, inspect, check up, investigate, search, research, inquire, scrutinize.
Untersuchung investigation, research, examination, probing, inspection, test.
Untersuchungsobjekt test specimen, t. sample, t. piece.
untersynchrone Resonanz subsynchronous resonance, submultiple resonance.
Unter-taste long key, white k. -tauchung immersion, submersion, deep dip. -teilung subdivision, submultiplication; partition, partitioning. -teilungspunkt tap (of a battery, etc.). -wagenantenne under-car antenna. -wasserschallempfänger hydrophone, submarine or subaqueous microphone, pickup or sound detector, asdic (submarine detector).
unterwegs en route, on the way.
Unterwind forced draft (introduced from below).
un-tilgbar indelible. -trennbar inseparable, indivisible, unseverable, undetachable. -übersehbar uncontrollable, hard to follow up or inspect, complex, intricate. -übertragbar nontransferrable, unassignable, inalienable. -ubertrefflich unexcelled, unsurpassable, unrivaled. -verbrennbarer Film slow-burning film, safety f. -verfaulbar unputretiable, imputrescible.
Unvergänglichkeit imperishableness, ever-lastingness.

unver-gängliches Lichtbild imperishable picture. **-hältnismässig** disproportionate.

Unverständlichkeit unintelligibility.

un-verträglich inconsistent, incompatible. **-verwechselbarer Stecker** non-interchangeable plug. **-verwüstlich** indestructible. **-verzeichnetes Bild,-verzerrtes Bild** orthoscopic image, undistorted picture.

Unvollkommenheit imperfection, imperfectness, faultiness, flaw, defect.

un-vollkommener Isolator imperfect insulation, leaky dielectric. **-wandelbar** immutable, invariable, unchangeable, imperishable. **-widerleglich** irrefutable, unrebuttable. **-wirksam** inactive, ineffective, inefficient, inert. **-wirksam machen** render inoperative or inactive, disable, deactivate, inactivate.

Unwucht unbalance, imbalance, desequilibrium. **-erreger,-masse** unbalance weight.

un-zeitgemäss untimely, unpropitious. **-zeitlich** untimely, immature. **-zerlegtes Licht** undispersed light. **-zerstörbar** indestructible, non-corrodible. **-ziehbar** non-ductile. **-zugänglich** inaccessible. **-zweideutig** unequivocal, unambiguous, clear.

Uran-tonbad uranium toning bath. **-verstärker** uranium intensifier.

Ur-atom primordial atom. **-eichkreis** master reference system. **-gewicht** standard weight, primary s.w.

Urheber author, originator, founder. **-recht** copyright, rights of an author or inventor.

Urkunde ausstellen execute a document.

Urkunde,Besitz- title deed. **Übertragungs-** deed of assignment.

Urmass primary standard.

ursächlich causal, causative.

Ursprung origin, source, provenience.

ursprünglich original, primitive, parent.., primary, first.

Ursprungs-bild original subject copy. **-type** prototype, original or primordial type or specimen.

Urstoff primary matter, element.

Urteil judgment, decision, verdict, opinion. **-fällen** render a decision, judgment, award. decree, sentence or verdict.

Urteil,Versäumnis- judgment by default.

Urteilsvollstreckung execution of a sentence, carrying out a judgment.

V

vagabondierende Ströme leakage currents, stray c., vagabond c.
Vakuole pinhole, vacuole.
vakuumdicht vacuum-tight, sealed (airtight) to insure a vacuous state.
Vakuum-dichtung vacuum plumbing, v. sealing, v. packing. **-fett** vacuum grease. **-gitterspektrograph** vacuum grating spectrograph. **-hahn** vacuum tap. **-mantelgefäss** thermos or Dewar vessel with exhaust jacket. **-messer,-meter** vacuometer, vacuum gage, ionization g. **-pumpe,Vor-** fore-pump, rough-vacuum p. **-schleuse** airlock (of electron microscope). **-thermoelement** vacuum thermo-couple. **-zelle** vacuum photo-electric cell.
Valenzkette,Haupt- primary valence chain.
Variationsmethode variational method.
Variometer,Ballon- balloon statoscope. **Klapp-** hinged-coil variometer.
Vater metal master (in disk manufacture).
Vektor,absoluter tensor. **Vierer-** four-component vector.
vektorielle Darstellung vectorial representation.
Vektorindikation vector response index.
Ventil,Ablass- discharge valve, drain v. **Etagen-** multi-seat valve. **Licht-** light valve, l. relay. **Rückschlag-** check valve.
Ventil-küken plug of a valve. **-stopfen** valve stopper, s. within v.
Venturidüse constricted or nozzle-like reducer, throat piece of Venturi meter.
ver-allgemeinern generalize. **-änderlich** variable, changeable, fluctuating, unstable, unsteady, non-constant. **-änderlich,laufend** running variable.
Veränderliche variable, variable quantity (math.)
veränderlich(er)Brennweite und Grösse,Linse von zooming lens for variable focus and variable magnification, "vario-objective." **-(e)Kopplung** vario-coupler.
Veränderung,Farben- change of color, discoloration.
Veranderungen,Seiten- turns and banks (in avigation).
verankern anchor, stay or guy (a pole).
Veranlassung von, auf at the request or suggestion of, on behalf of.
ver-antwortlich responsible, answerable, liable, chargeable. **-arbeiten** process, work.
Verarbeitung,Film- film processing.
veraschen incinerate.
Verband union, association; binding, bandage, fastening, tie, bond (in bricklaying).
Verbesserungs-mittel corrective. **-patent** patent of improvement.
verbiegen bend, buckle, warp.
Verbindlichkeiten eingehen incur liabilities.
Verbindung mit angeschärften Enden scarfed joint. **-,stumpfe** butt joint. **Dreiweg-** three-way connection. **Einlagerungs-** intercalation compound. **Glasschliff-** ground-glass joint. **Muffen-** sleeve joint. **Sauerstoff-** oxygen compound.
Verbindungs-fähigkeit affinity, combining power. **-glied** connecting link. **-klammer** brace, clip connector. **-klemme** connecting means, binding post, terminal. **-kraft** bonding strength, affinity, combining power. **-punkt** joining,

connecting or union point, junction, juncture.

Verbleiung leading, lead lining.

Verblitzen des Films dendriform exposure of film (due to static charges).

Verbot prohibition, inhibition, interdict, interdiction, injunction.

verboten(e) Linie forbidden line (spectr.) **-(er)Platz** forbidden place (cryst.) **-(er)Übergang** forbidden transition.

Verbrauch,Energie- energy consumption, e. dissipation. **Energie-, ohne** non-dissipative, wattless.

Verbraucher load, consuming device, consumer.

Verbrauchs-messer supply meter. **-zähler,Blind-** reactive volt-amp. hour meter, sine m., wattless component m., var m. **-zähler, Schein-** apparent power meter.

verbraucht spent, exhausted, used up, worn out, consumed, stale (of air), dissipated.

verbreitern widen, broaden, spread (out).

Verbreiterung spread (of image). **Druck-** pressure broadening (of gas-spectrum). **Linien-** line broadening (telev.) **Spektrallinien** broadening of spectral lines. **Stoss-** impact broadening. **Zeichen-** signal spread (due to echo effect).

verbrennlich inflammable, combustible, burnable.

Verbrennungsschiffchen combustion boat.

Verbund-glass compound glass, multilayer g. **-röhre** compound tube.

verdampfbar evaporable, vaporizable, volatile.

Verdampfer evaporator, drier, dehydrator, carburetter (water gas).

verdampfungsfähig evaporable, vaporizable, volatile.

Verdampfungsquelle source of evaporation, s. of volatilization.

verdecken cover (up), mask, conceal, hide, eclipse, occult.

verdeckte Zeichen,durch Störer signals swamped by interference.

Verdeckung masking(ac.), overriding (of noise).

Verdeckungserscheinung masking effect.

ver-derben spoil, damage, ruin, decay, rot. **-dichtbar** condensable, compressible. **-dichten** condense, condensate, concentrate, compress, compact, press,together, concrete, make denser, inspissate, consolidate.

Verdichter,Kreisel- rotary compressor, centrifugal c.

Verdichtungswelle compression wave.

ver-dicken thicken, concentrate, inspissate, coagulate, jell, become viscous. **-drahteter Empfänger** fully wired receiver set.

Verdrahtung,Blank- bare wiring.

ver-drallen twist, strand (together). **-drangen** displace, drive out, remove, expel, supplant, supersede.

Verdrängung,Strom- skin effect, current displacement.

Verdrängungsstrom displacement current, dielectric c., capacitance c.

verdrehen distort, twist, wrench, subject to torsional stress, stagger (rotationally), shift (a phase).

Verdrehung (cf. Zerdrehen) rotation, turn (of a picture); twist, warping, torsion.

Verdrehungs-dauerhaltbarkeit endurance under torsion stress. **-fähigkeit** torsibility. **-moment** torsional starting torque, moment of rotation. **-wechselfestigkeit** strength under alternating torsion stress. **-winkel** angle of torsion, twist a.

Verdreifachung,Frequenz- frequency tripling.

ver-drosseln provide with choke coil. **-dübeln** dowel.

Verdunkelung,Rücklauf- flyback

elimination, return-trace or return-path e., blackout or blanking of flyback trace (by gating and blanking pulse.

Verdunkelungs-kapazität shading condenser. **-phase** dark interval.

ver-dünnbar diluable, rarefiable, capable of being diluted. **-dünnen** dilute (liquid), rarefy (gas), attenuate, thin, de-concentrate, reduce or dim (light). **-dünntes Helium** low-pressure helium.

Verdünnungs-grad,Helligkeits- loss or reduction of brightness. **-mittel** diluent, attenuant. **-welle** rarefaction wave.

verdunstbar evaporable, vaporizable, capable of being evaporated.

Verdunstungsgefäss evaporimeter.

veredeln (cf **vergüten**) improve, elevate, refine, ennoble, purify, perfect; plate with a higher grade of metal.

Veredler,Ton- high-frequency filter, hf cutoff, sound clarifier, tone control, tonalizer.

Vereinfachung simplification, reduction (math.)

Vereinheitlichung standardization, normalization, make for uniformity, regularization, co-ordination.

Vereinigungs-koeffizient combination coefficient. **-stelle** junction, joining, meeting or union point.

Vereisung ice formation, icing (on wings).

ver-engen narrow, contract, constrict, squeeze. **-erzen** mineralize.

Verfahren process, method, procedure, proceeding, mode (of action), technique, processing (of film). **Beschwerde-** appeal action or procedure. **Billigkeits-** equity suit.

Verfärbung discoloration, shading.

verfestigen make firm or strong, solidify, consolidate, concrete, set, harden.

Verfestigungs-kurve plastic stress-strain curve (cryst.) **-zeit** time of set (of gels).

verfilzen felt, mat.

verflachen level off, flatten, become even, smooth (down), lose selectivity, become flatter or less steep, slope down, broaden or make broad-topped (as a resonance curve).

verflauter Übergang shaded-off transition (definition lacking).

ver-flechten inter-weave, involve. **-flüchtigen** volatilize, evaporate. **-flüssigen** liquefy.

Verfolgung,Ziel- tracking a target (by detecting and ranging means using a slewing motor).

Verfolgungs-aufnahme running shot. **-stativ** follow-shooting tripod, running t.

verformbar deformable, plastic, fictile, kneadable, moldable.

Verformung deformation, working, warping, strain.

Verformung unter Fliessen,Wissenschaft der rheology. **-,plastische** plastic deformation, p. working.

Verformungsvermögen degree of deformation, deformability.

Verfügung action (of Patent Office), order, decree, decision.

Vergangenheit,Wärme- thermal antecedent, prior heat treatment.

vergänglich perishable, transient.

vergasbar gasifiable, vaporizable.

Vergaser gasifier, vaporizer, carburetor.

Vergasung,Tieftemperatur- low-temperature distillation or carbonization, partial c.

Vergenz vergency (either convergence or divergence).

Vergiften poisoning, contamination, killing (of phosphor).

Vergitterung grating, gridding, lattice, grille.

verglasbar vitrifiable.

Verglasung vitrification, glazing, glaze.

Vergleich comparison, contrast, matching. **Farben-** color matching.

Vergleicher,Tonhöhe- tonvariator, pitch pipe.

Vergleichs-feld matching field (in photometry). **-masstab** standard of comparison, s. of reference, yardstick. **-prisma** comparison prism.

Vergleichung comparison, contrasting, matching.

Vergleichunterlage basis of comparison.

vergrössern enlarge, magnify, increase (in size).

Vergrösserung, lineare magnification in diameter. **Fernrohr-** telescopic magnification. **Über-** supplementary or extra magnification, over-m.

Verguss-kammer-Einführungsisolator pothead (lead-in) insulator. **-masse** filler or sealing compound.

vergüten refine, improve (quality), temper, coat (a lens).

vergütete Probe quenched or tempered test piece or specimen.

Vergütung, Glas- refining or treating of optical glass (e.g., by producing low-reflectance coating, dimming, etc., as by Bausch & Lomb Balcote magnesium-fluoride anti-reflection film on glass).

Verhältnis proportion, ratio, rate, relation, connection, circumstance. **Aufteilungs-** potentiometer ratio. **Molen-** mole ratio. **Öffnungs-** aperture ratio, relative a. **Stromstoss-** impulse ratio. **Winkel-** angular ratio.

Verhaltnis-anzeiger exponent (math.) **-arme** ratio arms (of a bridge).

verhältnismässig proportional, proportionate, commensurable, commensurate.

Verhandlung pleading, proceeding, hearing, argument, discussion. **zur mündlichen - gelangen** to be called for oral hearing or argument. **einer Sache, Termin zur - anberaumen** set a date for hearing a case.

verhärten harden, indurate, set (cement).

verharzen resinify, resining.

verheddern tangle (of antenna wire, etc.)

Verhinderung, Korrosions- corrosion inhibition.

verholzen lignify, turn woody.

Verhör hearing, interrogation, examination, trial, questioning.

Verjährungsgesetz law of limitations or superannuation.

ver-jüngen point, taper, thin, diminish, reduce, constrict, narrow, contract (a tensor). **-kabelte Leitung** cable line, stranded-wire l.

Verkabelung, Amts- central wiring, office w.

verkanten twist, tilt (of camera).

Verkehr, Fernsehsprech- video telephone traffic.

Verkehrsbezirk, Funk- radio controlled aerial navigation (or avigation) district or zone. **Nah-** approach or local control zone.

Verkehrs-flughafen commercial airport, civil a. **-kontrolle, Luft-** airways traffic control, avigation regulation. **-luftfahrt** commercial aviation, c. avigation.

verkehrt proportional inversely proportional.

Verkettung linking, linkage, concatenation, bonding.

Verkettung, Fluss- flux linkage.

Verkettungsspannung interlinked voltage.

ver-kitten cement, lute, seal. **-kittete Linse** cemented lens. **-kleben** cement, lute, paste, glue, cover, seal or close with adhesive or agglutinant. **-kleiden** face, case, line, disguise, mask. **-kleidet** faired (of airplane).

Verkleinerungskopie reduction print.

verkleistern paste (up), turn into p., agglutinate.

Verkleisterung gelatinization (of starch), agglutination, glutinization.

verklingen decay, die out (of waves, etc.).

Verkohlung carbonization, coking, dry distillation, destructive d., charring.

Verkokungszeit coking time.

Verkopplung spurious, stray or undesired coupling.

verkürzen shorten, abbreviate, condense, abridge, contract, curtail.

Verkürzungskondensator shortening condenser, padding c., aerial series c., antenna c.

Verlagerung displacement, shift, dislocation, misalignment. **Bildpunkt-** spot shift (causing plastic effect, in telev.). **Nullinien-** zero setting, electric biasing (in film recording). **Peilbearing** shift, error or distortion. **Zeilen-** migration or shift of line position.

verlängern lengthen, extend, prolong; produce (math.).

Verlängerung, Frist- extension of time, e. of term, respite. **Patent-** renewal of a patent.

Verlängerungsspule, Antennen- aerial load coil.

Verlauf pattern, plot, path, shape, trend, course, slope. **Feld-** field map, f. pattern, f. configuration. **Kopplungs-** coupling coefficient, c. factor.

Verlauf-filter sky filter.

Verlegung, Kabel- cable laying, burying of c. **Rohr-** pipe laying.

Verleihkopie release print, distributing p. (of film).

verletzen infringe, violate, transgress, contravene, prejudice, injure.

Verletzten, dem - eine Geldsumme zusprechen adjudicate a sum of money to aggrieved or injured party.

Verletzung offense, violation, infringement (upon patent rights).

Verletzungsklage, Patent- action or suit for infringement on patent rights.

Verlorener Kopf deadhead, wastehead, crop end.

Verlust loss, dissipation, drop (of potential), absorption (light).

Verlust, Belag- surface leakage loss (of condenser). **Einführungs-** insertion loss. **Erwärmungs-** Joule loss, joulean heat (dissipation). **Luftreibungs-** windage loss. **Nachwirkungs-** residual loss. **Spannungs-** voltage drop. **Sprüh-** corona discharge or brush d. loss. **Strahlungs-** corona loss. **Übergangs-** contact loss, transition l., junction l.

verlustarmes Kabel low-loss cable.

Verlust-dampfung damping (loss) resulting in dissipation of energy (by ohmic resistance). **-faktor** phase angle difference (of a condenser). **-kapazität** imperfect capacity, leaky c. **-konstante** attenuation constant. **-leistung, Anoden-** anode dissipation, effective power d. **-leistung, Auffangplatten-** collector dissipation (in multiplier).

verlustlos non-dissipative free from loss, lossless.

Verlust-strom watt current, active c., energy c. **-widerstand** nonreactive, effective, dissipative or ohmic resistance. **-winkel** phase angle difference, loss angle (of a condenser). **-ziffer** phase angle difference.

Verminderer, Nachleucht- killer, poison (of phosphors).

Vermittelung agency, intermediary, means, mediation, interposition, intervention, adjustment.

vermöge by or in virtue of, according to.

Vermögen, Abbildungs- resolving power. **Brems-** stopping power, retarding p., braking p. **Emissions-** emissivity, emitting power. **Reflexions-** reflectance, reflectivity, reflecting power.

Vermutung assumption, supposition, surmise, conjecture.

vernachlässigen disregard, neglect.

Vernachlässigung neglect, disregard, omission.

vernebeln atomize (under high

pressure).
verneinenden Falles in the negative case.
Verneinung negation, denial, disavowel, contradiction.
Vernetzung network, reticulation.
Vernichtungsstrahlung annihilation radiation.
ver-pflichtet obligated, bound, under obligation. **-puffen** puff off, deflagrate, explode, detonate; vaporize, atomize. **-regneter Film** rainy film. **-riegeln** block, lock, inter-lock, cut off, gang (condensers), bolt, latch. **-riegelnde Spannung** cutoff biasing potential.
Verriegler suppressor. **-relais** locking relay.
Verrückung displacement, disturbance, derangement, shift, dislocation, dislodgment **Atom-** atom displacement.
verrühren mix by stirring, stir up.
Versäumnisurteil judgment by default.
verschachteln nest, insert into one another.
verschiebbar displaceable, shiftable, movable, sliding, slidable. **-,posaunenartig** shiftable trombone-fashion, telescoping.
Verschiebe-fehler lack of synchronism, out of frame condition. **-zackenspur** bilateral variable-area sound track.
Verschiebung displacement, dislocation, drift, slip (in geology), shift (of phases), dephasing. **-,dielektrische** dielectric displacement.
Verschiebung, Bild- slow drift of picture, hunting of p. **Bild-, seitliche** lateral shift of image. **Druck-** pressure shift (of spectral lines). **Isotopie-** isotope effect, i. shift. **Längs-** axial adjustment (of electron microscope). **Parallel-** translatory motion, t. shift. **Phasen-** phase displacement, p. shift, dephased or out-of-phase condition. **Zeilen-** interlacing, interlaced scan, line-offset s. **Zeit-** time lag, t. shift.
Verschiebungs-konstante (Dielektizitätskonstante) displacement constant, specific inductive capacity (for vacuum). **-linie** line of displacement. **-satz** displacement law. **-strom** displacement current, dielectric d.c., capacitance c.
verschiedenfarbig varicolored.
verschieden gegliederte Bildteile picture portions of dissimilar make-up, nature or organization.
Verschlackung scorification.
Verschlag partition, compartment.
Verschlechterung impairment, deterioration, spoiling, debasement, vitiation.
verschleiern veil, make turbid, fogged, or cloudy, befog.
Verschleierung veiling glare (of projected image), foggy or fogged condition(of film).
Verschleiss wear and tear, abrasion, attrition. **-prüfung** abrasion test.
verschlungene Blitzbahn tortuous path of lightning flash.
Verschluss closing, closure, shutting, shutter, stopping, stopper, fastening, lock, clasp, trap, seal, zipper, seal-off (of ignition mercury pool, etc.).
Verschluss, elektro-optischer electro-optical shutter. **-,luftdichter** airtight seal, hermetic s. **End-** cable box, termination b. **Fallschieber-,** drop shutter. **Gas-** gas seal, ball and hopper, cup and cone arrangement (of furnace). **Reiss-** slide fastener, zipper. **Rouleau-** roller blind shutter. **Schlitz-** focal plane shutter. **Sektoren-** sector shutter. **Umlauf-** rotary shutter. **Wasser-** water seal.
Verschlussblende, flimmerfreie non-

non-flicker shutter. **Umlauf-, rotierende** rotary flicker shutter or light cutoff.

Verschlüsselung coding, cryptography.

Verschluss-kopf cable terminal. **-ring** locking collar (of magnifier). **-stück** plug, stopper, seal.

Verschmälerung des Tonstreifens squeezing of sound track, matting.

verschmelzen melt, smelt, fuse, melt together, blend, merge, alloy.

Verschmelzung merging (of pictures).

Verschmelzungsfrequenz critical flicker frequency, fusion or no-flicker f.

verschmieren smear, daub, lute.

verschmoren scorch, freeze together (of contacts).

ver-schobene Phase displaced phase, shifted p., dephased or out-of-p. condition. **-schränken** toggle, cross, interlace. **-schrumpfen** shrink, contract.

Verschweigung concealment.

ver-schwimmen blur, become indistinct; blend. **-schwinden** vanish, disappear, evanesce, fade (away). **-schwommen** indistinct, foggy, indefinite, blurred, bleary, woolly (of zero signal point in df apparatus); fuzzy (lack of sharpness of sound track). **-schwommenes Aussehen** bleary, blurred or foggy appearance (of image).

Versegelung distance sailed or covered (between two bearings); deviation from right course, yaw.

verseilen strand, cable, twist together.

Verseilung cable lay up, stranding (of wires or leads).

ver-sengen singe, scorch. **-senken** sink, submerge. **-senkter Schalter, halb-** semi-sunk switch, semi-recessed s. **-setzbar** capable of being mixed or treated. **-setzen** mix, treat, compound, alloy, add (one substance to another); stagger, displace, shift. **-setzt** staggered, offset, displaced, dislocated, shifted (of phases, in quadrature = 90°, in opposition = 180°), dephased, out of phase.

Versetzung dislocation (cryst.), permutation, alligation (math.), alloy (metal). **Flugzeug-** drift, lateral displacement or shift of airplane, deviation from course. **Metall-** alloying, alloy. **Perforations-** slipping of perforation, dislocation.

verseuchen vitiate, poison, infect.

Versicherung, eidesstattliche affirmation in lieu of oath, statutory declaration. **-, eidliche** affidavit, sworn statement, testimony or declaration.

versickern ooze or seep away.

versinnbildlichen symbolize, express or represent by symbols.

Versorgungs-gebiet service area, supply district. **-spannung** supply (source) potential.

verspannen stay, guy, span.

versperren obstruct, block, lock.

Verspieglung, Hochheim'sche Hochheim special mirror plating or "silvering" substance (of I.G. Farben).

verspleissen splice.

Versprödung embrittling.

Verständigung intelligence transmission, signal t. **-im Flugzeug** inter-phone. **-zwischen Flugzeugen** inter-aircraft communication or signaling. **Eigen-** interphone (on board airplane).

Verständlichkeit intelligibility (teleph.), transmission audibility, articulation (consonant, vowel, with the use of logatoms). **Silben-** syllable articulation, s. intelligibility.

Verstärkbarkeitsgrenze limiting amplifiableness, maximum (practical) amplification or gain.

verstärken amplify, multiply, intensify; re-inforce, strengthen, increase, fortify, concentrate, magnify, boost; truss (poles, etc.)

Verstärker amplifier, multipler, intensifier (RF stage between antenna and receiver input), repeater (in a telephone line).

Verstärker, A- class A amplifier, pre-amplifier (with operating angle of 360 degrees). **B-** class B amplifier (operating angle 180-360 degrees). **C-** class C amplifier (operating angle less than 180 degress). **-, ausgeglichener** balanced amplifier. **-, detonierender** heterodyne amplifier. **-mit negativem Widerstand** kallirotron, dynatron type of tube (with negative resistance). **-mit zwei gegeneinandergeschalteten Röhren** push-pull amplifier. **-, untersetzender** scaling down vacuum tube. **Aufnahme-** recording amplifier. **Bildfrequenz-, Bildpuls-** amplifier for frame synchronizing impulses, low-sync a. **Direkt-** straight-ahead amplifier, non-heterodyne a. **Druck-Zug-** push-pull amplifier. **End-** power amplifier, final a. **Geradeaus-** straight-ahead amplifier (in non-heterodyne receiver). **Gleichtakt-** in-phase amplifier. **Kaskaden-** cascade amplifier, repeating a. **Kathoden-** cathode follower (scheme). **Kraft-, Leistungs-** power amplifier. **Leitungs-** telephone (circuit) amplifier or repeater. **Mehrfach-** multi-stage amplifier **Mikrophon-** microphone pre-amplier, m. amplifier. **Nach-** amplifier above input stage, (mostly) power a. **Nadelton-** phonograph-type amplifier. **Netz-** mesh amplifier, m. multiplier (with electron-permeable targets of fine mesh or gauze). **Photozellen-** electron multiplier phototube; photo-cell amplifier. **Regel-** variable-gain amplifier, avc (automatic volume-control) a.; noise-reduction or silencing a.; any amplifier in which regulator or control potentials are operative. **Richt-** amplifying detector, plate-current detector. **Schirmgitter-Richt-** screen-grid tube with plate-current rectification. **Steuer-** amplifier between power and input stage (operates on pure voltage amplification). **Uranium-** uranium-intensifier (phot.) **Vor-** input amplifier, pre-amplifier (class A RF amplifier in superheterodyne receiver). **Zeilenimpuls-** amplifier to increase line synchronizing impulses, high sync a. **Zusatz-** booster amplifier. **Zwischen-** (inter-) stage amplifier.

Verstärker-brett amplifier rack, a. panel. **-bucht, -bunker** amplifier or repeater rack or bay. **-geräusch** amplifier noise, tube n. **-gestell** amplifier or repeater rack or bay. **-gleichrichter** amplifying detector. **-kette** amplifier cascade. **-netz** amplifier channel. **-stufe (zwischen Antenne und Verstarkereingang)** intensifier (tuned RF) stage, intermediate circuit. **-zug** amplifier channel.

Verstärkung, Wirk- transducer gain.

Verstärkungs-anlage, Sprach- (public) address system. **-begrenzer** output limiter, o. suppressor, gain spoiler. **-faktor** amplification factor, voltage f. **-mass** unit of gain, repeater gain, equivalent r. g. **-mittel, Schall-** sound reinforcing system (for halls, etc.). **-schirm** intensifying screen (x-ray work). **-unterdrücker** gain spoiler, output suppressor. **-zahl** amplification constant, a. coefficient, a. factor,

verstecken hide, conceal.

Versteifung stiffening, propping, bracing, strutting.

Versteilerung (Vergrösserung des Amplituden- verhaltnisses zwischen Zeilen oder Signalen) exaggeration of amplitude differences (emphasizing the stronger and supressing the weaker one of two or more signals), expansion. **-der Konturen** insuring higher definition (telev. pictures), improvement in resolution and black-white contrast (of an image), expansion.

Versteinung devitrification (glass); petrifaction.

verstellen shift, move, transpose, adjust.
Verstellpropeller feathered (variable-pitch) propeller.
Verstellung,Höhen- adjustment in elevation.
Verstellungs-knopf,Bild- framing knob (m.p.) **-welle,Bild-** framing shaft (m.p.)
verstimmt detuned, out of tune, dissonant, untuned, disharmonious, dissyntonized, tuned off resonance.
Verstimmungsschalter wave (length) change switch.
verstopfen stop, choke, clog, obstruct, stopper (up).
verstopfte Trompete stopped trumpet.
Verstopfung,Rohr- pipe clogging, stop-up, choke or obstruction in a pipe or piping.
verstreichen fill or stop up, caulk, spread or coat over; elapse, expire.
verstümmelt mutilated, obliterated, maimed, clipped (teleph., ac.).
Verstümmelung,wörter- clipping of words, obliteration of w.
verstürzen fall to pieces, scatter.
Versuch,Beregnungs- rain test, wet t. (of an insulator). **Drehschwingungs-** oscillation torsion test. **Rotbruch-** hot breaking test. **Verwindungs-** torsion test. **Vorführungs-** demonstration experiment. **Vorlesungs-** classroom experiment, lecture demonstration.
Versuchs-fehler experimental error. **-feld** proving ground, field for making tests and experiments. **-laboratorium** experimental or research laboratory. **-stab** test rod, t. bar.
Vertagung adjournment, postponement.
vertäubtes Ohr deafened ear.
vertauschen exchange, interchange, play reverse rôles, transpose.
Vertauschung exchange, interchange, permutation (math.), transposition (of wires).
Verteidungsschrift argued statement of defense, plea.
verteilen distribute, divide, disperse, diffuse, scatter, spread, space apart.
Verteiler,Bildpunkt- picture scanner (at receiving end). **Licht-** beam splitter (m.p.).
verteilte Induktion oder Induktivität distributed inductance, continuous loading (opposite of lumped or concentrated i.) **-Kapazität** distributed capacitance, self-c. (as distinguished from lumped c.) **-Wicklung** distributed winding.
Verteilung,räumliche spatial distribution, geometric d. **Empfindlichkeits-** sensitivity distribution. **Farbenempfindlichkeits-, des Auges** spectral response of eye, color sensitivity of eye. **Geschwindigkeits-** velocity distribution, d. in energy. **Geschwindigkeits-, Apparat zur Bestimmung der** velocity selector. **Gleich-** equipartition. **Lichtstärke-** light intensity distribution (on photo-cathode). **Maxwell'sche** Maxwellian distribution (law). **Raum-** spatial distribution, geometric d. **Reichweiten-** range distribution.
Verteilungs-anlage relay equipment, distribution gear. **-fläche,Licht-** isophotic curve. **-gesetz** distribution law, partition l. **-konstante** distribution or partition constant or coefficient. **-kurve,Fehler-** error distribution curve. **-satz** principle of distribution, d. law.
Vertikalausbreitung von Schall vertical diffusion of sound.
Vertikal-hinlauf vertical scansion. **-navigation** vertical guidance, v. avigation. **-wechsel** vertical or frame synchronizing pulse or cycle.
vertonte Oberfläche surface bearing sound track, s.-impressed surface or area.
Vertonung,Nach- dubbing.
vertraglich contractual, (bound)

by contract.
verträglich compatible.
verträglich verpflichtet contractually bound or obligated.
Vertragspartei party to contract, contractant.
Vertrauensbruch breach of confidence.
Vertrieb eines Filmes releasing or distribution of a film.
Vertrocknung drying (up), desiccation.
verunreinigen vitiate, soil, contaminate, pollute, adulterate, make impure.
Verunreinigungskerne impurity nuclei.
ver-ursachen cause, produce, occasion, result in. **-urteilen** pass judgment against, sentence, fine, condemn, convict, adjudicate. **-vielfachen** multiply.
Vervielfacher, Farnsworth- multipactor, F. multiplier. **Netz-** mesh multiplier, m. amplifier (works with secondary emission, cascaded electron-permeable mesh or gauze targets). **Pendel-** reciprocating and acceleration electron multiplier predicated on secondary emission. **Sekundäremissions-** electron multiplier (with plurality of secondary-emission targets or reflecting electrodes).
Vervielfacherzelle, photo-elektrische, mit Sekundäremission electron multiplier photo tube, photo-electric electron-m. tube (with reflecting electrodes or dynodes).
Vervielfältiger, Gitter- mesh-type multiplier.
Vervielfältigung copying, printing (of film).
Vervielfältigungs-faktor multiplication factor. **-kondensator** condensor-type d.c. multiplier. **-oberfläche** dynode (in Farnsworth multiplier). **-röhre** multiplier tube, multipactor.
Vervierfacher quadrupler.
Verwachsungen inter-growths, intercrescences.
Verwackeln des Bildes unsteadiness of frame.
verwandelbar transformable, convertible, transmutable.
verwandeln transform, convert, change, turn.
Verwandlung, Paschen-Back- P.B. magneto-optic effect.
Verwandschaft relation, affinity.
verwaschen obliterated, blurred, vitiated, bleary.
Verwaschungszone blurred zone, bleary z., z. lacking definition, confusion z. (telev.).
Verwehung draft of air, wind.
verweilen dwell, sojourn, stay.
verweisen, Berufung an den Gerichtshof refer appeal to the court.
verwerfen warp, distort, twist, repudiate, reject.
Verwerfung rejection, repudiation, warping, deformation; slow frequency drift (in radio apparatus, after tuning).
Verwertungsperiode evaluation period.
verwickelt complex, complicated, involved, intricate.
verwinden twist, subject to torsional stress.
Verwindungsversuch torsion test.
Verwirklichungsform practical embodiment (of an invention).
ver-wischen blot out, efface, obliterate, blur, make bleary. **-wittern** disintegrate, effloresce. **-würfeln** jumble, garble. **-zahnen** tooth, cog, indent, provide with (gear) teeth, serrate.
Verzahnung, Evolventen- involute tooth gear.
Verzahnungsgetriebe, Winkel- double helical gear.
verzapfen joint, mortise, dispense by tap.
verzehren consume, spend, dissipate, abate (smoke).
Verzeichnung faulty delineation or re-creation of a picture, distortion of a p. (e.g., pincushion, barrel, etc., distortion, in television).
Verzeichnung, anisotrope anisotropic

shear distortion, distortion of orientation (telev.) -, kissenförmige pincushion distortion, pillow d., negative d. -, nichtlineare non-linear distortion. -, tonnenförmige barrel distortion, positive d.
verzeichnungsfreie Linse orthoscopic lens, distortion-free l.
Verzerrer, Dynamik- dynamic compressor.
Verzerrung durch Ein- und Ausschwingung transient (non-linear) distortion, build-up and decay d. -durch Feldkrümmung distortion due to field curvature, curvilinear distortion.
Verzerrung, Ausschwing- decay distortion, non-linear d., overthrow or underthrow d. (in facsimile transient). Bild-(cf Verzeichnung) picture distortion (in facsimile: jiggers, f. transients known also as hangover, tailing, overthrow and underthrow distortion). Dampfungs- frequency distortion; tone d. Einschwing- build-up or transient (non-linear) distortion. Frequenz- attenuation-frequency distortion, amplitude f.d., attenuation d., frequency d. Laufzeit-transit-time distortion, phase d. Lichtfleck-spot distortion. Phasen- phase distortion.
verzerrungsfrei distortionless, d.-free, non-distorting. -(es) Okular orthoscopic eyepiece, distortion-free e.
Verzerrungs-leistung distortion power (in terms of d. volt-amperes). -toleranz tolerance of distortion.
Verzicht disclaimer, renunciation, waiver.
verzichten (auf Anmeldung, Patent, etc.) abandon, drop or relinquish (an application, patent, rights, etc.), disclaim, renounce, waive rights.
Verzichtleistung disclaimer, renunciation, waiver, abandonment.
verziehen warp, buckle.
Verziehung warpage, buckling.
Verzinkung, Feuer- hot galvanization, pot g.
Verzögerer, Vielfach- Barkhausen-Kurz type of tube (retarding-field t., magnetron, etc.)
Verzögerung retardation, deceleration, braking, delay, lag, checking. Funken- spark lag. Phasen- phase delay, phase lag, (phase) angle of lag.
Verzogerungsbad restraining bath.
verzögerungsfrei inertialess, free from lag, delay or sluggishness.
Verzögerungs-kette delay network. -linse cutoff or stopping (electron-optic) lens. -relais slow-acting relay, time-lag r. -schalter time-lag switch. -spannung delay voltage, delayed-action v. (in avc circuits).
Verzug time delay, t. lag. Ausschwingungs- hangover, tailings (in facsimile). Entladungs- discharge delay, d. lag. Fliess- yield or flow distortion or deformation.
Verzugsrechner acoustic corrector (sound location).
verzweigte Ströme branched currents, branch c. -Stromkreise divided circuits, branched c.
Verzweigung by-pass, shunt, branching (of transmutation products), ramification.
Verzweigungs-struktur lineage structure (cryst.). -verhältnis branching ratio.
Verzwillingen twinning.
Vibrations-galvanometer, kompensiertes Abraham's rheograph (inertia and damping compensated). -schwelle vibrational threshold. -tastgerät vibrato-tactile device.
vieldeutig ambiguous, non-equivocal, capable of various interpretations.
Viel-doppelzackenschrift multiple double-edged variable-width track. -ebene multiplane. -eck polygon.
vieleckig polygonal, multiangular.
vielerlei various, multifarious, variegated.
vielfach manifold, multiple, poly...,

multi..., various, frequent.
gleich- equimultiple.
Vielfach-abtastung multiple scanning. -beschleuniger cyclotron (type of) tube (with two dees and spiral beam), betatron (of G.E.Co.) -echo multiple echo, flutter e.
Vielfaches, ganzes integral multiple, whole m. -, gerades even multiple. -, ungerades odd multiple.
Vielfach-konturen multiple outlines along edges of objects or of lines (in pictures or images), general masking of details. -schaltung connection in multiple. -schwingungserzeuger multivibrator. -streuung multiple scattering. -telegraphie multiplex telegraphy. -verzögerer Barkhausen-Kurz type of tube (retarding-field, magnetron, etc. tubes). -zackenblende shutter or light stop for making multilateral sound track. -zackenschrift multilateral sound track or s. recording.
Vielfall multiplicity.
vielfarbig many-colored, multi-c., variegated, polychromatic.
Vielflach polyhedron.
viel-flächig polyhedral. -gestaltig multiform, manifold, diverse. -gliedrig polynomial (math.). -gliedriger Kettenleiter multi-mesh network.
Viel-kanalaufzeichnung multi-channel recording. -kantscheibe polyhedral mirror (of scanner).
vielkernig multinuclear, polynuclear, polynucleate; polycyclic.
Viellinien-spektrum discrete band spectrum, manyline spectrum. -system many-lined system.
vielseitig many-sided, polyhedral, multilateral; having diversity, versatile.
Vielspiegelung multiple of image.
vielstimmig polyphonous, polyphonic (of organ).
Vielstufenwärmeaustauscher multipass heat-exchanger.
vielteilig multipartite, of many parts; polynomial (math.)
Vielwegehahn multiple-way stopcock.
vielwertig multivalent, polyvalent.
Vier-bandlichtschleuse four-band light valve, four-ribbon l.v. -eck, Gelenk- linked or articulated quadrilateral or quadrangle.
viereckig four-cornered, square, quadrangular, oblong (of reading glass).
Vierervektor four-component vector.
Vierfachspirallochscheibe quadruple-spiral scanning disk.
Vierflach tetrahedron.
vier-flächig tetrahedral, 4-faced. -gliedrig four-membered, tetragonal (cryst.); quadrinomial (math.). -gliedrige Farbgleichung four-color equation, four-stimulus e.
Vierpol quadripole, four-terminal network, transducer, ideal artificial line (with two input and two output terminals). Teil- section of recurrent structure.
Vierpol-dämpfung image attenuation constant. -lautsprecher balanced armature loudspeaker (with permanent magnet). -winkelmass image phase constant.
Vierseit quadrilateral.
vierstellige Zahl four-figure number.
Vier-stoffsystem quaternary system. -teilung quadripartition. -telkreis quadrant.
viertelkreisige D-Werte quadrantal error (Q.E.) values (due to airplane fuselage, etc., in d.f.)
Vier-telperiode quarter period. -weghahn four-way cock.
Vignettierapparat vignetter.
Vignettieren des Lichtbündels vignetting of cone of light.
Vignettier-maske vignetting mask. -wirkung durch Linsenfassung trimming of pencil intensity caused by lens mount, vignetting effect.

Viola viola, tenor violin.
Violin E violin E string.
Visier sight (of gun, etc.), visor. **Abdrift-** drift meter. **Bomben-** bomb sight.
visieren sight, aim, gage.
Visier-einrichtung view-finder. **-linie** line of sight. **-mattscheibe** focusing screen. **-richtung** line of sight, direction of vision. **-scheibe** focusing screen. **-vorrichtung** (visual) sight for taking direct bearings. **-zirkel** gaging calipers.
Vogel-augenahorn bird's eye maple. **-orgel** bird's organ. **-perspektive** bird's eye view. **-pfeife** bird whistle.
Vokale,Halb- liquid consonants, semi-vowels, liquidae.
Vokaleinsatz beginning of vowel sound.
Voll-anode unsplit anode (of magnetron). **-bogen** semi-circular rail (of microscope).
vollfarbig of full (saturated) color.
Voll-flächner holohedron. **-leiter** compact, unstranded or solid conductor. **-macht ausstellen** execute a power of attorney. **-machtgeber** principal, mandator. **-netzanschluss** all-electric supply, mains s. **-niete** solid rivet. **-opalscheibe** pot opal diffuser disk.
vollstreckbar enforceable, executable, executory.
Voll-weggleichrichter full-wave rectifier. **-ziehungsbefehl** writ of execution warrant.
Voltampere(Scheinleistung) var (unit of reactive power).
Voltapotential Volta potential, V. emf, V. effect.
Voltmesser,Gleichrichter- rectifying voltmeter. **Ionenwind-** ionic wind voltmeter. **Röhren-,selbstgleichrichtendes** self-rectifying tube voltmeter. **Scheitelspannungs-** peak voltmeter, crest v.
Voltsekunde weber (magnetic flux unit).
Volumen,Mol- molar volume.
Volumen-gleichrichtung volume rectification. **-schwankung** volume flutter. **-verlust** loss in volume.
Volum-gewicht volume weight, weight of unit v. **-prozent** percent by volume. **-verhältnis** volume relation, proportion by v.
Vomhundertsatz percentage.
Voramt,Hilfs- sub-control station.
vorangehend preceding, leading, antecedent.
Voranode first anode (c.-r. tube); screen grid connected with cathode and acting as "fore-anode" in power pentode (Pierce circuit).
voraus,Backbord port bow. **-,recht** right ahead. **-,Steuerbord** starboard bow.
voraus-gehend preceding, leading, previous, prior, antecedent. **-nehmend** anticipatory (patent law).
Voraussetzung postulate, premise, assumption, supposition, presumption, hypothesis.
Vorbedingung prerequisite, precondition, conditio sine qua non.
vorbei-führen (cf **vorbeistreifen**) skirt, conduct or direct past (something), lead past at a glancing angle. **-sehen** not to look straight at each other (in two-way television telephone). **-streichen,-streifen** skirt, glance, graze, pass by, brush against (in more or less close contact with something).
Vor-beladung pre-charge, preliminary c. (of sorptive in sorbent). **-belichtung** priming exposure or illumination, pre-exposure. **--, Zellen-** primary illumination of a cell. **-benutzung,offenkundige oder öffentliche** prior use, public use. **-bereitung** preparation, line-up, readying. **-bereitung für Bildaufnahme** lining up for shooting pictures. **-bescheid** preliminary action (of

Patent Office Examiner), interim decision. **-bild** pattern, model, copy, prototype, type, exemplar.
vorbringen argue, plea, reason.
Vorder-ansicht front elevation, f. view. **-blende** anterior stop, field s.
vordere Brennebene front focal plane. **-(r) Brennpunkt** first focal point.
Vorder-glied predecessor, antecedent. **-licht** front light (studio). **-linse** front lens, field l. **-mann stehen,auf** lie in the shadow of. **-satz** premise, antecedent. **-seite** front side, front, obverse, face, panel.
Vordruck first impression, proof.
voreilen lead.
Vor-eilung,Phasen-phase lead. **-eilungswinkel** angle of lead (of phases). **-einflugzeichen** fore marker, outer m. (beacon) (of air-field). **-entladung** pre-discharge. **-entladungskanal** pre-discharge track. **-form** gathering mold (in glass manufacture).
vorführen demonstrate, display, bring out, produce.
Vorfuhrer projectionist, motion-picture operator.
Vorfuhrung,Film- film projection, exposure of f. (in gate).
Vorführungs-film display film. **-kabine,-raum** projection room or booth (m.p.) **-versuch** demonstration experiment.
Vorgang process, action, event, phenomenon, reaction, proceeding, procedure, occurrence, happening.
Vorgang,einmaliger non-recurrent action, unique a., singular a., event or phenomenon. **-,unperiodischer** non-recurrent action, non-periodic a. or phenomenon. **-,zeitlich veränderlicher** action subject to time variation, a. variable with time. **Ausgleich-** transient. **Bewegungs-** motional action, cinematographic a. **Einzel-**single or separate process, action, phenomenon or event. **Schalt-** switching operation, process or action.
Vor-gänger predecessor, antecessor, antecedent, progenitor. **-gelege** transmission gearing, connecting gear. **-geschichte** previous history, antecedents.
vorgespannt pre-stressed (mech.), initially tensioned; biased (electrical potential).
vorgetäuschte Sehschwäche simulated amblyopia.
Vorglimmlicht-gebiet sustaining voltage range, pre-photoglow region (arises at a voltage slightly below striking v.). **-zelle** photo-emissive gas-filled cell operating at sustaining voltage (in pre-glow region).
vorhallen over-emphasize; preverberate, pre-echo (ac.)
Vor-haltewinkel drift angle, a. of lead. **-hangblende** curtain fading shutter. **-hängeaufblendung** curtain fade-in. **-hof,Trichter-** air chamber (adjacent horn of loudspeaker).
vorionisierte Spur pre-ionized track.
Vor-kammer pre-chamber, ante-chamber; forehearth (oven). **-kehrung** precaution, provision, precautionary measure or step. **-kehrungen,sicherstellende** preventive measures.
vorkommen occur, happen.
Vorkonzentration (cf Vorsammellinse) preliminary focusing, pre-focus, first focus action (in c.-r. tube).
vorladen cite, serve, subpoena, summon,
Vor-ladung appointment, summons, subpoena, writ, citation, notice, notification. **-lage** original copy, copy, pattern, receiver, condenser (zinc smelting). **-lauf** lead or precession of sound (in relation to picture = 19 frames = 361 mm). **-läufer** forerunner, precursor; sign, indication, clue.
vorläufig provisional, preliminary, advance (e.g., notice).
Vor-legierung key alloy.

-lesungsversuch classroom experiment or study, lecture demonstration.
vorliche Kurslinie ahead course or direction, "go-to" indication.
Vorlicht priming or biasing illumination.
vormagnetisierte Spule, gleichstrom- d.c. controlled saturable coil, three-legged reactor.
Vormagnetisierung bias magnetization, magnetic polarization.
Vormagnetisierungsstrom biasing current.
vorn in front, anteriorly.
vornherein, von from the outset, from the first, to begin with.
Vor-prüfer examiner (in Patent Office). **-prufungsverfahren** preliminary examination, preliminary search or action. **-pumpe** fore-pump, backing p. (for vacuum). **-rang** priority, precedence, superiority, first rank. **-rat, Metall-, Quecksilber-** mercury pool (in rectifier).
Vorrats-flasche "Winchester", stock bottle. **-leitung** spare circuit, standby c. **-lösung** stock solution. **-röhre** input tube. **-rolle, -spule** magazine, m. roll.
Vorraum ante-chanber, fore-c., outer c.
Vorröhre input tube.
vorrücken feed (forward), advance, progress, step forward, notch forward (e.g., a switch or controller).
Vor-sammellinse first focusing lens, cathode l. (like Wehnelt cylinder, apertured disk, etc., in c.-r. tube). **-satz, Peil-** direction-finder means added to standard receiver apparatus. **-satzgerät** attached or accessory device, head (m.p.) **-satzlinse** anteriorly attached lens, front l. attachment, magnascope (to enlarge projector image).
vorschalten cut in circuit, connect in series.
Vorschaltwiderstand series resistance; multiplier (of a voltmeter).
vorschlagbar capable of being moved, dropped, let or folded down on hinges.
Vorschrift, Bedienungs- service direction, working instructions, directions for use. **Dienst-** service or working rules and regulations. **Gebrauchs-** directions for use.
Vorschub feed, conveyance. **Papier-** paper feed.
Vorschub-bewegung, Zeilen- line sweep, l. traversing motion. **-rad** feed wheel.
Vor-selektion pre-selection. **-sichtsmassregel** precaution, precautionary step or measure. **-signal** first or outer marking signal, caution or warning s. **-spann** label (of a print), leader (of a film). **-spannprogramm** prologue.
Vorspannung biasing potential (of a grid, magnet, etc.), bias, polarizing p., inherent bias tension (as in a spring) to restore normal position, normally tensed or tensioned condition, initial tension, pre-stress (in metal, etc.); anode, drive or positive p. (of a photo-cell). **-verriegelnde** cut-off biasing potential. **Gitter-, automatische** automatic bias or self-biasing of grid.
Vor-spektrum preliminary spectrum (of Goldstein).
vorspringen project, protrude, be salient or prominent.
Vor-sprung projection, salient, lug, shoulder, protrusion, prominence. **-stellung** mental picture, conception, ideas, notion; representation, demonstration, performance, display. **-stoss** adapter, lap (of a tile), edging. **-strom** dark pre-sparking current. **-stufe** first or preliminary step, input or first stage (of an amplifier); primer. **-synchronisieren** pre-scoring (first sound, then picture).
vortauschen simulate, mislead,

delude, deceive, cause an illusion.
Vortexring vortex (ring).
Vor-transport supply sprocket.
-trieb positive drive, forward d., torque. **-vakuumpumpe** fore-pump, rough-vacuum p., backing p. **-verfahren** preliminary procedure or proceedings. **-veröffentlichung** anticipatory reference, anticipation (in patent or magazine literature), earlier publication. **-verstärker** input amplifier, pre-a. (class A RF amplifier, in a superheterodyne). **-verstarkung** gain amplification.
vorwalzen rough down (rolling mill work).
Vor-warmung pre-heating, fore-warming. **-warnung** alert signal.
Vorwarts-bewegung active stroke (in film feed), forward motion, translatory movement. **-gang** direct action (camera). **-regelung** direct control, forward-acting regulation.
vorwegnehmend anticipatory (in the form of a disclosure in prior art).
Vorwickel pull-down sprocket. **-rolle,-trommel** supply reel, (upper) feed sprocket, pull-down sprocket.
Vorwickler upper feed sprocket.
Vorwiderstand series resistance.
vorwiegend preponderant, predominant, especially, chiefly.
Vor-wolbung protrusion, anterior curvature. **-wurf** subject (phot.) **-zeichen** indication, symptom, sign (math.). **-zerleger** filter (for ultraviolet) mounted anteriorly of monochromator entrance slit). **-zugsrichtung** privileged direction.

W

wabenartig honeycombed, pitted (of structure).
Wabenspule honeycomb coil, lattice-wound c., duo-lateral c.
Wachfrequenz watch frequency, f. of international automatic alarm signal, distress f., mayday (in radiophony).
Wachs-abdruck impression in wax. **-paste** cerate or encaustic paste. **-platte** (soft) wax disk, w. master (in disk manufacture). **-tuch** oilcloth, cerecloth (obs.). **-tum,Kristallisations-** germination, crystal growth. **-tumkurve** growth curve. **-tumzentrum** nucleus, center of growth, kernel.
Wackelkontakt loose contact, variable or defective c.
wackeln shake, totter, rock, sway, quake.
Wackelung,Ver- frame unsteadiness.
Wage,Dreh(feder)-torsion balance. **Feder-** spring balance, s. scales. **Fein-** precision balance, micro-b. **Hebel-** beam scale, lever s. **Mikro-**micro balance, precision b. **Stoss-** ballistic pendulum. **Strom-** -ampere balance.
Wagebalken beam of balance, scale b.
Wagen,Aufnahme- pickup camera truck, dolly. **Fernsehaufnahme-**television pickup camera truck, video bus. **Lautsprecher-** loudspeaker truck, sound t. **Ton-(film)-** sound truck, location t., lorry set, sound van.
Wagenfett axle grease.
Wägeschaltung,Ab- comparator circuit organization.
Wagschale weighing-dish, scale or balance pan.
Wahl,Nummern- impulse action, stepping (automatic telephony).
Wähler,Schrittschalt- step-by-step selector.
Wähler-bucht bay of selector. **-hebeschritt** vertical step of selector. **-scheibe** dial type selector switch. **-segment** bank of (stationary) contacts.
wahllos at random, haphazardly, without choice or selection, non-selective.
Wahlschalter selector switch.
wahlweise selective, directional (in signal transmission.
wahre Peilung true bearing.
Wahr-heitsbekräftigung affirmation. **-nehmbarkeit** perceptibility, noticeability, observability, visibility, discernibility, audibility.
wahrnehmen perceive, sense, notice, observe, be sentient.
Wahrnehmung perception, sensation, observation, percipience, sentience. **Farben-** color vision, c. perception. **Schall-,des menschlichen Ohres** human sound or tone perception. **Ton-** acoustical perception, auditory or sound sensation.
Wahrscheinlichkeit,Ansprech-,eines Zählrohres efficiency of counter tube (probability to respond or register).
Wahrscheinlichkeits-gesetz probability law. **-rechnung** calculus of probability.
Waldhorn French horn, concert h.
Walkerde fuller's earth.
Walze roller, roll, cylinder, drum, wheel, barrel. **Bild-** picture cylinder (in Bakewell's p. transmitter).
walzen roll, mill, laminate. **vor-** rough down (by rolling mill).
walzenformig cylindrical, barrelshaped.
Walzen-mühle roller mill. **-strecke** roll train.
Walzwerk,Draht- wire rod mill.
Wand wall, partition, side, cheek, shell, panel, screen, baffle.

wand, Bild- picture (projection) screen. **Durchprojektions-** translucent screen. **Ganzmetall-** all-metal (projection) screen. **Schall-** baffle (board) of a loudspeaker; sound panel (for s. absorption or reflection). **Signal-** wall of signals, curtain of s. **Tonbild-** transoral screen (m.p.).

Wand-bekleidung, schallschluckende wall draping, w. lining, baffle, blankets, sound absorbing material put on walls or panels, often portable, called "gobos" or "tormentors" (m.p. and sound stage). **-durchführungsisolator** wall lead-in insulator, bushing i.

Wander-licht (Personenabtaster) splotlight scanner for persons. **-marke** wander mark, measuring m. (in range finder).

wandern migrate, shift, crawl, creep, wander (of beam), travel (of waves or surges, over electric lines).

wandernder Lichtstrahl flying spot, scanning spotlight.

Wanderung, Material- creep or flow of material (or metal).

wanderungs-geschwindigkeit rate of migration, crawl or creep (in telev., ions), m. velocity. **-sinn** direction of migration (of ions).

wander-welle traveling wave, transient w., surge. **-wellengenerator** surge generator, impulse g., "lightning" g. **-wellenmesser** klydonograph, surge indicator, s. recorder.

Wandler transformer, converter (of picture or image), transducer. **-, lichtelektrischer** photo-cell. **Bild-** image converter, picture transformer, transducer (changes optical into electric images or patterns, etc.) **Durchfuhrungs-** bush-type transformer. **Fernseh-** photo-cells, light relays and similar devices(changing light into currents and back), television transducer. **Frequenz-** frequency transformer, f. changer. **Kurzwellen-** short-wave converter. **Mess-** measuring transformer, instrument t.

Wandlerspeicherröhre, Bild- super-iconoscope.

Wand-stoffbekleidung draping, baffle, blanket, sound-absorbing material, called gobo or tormentor when in form of portable wall (m.p. studio). **-strom** wall current (of a tube).

Wange cheek, side piece, end p.

Wanne trough, tub, casing (of a multiple condenser and its plates). **-, pneumatische** pneumatic trough (in gas drying).

warmblasen hot blasting (water gas manufacture).

warmbrüchig hot-short, brittle when hot.

Wärme, gebundene latent heat. **Arbeitswert der** mechanical equivalent of heat. **Bildungs-** enthalpy, heat of formation. **Eigen-** specific heat, body heat, animal n.

Wärme-abführung, -abgabe, -ableitung, heat dissipation, h. evacuation, loss of h., h. "abduction", the carrying or conducting away of h. **-arbeitswert** mechanical equivalent of heat. **-austauscher, Vielstufen-** multi-pass heat exchanger. **-bewegung** thermal agitation, heat motion.

wärmedurchlässig diathermic, thermanous.

Wärme-durchschlag breakdown due to thermal instability. **-einheit (W.E.)** thermal unit, heat u. **-elektrizität** thermo-electricity. **-entbindung, -entwicklung** development, disengagement or evolution of heat, loss of h. **-fluss** heat flow, thermal f. or flux. **-geräusch** thermal noise (sort of Brownian motion of electrons in input circuit).

wärmegleich isothermal.

Wärme-ionisation thermal ionization, temperature i. **-kapazität** thermal capacity. **-kraftlehre, -mechanik** thermodynamics. **-leitvermögen, -leitzahl** thermal conductivity.

Wärmemesser thermometer, calorimeter. **-, thermoelektrischer** thermel. **Sonnen-** solarimeter, pyroheliometer.
Warme-rauschen thermal noise (thermal agitation or motion of electrons). **-schwingung** heat vibration. **-speicher** heat accumulator, regenerator. **-stoff** caloric, thermogen. **-strahlung** thermal radiation, temperature r., heat r., radiant heat. **-tod** heat death. **-tonung** heat tone (in thermochemistry), heat quantity, h. effect.
wärmeundurchlassig athermanous, impervious to heat, heat insulating.
Wärme-vergangenheit thermal antecedents, prior heat treatment. **-verlust** heat loss, thermal l. **-wirkung** temperature effect, heat action, thermal a.
warm-fest heat resistant. **-gepresst** hot-pressed, hot-press-fitted, subjected to thermo-plastic treatment.
Warnsignal (cf **Entwarnen**) air-raid alarm, caution signal.
Warnung, Vor- alert signal.
Warze wart, nipple, pin, knob, boss.
Wasser, Ultra- optically empty water.
Wasser-abguss, Fagott- bassoon syphon. **-ablass, -ableitung** water drain, w. drainage, w. outlet.
wasser-abstossend repelling water, w. repellent, non-hygroscopic. **-anziehend** water or moisture attracting, hygroscopic. **-beständig** stable in water or towards action of w. **-bindend** hydrophylic, water-absorbent, w. imbibent.
Wasser-blase bubble of water; vessel or container for water heating. **-druck** hydraulic pressure. **-entziehung** dehydration, removal of water, desiccation.
wasser-fest water-tight, w.-proof, resistant to w. **-frei** dehydrated, desiccated, anhydrous.
Wasserhahn water cock, w. tap, faucet.
wasserhaltig hydrous, hydrated, aqueous, containing water, w. bearing.
wässerig watery, hydrous, aqueous.
Wasser-kraftlehre hydrodynamics. **-kuhlröhre** water-cooled tube. **-luftpumpe** water vacuum pump. **-mantel** water jacket.
wassern airplane traveling in contact with water. **ab-** rise or start from water. **an-** "land", alight or descend on water.
Wasser-schalter hydroblast switch. **-schluss** water seal, trap. **-siedemesser** hypsometer.
Wasserstoff, schwerer deuterium (mass of 2), tritium (mass of 3). **-ionenkonzentration** pH value, hydrogen-ion concentration. **-kern** hydrogen or H particle or nucleus. **-knallgas** detonating gas (explosive mixture of hydrogen and oxygen). **-strahl** H ray (consisting of H particle or positive hydrogen ion). **-zahl** hydrogen-ion concentration, pH value.
Wässerung washing (of film, etc.)
Wässerungsgestell washing rack (phot.)
wasserunlöslich insoluble in water.
Wasser-verschluss water seal. **-wert** water equivalent (of a calorimeter).
Watte absorbent cotton, wadding (of cotton, glass wool, etc.), pad.
Watt-leistung, -zahl wattage, power in terms of watt.
W.E. (Warmeeinheit) heat unit, thermal u.
Wechsel change, shift, alternation, cycle. **Bild-** (cf **Vertikal-**) picture cycle, moving period, feed stroke (m.p.) **Horizontal-** horizontal or line sync impulse or cycle. **Last-** cyclic stress, stress application cycle. **Licht-** variation of exposure. **Pol-** reversal of polarity, change of p. **Vertikal-** vertical or frame synchronizing pulse of cycle. **Zeilen-** s. **Horizontal-**
Wechsel-beziehung inter-relation, correlation, mutual or reciprocal relationship. **-blende** alternating shutter (in stereoscopy); masking disk (with spiral slot

cyclically cooperating with quadruple scanner disk or drum apertures), auxiliary rotary shutter disk with spiral slot.
-druck alternating pressure.
-festigkeit, Biege- alternating bending strength. **-festigkeit, Verdrehungs-** strength under alternating torsion stress.
-frequenz, Bild- picture or frame frequency, repetition f.
-frequenz, Zeilen- line frequency.
-getriebe, Geschwindigkeits- change-speed gear. **-hahn** change cock. **-induktion** mutual inductance. **-kardioide** switched cardioid (d.f.) **-kontakt** make-and-break contact. **-licht** light intensity variations (striking photo-cell). **-note** warble note, warble sound (in airplane alarms); appogiatura note (mus.)
-periode, Zeilenzug- field frequenry (= 2 x frame f.) **-richter** inverter, inverse or inverted (d.c. to a.c.) rectifier, d.c. to a.c. transverter (mostly combined with transformer).
-richter, Pendel- vibratory inverter, vibrator. **-satz** exchange principle. **-schalter** double-throw switch.

wechselseitig reciprocal, mutual, interchangeable, alternate.

Wechselsprechen, Funk- alternate two-way radiophone communication.

Wechselstrom alternating current, indirect c., a.c. (specifically, single-phase current is mostly meant), any periodic c. **Leitungs-** conduction alternating current.

Wechsel-stromempfanger a.c. electric radio set or receiver apparatus. **-stromton** hum in a telev. picture (causing wavy edge or sides, pairing of lines, etc.)
-tastverfahren reversing switch method (in d. f. work), make-and-break switch or keying method.
-ton warble note or sound (used in airplane alarms), appoggiatura note (mus.) **-verhältnis** reciprocal relation, r. proportion.
-winkel alternate angle.
-wirkung inter-action, mutual or cooperative a., reciprocal effect.

Wechselwirkungs-energie mutual potential energy. **-glieder** interaction terms. **-potential** interaction potential.

Wechsel-zahl, Bild- frame frequency, picture f., repetition f. **-zahl, Last-** alternating stress number.
-zeit, Bild-frame frequency, picture f., repetition f.

Wechsler, Objektiv- revolving nosepiece (in microscope). **Platten-** record changer (in phonograph). **Strom-** commutator, reverser.

Wechslung, Fallsystem- drop system record changer.

Wecker, Schnarr- buzzer alarm.

Weg (in Mechanik) displacement (in centimeters, in mechanics, equivalent to charge, in coulombs, in electricity, and to volume displacement, in cubic centimeters, in ac.)

Weg, Eisen- iron path, magnetic or ferro- m. circuit. **Gleis-** rail track. **Gleit-** glide path, landing curve (of an airplane), landing beam. **Neben-**, by-path, by-pass. **Ubertragungs-** transmission channel, t. path, t. medium.

Weg-ablenkung, Peilstrahl- distortion of bearing, deviation (due to diurnal or seasonal factors, weather, terrain, local conditions, etc.), lateral deviation.
-abschnitt element of path, length or distance, portion, fraction or subdivision of length.

weg-diffundieren diffuse away.
-heben, sich cancel out (math.)

wegig, ein- single-channel, one-way, simplex. **mehr-** of multi-channel or m.-path nature.

Weglänge, mittle freie mean free path. **Schall-** sound path length.

Weg-leitung directions for tracking trouble. **-strecke** distance, stretch, (element of) length.
-tasten des Trägers, Synchronierung durch carrier interruption, suppression or gating of c.,

synchronizing impulse by c. gating. -überführung overpass, overhead crossing. -unterführung underpass. -winkel transit angle (of klystron and other beam tubes). -zeitkurve trajectory.

Wehneltzylinder Wehnelt cylinder, W. grid, W. shield (telev.)

weich gentle, smooth, yielding, soft; weak or of low contrast (of picture or image).

weich,schall- sound absorbent.

weiche Röhre soft valve, ionic tube, gas-filled t., gassy t.

Weiche dividing network, d. filter, separator (for sync and video signals) comprising limiter tube and network.

Weiche,Strom- current divider (filter).

weichen soak, steep; yield, give way, soften, plasticize.

weichlöten soft-solder.

Weichzeichner soft-focus lens.

weigern refuse, decline.

Weihnachtsbaumantenne christmas tree antenna.

Weiser pointer, indicator, hand, guide. Kurs-,Richt- radio beacon, r. range.

weissbrüchig white fracture, pale f.

Weissbuche white beech.

weisser Elektronenstrahl heterogeneous beam (of electrons).

weissglühend white-hot, incandescent.

Weiss-guss white metal (cast). -lot soft solder. -metallfutter, Lager mit babbitted bearing. -wert bright level value (telev.)

Weisung direction-finding, guiding (by radio compass, etc.). Fehl- directional or d.f. error, quadrantal e. (Q.E.), compass deviation.

Weite,lichte inside diameter, i. width, lumen. Bild- distance between screen and lens, image intercept, i. distance. Brenn- focal distance, f. length, convergence d. Einstell- focal range, focusing r. Gegenstands- distance from object to lens, o. distance, d. from electron source (or from stop or crossover) to main l. Schnitt- distance from back lens to image, intercept length. Schwingungs- amplitude of vibration, a. of oscillation. Seh- distance of vision, visual range. Sicht- optical distance, line of sight, direct-sight range or distance. Start- (der Elektronen) distance from origin, d. from object to lens, object d.

Weiter-bildung further development (of an invention), evolutionary development or progress. -drehen des Schaltarmes stepping around or notching forward of the wiper, controller handle, etc. -führung einer Anmeldung prosecution of an application (for letters patent).

weit-gehend far-reaching, substantial, thorough, appreciable; broad (of a patent specification, its disclosure or claim). -geöffneter Konus wide-open cone or pencil. -maschiges Gitter open or wide-meshed grid, coarse g. -sichtig far-sighted, long-s., hyperopic.

Weitwinkelbild wide-angle picture, "Wide-scope" p.

weitwinklig wide-angled (phot., etc.) -(e) Aufnahme magnascopic picture. -(en) Kohlenstiften, Bogenlampe mit scissors-type arc lamp.

Welle (cf. wellenabgrenzung) wave; shaft, arbor, axle, spindle, roll, roller.

Welle,direkte ground wave, direct w. ("hugging" curvature of earth). -,ebene plane wave. -,extremkurze ultrashort wave. -,fortschreitende progressing or progressive wave. -,gedämpfte discontinuous, damped or decadent wave. -,gleichstromüberlagerte wave superposed on or with d.c. -,impulsmodulierte pulse-modulated wave (used in radar). -,linkslaufende sinistro-propagating wave or surge.

-, rechteckige square wave. -,rechtslaufende dextro-propagating wave or surge. -,reflektierte reflected wave. -,rucklaufende retrogressive wave, reflected w. -, stehende standing wave, stationary w. -,ultrakurze ultra-short wave, quasi-optic w. -,ungedampfte undamped wave, persistent w., continuous w. (c.w.).

Welle,Anregungs- exciton (strictly an electron in a crystal moving in a field of a positive hole). Antriebszahntrommel- drive sprocket shaft. Boden- direct or ground wave ("hugging" curvature of earth). Bug- (eines Geschosses) nose wave (of a projectile). de Broglie- de Broglie wave, electron w., phase w. Dehnungs- dilational wave. Dezimeter- micro-wave, decimeter w. Eigen- natural wave. Elektronen- de Broglie electron wave, phase w. Elementar- wavelet (of Huygens). Feld- field wave. Front- onde de choc, impact wave, bow w. (preceding a projectile's nose). Hohl (metall) rohr- metal-sheath conducting wave (in tubular or hollow pipe wave guide). Huygens'sche Elementar- Huygen wavelet. Knall- shock wave. Kopf- onde de choc, impact wave, bow w., shell w., ballistic w. (ballistics). Kreis- circular (electric) wave. Kugel- spherical wave. Leitbahn- transit-time or electron-path oscillation. Luft- sky wave, space w., indirect w. Mantel- wave on outer surface of co-axial cable, "shell" wave. Materie- de Broglie elementary particle (electron or proton) wave, elementary w., phase w. Mikro- micro wave, hyper-frequency w. (1-100 centimeters). Nuten- slot ripple. Phasen- de Broglie wave, phase w., electron w. Raum- space wave, sky w., indirect w.; spherical w. Raumladungs- space-charge wave. Rohr- dielectric or conducting wave (solid or hollow pipe wave guide). Rundfunk-(cf Wellenabgrenzung) broadcast wave, medium-frequency w. Schalt- wiper shaft, switch s. Spiegel- reflected wave. Staub-,Kundt'sche acoustic dust pattern (showing nodes and antinodes). Steuer- control wave (in chain broadcast). Stirn- onde de choc, impact w., bow w. Störungs- garbling wave (in secret or coded telephony); parasitic w., interfering w., jamming w. Stoss- percussion wave, impact w., impulse w., wave impacting upon a joint (say, in a pipe). Überschall- supersonic wave, ultra-audio w., ultrasonic w. Ultrakurz-(cf. Wellenabgrenzung) ultra short wave, uhf w., quasi-optical w. Unterschall- subaudio wave, infrasonic w. Verdichtungs- compression wave. Verdünnungs- rarefaction wave. Wander- traveling wave, transient (wave), surge. Zeichen- marking wave, signal w., keying w. Zwerg(cf Wellenabgrenzung) micro wave, midget w., dwarf w. Zwischenzeichen- spacing wave, back w.

Wellenabgrenzung und -Einteilung frequency and wave-band designation suggested by Fed. Communic. Comm. (FCC): very low-frequency (vlf) 10-30 kc; low-f. (lf) 30-300 kc, medium-f. (mf) 300-3000 kc; high-f. (hf) 3-30 mc; very-high f. (vhf) 30-300 mc; ultra-high f. (uhf) 300-3000 mc; super-high f. (shf) 3,000-30,000 mc.

Wellen-antenne wave antenna, Beverage a. -anzeiger cymoscope (obs.), wave detector, oscillation detector, coherer. -aufzeichner, Schall- phonodeik (record on film), phonautograph. -ausbreitung in Hohlmetallrohr wave propagation in hollow metal pipe, metal-sheath conducting wave p. in wave guide. -bereichmelder wave band indicator. -bereichschalter wave band switch. -berg

peak, crest or hump of a wave.
-berglochweite distance separating peaks, intercrest d. -betrieb, Gleich- operation of stations on one wave, chain operation (from key station), network operation.
-bündel wave packets, beam of electrons and ions. -bündelung beaming, directional effect of waves. -dampfungskonstante wave attenuation constant. -detektor cymoscope, wave detector, oscillation d., coherer. -echo radio (signal or wave) echo (involving single-, double-or multiple-hop).
-feld wave field. -flache, Fresnel'sche Fresnel zone (of half-period elements).
wellenförmig undulatory, wavy, corrugated, rippled.
Wellen-front, geneigte tilted wave front, t. wave head. -frontwinkel wave front angle, w. tilt.
-führung wave guide (either dielectric or conducting). -funktion wave function.
wellengerader Kondensator straight-line wave-length condenser.
Wellen-geschwindigkeit wave velocity, phase v. -gipfel peak, crest, or hump of a wave. -gleichung, d'Alembert'sche d'Alembertian wave equation. -konstante wave constant (meters per millimeter of piezo-electric resonator).
-länge, Grenz- limiting wave-length. -leitung wave-guide (for dielectric or conducting waves).
-linie wavy line, undulatory l., sinuous l., rippled l. -mechanik, Geometrisierung der geometric derivation of wave equation.
wellenmechanisch wave mechanical.
Wellenmesser (cf Frequenzmesser) wave meter, Fleming's cymometer, frequency meter, ondometer. Schwingaudion- autodyne wave meter or frequency m. Summer- buzzer wave meter.
Wellen-rückstrahlung wave reflection (by hop to ionosphere and back). -sauger smoothing choke, series reactor, wave trap.
-schalter wave-band switch.
-schlucker(cf-sauger) wave trap (of absorptive or of impedance type). -schwanz wave tail (in facsimile). -spannung, Bild- video wave potential (telev.)
-spannung, Ton- audio wave potential (telev.) -spektrum(cf -abgrenzung) spectrum of electromagnetic waves, radio wave spectrum. -stirn wave face, w. front, w. head. -strom pulsating current, ripple c. -stromfilter ripple filter. -stumpf butt end of a shaft. -tal trough or valley of a wave. -telegraphie, Draht- wired-wave telegraphy, wired-radio t. -widerstand characteristic impedance, surge i. (distinguished by some authors).
-zahl wave number. -zugfrequenz group or wave-train frequency, spark f. -zugkohärenzlänge coherence length of wave trains.
wellig wavy, rippled, undulated, pulsating. -, lang- long-wave, of great w.-length. -(e) Gleichspannung ripple potential, pulsating p.
Welligkeit ripple, rippled condition, humpiness (of resonance curve), pulsation factor, lack of linearity. Mehr- multi-wave property (of a crystal).
Wellrohr corrugated tube or tubing.
Wellung rippling, corrugation, undulation.
Welt, Aussen- external world.
Welt-beschreibung cosmography.
-raumstrahlen cosmic rays.
weltzeitlicher Anteil des Potentialgefälles universal diurnal variation component of potential gradient.
Wendekurve turn in landing (in ZZ method), U turn.
Wendel spiral, helix, coil. Kehrdoppel- reversed double loop.
Wendelsystem, Doppel- biplane filament system.
Wendepunkt inflection point, reversing p., cusp, turning p.
Wender, Strom- current reverser,

commutator. **Ton-** tone converter.
Wende-schalter reversing switch, r. key. **-tangente** tangent to reversing point. **-zeiger** turn meter, t. indicator (of airplane).
Wendigkeit maneuvrability, manageableness, ease of handling or operating.
Werbung, Licht- light-advertising, propaganda using illumination effects.
werfen, sich warp, distort, deform.
werfer, bild- picture projector. **Strahlungs-** beacon or beam station.
Werg oakum, tow.
Werk, Lauf- feed mechanism, drive m. **Schalt-** film feed mechanism, intermittent m. (m.p.)
Werk-blei raw lead. **-photo** studio still. **-stoff, technischer** industrial, technical or commercial material. **-stoff, Austausch-** substitute material. **-stück** work, piece of material or metal (to work on), blank to be machined or tooled, etc.
Wert, quadratischer Mittel- root-mean-square (rms) value, virtual v. **Behaglichkeits-** comfort value. **Bei-** constant, coefficient, factor, parameter, co-ordinate. **Effektiv-** effective value, root-mean-square v. (rms), virtual v. **Nenn-** nominal, normal, assessed, face or rated value. **Schwarz-** black-level value. **Unterhaltungs-** entertainment value. **Weiss-** bright-level value (telev.)
Wertanzeiger, Maximum- peak indicator, crest i. **Minimum-** minimum indicator (for instance, of voltage).
Wertbestimmung valuation, evaluation, appraisal, estimate.
Werte, auseinandergehende divergent or discrepant values.
wertig, gerad- of even value.
Wertigkeitsformel valence formula, linkage f.
Wertung valuation, evaluation, appraisal, estimate.
Wesen (einer Erfindung) essence or essential feature of an invention, (main) object of invention.
Wettbewerb, unlauterer unfair competition.
Wetter-forschung, Radiosonde für radiometeorographic sonde, radiosonde. **-messinstrumente mit funkentelegraphischer Fernübertragung** radio-telemeteorologic or telemetric instruments.
Wetzstein whetstone.
Wichte specific gravity (sp.gr.), unit of weight.
Wickel, Kondensator- tubular condenser, paper capacitor, rolled or wrapped fixed condenser (of metal foils and paper or mica insulation), Mansbridge c. **Vor-** pull-down sprocket.
Wickel-blockkondensator tubular condenser, paper capacitor, wound or wrapped c. of fixed value (with metal foils and paper or mica insulation), Mansbridge c. **-drähte, Gitter-** grid coil wires. **-korper** coil form, former.
wickeln wind, wrap, roll, twist.
Wickelturm mandrel.
Wickler, Nach- lower take - up sprocket. **Vor-** pull-down sprocket.
Wicklung, angezapfte, in der Mitte center-tapped winding. **-, gesehnte** short-pitch winding, chord w. **-, kleine** bobbin. **-, schrittverkurzte** short-pitch winding. **-, verteilte** distributed winding. **Bifilar-** bifalar winding, Ayrton-Perry w. **Dampfer-** amortisseur winding, damping w. **Dritt-** tertiary winding (of a transformer). **Durchmesser** full-pitch winding. **Erst-** primary winding. **Korb-** basket winding, low-capacitance w. **Stufen-** banked winding, pile w. **Zweidraht-** bifilar winding, Ayrton-Perry w. **Zweit-** secondary winding.
Wicklungs-faktor space factor, copper f. **-halter, -träger** winding form, coil form, c. support, skeleton form, former.
Wider-beklagter cross-claim

defendant, counteraction d. -druck counterpressure, reaction. -haken barb. -hall repeated echo, e. reflection, reverberation, resonance. -hall,Flatter- flutter echo, multiple e. (between two parallel surfaces). -hallzeit reverberation period, r. time. -kläger cross-claim plaintiff, counteraction p. -lager abutment (immovable point or surface sustaining pressure or reaction). -legung rebuttal, refutation, disproof.

wider-rechtlich illegal, unlawful, contrary to law. -sinnig in or of opposite or contrary sense or direction (say, of rotation); devoid of sense or logic, nonsensical, illogical, irrational, absurd. -spiegeln reflect. -sprechend contrary, contradictory, inconsistent, opposite.

widerspruch opposition, protest, objection, inconsistency, in opposition or conflict with.

widerstand ein- und ausschalten cut in and out resistance. -, induktionsfreier non-inductive resistance, ohm r. -, innerer(im Innern) volume resistance, insulation r. -, magnetischer reluctance, magnetic resistance. -, ohm'scher ohmic resistance, d. c. resistance. -, spezifischer resistivity (mass or volume), specific resistance. -, winkelfreier resistance with zero phase angle.

widerstand, Ableitungs- leak resistance, leakance, resistance leak (of a tube). Abschluss- termination or terminal resistance or impedance. Abzweig- leak coil, l. resistance (of a telegraph repeater). Ausbreitungs- diffusion resistance. Ausgleich- compensating resistance, balancing r., build-out r. Ballast- ballasting resistance, ballast resistor. Beruhigungs- steadying resistance, ballasting r. Bewegungs- motional impedance. Blind-, induktiver inductive reactance, positive r., magnetic r. Blind-, kapazitiver capacitive reactance, negative r., condensance. Dämpfungs- nonreactive, active, dissipative or ohmic resistance. Dunkel- dark resistance (of a cell). Durchgangs-volume resistance (exclusive of surface r.), insulation r. Durchlass- forward resistance (of a rectifier). Eisenwasserstoff- iron-hydrogen resistance. Entknurrungs- anti-growl resistance, r. to suppress growling noise due to RF or tickler coil in regenerative path. Erder- grounder resistance. Fahr- aerodynamic drag (of an airplane). Falz- folding resistance. Federungs- compliance (ac.). Fusspunkt- terminating impedance, base-loading i. (of antenna). Haut- dermal or skin resistance; s. effect r. Isolations- insulation resistance, insulance (reciprocal of leakance). Kenn- characteristic impedance, image i., indicial i. Kenndämpfungs-, Ketten- iterative impedance. Klemmen- terminal impedance, t. resistance. Kopplungs- coupling resistance. Länge- line resistance, series r. (of a filter or network). Leerlauf- open-circuit impedance. Luft- air resistance, aerodynamic drag (of airplane). Massen- inertance. Nutz- (cf Wirkwiderstand) useful resistance, signal r. Oberflächen- surface resistance; skin effect r. Quer- cross resistance, shunt r. (of a network). Reibungs- (Schall) acoustic resistance (sound). Reihenverlust- equivalent series resistance (of condenser). Richt- unidirectional resistance, valve effect. Schall- ratio pressure amplitude: velocity a. Schallreibungs- acoustic resistance. Schallwellen-

acoustic impedance (comprising a. resistance and reactance).
Schallwirk- acoustic resistance. **Schein-** impedance, apparent resistance; impedor (part having i.) **Schein-,Blindkomponente** reactive component of impedance. - **Wirkkomponente** active or dissipative component of impedance. **Schieber-** slide rheostat. **Schwächungs-** gain controller, g. regulator, potentiometer. **Sieb-** contact resistance (in barrier cell). **Stirn-** (front) end resistance, leading end r. **Strahlungs-** effective resistance (of loudspeaker, antenna, etc.), radiation r., characteristic impedance. **Strömungs-** aerodynamic resistance, drag (of air). **Übergangs-** contact resistance. **Übergangs-,einer Probe** flow resistance of test specimen. **Verlust-** non-reactive, effective, ohmic or dissipative resistance. **Vor-** series resistance. **Vorschalt-** series resistance; multiplier (of voltmeter). **Wellen-** characteristic impedance, surge i. **Wirk-** non-reactive, active, effective, dissipative or ohm resistance. **Zersetzungs-** electrolytic resistance.

Widerstand-Reaktanzverhältnis Q factor, magnification f.

Widerstands-ballast resistance ballast, ballasting r. **-brücke** resistance bridge, Wheatstone b. **-büchse** resistance box. **-fähigkeit,Knick-** folding endurance (of film). **-falschanpassung** mismatching of impedances. **-kette** voltage divider, potentiometer. **-körper** baffle, damper.

widerstandsloser Kreis resistanceless circuit (devoid of resistance or impedance).

Widerstands-messer ohm meter, megger. **-moment** section modulus (metal testing). **-rauschen** circuit noise (due to Brownian movement). **-reziprok** inverse resistance, r. reciprocal. **-überanpassung** overmatching of impedance (load resistance exceeds internal tube r.) **-unteranpassung** under-matching of impedance. **-verteilung (im Potentiometer)** taper (called linear t. if distribution of resistance is uniform per unit of length).

widerstrahlen reflect, re-radiate.

widrigenfalls in default of, failing which.

Wieder-abtretung re-assignment. **-aufbau** re-construction (of a facsimile picture). **-aufnahme** re-opening (of a legal case); re-take (m.p.). **-belebung** re-activation, rejuvenation (of a thoriated filament, etc.). **-einführung** re-insertion, restoration.

wiedereinschalten re-close (a switch), cut or connect in circuit again.

Wieder-erlangung recovery. **-eröffnung** re-opening (of a law suit).

wiedererstatten return, re-imburse, restitute.

Wiedergabe,flache flat reproduction. **-,flimmerfreie** flicker-free reproduction. **-,getreue** good definition (of image by a lens, by telev. or facsimile, etc., receiver); orthophonic reproduction (of loudspeaker). **-,hohle** boomy reproduction (hf cut off.) **-,kontinuierliche** flicker-free reproduction. **Farben-** color reproduction. **Frequenz-** frequency response (characteristic).

Wiedergabe-brillianz brilliance of sound reproduction, "bounce." **-dose** sound pickup (head). **-natürlichkeit** faithfulness, fidelity or realism of reproduction, brilliance. **-optik,Ton-** soundhead lens, s. optic. **-qualität** quality or faithfulness of reproduction (with "atmosphere" or "room tone"). **-röhre,Braun'sche** Braun c.-r. tube or electronic picture reproducing tube.

wiedergewinnen recover, recuperate, re-claim.

Wiederhall (repeated) echo, r. reflection, reverberation, resonance. **Flatter-** flutter echo, multiple e. **-formel** Eyring formula.

wiederholbar reproducible, repeatable, reiterable.

Wieder-holungsaufnahme re-take. **-holungsfrequenz** repetition frequency, f. of recurrence. **-instandsetzung** re-conditioning, restoration, renovation.

wieder-kehrend recurrent (periodically or not).

Wieder-kehrspannung recovery voltage. **-schein** reflection, reflex.

wieder-spiegeln,-strahlen reflect.

Wieder-vereinigung re-combination (of ions in gas); focusing (of rays, electrons, etc.). **-zündung** re-striking, re-ignition.

wilde Kopplung stray or spurious coupling, undesired c.

willkürlich arbitrary, haphazard, (at) random. **-verteilt** randomly distributed, scattered at random.

Wilson-kammer cloud chamber, expansion c., fog c. **-nebelspurmethode** Wilson cloud track method (using streak of droplets and ionizing particle).

Wind, dem - entgegendrehen head into the wind, crab. **gegen den - fliegen** fly up wind. **mit dem - fliegen** fly with tail wind.

Wind, Gegen- head wind, contrary w. **Rücken-** tail wind. **Schall-** flow or draft of matter particles. **Seiten-** side or cross wind, wind on the beam, wind blowing across course.

Winddreiecksrechnung airdrift triangulation.

winden wind, coil, wrap.

Wind-geschwindigkeit air velocity. **-geschwindigkeitsmesser** anemometer, wind meter. **-kurs** course heading in wind (coincides with map or head-c. in absence of wind). **-kurs, rechtweisender** true heading. **-lade** wind chest, sound box (of organ). **-maschine** fan, blower. **-sack** wind cone, w. hose (airport). **-sammler** air reservoir, compressed-air tank.

wind-schief warped, deformed or twisted (out of shape). **schnittig** streamlined.

Windstärkemesser, Schalen- cup anemometer.

Windungen, Ampere- ampere turns (ats).

Windungs-fluss turn-flux, flux-turns, magnetic linkage. **-ganghöhe** pitch (of turns). **-kapazität** internal capacitance, inter-turn c., self-capacitance. **-schlussprüfer** inter-turn short-circuit tester or continuity tester.

Windwinkel angle of the wind, angle between head-on course and direction of w., yaw angle.

Winkel, einspringender re-entrant angle. **-, räumlicher** solid angle (measured in steradian units).

Winkel, Abgangs- angle of departure. **Ablenkungs-** angle of deflection, a. of refraction. **Abprall-** angle of ricochet. **Achsen-** axial angle. **Anlauf-** angle of approach. **Anstell-** angle of attack, a. of pitch, blade a. **Aufnahme-** shooting angle. **Ausfall-** angle of emergence, a. of reflection. **Aussprung-** angle of reflection. **Beugungs-** diffraction angle. **Bild-** angle of image, a. of view. **Blick-** angle of view. **Brechungs-** angle of refraction. **Dämpfungs-** phase angle difference, loss a. **Drall-** angle of twist. **Einfall-** angle of incidence. **Einspring-** re-entrant angle. **Ergänzungs-** complementary angle. **Erhebungs-** angle of elevation. **Fall-** angle of inclination. **Fehl-** phase displacement angle, shift a. **Flächen-** plane angle. **Gesichts-** visual angle, optic a., facial a., camera or viewing angle. **Glanz-** glancing angle. **Gleit-** glide angle. **Grenz-** critical angle, limiting a. (of refractometer). **Grenzflächen-** interfacial angle (between two adjacent crystal faces). Grundrichtungs- base angle. Halb- half angle. Halböffnungs- semi-apertural angle. Haupteinfall- angle of principal incidence. Höhen- angle of elevation, azimuth a., angular height, vertical visual a. Inklinations- magnetic inclination, m. dip. Komplement- complementary angle. Konvergenz- angle of convergence. Kurs- azimuth of target (gunnery), magnetic a. Lauf- transit angle. Leitstrahl- angle of beam. Luft- angle of the wind. Luv- drift angle, a. of lead. Nacheil- angle of lag. Neben- adjacent angle, adjoining a. Neigungs- inclination (angle), a. of slope, a. of dip. Öffnungs- aperture angle, apertural a. Phasen- impedance angle, phase a., G factor, quality f. Polarisations- polarizing angle, Brewster a. Prall- angle of reflection. Rand- angle of contact (between wall and liquid), wetting angle. Raum- solid angle. Reflexions- angle of reflection. Ruhe- angle of repose, angle of friction (rarely used). Schärfen- angular resolving power, focal angle. Scheitel- opposite angle, vertex a., vertical a. Scherungs-, Schiebungs- angle of shear. Schiel- angle of strabism, a. of squint. Schneid- cutting angle (called dig-in a. or drag angle when other than 90 degrees, in record making). Schütt- angle of repose, a. of friction (rare). Schwenkungs- angle of traverse (in gunnery). Seh- visual angle. Sehschärfengrenz- angular resolving power, critical angle of visual acuity. Senkungs- depression angle. Sicht- angle of sight. Sinus- sine of angle. Steig- rise angle. Stoss- butting angle, a. of joint. Streu- angle of scattering. Tiefen- angle of elevation below horizontal, a. of depression. Trennungs- angle of separation (in solid moving through fluid). Umschlingungs- looping angle (m.p.). Verdrehungs- angle of torsion, twist a. Verlust- phase angle difference (of condenser), loss a. Voreilungs- angle of lead (of airplane). Vorhalte- drift angle, angle of lead. Wechsel- alternate angle. Weg- transit angle (of klystron). Wellenfront- wave front angle. Wind- angle of the wind, angle between head-on course and wind direction. Zentri- sector angle, center a.

Winkel-abhängigkeit angle dependence, as function of an angle. **-abweichung** angular deviation. **-aufblendung** angle fade-in. **-bild, Weit-** wide-angle picture, "Wide-scope" p. **-eikonal** angle iconal (opt.).

winkelfreier Widerstand resistance with zero phase angle.

Winkel-frequenz angular frequency, radian f., pulsatance. **-funktion** trigonometric function, angular f. **-geschwindigkeit** pulsatance (ω), angular velocity, frequency in radians. **-grad** degree of angle, radian (unit of circular measure of angle). **-halbierende** bisector, bisectrix (of angle). **-hebel** bell crank lever, angle l. **-konvergenz** half-convergence error. **-linie** diagonal. **-mass** phase constant, wave-length c. (denotes phase shift between input and output potential, and is the imaginary component of image transfer constant); set square. **-mass, Kenn-** iterative phase constant. **-mass, Vierpol-** image phase constant.

Winkelmesser protractor, goniometer, angle gage (for eyes), theodolite (for surveys), angleometer (for measuring external angles). Karten- map protractor. Neigungsclinometer. **Öffnungs-** apertometer.

Winkel-messung,Kristall- crystallographic goniometry. **-minute** one-sixtieth of one angular degree. **-punkt** infection point (of a curve).

winkel-randig with beveled edge. **-recht** right-angled, rectangular, orthogonal.

Winkelstellung angular position

winkeltreu of true angle or bearing.

Winkel-unterschied,Konvergenz- binocular parallax difference. **-verhältnis** angular ratio. **-verteilung** angular distribution, d. in angle. **-verzahnungsgetriebe** double helical gear.

winklig,gleich- equi-angular, equi-angled, isogonal. **scharf-** acute-angled. **schief-** oblique-angled. **spitz-** acute-angled. **stumpf-** obtuse-angled. **weit-** wide-angled (phot.)

winzige Gewebegebilde ultramicrons (of eye tissue).

Wippe rocker, balance, counterpoise b. **Pohl'sche -** Pohl commutator (double-pole, d.-throw switch). **Röhren-** Kipp relay, trigger circuit.

Wirbel vortex, whorl, eddy, turbulence; spigot, button, collar, swivel, peg (in string instrument). **Stimm-** wrest pin.

Wirbelbewegung vortex motion, eddy m., eddying, whirl.

wirbel-frei irrotational. **-lose Strömung** flow free from vortices, eddies or turbulence, irrotational f.

wirbeln whirl, spin, warble.

Wirbel-strasse vortex avenue. **-transporttheorie** vorticity transport theory. **-trommel** high side drum.

Wirkdämpfung transducer loss.

wirken act, work, operate, effectuate, effect, result in.

Wirk-komponente energy, active, watt, dissipative or wattful component, in-phase c. **-leistung** active power, actual p., true p. (in watts). **-leitwert** active admittance, conductance, **-samkeit** activity, effectiveness, efficacy, efficiency. **in - setzen** render operative, throw in gear, start. **-schema** actual or practical operating or working diagram. **-strom** active current, energy c., watt or wattful component.

Wirkung,Fern- telemetric action, distance a., remote effect.

Wirkungs-grad,Gesamt- overall efficiency, total e., commercial e. **-grösse** action magnitude, a. quantity. **-kreis** sphere of action, s. of influence, action radius.

wirkungslos inactive, ineffectual, inefficient, ineffective.

Wirkungs-quantum,Planck's Planck's constant, quantity of action. **-querschnitt** effective cross-section or c.-sectional area.

Wirk-verbrauchsmesser active power meter. **-verstärkung** transducer gain. **-widerstand** non-reactive, active, effective, dissipative or ohm resistance. **-widerstand-Blindwiderstandverhältnis** Q factor, quality f.

Wischkontakt self-cleaning or -wiping contact.

wissentlich deliberately, wilfully, knowingly.

wobbeln wobble (intentional frequency variation, as by wobbulator, used in f.-response test of radio apparatus); warble (as in secret telephony).

Wohnsitz legal or postal address, residence, place of living, domicile, office (in case of a company).

wölben vault, arch, make arcuate, curve, camber.

Wölbung buckling, curvature (of image); arching, vaulting. **Bildfeld-** curvature of image field, lack of flatness. **Film-** film buckling.

wölbungsfrei flat, uncurved, non-buckled, flattened (of image or field).

Wolfram-bogenlampe tungsten arc-lamp, pointolite. **-faden, thoriumhaltiger** thoriated tungsten wire or filament. **-faden,wiederbelebter** rejuvenated or re-activated tungsten filament,(subjected to flashing).
Wolframsaures Calzium calcium tungstate.
Wolke,Elektronen- electron cloud; e. shell. **Ionen-** ionic atmosphere. **Raumladungs-** concentration or accumulation of space-charges.
Wolkenbildung formation of clouds, clouding.
Wollastonprisma Wollaston prism, double-image p.
Wollfilz wool felt.
Wortabschneidung clipping, obliteration or mutilation of words (in telephony).
Wörter-buch,Sach- encyclopedia. **-verstummelung** clipping, obliteration (of words).
wortlich verbal, verbatim, literal, literatim.
wubbeln wobble, warble (by wobbulator, in coded telephony).
Wucht,Elektronen- collision or bombarding force of electrons, impact of e. **Mundungs-** kinetic energy at the muzzle. **Über-** unbalance, imbalance.
wuchtsteigerer dynamic expander.
Wulst,Ring- tore, torus, toroid, "doughnut." **Röhren-** tubular tore. **Spruhschutz-** guard ring (against corona).
wulstförmig toroidal, doughnut-shaped.
wunder Punkt snag, trouble, difficulty, moot point.
Wunschelrute divining rod, dousing r.
Wünschelrutengänger douser.
Wurf throw, cast, projection (of pictures). **Schatten-** sound shadows.
Würfel,Lummer-Brodhun'scher L.-B. contrast photometer, cube p. (with cubical cavity). **Pyramiden-**tetrahexahedron.
Würfelchen,Abbe'sches Abbe drawing cube.
Würfeleck corner of cube, cubic summit (cryst.)
Wurfelektrode,Bild- target electrode.
würfelförmig cubic, cubical.
Wurf-kraft projectile force. **-lehre** ballistics. **-linie** line of projection, curve of p., trajectory. **-parabel** trajectory parabola.
Wurmerkriechen swarming (indicative of excessive graininess, m.p.)
wurstförmig toric, tire-shaped.
Wurzel,Kubik- cube root. **Quadrat-** square root.
Wurzel-zeichen radical, root sign. **-ziehen** extracting a root, process of evolution.

X

X-Einheit unit of wave-length for Rontgen rays, etc., Siegbahn unit.
x-geschnittene Platte X-cut plate or crystal.

Y

y-geschnittete Platte Y-cut plate or crystal.
Yucon (cf. schweres Elektron) Yukawa particle or heavy electron (barytron, now known as mesotron).

Z

Zacke prong, tine, tooth, jag, serration.

Zackenblende vane with serrated or triangular edge, t. aperture (used in sound recording). Vielfach- light stop for making multi-lateral sound track.

Zackenrolle sprocket wheel.

Zackenschrift variable-area sound track, v.-width s. recording. Mehrfach- multilateral sound track. Zweifach- bilateral track.

Zackenspur, Abdeckdoppel- duplex variable-area track. Abdeckeinfach-, Einfach- unilateral variable-area sound track. Gegentakt- push-pull variable-area track. Verschiebe- bilateral variable-area track.

Zacken-tonspurbreite breadth of variable-width or variable-area sound track. -trommel sprocket drum, s. wheel.

zackig jagged, toothed, indented, serrated.

zäh tough, tenacious, viscous, viscid (liquids).

Zähflüssigkeit viscosity; refractoriness.

Zähigkeits-einheit (poise) unit of viscosity. -kehrwert fluidity (inverse of viscosity). -messer mit Fallkörper viscosimeter, viscometer of balldrop type.

Zahl, dreistellige three-figure number. -, ganze integer, integral number. -, gerade even number. -, natürliche natural number. -, unbenannte indefinite number. -, ungerade odd number. -, vierstellige four-figure number, four-place n.

Zahl, Bild- picture frequency, frame f., repetition f. Dehnungs- coefficient of expansion, c. of extension. Güte- figure of merit, quality of Q factor (ratio reactance-resistance of coil). Hallraum- chamber coefficient (ac). Haupt- cardinal number. Kenn- characteristic factor; office code. Kerbwirkungs- fatigue stress concentration figure. Kernladungs- nuclear charge number. Koordinations- co-ordination number. Massen- mass number. Ordnungs- ordinal number, atomic n., serial n. Stellen- position number, atomic n., index (math.). Überführungs- transport number (of ions), transference n., Hittorf n. Übergangs- transfer number (of solutions).

Zählader meter wire, pilot w., marked w.

Zahlen-blank figure space, f. blank. -lehre arithmology, arithmetic, numerology.

zahlenmässig numerically, as far as numbers are concerned. -erfassbar numerically evaluable.

Zahlentafel table of figures, numerical table of tabulation.

Zähler, Bild- frame indicator, f. counter. Blindverbrauchs- reactive volt-amp.-hour meter, sine m., wattless component m., var-hour m. Blutkörper- haemacytometer, blood counter. Faden- thread counter, linen tester. Film- footage counter (m.p.) Funken- spark counter. Meter- footage counter (m.p.). Scheinverbrauch- apparent power meter, trivector. Spitzen- needle counter (e.g., of Geiger). Staub- dust counter.

..zählig ...nary (binary, ternary, etc.) ...fold.

Zahn und Trieb rack and pinion.

Zahn, Sperr- ratchet tooth, pawl, detent.

zähneln, zahnen tooth, indent, denticulate, notch, mill, knurl.

Zahn-getriebe, Schrauben- helical gear, spiral g., worm. -kranz tooth-wheel, gear rim, toothed r. -motor mit Stimmgabelsteuerung phonic drum motor, p. wheel (of La Cour). **-radfeinbewegung** slow-motion gear. **-sirene** tooth-wheel synchronizer, tone-wheel. **-stange** rack (rod), tooth rack. **-trieb** rack and pinion drive, tooth-wheel d. **-trommelwelle, Antriebs-** drive sprocket shaft. **-werk** gearing.

Zain ingot, bar, rod, pig.
Zängchen small forceps or tweezers.
Zange (pair of) tongs, pincers, nippers, pliers, forceps, tweezers.
Zange, Feder- spring pliers.
Zapfchen, farbenempfindliche color distinguishing cones (of eye).
Zapfen peg, tenon, pin, plug, pivot, tap, spigot, bung; axle, journal, trunnion.
Zapfen, Dreh- pivot, trunnion. **Lager-** journal. **Lauf-** journal, neck (of rolling mill). **Stirn-** trunnion, end journal.
Zapfen-kurve spectral response curve of eye (cones). **-lager** plain bearing, socket, bush, collar.
Zapfstelle tap, tapping point, tap connection.
Zapon varnish.
Zaser fiber, filament.
Zeemanstruktur, aufgelöste resolved Zeeman pattern.
Zehneck decagon.
Zehner-logarithmus logarithm to the base 10. **-system** decimal base, d. system.
Zehntellösung tenth-normal solution.
zehntelnormale Lösung decinormal solution.
Zeichen signal, sign, symbol, mark, stamp, imprint.
Zeichen, durch Störer verdeckte swamped signals.
Zeichen, Akten- serial number, file n., docket n. **betrachtungs-** reference letter, r. numeral, symbol. **Bezugs-** reference symbols (numerals or letters, used in drawings). **Bodenfunkstellenruf-** ground station call signal or sign. **Einflug-** boundary marker signal. **Funkruf-** radio call signal or sign, code signal, "signature". **Gleichheits-** equality sign. **Gleichlauf-** synchronizing pulse, sync signal. **Haupteinflug-** inner marker signal, main entrance s. **Kenn-** characteristic feature, mark, sign, indication, clue, symptom; object (of an invention or patent). **Kurz-** symbol, symbolic denotation. **Mehrfach-** multiple (or multiplied echo) signal. **Not-** distress signal, SOS s. or call. **Peil-** call sign, code signal (e.g., of ground station). **Ruf-** call or code signal or sign, "signature" (of a station). **Schalt-** symbols of circuit elements. **Schau-** optical signal, visual s., indicator s., telltale means. **Sender-** transmission signal, transmitter s., station s. **Voreinflug-** fore or outer marker (of beacon). **Wurzel-** radical sign, root s., radical. **Zeit-** time signal.
Zeichen-frequenz signal frequency. **-schiefer** pencil slate. **-spannung** signal potential. **-stromrohre** marking valve. **-trickfilm** animated cartoon. **-verbreiterung** signal spread (due to echo effect).
zeichnen, eine Kurve plot a curve, map a graph.
zeichnende Linse, scharf- achromatic lens.
Zeichner, Schall- sound recorder. **Trickfilm-** animator (of animated cartoons). **Weich-** soft-focus lens.
Zeichnung drawing, sketch, cut, view, illustration, delineation or recreation (of an image, etc.), tracing, graphic representation.
Zeichnung, Akten- drawing in docket, in case-records or file wrapper (of a patent application or legal case). **Block-** block diagram, schematic drawing, diagrammatic illustration. **Fern-** perspective drawing. **Schnitt-** cross-sectional drawing, illustration, view or picture. **Umriss-** linear drawing, sketch or outline.
Zeiger pointer, hand, index hand,

needle, indicator, measuring instrument, meter (e.g., wattmeter, etc), index, vector (math.).
Zeiger,Kurs- navigator's compass. **Leucht-** dial light resonance or tuning indicator. **Licht-** spot lighting, point l. **Maximumwert-** peak indicator, crest i. **Messer-** knife-edge pointer. **Peil-** pointer of d.f. apparatus. **Schatten-** shadowgraph (tuning indicator). **Sekunden-** seconds hand. **Spannungs-** voltmeter; voltage vector. **Strom-** ammeter; current vector. **Wende-** turn indicator, t. meter (of airplane).
Zeiger-diagramm vector diagram. **-instrument** pointer-type instrument. **-länge** length of brush or pencil (in cathode-ray tube). **-synchronisator** dial scoring machine (m.p.)
Zeile,Dipol- dipole array.
Zeilen-ablenkung line scan, l. scansion, l. sweep. **-abstand** separation between lines, pitch, inter-line distance (telev.) **-bewegung** line scan (telev.) **-breite** line width, strip w., spot diameter (telev.). **-durchlauf, Bild-** line traversal (telev.) **-flimmer,Zwischen-** inter-line flicker, shimmer, weave (telev.). **-folgeimpulse** fractional scan or partial s. impulses, line-sequence or l. - set i. (in interlaced scanning). **-frequenz** line frequency, strip f., horizontal (scan) f. **-hinlauf** (cf **Rücklauf**) active line trace, scansion (of pencil or spot). **-kippeinsatz** incipient flyback. **-kipper** line time base. **-kippschwingung** line time base impulse. **-raster, Zwischen-**interlaced scan. **-raster,Zwischen-,ungerader** odd-line interlaced scanning. **-rücklauf** flyback, return, retrace (of pencil or spot). **-schaltung** picture traversing, vertical stepping down movement. **-spiegelrad** line scanning mirror drum (of Scophony) **-sprungfaktor** interlace factor or ratio. **-sprungmethode** interlaced, intermeshed or interleaved scanning method. **-spulen** line scan coils, line sweep c. **-steuerung** velocity modulation, variable speed m.
Zeilensynchronisierungs-impuls line synchronizing impulse, high sync i. **-lucke** (cf **Lückensynchronis.**) line gap or interval (for introduction of line sync pulse).
Zeilen-teilung line pitch (telev.). **-überdeckung** line overlap, spot o. **-verlagerung** migration or shift of line position. **-verschiebung** interlacing; line offset scan. **-verstärker** line or high sync impulse amplifier (telev.). **-vorschubbewegung** line sweep, line traversing motion. **-wechselfrequenz** line frequency. **-zug** line sequence, l. series. **-zugwechselperiode** field frequency (= 2 x frame f., telev.)
Zeit,Abkling- decay period, die-out p., quench p. (in superregeneration). **Anfach-** building-up period (time required for oscillating to start). **Anheiz-** (cathode) heating time (till stable temperature is reached), warm-up period, thermal time constant. **Ankling-** starting time, onset t. (of a tone). **Anschwing-** build-up period (time required for oscillating to start). **Aufbau-** formation time (in spark discharge). **Aufschaukel-** time constant of resonant amplification. **Ausregel-** decline period (of control potential). **Bildspur-** Watkins factor, development f. (film). **Dunkel-** dark interval, d. period. **Einregelungs-** building-up or waxing period (of control potential). **Einschwing-** build-up period (required for starting of oscillations). **Eintauch-** immersion period. **Nachhall-** reverberation period. **Nachwirk-** hang-over period, decay p. **Zerfall-** decay time.

Zeit-ablenkung time base. **-abschnitt** element of time, interval of t. **-aufnahme** time exposure. **-blinker** chronograph, time recording camera. **-dauer, Abstoppen von** stopwatch measurement of time values or intervals of time. **-dehnaufnahme** high-speed camera shooting or picture (for slow-motion projection). **-dehner** slow-motion device. **-dehnung** high-speed camera shooting; time scale. **-dehnung, Ton-** acoustic slow motion. **-festigkeit** time strength, endurance s. (found by time test). **-folge** time sequence, consecution. **-funk** time topics program, timely t. **-härtung** time hardening. **-kreiskondensator** time-base condenser.

zeitlich temporal, concerning time, temporary. **-veränderlicher Trägerstrom** wobbled carrier current. **-veränderlicher Vorgang** action variable with time, a. subject to time variation. **-(er) Verlauf** shape of curve as function of time, t. slope.

Zeit-lupenverfahren high-speed take and low-speed projection method, slow-motion or retarded-action m. **-marke, -mass** time scale, marking of time (for synchronization). **-masstab** time scale. **-messer** chronometer, time piece, watch, clock. **-messer, registrierender** chronoscope, chronograph. **-messgerät** chronographoscope. **-messgerät, Mikro-** microchronograph. **-modulation** velocity modulation, variable-speed m., variable-speed scanning. **-raffaufnahmeverfahren** time-lapse photography, low-speed shooting for high-speed projection. **-raffer** time-lapse-motion camera, stop-motion device.

zeitraubend time-consuming, tedious.

Zeit-rechnung chronology. **-regulierband** timing tape. **-relais, Invert-** inverse-time relay. **-schnellschalter** high-speed circuit breaker or switch. **-schriftenliteratur** journal literature, magazine l. **-signal** time signal. **-skaleninstrument** time-scale instrument. **-spirale** time spiral. **-verschiebung** time lag, t. shift. **-zeichen** time signal.

zellähnlich cell-like, celloid, cellular, pitted, honeycomb-like.

Zelle, lichtelektrische od. photoelektrische photo-electric cell, electric eye (either photoconductive, p.-voltaic or p.-emissive, the latter preferably called photo-tube). **-, photoelektrische, mit Sekundärelektronen-Vervielfachung** photo-electric electron-multiplier tube, e.-m. with photo-cathode.

Zelle, Alkali- alkali-metal (photoemissive) cell. **Einheits-** unit cell, elementary c., unit crystal, lattice u. **Elektrophorese-** electrophoresis cell. **Elementar-** unit cell, elementary c., u. crystal, lattice u. **Gas-** gas-filled (photo-emissive) cell. **Gleichrichter-** rectifier cell, barrier-layer c. **Glimm-** photoemissive gas-filled cell (operated at potential slightly below critical discharge p.), photo-glow tube. **Haar-** hair cell, capillary c. **Hinterwand-** photovoltaic barrier-layer cell with posterior metallic l. **Licht-** photo (-electric) cell. **Lichtsteuer- (Debye-Sears)** supersonic light valve, D.-S. light modulator. **Mehrfachplatten-, von Karolus-Kerr** Karolus multiple-plate Kerr cell. **Piezoquarz-** supersonic light valve (predicated on Brillouin and Debye-Sears effect). **Selenium-** selenium (photoconductive) cell. **Sperr(schicht)-** barrier-layer photo-voltaic or photronic cell, rectifier c. (operating with a blocking or stopping layer). **Streu-** scattering cell. **Thallium (sulfid)-** thalofide cell. **Überschall-** supersonic light valve. **Vakuum-** vacuum photo (-electric) cell, photo-emissive (alkali) c. **Vervielfacher-, mit Sekundäremission**

electron multiplier photo-tube (working with dynodes). **Vorderwand-** photo voltaic barrier-layer cell with anterior metallic layer. **Vorglimmlicht-** photo-emissive gas-filled cell operated at a sustaining potential in pre-glow region (slightly below striking or critical discharge p.).

zellenartige Struktur honeycomb, pitted or cellular structure.

Zellen-kabel, Photo- low-capacitance shielded cable (connecting photocell and amplifier), video cable, concentric c., co-axial c. **-verstärker, Photo-** electron multiplier phototube; photocell amplifier. **-vorbelichtung** priming illumination of a cell.

zellig, zwischen- intercellular.

Zellstoff cellulose, pulp (paper). **-pappe** pulp board, carton, mill-board.

Zentralstrahl central ray.

zentralsymmetrisch centro-symmetrical.

Zentren, Farb- F centers. **Keim-** grain centers.

zentriert, flächen- face-centered (cryst.). **raum-** body-centered, space-c. (cryst.)

Zentrierungsfeder spider (of loud-speaker coil).

Zentrifuge, Ultra- high-speed centrifuge.

zentrifugieren centrifuge, centrifugalize.

zentrisch centric, central, co-centric.

Zentriwinkel center angle, sector a.

Zentrum, Kurbel- crank center. **Massen-** center of mass, c. of inertia, centroid. **Ton-** tonal center, acoustic c. (of "gravity") of frequencies. **Wachstum-** center of growth, nucleus kernel.

zerbrechlich breakable, fragile, brittle.

Zerbrechungsfestigkeit resistance to breaking strain, b. strength.

zerbrockelnd crumbling, crumbly, friable.

Zerdehnung, Bild- radial distortion of picture or image (pincushion effect).

Zerdrehung, Bild- rotational or tangential distortion, twist (barrel effect; telev. pictures).

Zerdrückfestigkeit resistance to crushing or compression strain, crush strength.

Zerfall decomposition, dissociation, disintegration, decay. **Atom-** disintegration of atom, splitting or fission of a. **Korn-** grain disintegration. **Mesonen-** meson decay.

Zerfallbarkeit, Nicht- non-disintegrability.

Zerfallelektronen electrons resulting from disintegration (say, mesotrons), decay e.

zerfallen decompose, dissociate, disintegrate, undergo fission, break up, divide, crumble, fall apart, decay, breakup.

Zerfall-kurve decay curve. **-produkte, radioaktive** metabolons, successive disintegration products of radioactivity. **-prozess** decomposition process.

Zerfalls-elektronen decay electrons (resulting from disintegration, say, mesotrons). **-konstante** disintegration constant, transformation c., radio-active c., decay coefficient.

Zerfallzeit decay time, disintegration t.

zerfliessen deliquesce, melt, dissolve, run.

Zerhacker chopper. **Licht-** light chopper, episcotister.

zerhackt flicked off, chopped.

zerlegen decompose or split up (light beam), refract, diffract, take apart, demount, cut up, dissect, analyze, resolve, separate, knock down.

Zerleger scanner, dissector (of Farnsworth), exploring means. **Bildfeld-** picture scanner, image dissector. **Vor-** filter mounted anteriorly of monochromator (for ultraviolet rays). **-blende** dissector aperture, scanning hole.

zerlegtes Licht, spektral- spectroscopically dispersed or separated light.

Zerlegung (cf **zerlegen**) analysis, decomposition, dissection, dispersion, resolution, definition. **Bildfeld-** image (field) definition. **Licht-** light dispersion, l. decomposition, l. refraction. **Ton-** sound analysis.

Zerlegungsvorrichtung scanner, exploring means.

Zerplatzen, Kern- nuclear fission, n. disintegration. **Vakuumröhren-, nach innen** implosion, inward collapse or burst of a tube.

Zerrbild distorted picture, p. made with an anamorphotic lens.

zerreiben pulverize, powder, triturate, disintegrate or comminute by grinding or abrasive action.

zerreiblich friable, crumbly, triturable.

Zerreibung pulverization, trituration, attrition, comminution (by abrasion).

zerreissen tear, lacerate, break, rend, rupture, wear out, shred.

Zerreissprobe breaking test, tensile t., rending t.

Zerrlinse anamorphosing lens, distorting lens.

Zerrungskreis annulating network; suppressing n.

zerschlagen break in pieces, smash, shatter, crush, batter.

zersetzen dissociate, decompose, disintegrate, analyze.

Zersetzung, Ton- analysis of sound.

Zersetzungs-kunst analysis, art of analyzing. **-widerstand** electrolytic resistance.

Zerspanbarkeit free cutting and machining property.

Zerspanung removal of metal by cutting tool (chipping, turning, shaving, etc.).

zerstäuben spatter, sputter, disintegrate, vaporize, atomize, reduce to dust or mist, spray.

Zerstäubung disintegration, sputter, splutter or spatter (of cathode), spraying, atomization.

zerstörend, holz- lignicidal.

Zerstörung, lochfrassähnliche pitting, destruction in form of pits, honeycombing.

zerstörungsfreie Prüfung non-destructive test.

zerstossen pound (to pieces), bray, bruise, powder, pulverize.

Zerstrahlung annihilation radiation (due to collision and mutual annihilation of atoms, electrons and positrons).

zerstreuen disperse, scatter, dissipate, diffuse.

zerstreuend, licht- light dispersive, l. diffusive, l. scattering.

Zerstreuung dispersion, diffusion. **Energie-** scatter of energy. **Farben-** chromatic aberration, c. dispersion, chromatism. **Richtungs-** scatter of direction (of beam).

Zerstreuungs-bild image formed by divergent lens. **-bilder des Auges** blur circles of eye. **-kreis** circle of diffusion, blur c. **-linse** dispersive lens, divergent l., negative l. **-scheibchen** circle of (least) diffusion, blur c. **-vermögen** dispersive power (opt.) **-wirkung** scattering (of electrons, etc.)

zerteilt, fein- finely divided.

zertrümmerbares Material fissionable or disintegrable substance.

Zertrümmerbarkeit der Materie disintegrability of matter.

Zertrümmerung fragmentation, breaking up into fragments, shattering, smashing, disintegration, fission (of atom).

Zertrümmerungswahrscheinlichkeit probability of transmutation or disintegration.

Zessionar assign, assignee, transferee.

Zettel label, slip of paper, card, ticket, check, note.

Zeugelinie generatrix, generating line

zeugen testify, give or bear testimony, produce, create, generate.

Zeugengebühren witness fees.

Zeugnis testimony, testimonial, evidence, witness. **- ablegen** bear testimony or witness.

Zickzack-Reflexion staggered reflection.

ziegelrot brick-red.

Ziegenleder kid leather.

ziehbar ductile.

Zieh-diamant diamond die. **-eisen** drawing die, dieblock, draw plate, die p. (for drawing wire).

ziehen (cf **Ziehvorgänge**) draw, pull, tug, drag; warp; describing a phenomenon in which oscillating suddenly ceases on loosening feedback coupling.

Ziehen eines Bildes photographic travel ghost.

Ziehen,Kubikwurzel- extraction of cubic root. **Quadratwurzel-** extraction of square root. **Wurzel-** root extraction, process of evolution.

Zieh-fähigkeit,Tief- cuppability, capability of being drawn or cupped in drop press. **-grad** ductility. **-kraft** drawing or tractive power, traction. **-presse** drop (forge) press (for drawing and forming sheet metal articles, etc.). **-probe** tensile test, t. t. specimen. **-riefen** fissures, drawing grooves or scratches. **-vorgänge** pulling, oscillation hysteresis, resonance discontinuity,protraction or instability phenomena.

Ziel goal, aim, object, end, destination, mark, target, objective, boundary, limit.

Ziel-abflug flight away from object. **-anflug** flight towards object. **-aufsuchen** detection of target or object (by radar or sonar, etc.).

zielen auf aim at, strive for, direct or tend to, adjust (sight).

Zielender,Geschütz- gunlayer.

Zielfernrohr sight telescope, bombsight, rifle s., telescopic s. **Doppelblick-** double direct sighting telescope.

Zielflug,Gehör- aural homing flight. **Sicht-** visual homing flight. **-empfänger** homing receiver apparatus. **-verfahren** point to point avigation method, homing method.

Ziel-gerät course indicator. **-kontrollapparat** radio timing apparatus.

Ziel-kurs,ablaufender course away from transmitter or beacon. **-,anlaufender** course towards transmitter or beacon. **-,missweisender** magnetic course to steer. **-,rechtweisender** true course to steer.

Ziel-linie finish line (in racing); line of sight, l. of collimation (opt.). **-marke** sight graticule or reticle.

Zielpeilung,rechtweisende Richtung vom Flugzeug zur Bodenpeilstelle true reciprocal course (from airplane to ground station). **-,missweisende Richtung vom Flugzeug zur Bodenpeilstelle** magnetic reciprocal course (from airplane to ground station).

Zielpunktes,Anzielen des sighting of target.

zielstrebig purposive.

Ziel-sucher radar (radio detection and ranging apparatus), sonar (under-water sound navigation and ranging). **-verfolgung** tracking a target (with detecting and ranging equipment and with directors, using a slewing motor).

Zierschiene grille bar.

Ziffer figure, digit, number, cipher. **Anteil-** percentage share. **Form-** shape factor, theoretical stress concentration f. **Kenn-** index (of a log.), characteristic, code number. **Verlust-** phase angle difference (tan δ).

Zifferblatt,Leucht- luminous dial.

Zinkbecher zinc case or cylinder (of dry cell).

Zinken (cf **Zacke**) prong, tine (of tuning fork); spike, tooth, jag, lug.

Zink-pfeife zinc pipe. **-pol** zinc pole, cathode.

Zinnpfeife tin pipe.

Zipfel lobes, ears, secondary beams, stray or side radiations (of a beacon station).
Zipfel,Haupt- major lobe (of space pattern). **Neben-** minor lobe or ear (of space pattern).
Zirkel circle; pair of compasses. **Spitzen-**compasses, dividers. **Taster-, Visier-** gaging calipers.
Zirkonbrenner zirconium burner, z. filament.
zischen hiss, sizz, fizz, sizzle.
Zischlaute sibilants, fricatives.
Zither,Akkord- auto harp.
Zitter-bewegung circular fluctuation movement (of spin). **-elektrode** vibrating electrode, vibratory e.
zittern tremble, shake, quiver, vibrate, quake.
Zittern,Bild- unsteadiness of picture, jumping of p.
Zone,stille oder tote skip zone, dead z. (radio reception).
Zone,Anhäufungs- zone of accumulation (of ultrasonic waves). **Auslösch-** dead spot, skip distance, skip region (of silence). **Leit-** equisignal zone, glide z. or path. **Schlieren-** dead zone, skip z., shadow z., shadow. **Schweige-** skip distance, shadow region, zone of silence (airplane landing). **Verwaschungs-** confusion zone, z. lacking definition, blurred or bleary z. (telev. pictures).
Zonen-blende limiting stop (diaphragming out a given zone or area). **-fehler** zonal aberration.
zu dicht overdense. **-flach** too low a gamma.
Zubehörteile accessories, spare parts.
zubereiten prepare, dress, finish, make ready.
züchten grow, raise, produce, cultivate.
Züchtung growth or growing (of crystals).
Züchtungsgefäss growing vessel.
Zuckanzeige ballistic, kick or flash (instrument) reading or indication (d.f., etc.)
Zuckerbestimmung saccharimetry, determination or analysis of sugar (with polarimeter).
zuerkennen adjudge, adjudicate, award. **-,eine Entschädigung** award damages.
zuerteillen award, allot.
zufällig accidental, incidental, haphazard, casual, random, by chance, fortuitous.
Zufälligkeit accidentalness, casualness, chance, probability, vagary, fortuity.
Zufalls-ergebnisse fortuitous results, f. findings. **-gesetz** law of chance, l. of probability.
Zufluss flow, flux, afflux, influx, inlet, admission.
Zufuhr admission, inlet (for instance, in water-cooled tube), feed, supply, afferrence.
Zug drawing, pull, traction, draft, train, stroke, flue, drawtube (of microscope).
Zug(e), im - des Verstärkers in the channel or cascade of an amplifier. **im - des Tonspaltes** along the slit, in line with s.
Zug,künstlicher forced draft. **Banden-** band progression. **Haupt-** main, principal or outstanding feature or trait, main object (of an invention or a patent). **Kurven-,abklingender** decay train. **Verstärker-** amplifier channel. **Zeilen-** line sequence, l. series.
zugänglich accessible, amenable.
Zügen, in rohen in rough outline, roughly, approximately, practically, substantially, closely.
Zugentlastung traction relief.
zugestandenermassen admittedly, concededly.
Zug-feder tension spring, traction s., retractile s. **-festigkeit** tensile strength, tenacity. **-gitter** positive grid; space charge g.
zügiger Gang intimate or positive threading (without play or backlash).
Zug-kraft traction, tractive force or effort, tension, attraction.

-linie tractrix. -loch vent hole, air h. -luft (air) draft. -posaune slide trombone. -rohr air pipe, vent p. or tube, stack. -wechselperiode, Zeilen- field frequency (=2 x frame f.)

Zuhörer auditor, audience (plurality of listeners).

Zuleitung, Antennen- antenna lead, downlead, feeder line, transmission l.

Zuleitungsinduktivität lead inductance.

zumischen admix, mix with, add.

Zündelektrode ignitor (in ignitron).

zünden ignite, start (an arc), set afire, kindle, light.

Zünder lighter, igniter, fuse, tinder.

Zünder, Dampfdruck- vapor-pressure igniter. Doppel- time and percussion fuse. Druck- pressure igniter. Selbst- (pyrophorous) self-igniter, automatic lighter.

zunderfest tinder-proof.

Zünd-flämmchen pilot flame. -kirsche ignition pellet. -löschspannungsdifferenz striking-extinction potential difference. -papier ignition paper (phot.). -spannung firing potential, striking p. (of arc-lamp, rectifier, thyratron, ignitron, etc.), break-down p. (of insulator), flashover (of spark-gap), glow p. (at which initial glow-discharge begins, in photocell), p. at which, in glow-tube, discharge changes from assisted to self-sustained form. -stab ignitor, ignition rod, striking r. (in ignitron).

Zündung ignition, priming. Fehl- failure to fire or ignite (in rectifier, ignitron, etc.); misfiring (of engine). Früh- pre-ignition, premature i. Rück- backlighting, arc-back, back-firing, backlash (in imperfect valve rectification).

Zündungseinsetzen initiation or incipience of striking or firing.

Zunge blade, reed, ditton, tongue, vibrator, keeper (of a magnet). Kontakt- spring contact, reed, blade, ditton.

Zungen-frequenzmesser vibrating reed frequency meter. -öffner latch opener.

Zungenfeife reed pipe or flute. Doppel- double-tongued flute. Orgel- reed organ pipe.

Zungentelephom reed telephone receiver, Brown-type t.r.

Zuordnung correlation, co-ordination, assignment, allotment.

zupassen fit, adjust.

zupfen pick, pluck, tug, pull.

zurechtmachen get ready, make ready, ready, prepare, adjust, line-up.

Zurichten prepare, make ready, dress, finish, straighten, adjust.

zurückdrehen turn off, turn down (as in a lathe).

Zurückfallen in Grundzustand return or re-transition to ground state or lowermost energy level (of an atom).

zurück-führen trace (back), ascribe, attribute, fly or lead back, re-convene, reduce, return, restore. -halten retain, hold back, retard, stop. -prallen rebound, recoil, be reflected. -weisen dismiss, reject, refuse, disallow.

Zurückweisung overruling, rebuttal, dismissal, disallowance, rejection (of an application, petition, etc.)

zurückwerfen throw back, reflect (light waves), reverberate (sound waves), echo back, bend back (a ray, by ionosphere).

zusammenbacken cake, stick together, conglomerate, agglomerate.

Zusammenballen bunching of electrons in groups (by phase focusing, in drift space of beam tubes); agglomeration, conglomeration, coagulation (of colloids and smoke particles).

Zusammen-ballungen bunching (of charges or electrons). -bau erection, assembly or mounting work.

zusammenbrechen collapse, break down, implode (as an inwardly

collapsing vacuum vessel).
Zusammendrückbarkeit compressibility.
zusammen-drücken compress, squeeze, crush, compact, consolidate. -fallen collapse (as a field), implode (as a vacuum tube); coincide, register, come together. -fallend coincident, be in congruity or in registration. -fassender Bericht survey, summarizing report or article.
Zusammen-fassung summary, résumé, compilation, abstraction, recapitulation. -gehorigkeit correlation (of things).
zusammengesetzt(er) Klang complex musical sound. -(e) Leitung compound circuit. -(e) Schwingung complex harmonic wave.
zusammengesprengte Prismen broken-contact prisms (closely adjoined without cement, as in Cornu p.)
Zusammenhalt cohesion, unity.
zusammen-hängende Photokathode plain mirror or continuous (non-mosaic) photo-cathode (as in Farnsworth dissector and super-emitron). -häufen heap or pile up, accumulate, aggregate, gather together. -klappbar collapsible, foldable, knock-down. -laufen run together, converge, blend, coagulate, concur.
Zusammen-pressung der Luftteilchen compression or condensation of air particles (in sound wave). -schmoren von Kontakten scorching, freezing or melting together of contacts. -setzblende scanning hole, picture recreator aperture (to focus beam, in Farnsworth oscillight receiver).
zusammensetzen compose, compound, assemble, fit together.
Zusammensetzungsvorrichtung, Bild- picture scanning disk (of telev. receiver), picture reproducer, recreator or delineator.
zusammensprengen join or fit together, without cement, as prims or objectives, in a broken-contact assembly.
Zusammen-stellung, Linsen- lens combination. -ziehung, Quer- lateral contraction. -ziehung, Querschnitts-contraction or reduction of cross-sectional area, necking, formation of a waist (in piece under tensile test).
Zusatz addition, addendum, appendix. Lichtton- sound-on-film head or attachment. Nadelton- sound-on-disk attachment.
Zusatz-einrichtung accessory, ancillary or auxiliary means or equipment. -gerat, Tonfilm- sound film head or attachment.
zusätzlich additional, supplemental, accessory, auxiliary, boosting.
Zusatz-linse supplementary lens. -nachbildung building-out section (of transmission line). -patent patent of addition (to parent p.) -spannung booster potential, addition p. -träger suppressed carrier re-inserted in receiver (in side-band transmission of signals). -verluste stray load losses. -verstärker booster amplifier.
Zuschauer audience, (movie) patrons, spectators, viewers and/or listeners (sound film). -raum auditorium.
Zuschlag flux.
Zuschlagsgebühr extra tax or fee, supplementary fees, dues or charges.
zuschmelzen seal, melt together.
zusetzen alloy, admix, add.
zuspitzen point, sharpen, taper, tip.
zusprechen, dem Verletzten eine Geldsumme adjudicate or award a sum of money to injured or aggrieved party.
Zustand state, condition, situation, circumstance. -, eingeschwungener steady state oscillation. -, geordneter ordered state or arrangement (of a structure), preferred configuration (of units in crystal lattice). Augenadaptations- (state of) retinal adaptation. Dauer- steady state. Elektronen-, entarteter degenerate electron

state. **Entstehungs-** nascent state, status nascendi. **Gleichgewichts-** state of equilibrium, state of balance, stable s. **Grund-** ground state (of nucleus). **Grund-, Zurückfallen in** return or re-transition into ground state or lowermost energy level. **Ruhe-** state of rest, repose or quiescence, steady, stationary or neutral s. or condition, no-signal or Q s. (of a tube). (For additional terms of Ruhe...)

zuständig sein als Berufungsinstanz to have appellate jurisdiction.

zuständiger Gerichtshof court of competent jurisdiction or venue.

Zustands-diagramm phase diagram. **-formel** state formula. **-gleichung** equation of state. **-grössen** variables of state. **-summe** partial function. **-verschiebung** change in state, displacement of s.

Zustimmung assent, accession, acquiescence, consent, agreement, permission.

zuteilen allot, assign, delegate, give over, allocate; adjudicate, award.

Zutritt, Luft- access or admission of air, in the presence of a.

Zuweisung der Wellenlängen frequency allocation or assignment.

Zuwiderhandlung act in contravention of.

Zwang(es), Prinzip des geringsten principle of least constraint or resistance.

Zwang, Ausführungs- compulsory working (of a patent). **Synchronisier-** locked synchronism.

zwangläufig (cf **kraftschlüssig**) positive, by positively acting means, by constraint, in a locked manner, non-slip (in drive).

Zwangs-lizenz compulsory license. **-schwingung** forced oscillation. **-vollstreckungsbefehl** writ of (compulsory) execution.

zwangsweise by mandamus, by compulsion, compulsorily.

Zweckbeleuchtung, Zwei- dual-purpose illumination.

Zwecke tack, pin, peg. **Kamm-** tack.

zweckmässig suitable, appropriate, expedient, adapted, fit.

zwei-achsig biaxial. **-adrig** bifilar. **-atomig** diatomic. **-äugiges Sehen** binocular vision, stereoscopic v.

Zweibandspieler two-unit dubber.

zwei-chöriges Piano bichord piano, two-string p. **-deutig** ambiguous, equivocal, double meaning, double entendre.

Zwei-drahtwicklung bifilar winding, Ayrton-Perry w. **-elementkristall** bimorph crystal (either of bender or twister type).

Zweier-schalen duplet rings. **-verfahren** two-line-series or two fractional scans per frame (interlaced scanning method).

zweifach dual, double, two-fold.

Zweifach-diodengleichrichter, -zweipolröhre duodiode rectifier, d. valve.

zwei-fädig bifilar. **-farbiger Schleier** dichroic fog, dichromatic f., silver or red f.

Zweifeldröhre drift or beam tube working with two fields (of Heil, etc.)

zweiflächig dihedral, two-faced.

Zweig branch, leg, arm (of a bridge or filter). **Längs-** line arm, series arm (of a filter or network). **Mess-** measuring arm. **P-** P branch (spectrum lines). **Quer-** cross arm, shunt a. (of network).

zweigliedrig binary, two-membered, binomial (math.) **-(er) Kettenleiter** two-section network.

Zweig-linien, Null- zero branch lines. **-stromkreis** divided circuit, branch c.

Zweiheit duality, dyade, couple.

zweihöckrige Resonanzkurve double-hump resonance curve.

Zwei-koffergerät outfit in two carrying cases. **-kreisempfänger** two-circuit receiver. **-kristallspektrometer** double-crystal spectrometer.

zweiohriges Hören stereophonic or binaural hearing, plastic audition or hearing, two-channel listening (with auditory perspective).

Zweipackverfahren bipack method.

zweipoliger Hebelschalter double-lever switch.

Zweipol-röhre, Zweifach- duodiode (tube). **-strecke** dipole tube, dipole path.

zweischenklig having two limbs, legs or branches, two-legged.

Zwei-schleifenoszillograph two-string oscillograph. **-schlitz-magnetron** two-split magnetron, two-segment m.

zweischneidig double-edged, two-e.

Zweiseitenbandüberträger side-band transmitter with suppressed carrier.

zweiseitig two-sided, bilateral. **-(es) Arbeiten** two-way operation, double action.

zweispitzige Kurve double-peak curve, double-hump c.

Zwei-stärkenbrille bifocal spectacles. **-stoffsystem** binary system.

zweistufiger Nocken double-lift cam (with two lobes).

zweiteilig bipartite, consisting of two parts.

Zwei-teilung bipartition, bisection. **-tellergerät** two-turntable type of player.

Zweitwicklung secondary winding.

zweiwellige Resonanzkurve two-hump resonance curve, double-hump r.c.

Zweiwelligkeit two-wave property (of a tube, crystal, etc.).

Zwei-zackenschrift bilateral track (sound film). **-zenterproblem** two-center problem. **-zweckbeleuchtung** dual-purpose illumination.

Zwergwelle (cf Wellenabgrenzung) micro wave, midget w., dwarf w.

Zwickbohrer twist drill.

Zwickel wedge, trycock. **Balg-** gusset.

Zwiebellinse biconvex lens.

Zwillings-bildung twinning, twin formation. **-gleitung** twin slipping. **-kerne** twin nuclei. **-kristall** twin crystal. **-prisma** biprism.

Zwinge clamp, cramp, vise, ferrule, hoop.

Zwischen-bild intermediate image, first i. (in super-microscopy). **-bürstenmaschine** metadyne, metadynamo. **-fehler** zonal aberration. **-filmmethode** intermediate-film method (telev.) **-flügel** intermediary blade, anti-flicker b., balancing b. (m.p.). **-fluoreszenzschirm** intermediate fluorescence screen (supermicroscopy). **-flüssigkeit** intermediate liquid. **-form** passage type. **-frequenz** intermediate frequency, i. beat or transfer f. (in superhet. receiver). **-frequenzempfänger mit Kristallsteuerung** crystal stabilized superheterodyne receiver, stenode radiostat. **-gefäss** intermediate vessel, i. receptacle.

zwischengeschichtet inter-stratified.

Zwischengitter-ion interstitial ion. **-platzmechanismus** interstitial mechanism (of alloys).

Zwischen-glied intermediate member, connecting m., link, intermediate. **-hörbetrieb** duplex operation, transmitter and receiver alternately connected with aerial, alternate two-way communication. **-kern** compound core, intermediate. **-kopie** lavender copy. **-körper** intermediate medium or substance, interposed m. or means. **-lagscheibe** washer.

zwischenlegen interpose, interlay, intercalate, insert.

Zwischen-linienflimmer interline flicker, shimmer, weave. **-masse** interstitial matter, cement (geology). **-mittel** intermediate medium or substance, interposed m. or means. **-modulation** intermodulation.

zwischenmolekulare Kräfte intermolecular forces.

Zwischen-optik intermediate optic, interposed o. or optical means. **-phase** inter-phase (liquid dielectric). **-positiv** master positive. **-raum** intermediate space, intervening s., interstice, interspace, interstitial s., interval (of space). **-raum,Pol-** pole clearance, interpolar space or gap. **-ring** adapter (of microscope).

zwischenschalten interpose, interpolate, cut in between, cut in circuit, insert, include, intercalate.

Zwischen-sender re-transmitter, repeater station. **-stecker** adapter plug, socket a. **-stecker,Röhren-** tube adapter, valve a. **-stromröhre** spacing valve. **-ton** medium tone, intermediate t. or shade (of colors). **-träger** sub-carrier. **-transformator** inter-stage transformer. **-tubus** intermediate tube (supermicroscope). **-verstärker** stage amplifier. **-wand** partition, separating wall. **-zeichenwelle** spacing wave, back w.

Zwischenzeilen-flimmer inter-line flicker, shimmer, weave (telev.). **-raster** interlaced scanning. **-raster, ungerader** odd-line scanning.

Zwischenzeit time interval, interim, meanwhile, intervening t.

zwischenzellig intercellular.

Zwitschern birdies, canaries (extraneous high-frequency noise).

Zwitscherpfeife bird warbler.

Zwitter-apertur split focus (Walton scanning arrangement). **-ion** dual ion, amphoteric i., hybrid i.

Zwölf-eck dodecagon. **-flach** dodecahedron.

zwölfstufiges System duodecimal scale (tonal system).

Zyklotron (cf Resonanzbeschleuniger) cyclotron comprising two dees or half hollow cylinders and spiral beam. (Other models of resonance accelerators using a bent beam are known as induction electron accelerators, rheotron, synchrotron, etc., all of them to be distinguished from a late 1946 model of accelerator which is of the linear, non-spiral type of accelerator gun).

Zylinder,Sammel- focusing cylinder. **Steuer-** Wehnelt grid or cylinder, focusing g., control electrode, shield (c.-r. tube).

Zymbal dulcimer, cymbal.

ZZ Verfahren ZZ (airplane) landing method.

Lightning Source UK Ltd.
Milton Keynes UK
UKHW022317060223
416579UK00001B/404